THE C# WORKSHOP

Kickstart your career as a software developer with C#

Jason Hales, Almantas Karpavicius, and Mateus Viegas

THE C# WORKSHOP

Authors: Jason Hales, Almantas Karpavicius, and Mateus Viegas

Reviewers: Omprakash Pandey and Dara Oladapo

Development Editor: M Keerthi Nair

Acquisitions Editors: Royluis Rodrigues, Kunal Sawant, and Anindya Sil

Production Editor: Shantanu Zagade

First published: September 2022

Production reference: 2281022

ISBN: 978-1-80056-649-1

Published by Packt Publishing Ltd.
Livery Place, 35 Livery Street
Birmingham B3 2PB, UK

Table of Contents

Chapter 2: Building Quality Object-Oriented Code 61

Chapter 3: Delegates, Events, and Lambdas 157

Chapter 4: Data Structures and LINQ 225

Chapter 5: Concurrency: Multithreading Parallel and Async Code 313

Chapter 6: Entity Framework with SQL Server 409

Chapter 7: Creating Modern Web Applications with ASP.NET 483

Chapter 8: Creating and Using Web API Clients 535

Chapter 9: Creating API Services 605

PREFACE

ABOUT THE BOOK

C# is a powerful and versatile **Object-Oriented Programming** (**OOP**) language that can unlock a variety of career paths. But, as with any programming language, learning C# can be challenging. With a wide range of different resources available, it's difficult to know where to start.

That's where *The C# Workshop* comes in. Written and reviewed by industry experts, it provides a fast-paced, supportive learning experience that will quickly get you writing C# code and building applications. Unlike other software development books that focus on dry, technical explanations of the underlying theory, this workshop cuts through the noise and uses engaging examples to help you learn how each concept is applied in the real world.

As you work through the book, you'll tackle realistic exercises that simulate the types of problems software developers work on every day. These mini-projects include building a random-number guessing game, using the publisher-subscriber model to design a web file downloader, creating a to-do list using Razor Pages, generating images from the Fibonacci sequence using async/await tasks, and developing a temperature unit conversion app that you will then deploy to a production server.

By the end of this book, you'll have the knowledge, skills, and confidence required to advance your career and tackle your ambitious projects with C#.

AUDIENCE

This book is for aspiring C# developers. It is recommended that you have a basic knowledge of core programming concepts before you start. Prior experience with another programming language would be beneficial, though it is not absolutely necessary.

ABOUT THE AUTHORS

Jason Hales has been developing low-latency, real-time applications using various Microsoft technologies since the first release of C# in 2001. He is a keen advocate of design patterns, OO principles, and test-driven practices. When he's not dabbling with code, he likes to spend time with his wife, Ann, and their three daughters in Cambridgeshire, UK.

Almantas Karpavicius is a lead software engineer working in the information and technology company, TransUnion. He has been a professional programmer for over five years. On top of his full-time programming career, Almantas has spent three years teaching programming for free in his free time on Twitch.tv. He is a founder of a C# programming community called C# Inn that boasts over 7000 members and the creator of two free C# boot camps in which he has helped hundreds of people get a start in their careers. He has taken interviews with programming celebrities, such as Jon Skeet, Robert C. Martin (Uncle Bob), Mark Seemann, and was also a part-time Java teacher for a time. Almantas likes talking about software design, clean code, and architecture. He is also interested in Agile (Scrum, in particular) and is a big fan of automated tests, especially those done using BDD. He also holds a two-year Microsoft MVP (https://packt.link/2qUJp).

Mateus Viegas has been working in Software Engineering and Architecture for over a decade, dedicating the last few years to Leadership and Management roles. His main interests in technology are C#, Distributed Systems, and Product Development. A lover of the outdoors, when not working he likes to spend his time either exploring nature with his family, taking photographs, or running.

ABOUT THE CHAPTERS

Chapter 1, *Hello C#*, introduces the fundamental concepts of the language, such as variables, constants, loops, and arithmetic and logical operators.

Chapter 2, *Building Quality Object-Oriented Code*, covers the basics of Object-oriented programming and its four pillars, before introducing the five main principles of clean coding—SOLID. This chapter also covers the latest features in the C# language.

Chapter 3, *Delegates, Events, and Lambdas*, introduces delegates and events, which form the core mechanism for communicating between objects, and lambda syntax, which offers a way to clearly express the intent of code.

Chapter 4, *Data Structures and LINQ*, covers the common collection classes that are used to store multiple values and the integrated language LINQ which is designed for querying collections in memory.

Chapter 5, *Concurrency: Multithreading Parallel and Async Code*, provides an introduction to writing efficient code that is high performing across different scenarios and how to avoid common pitfalls and mistakes.

Chapter 6, Entity Framework with SQL Server, introduces database design and storage using SQL and C# and provides an in-depth look at object-relational mapping using Entity Framework. The chapter also teaches common design patterns for working with databases.

> **NOTE**
>
> For those who are interested in learning the basics of databases and how to work with PostgreSQL, a reference chapter has been included in the GitHub repository of this book. You can access it at https://packt.link/oLQsL.

Chapter 7, Creating Modern Web Applications with ASP.NET, looks at how to write simple ASP.NET applications and how to use approaches such as server-side rendering and single-page applications to create web applications.

Chapter 8, Creating and Using Web API Clients, introduces APIs and teaches you how to access and consume Web APIs from ASP.NET code.

Chapter 9, Creating API Services, continues with the topic of APIs and teaches you how to create your API services for consumption, and how to secure it. The chapter also introduces you to the concept of microservices.

> **NOTE**
>
> There are also two bonus chapters (*Chapter 10*, *Automated Testing*, and *Chapter 11*, *Production-Ready C#: From Development to Deployment*) which you can find at https://packt.link/44j2X and https://packt.link/39qQA, respectively.
>
> You can also find solutions for all activities in this Workshop online at https://packt.link/qclbF.

This book has some conventions set to arrange content efficiently. Read about them in the next section.

CONVENTIONS

BLOCK OF CODE

In the book, a block of code is set as follows:

```
using System;

namespace Exercise1_01
{
    class Program
    {
        static void Main(string[] args)
        {
            Console.WriteLine("Hello World!");
        }
    }
}
```

In cases where inputting and executing some code gives an immediate output, this is shown as follows:

```
dotnet run
Hello World!
Good morning Mars!
```

EMPHASIS

Definitions, new terms, and important words are shown like this:

Multithreading is a form of concurrency whereby **multiple** threads are used to perform operations.

TECHNICAL TERMS

Language commands within the body of the chapter are indicated in the following manner:

Here, the simplest **`Task`** constructor is passed an **`Action`** lambda statement, which is the actual target code that you want to execute. The target code writes the message **`Inside taskA`** to the console.

ADDED INFORMATION

Essential information is indicated in the following way:

> **NOTE**
>
> The term **Factory** is often used in software development to represent methods that help create objects.

TRUNCATION

Long code snippets are truncated and the corresponding names of the code files on GitHub are placed at the top of the truncated code. Permalinks to the entire code are placed below the code snippet, as follows:

HashSetExamples.cs

```
using System;
using System.Collections.Generic;
namespace Chapter04.Examples
{
}
```

You can find the complete code here: http://packt.link/ZdNbS.

Before you dive into the power of the C# language, you will need to install the .NET runtime and the C# development and debugging tools.

BEFORE YOU BEGIN

You can either install the full Visual Studio **Integrated Development Environment** (**IDE**), which offers a fully featured code editor (this is a costly license) or you can install **Visual Studio Code** (**VS Code**), Microsoft's lightweight cross-platform editor. *The C# Workshop* targets the VS Code editor as this does not require a license fee and works seamlessly across multiple platforms.

INSTALLING VS CODE

Visit the VS Code site at https://code.visualstudio.com and download it for Windows, macOS, or Linux, following the installation instructions for your preferred platform.

> **NOTE**
>
> It is better to check the **`Create a Desktop Icon`** checkbox for ease of use.

VS Code is free and open source. It supports multiple languages and needs to be configured for the C# language. Once VS Code is installed, you will need to add the **`C# for Visual Studio Code`** (powered by OmniSharp) extension to support C#. This can be found at https://marketplace.visualstudio.com/items?itemName=ms-dotnettools.csharp. To install the C# extension, follow the per-platform instructions:

1. Open the **`Extension`** tab and type **`C#`**.

> **NOTE**
>
> If you do not want to directly install the C# extension from the website, install it from VS code itself.

2. Select the first selection, that is, **`C# for Visual Studio Code (powered by OmniSharp)`**.
3. Click on the **`Install`** button.

4. Restart **VS Code**:

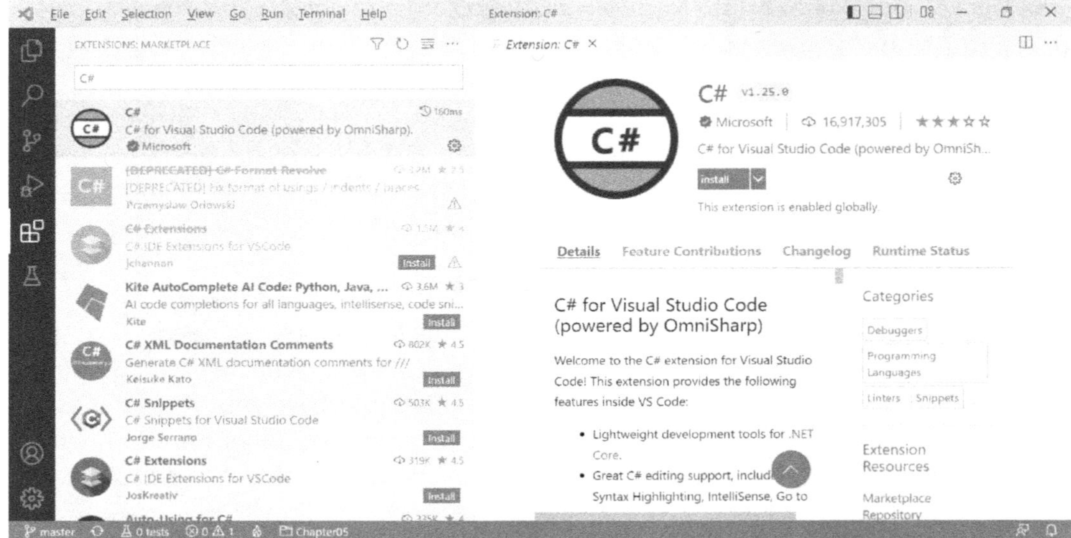

Figure 0.1: Installing the C# extension for VS Code

You will see that the C# extension gets successfully installed on VS Code. You have now installed VS Code on your system.

The next section will cover how VS Code can be used as you move between the book chapters.

MOVING BETWEEN CHAPTERS IN VS CODE

To change the default project to build (whether it is an activity, exercise, or demo), you will need to point to these exercise files:

- **`tasks.json`** / **`tasks.args`**
- **`launch.json`** / **`configurations.program`**

There are two different patterns of exercise that you should be aware of. Some exercises have a project of their own. Others have a different main method. The main method of a single project per exercise can be configured like this (in this example for *Chapter 3, Delegates, Events, and Lambdas*, you are configuring *Exercise02* to be the build and launch points):

launch.json

```
{
    "version": "0.2.0",
    "configurations": [
        {
            "name": ".NET Core Launch (console)",
            "type": "coreclr",
            "request": "launch",
            "preLaunchTask": "build",
            "program": "${workspaceFolder}/Exercises/ /Exercise02/bin/
Debug/net6.0/Exercise02.exe",
            "args": [],
            "cwd": "${workspaceFolder}",
            "stopAtEntry": false,
            "console": "internalConsole"
        }

    ]
}
```

tasks.json

```
{
    "version": "2.0.0",
    "tasks": [
        {
            "label": "build",
            "command": "dotnet",
            "type": "process",
            "args": [
                "build",
                "${workspaceFolder}/Chapter05.csproj",
                "/property:GenerateFullPaths=true",
                "/consoleloggerparameters:NoSummary"
            ],
            "problemMatcher": "$msCompile"
        },

    ]
}
```

One project for each exercise (for example, **Chapter05 Exercise02**) can be configured like this:

launch.json

```
{
    "version": "0.2.0",
    "configurations": [
        {
            "name": ".NET Core Launch (console)",
            "type": "coreclr",
            "request": "launch",
            "preLaunchTask": "build",
            "program": "${workspaceFolder}/bin/Debug/net6.0/Chapter05.
exe",
            "args": [],
            "cwd": "${workspaceFolder}",
            "stopAtEntry": false,
            "console": "internalConsole"
        }

    ]
}
```

tasks.json

```
{
    "version": "2.0.0",
    "tasks": [
        {
            "label": "build",
            "command": "dotnet",
            "type": "process",
            "args": [
              "build",
              "${workspaceFolder}/Chapter05.csproj",
              "/property:GenerateFullPaths=true",
              "/consoleloggerparameters:NoSummary",
              "-p:StartupObject=Chapter05.Exercises.Exercise02.
Program",
            ],
            "problemMatcher": "$msCompile"
        },
```

```
        ]
    }
```

Now that you are aware of **launch.json** and **tasks.json**, you can proceed to the next section which details the installation of the .NET developer platform.

INSTALLING THE .NET DEVELOPER PLATFORM

The .NET developer platform can be downloaded from https://dotnet.microsoft.com/download. There are variants for Windows, macOS, and Docker on Linux. *The C# Workshop* book uses .NET 6.0.

Follow the steps to install the .NET 6.0 platform on Windows:

1. Select the **Windows** platform tab:

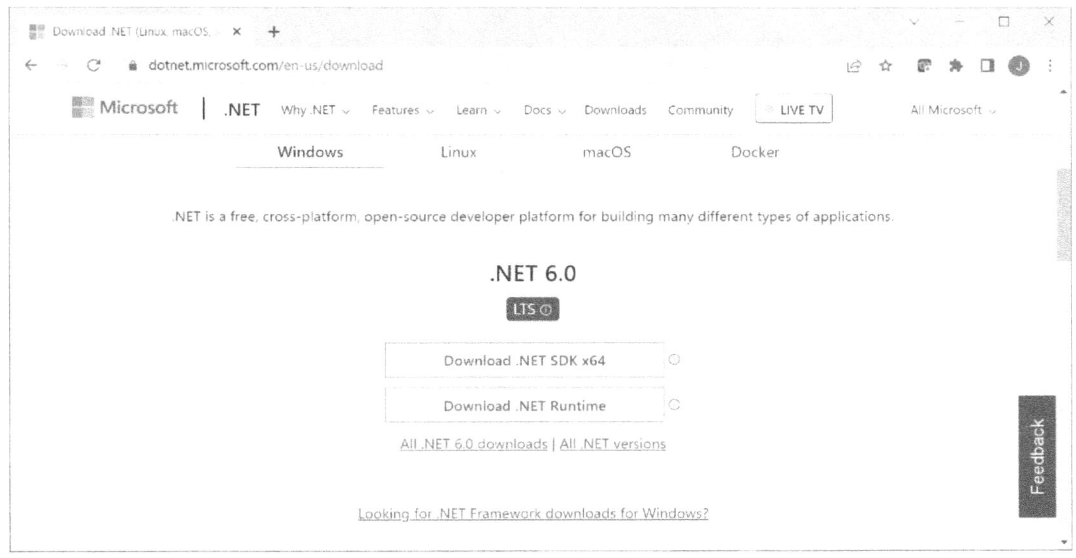

Figure 0.2: .NET 6.0 download window

2. Click on the **Download .NET SDK x64** option.

> **NOTE**
>
> The screen shown in *Figure 0.2* may change depending on the latest release from Microsoft.

3. Open and complete the installation according to the respective OS installed on your system.
4. Restart the computer after the installation.

Follow the steps to install the .NET 6.0 platform on macOS:

1. Select the **`macOS`** platform tab (*Figure 0.2*).
2. Click on the **`Download .NET SDK x64`** option.

   After the download is complete, open the installer file. You should have a screen similar to *Figure 0.3*:

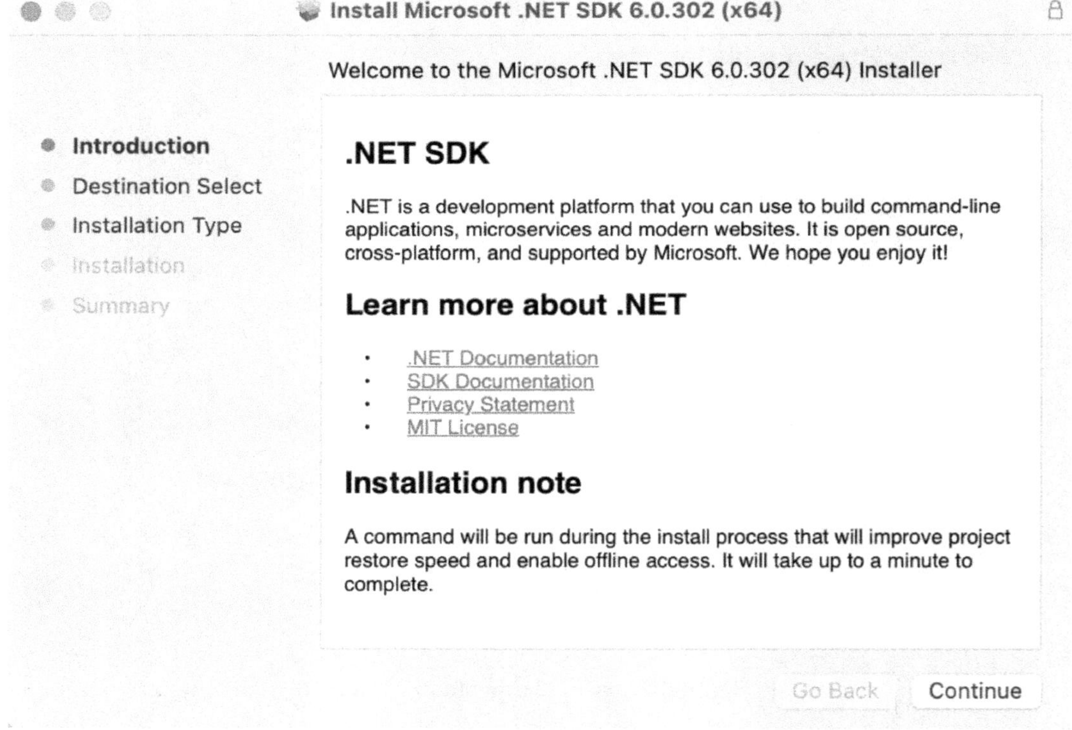

Figure 0.3: The macOS installation starting screen

3. Click on the **`Continue`** button.

   The following screen will confirm the amount of space that will be required for the installation:

4. Click on the **`Install`** button to continue:

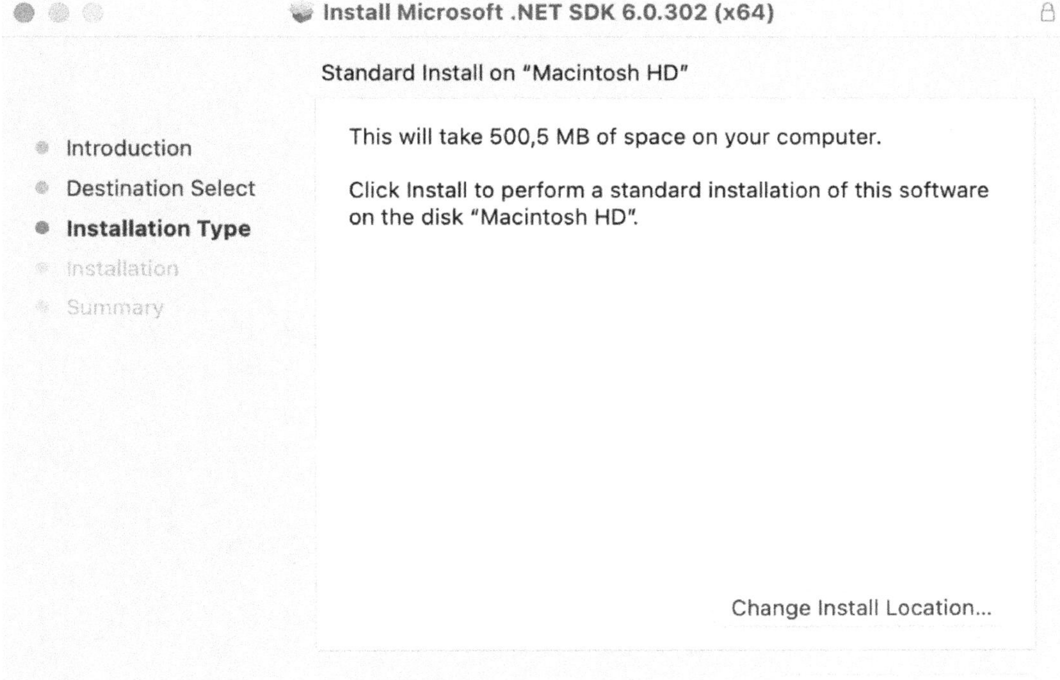

Figure 0.4: Window displaying the disk space required for installation

You will see a progress bar moving on the next screen:

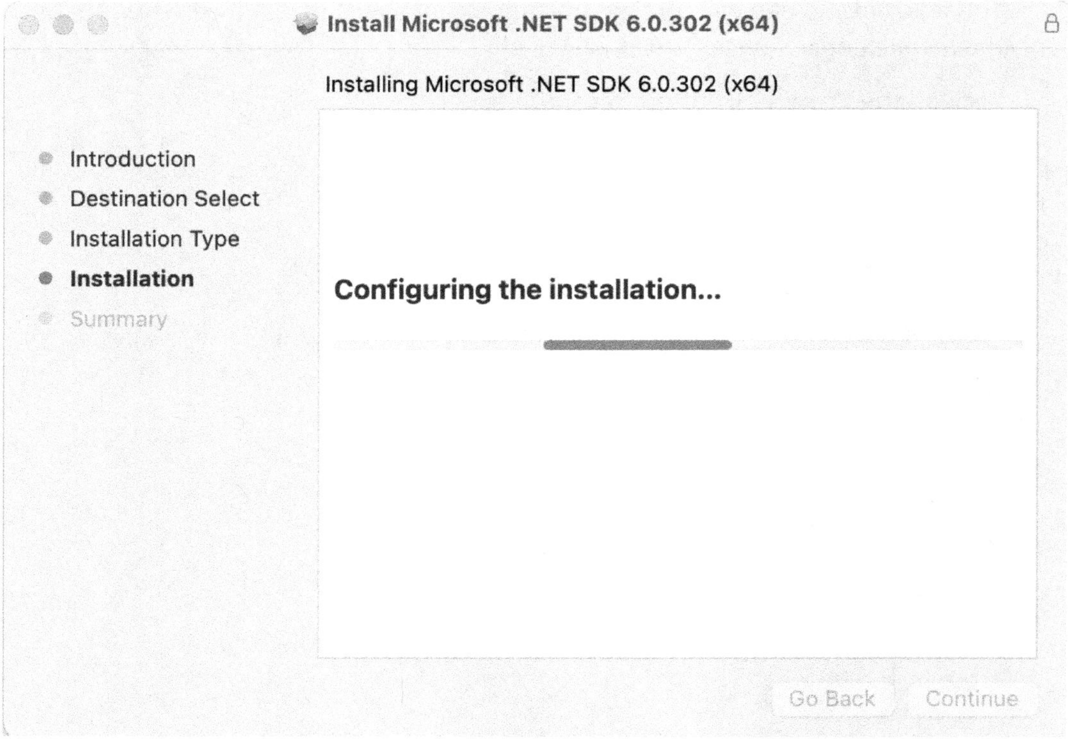

Figure 0.5: Window showing the Installation progress

Soon after the installation is finalized, you'll have a success screen (*Figure 0.6*):

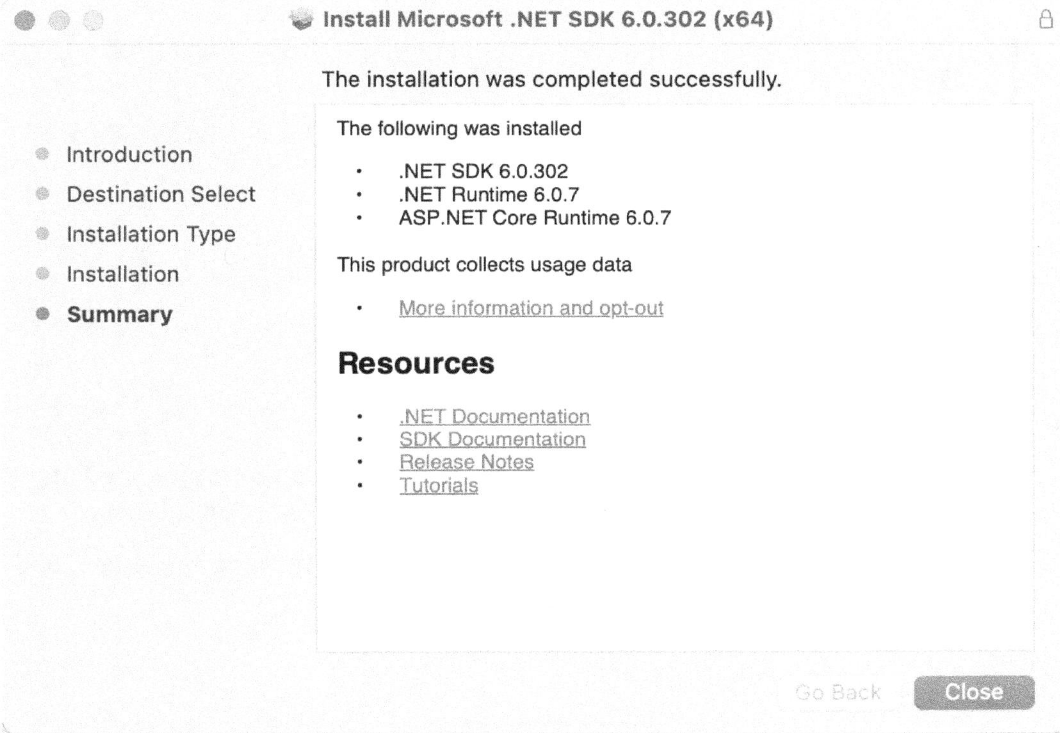

Figure 0.6: Window showing the installation as complete

5. In order to check whether the installation was a success, open your Terminal app and type:

```
dotnet -list-sdks
```

This will check the version of .NET installed on your machine. *Figure 0.7* shows the output where your installed SDKs will be listed:

```
mateusviegas@Mateuss-MacBook-Pro:~                    ⌥⌘1
❯ dotnet --list-sdks
5.0.100 [/usr/local/share/dotnet/sdk]
6.0.102 [/usr/local/share/dotnet/sdk]
6.0.200 [/usr/local/share/dotnet/sdk]
6.0.302 [/usr/local/share/dotnet/sdk]
~ ❯                                                   13:15:04
```

Figure 0.7: Checking the installed .NET SDKs in Terminal

With these steps, you can install the .NET 6.0 SDK on your machine and check the installed version.

> **NOTE**
>
> Net 6.0 installation steps for Linux are not included as they are like Windows and macOS.

Before proceeding further, it is important to know about .NET 6.0 features.

THE .NET 6.0 FEATURES FOUND IN WINDOWS, MACOS, AND LINUX

WINDOWS

- .NET 6.0: This is the latest **long-term support** (**LTS**) version recommended for Windows. It can be used for building many different types of applications.
- .NET Framework 4.8: This is a Windows-only version for building any type of app to run on Windows only.

MACOS

- .NET 6.0: This is the LTS version recommended for macOS. It can be used for building many different types of applications. Choose the version that is compatible with the processor of your Apple Computer—x64 for Intel chips and ARM64 for Apple chips.

LINUX

- .NET 6.0: This is the LTS version recommended for Linux. It can be used for building many different types of applications.

DOCKER

- .NET images: This developer platform can be used for building different types of applications.
- .NET Core images: This offers lifetime support for building many types of applications.
- .NET framework images: These are Windows-only versions of .NET for building any type of app that runs on Windows.

With .NET 6.0 installed on your system, the next step is to configure projects using CLI.

.NET COMMAND-LINE INTERFACE (CLI)

Once you have installed .NET, the CLI can be used to create and configure projects for use with VS Code. To launch the .NET CLI, run the following at the command prompt:

```
dotnet
```

If .NET is installed correctly, you will see the following message on your screen:

```
Usage: dotnet [options]
Usage: dotnet [path-to-application]
```

Once you have the CLI installed to configure projects with VS Code, you need to know about the powerful open source object-relational database system that uses and extends the SQL language that is, PostgreSQL.

> **NOTE**
>
> You will first go through the instructions to install PostgreSQL for Windows followed by macOS, and then by Linux.

POSTGRESQL INSTALLATION FOR WINDOWS

PostgreSQL has been used in *Chapter 6*, *Entity Framework with SQL Server*. Before you proceed with that chapter, you must install PostgreSQL on your system using the following steps:

1. Go to https://www.enterprisedb.com/downloads/postgres-postgresql-downloads and download the latest version installer for Windows:

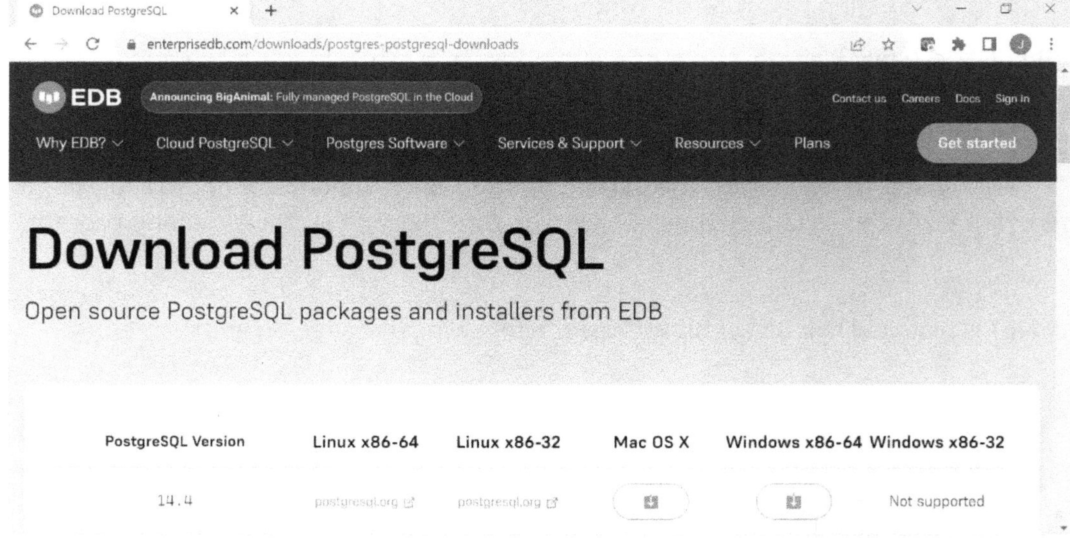

Figure 0.8: Latest PostgreSQL versions for each platform

> **NOTE**
>
> The screen shown in *Figure 0.8* may change depending upon the latest release from the vendor.

2. Open the downloaded interactive installer and click the **`Next`** button. The **`Setup PostgreSQL`** screen gets displayed:

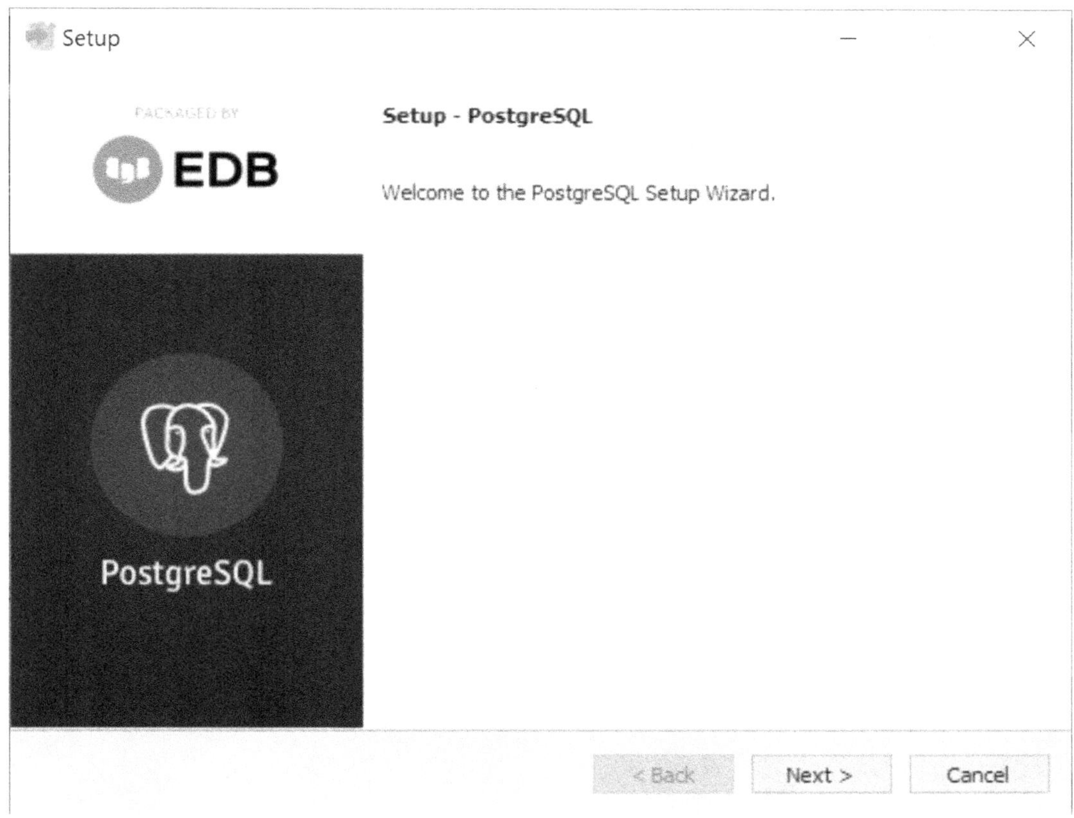

Figure 0.9: Welcome screen for PostgreSQL upload

3. Click on the **`Next`** button to move to the next screen which asks for the installation directory details:

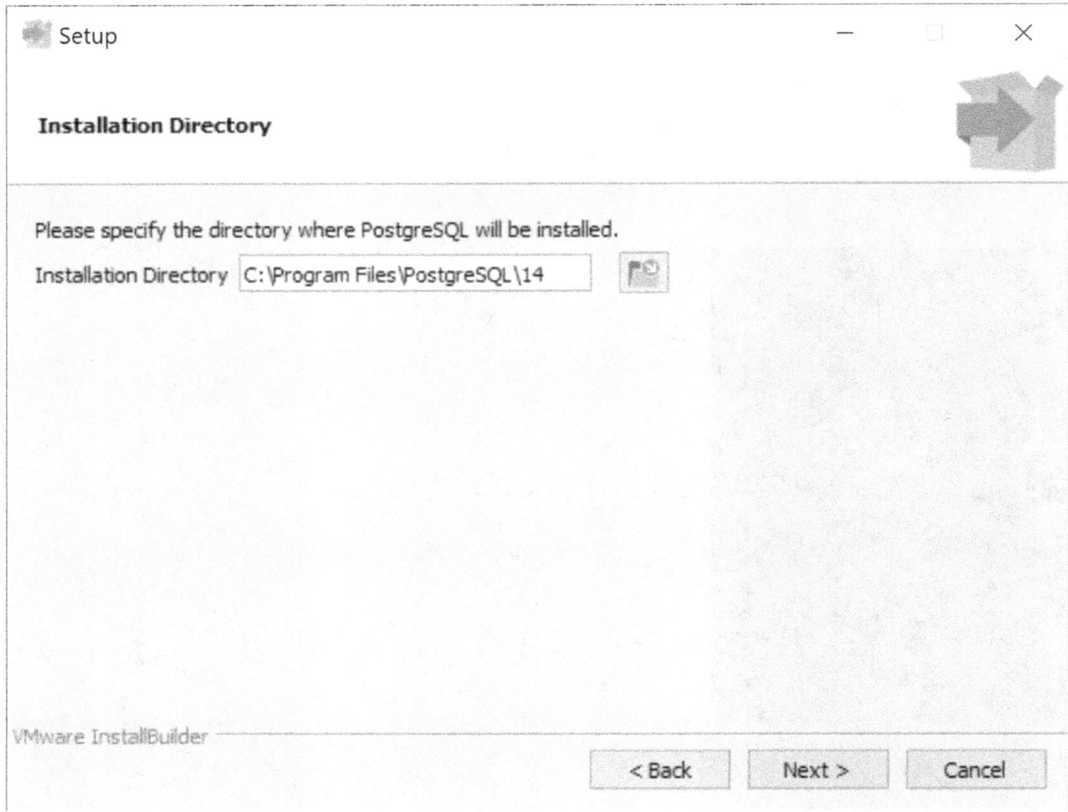

Figure 0.10: PostgreSQL default installation directory

4. Leave the default **`Installation Directory`** unchanged and click the **`Next`** button.

5. Select the following from the list in *Figure 0.11*:

    - **`PostgreSQL Server`** refers to the database.
    - **`pgAdmin 4`** is the database management tool.

- **`Stack Builder`** is the PostgreSQL environment builder (optional).
- **`Command Line Tools`** work with the database using a command line.

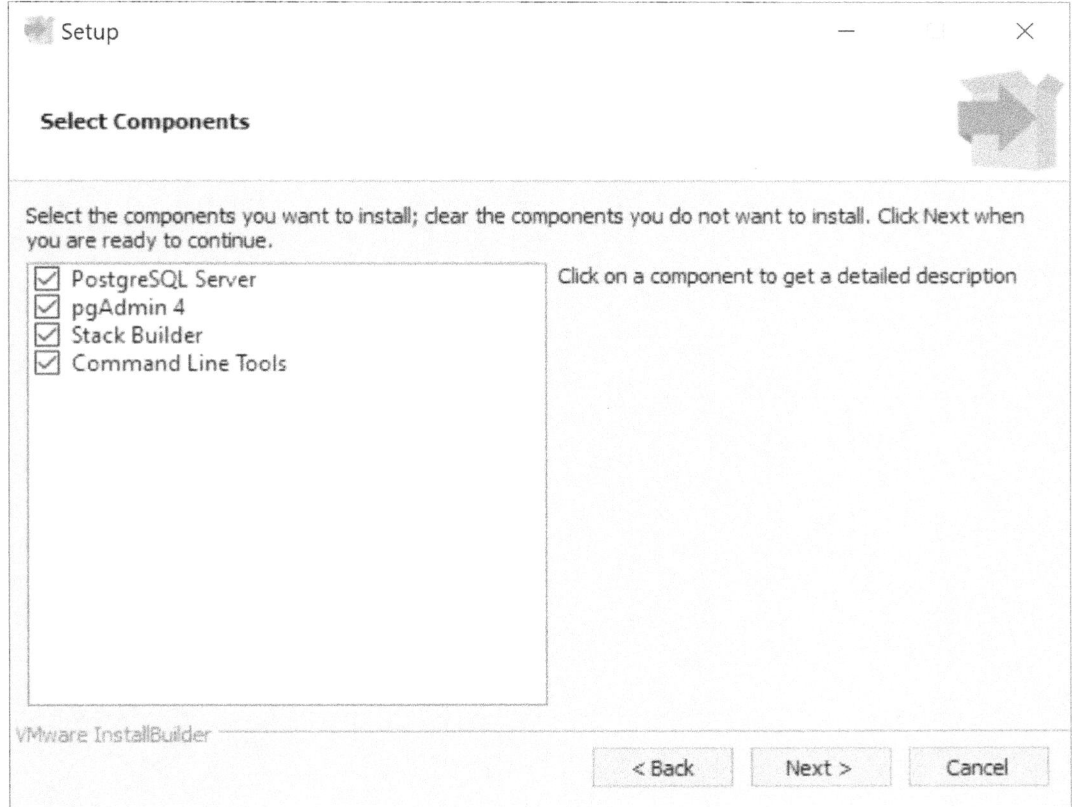

Figure 0.11: The PostgreSQL components selected to proceed

6. Then click the **`Next`** button.

7. In the next screen, the Data Directory screen asks you to enter the directory for storing your data. So, enter the data directory name:

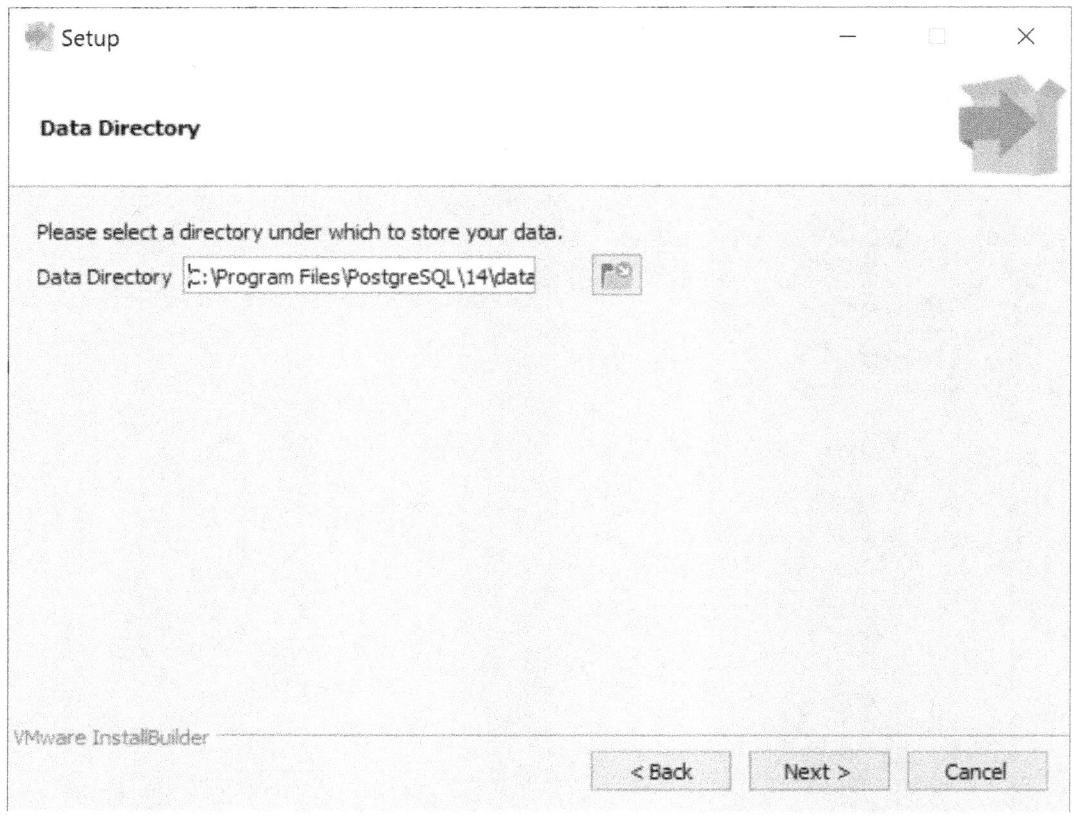

Figure 0.12: The directory for storing data

8. Once you have entered the data directory, click on the **`Next`** button to continue. The next screen asks you to enter the password.
9. Enter the new **`Password.`**
10. Retype the password beside **`Retype password`** for the database superuser:

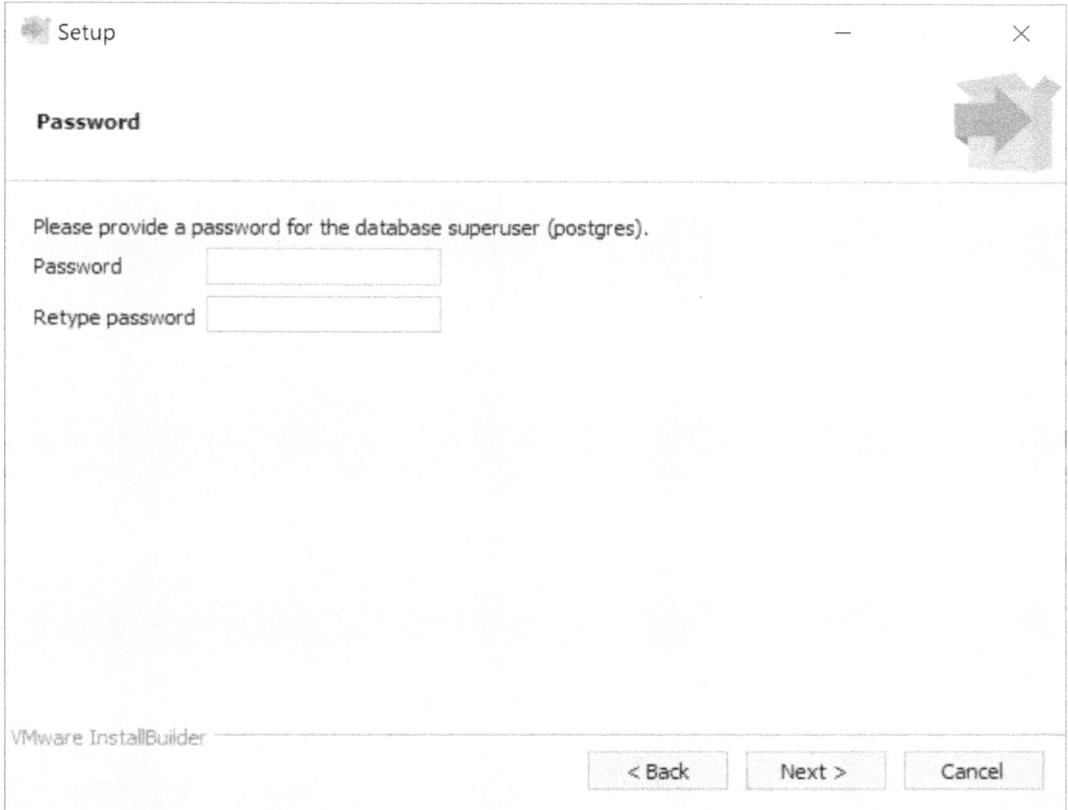

Figure 0.13: Providing password for database superuser

11. Then click the **`Next`** button to continue.

12. The next screen displays the Port as **5432**. Use the default port—that is, **5432**:

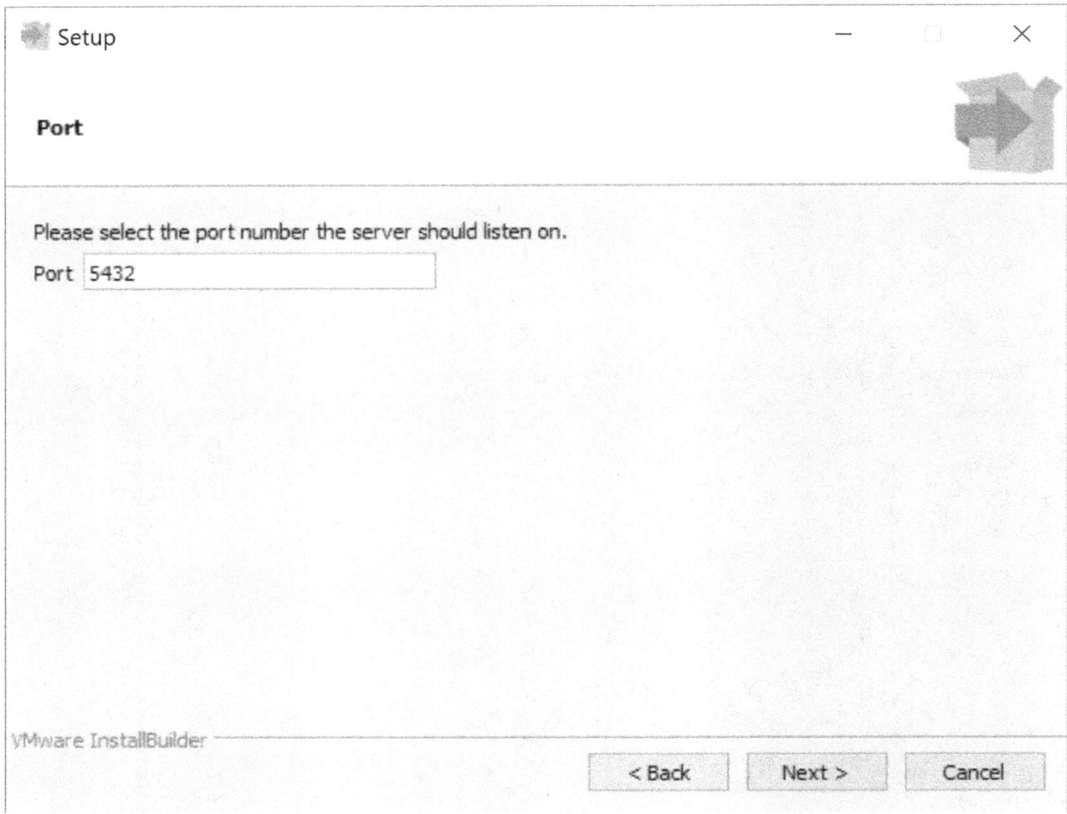

Figure 0.14: Selecting Port

13. Click the **Next** button.

14. The Advanced Options screen asks you to type the locale for the Database cluster. Leave it as **`[Default locale]`**:

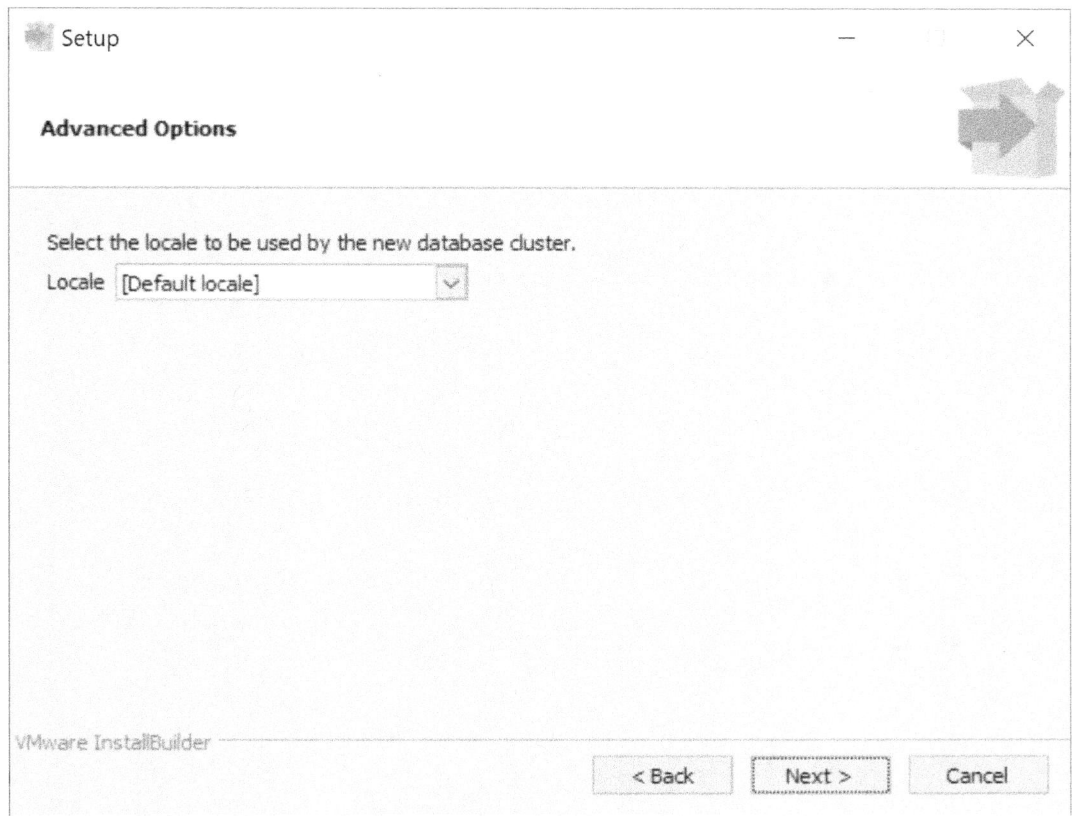

Figure 0.15: Selecting the locale for the Database cluster

15. Then click the **`Next`** button.
16. When the Preinstallation Summary screen gets displayed, click the **`Next`** button to go ahead:

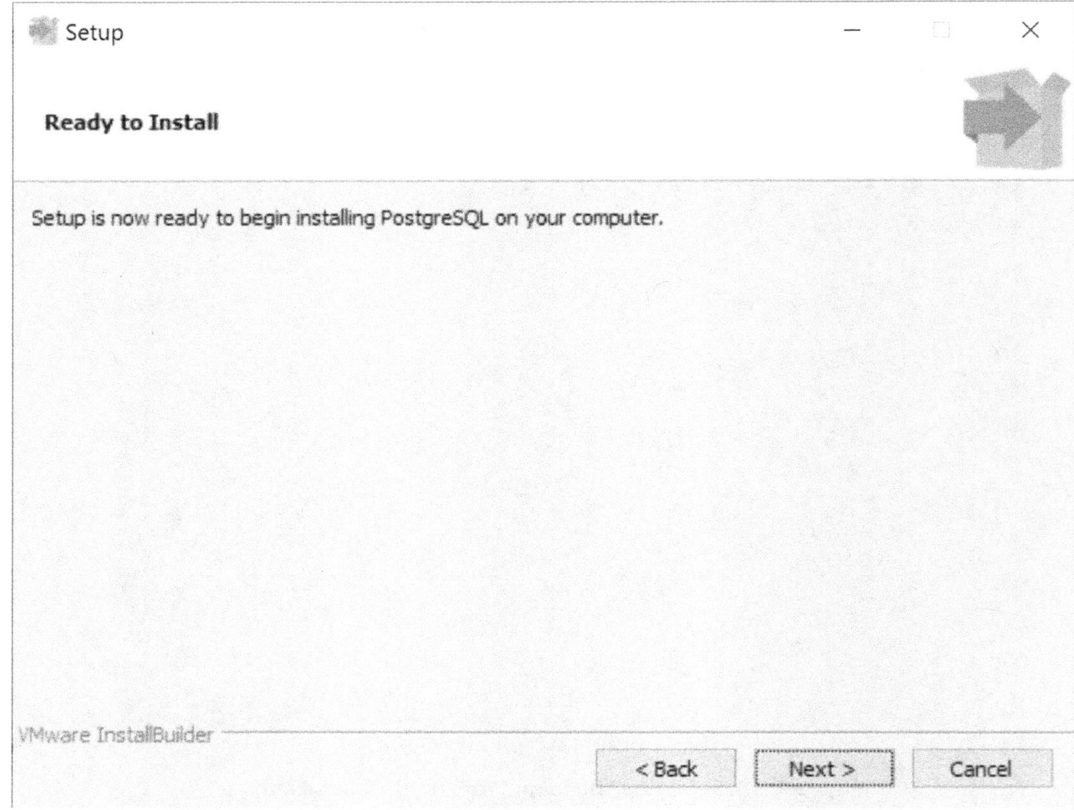

Figure 0.16: Setup window showing ready to install message

17. Continue selecting the **`Next`** button (leaving the default settings unchanged) until the installation process begins.

18. Wait for it to complete. On completion, the Completing the PostgreSQL Setup Wizard screen gets displayed.
19. Uncheck the **`Launch Stack Builder at exit`** option:

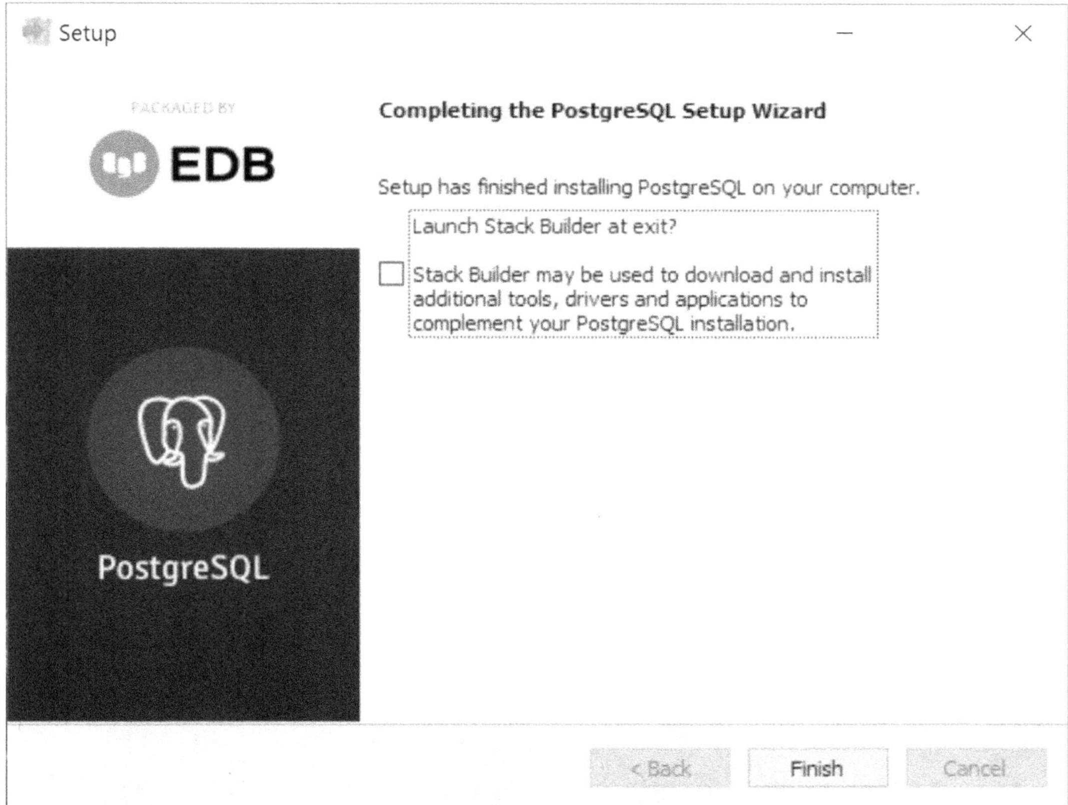

Figure 0.17: Installation complete with Stack Builder unchecked

The Stack Builder is used to download and install additional tools. The default installation contains all tools needed for the exercises and activities.

20. Finally, click the **Finish** button.
21. Now open **pgAdmin4** from Windows.
22. Enter a master **Password** for connecting to any database inside PostgreSQL in the Set Master Password window:

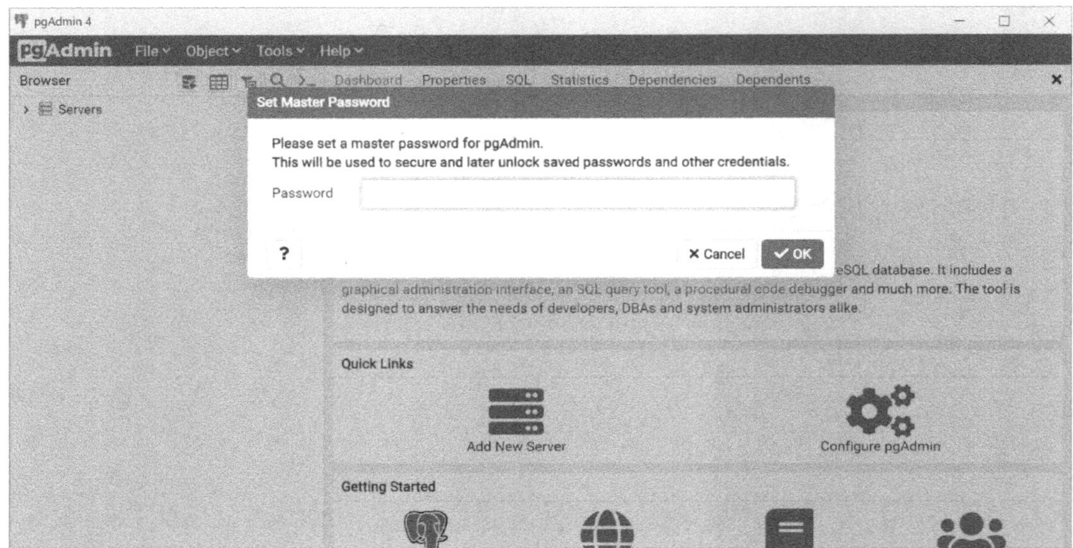

Figure 0.18: Setting Master Password for connecting to a PostgreSQL server

> **NOTE**
>
> It is better to type a password that you can easily memorize as it will be used to manage all your other credentials.

23. Next click the **OK** button.
24. On the left side of the pgadmin window, expand the **Server** by clicking the arrow beside it.
25. You will be asked to enter your PostgreSQL server password. Type the same password that you entered in *Step 22*.

26. Do not click **`Save password`** for security reasons:

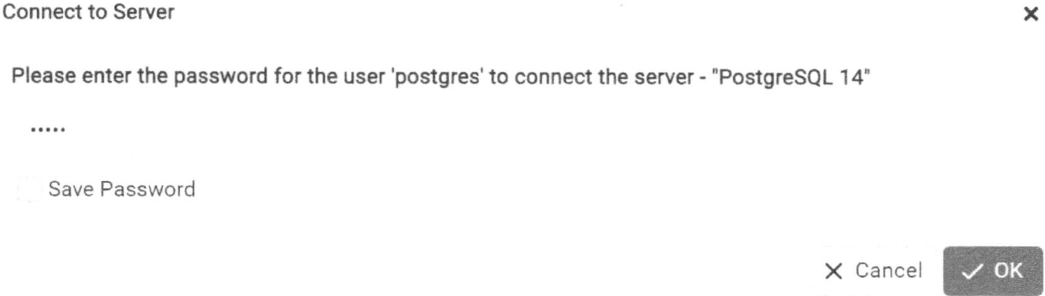

Figure 0.19: Setting the postgres user password for the PostgreSQL server

PostgreSQL server password is the password you will use when connecting to the PostgreSQL server and using the **`postgres`** user.

27. Finally click the **`OK`** button. You will see the pgAdmin dashboard:

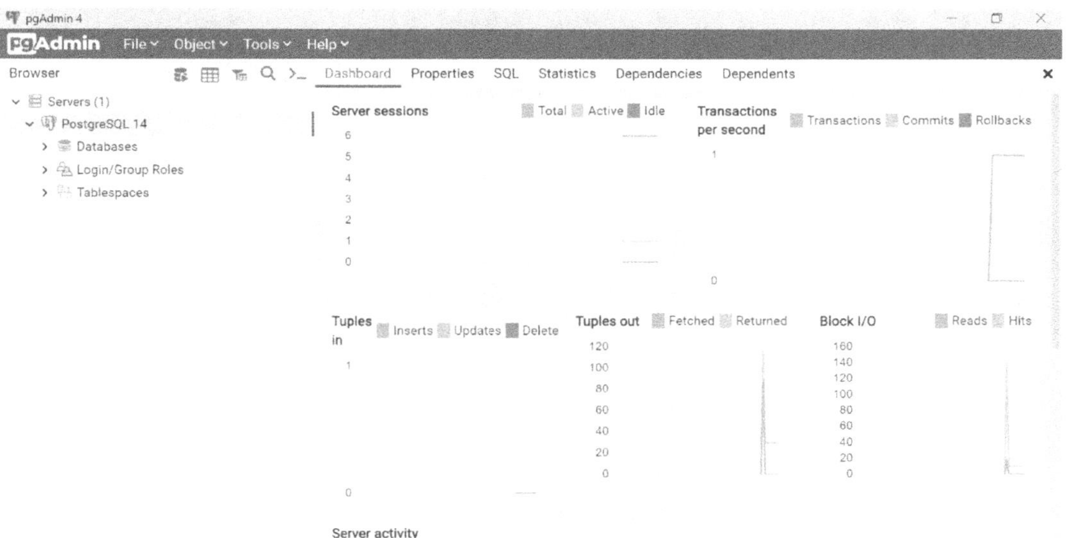

Figure 0.20: pgAdmin 4 dashboard window

In order to explore the pgAdmin dashboard, move to the *Exploring pgAdmin Dashboard* section.

POSTGRESQL INSTALLATION FOR MACOS

Install PostgreSQL on your macOS using the following steps:

1. Visit the official site of the Postgres app to download and install PostgreSQL on your mac platform: https://www.enterprisedb.com/downloads/postgres-postgresql-downloads.

2. Download the latest PostgreSQL for macOS:

> **NOTE**
>
> The following screenshots were taken for version 14.4 on macOS Monterey (version 12.2).

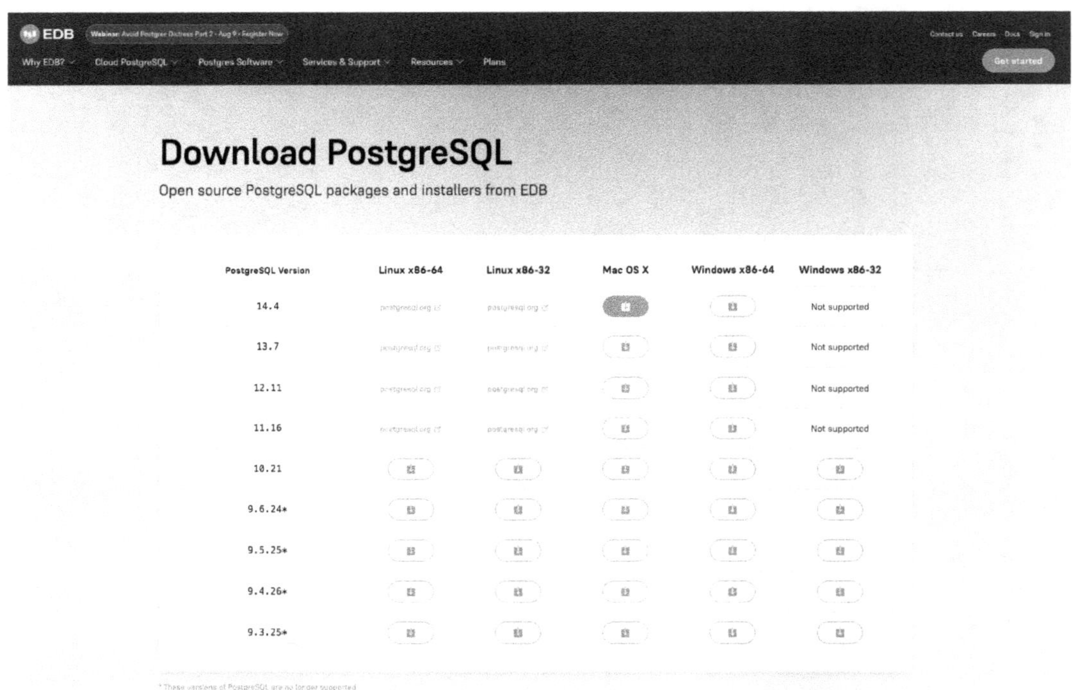

Figure 0.21: Installation page for PostgreSQL

3. Once you have downloaded the installer file for macOS, double-click the `installer file` to launch the PostgreSQL Setup Wizard:

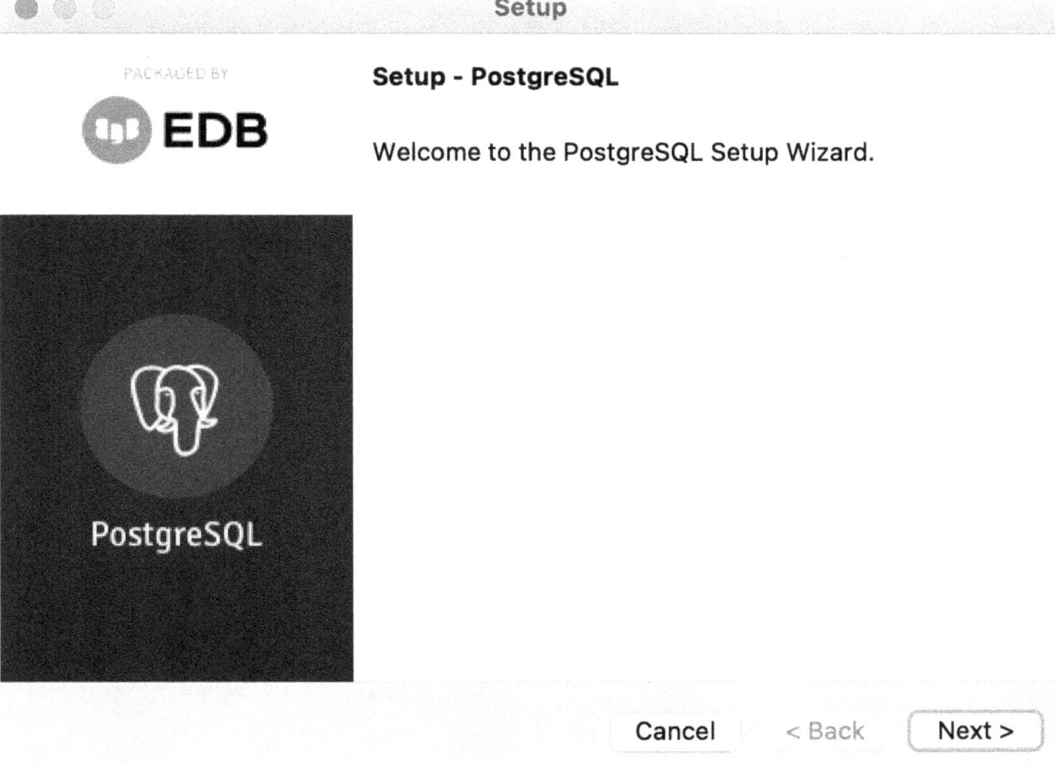

Figure 0.22: Launching the PostgreSQL setup wizard

4. Select the location where you want PostgreSQL installed:

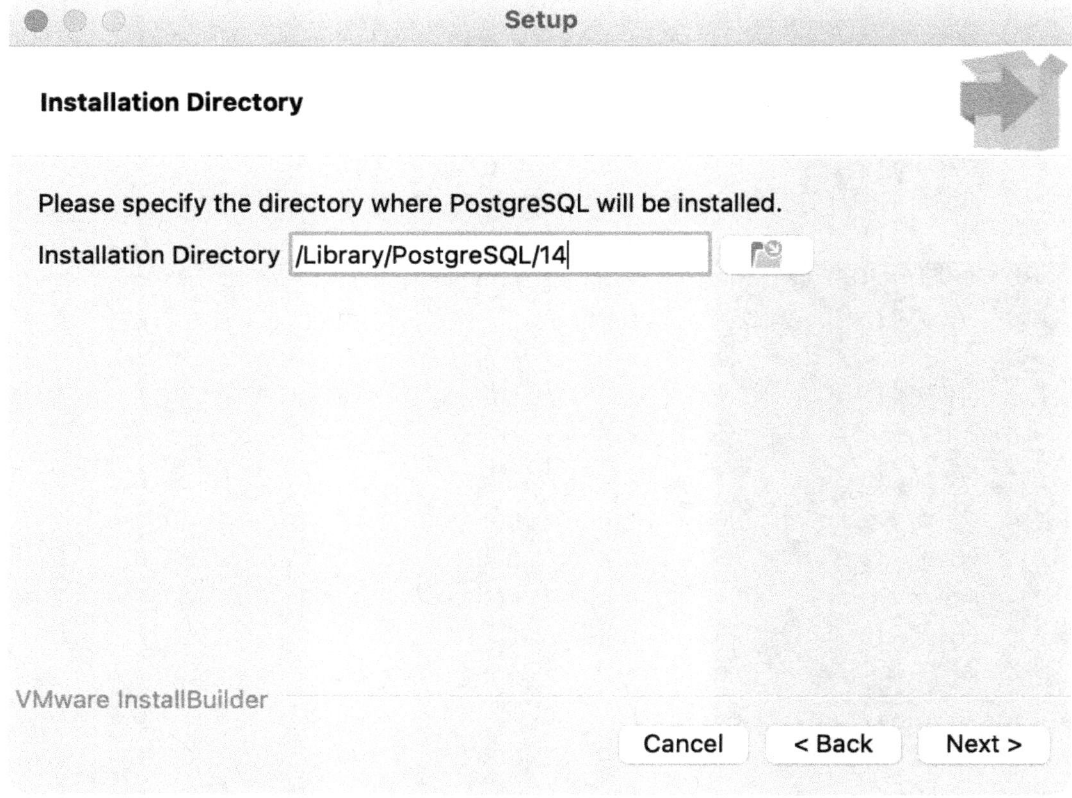

Figure 0.23: Selecting the installation directory

5. Click on the **Next** button.
6. In the next screen, select the following components for installation:
   - PostgreSQL Server
   - **pgAdmin 4**
   - Command Line Tools

7. Uncheck the **`Stack Builder`** component:

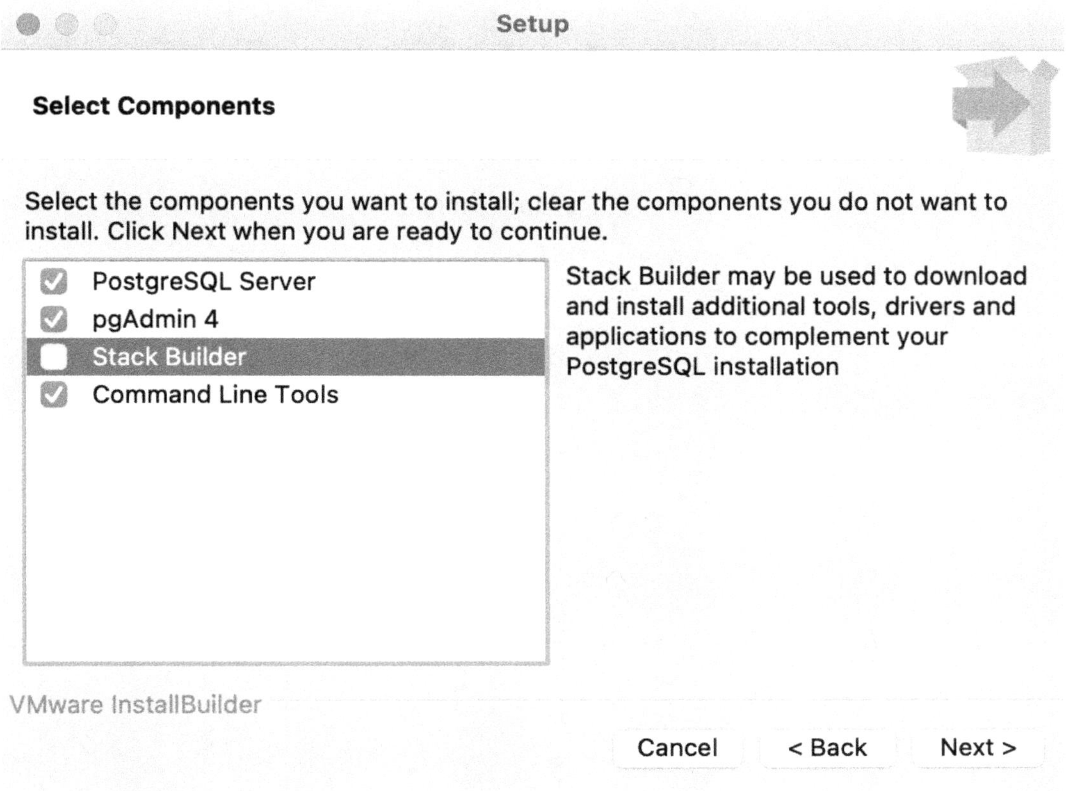

Figure 0.24: Selecting the components for installation

8. Once you have selected the options, click on the **`Next`** button.

9. Specify the data directory in which PostgreSQL will store the data:

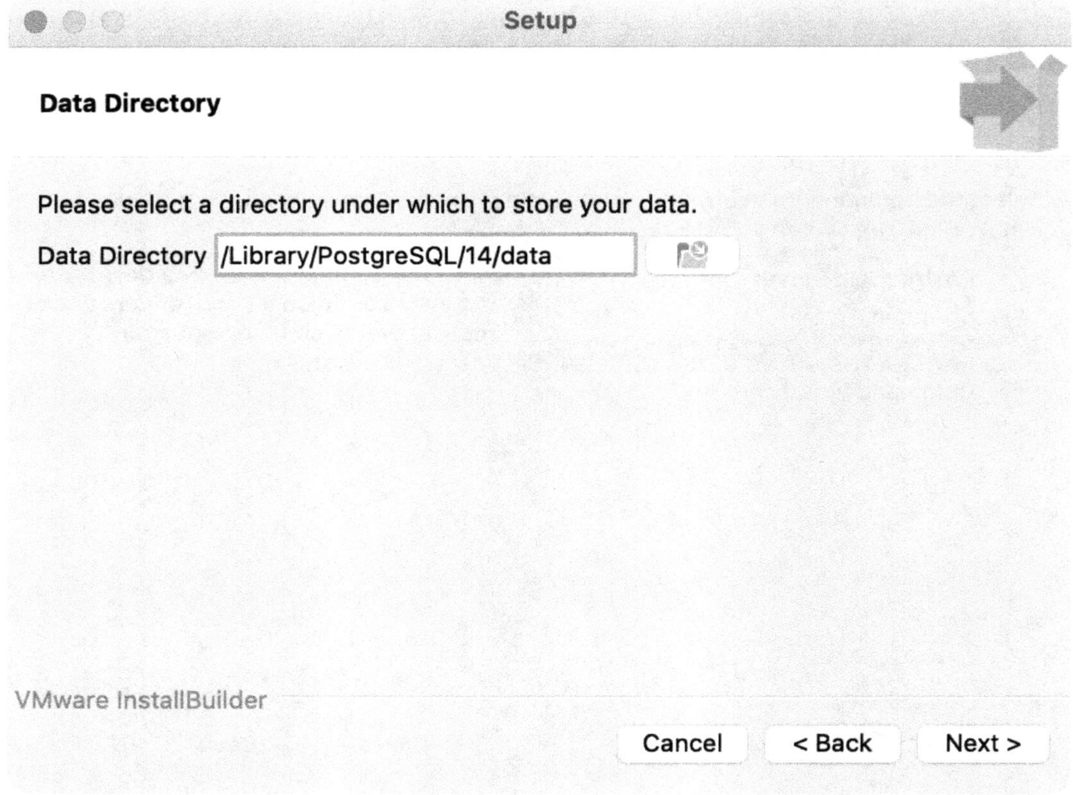

Figure 0.25: Specifying the data directory

10. Click on the **Next** button.

11. Now set a **`Password`** for the Postgres database superuser:

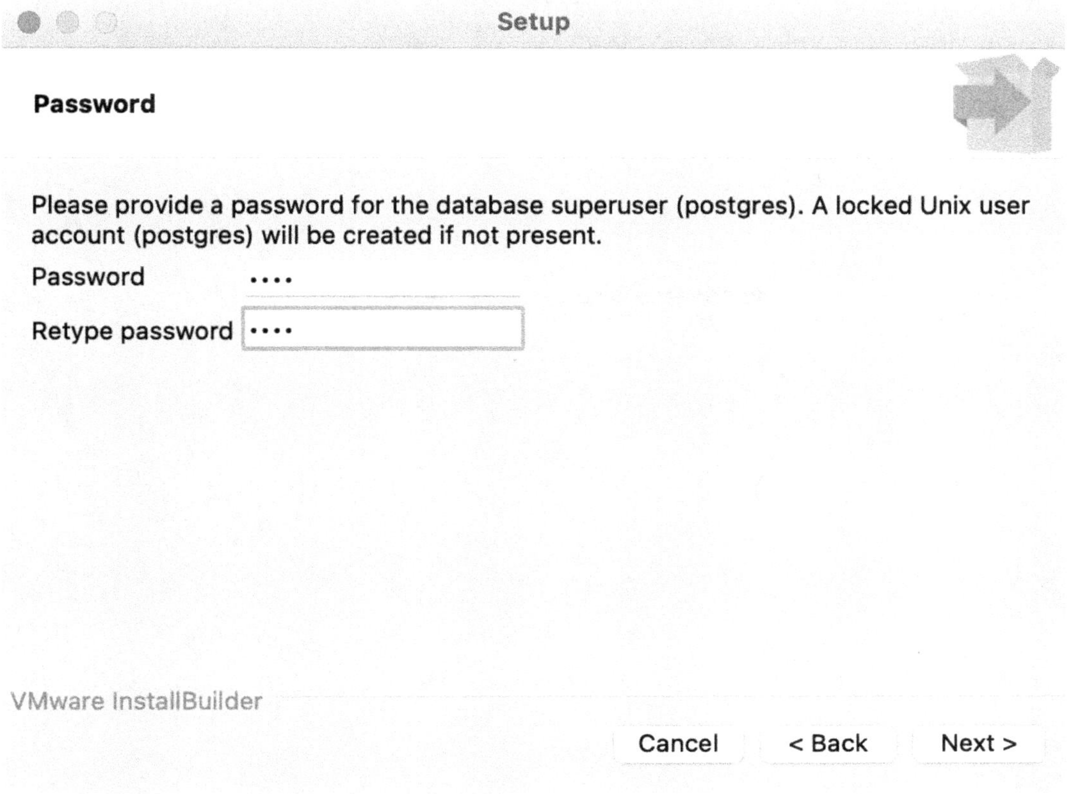

Setup

Password

Please provide a password for the database superuser (postgres). A locked Unix user account (postgres) will be created if not present.

Password ••••

Retype password ••••

VMware InstallBuilder

Cancel < Back Next >

Figure 0.26: Setting the password

Make sure to note down the password safely for logging in to the PostgreSQL database.

12. Click on the **`Next`** button.

    Set the port number where you want to run the PostgreSQL server. Here the default Port number is set as **`5432`**:

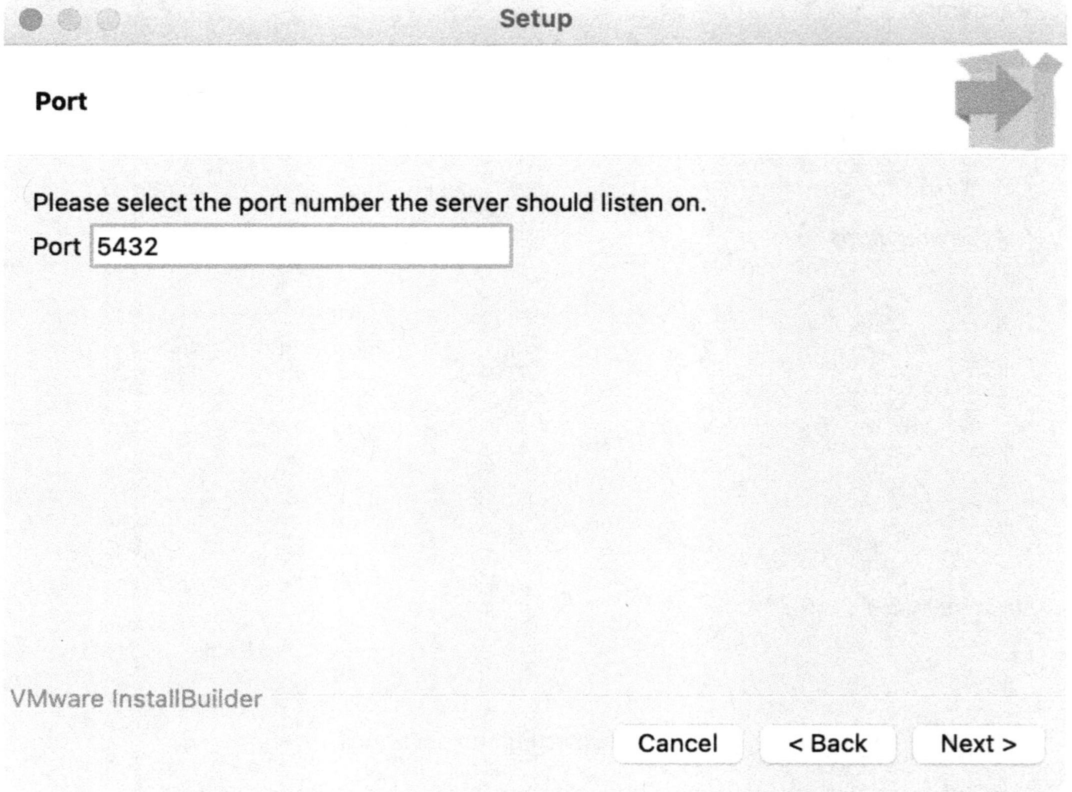

Figure 0.27: Specifying the port number

13. Click on the **`Next`** button.

14. Select the locale to be used by PostgreSQL. Here, **[Default locale]** is the locale selected for macOS:

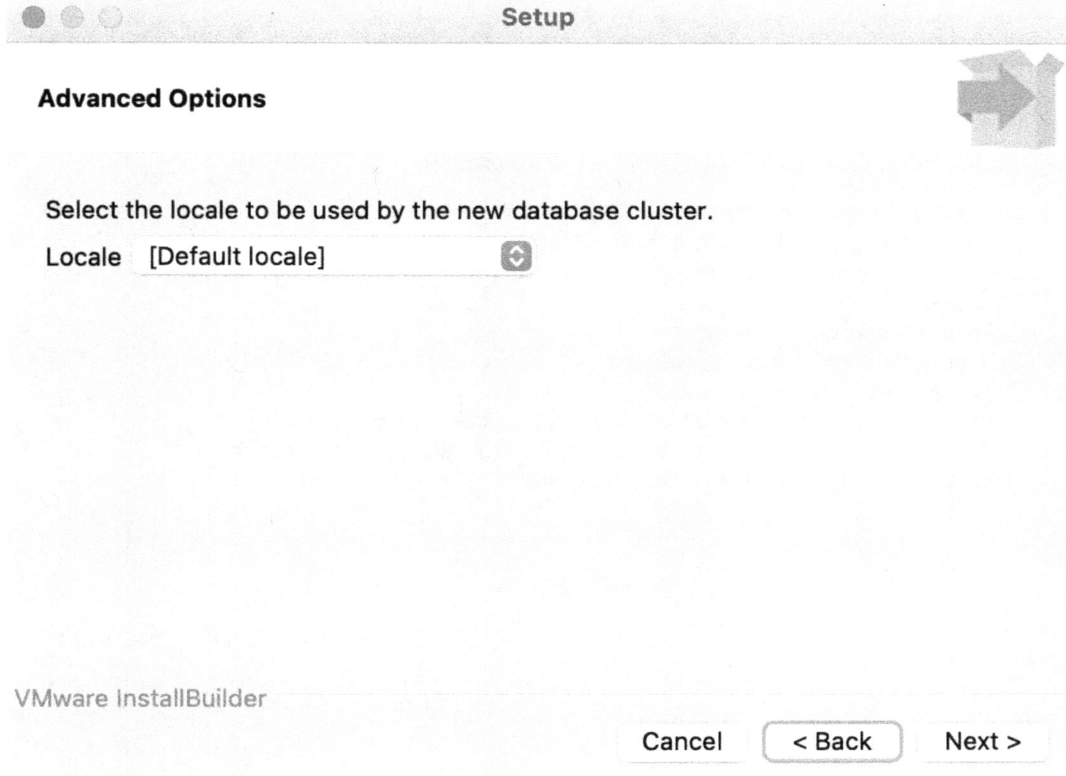

Figure 0.28: Selecting the locale specification

15. Click on the **Next** button.

16. In the next screen, check the installation details:

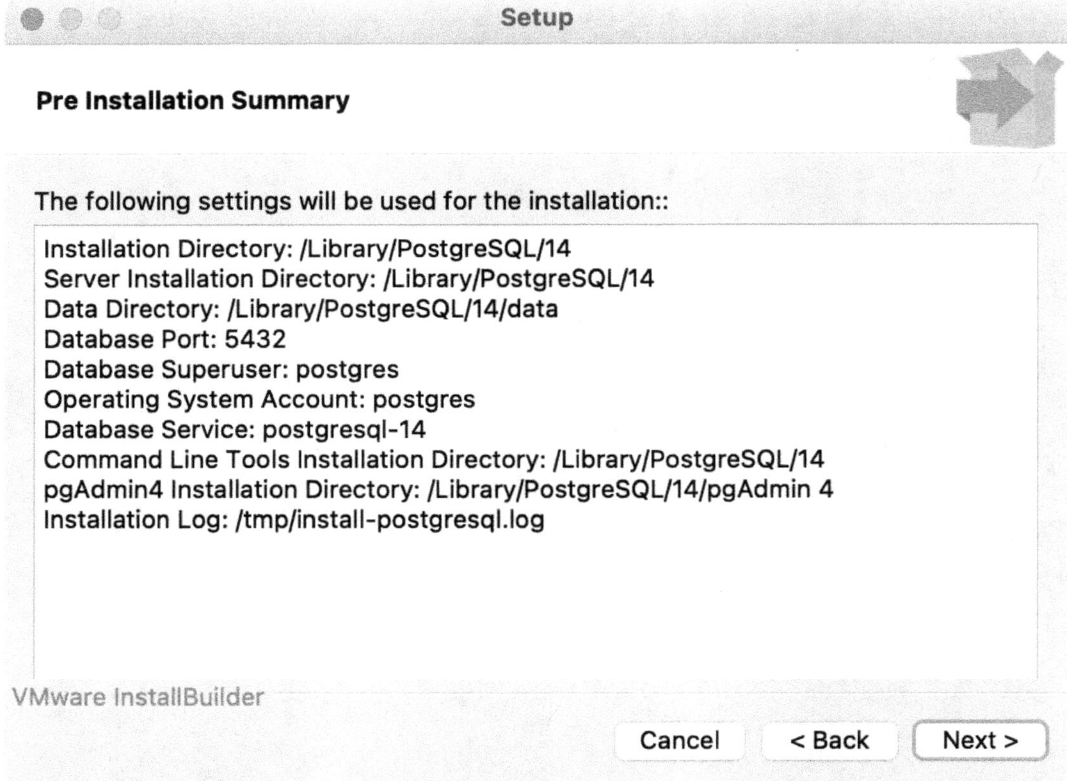

Figure 0.29: Pre Installation summary page

Finally, click on the **`Next`** button to start the installation process of the PostgreSQL database server on your system:

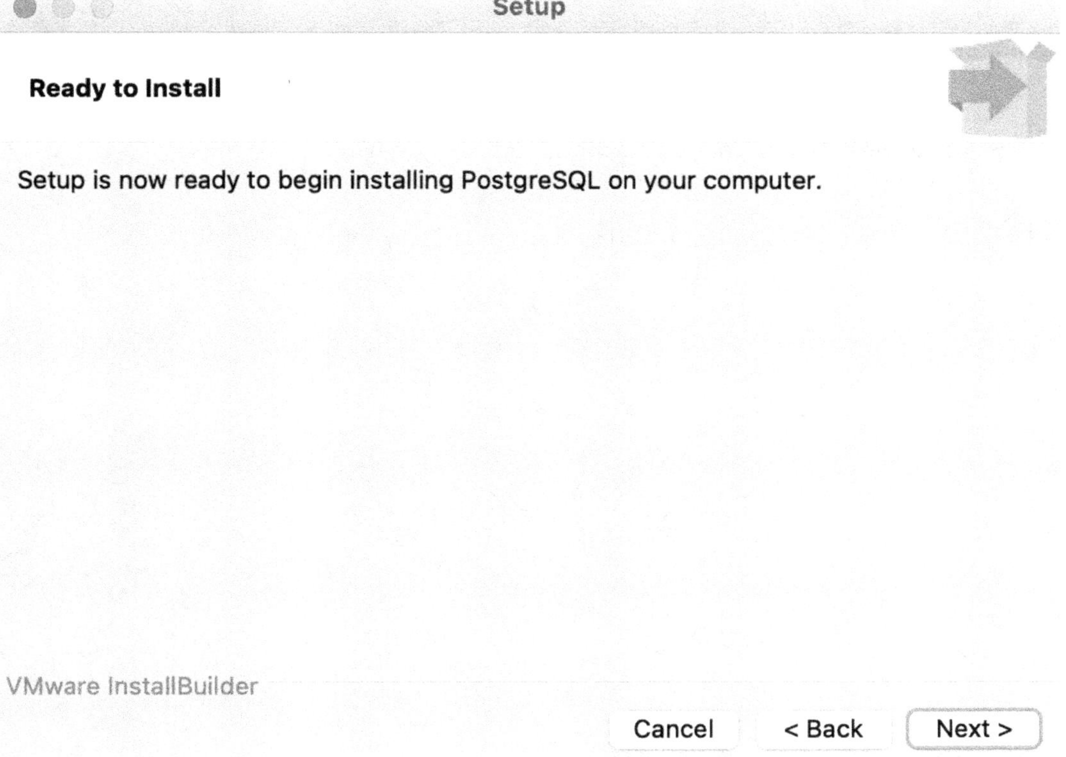

Figure 0.30: Ready to Install page before starting the installation process

17. Wait for a few moments for the installation process to complete:

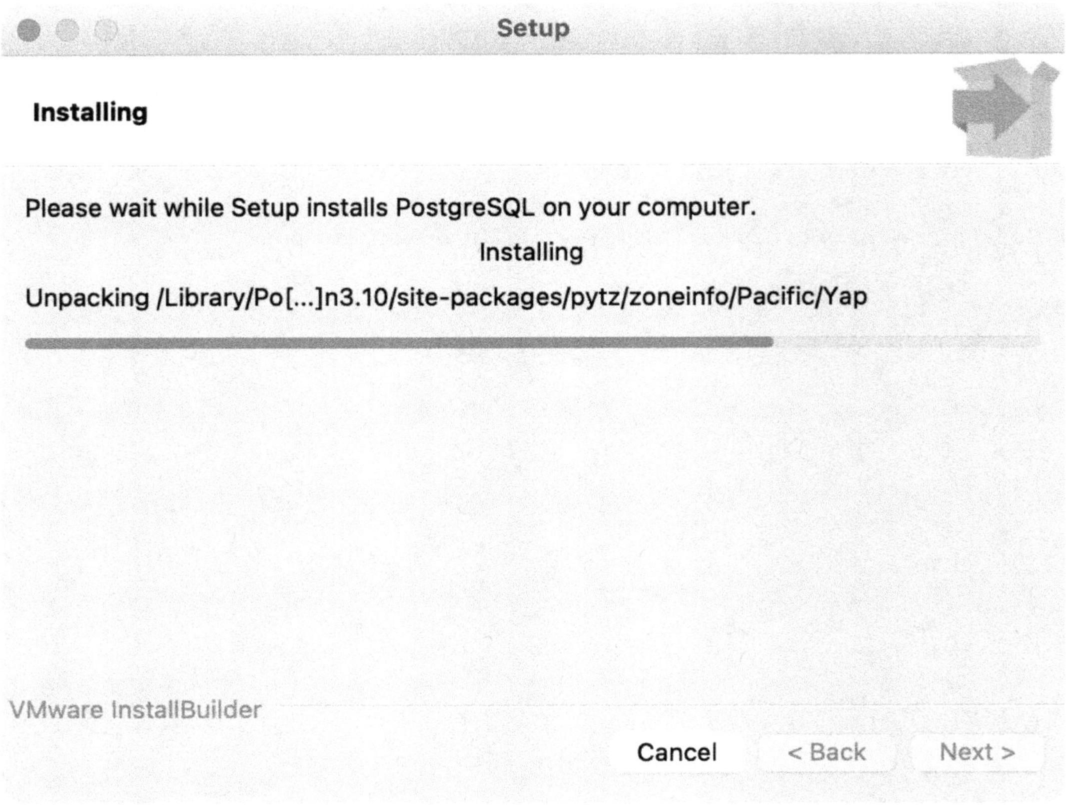

Figure 0.31: Setup Installation in progress

18. When prompted, click the **`Next`** button. The next screen displays the message that the PostgreSQL installation is complete on your system:

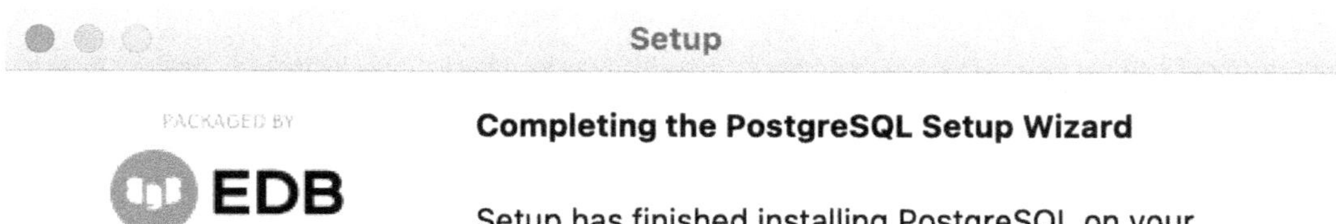

Figure 0.32: Success message showing the setup as complete

19. Click the **`Finish`** button once the installation gets complete.

20. Now load the database in the PostgreSQL server.

21. Double-click on the **`pgAdmin 4`** icon to launch it from your Launchpad.

22. Enter the password for the PostgreSQL user that you had set during the installation process.

23. Then click the **OK** button. You will now see the pgAdmin dashboard.

This completes the installation of PostgreSQL for the macOS. The next section will familiarize you with the PostgreSQL interface.

EXPLORING PGADMIN DASHBOARD

Once you have installed PostgreSQL in Windows and macOS, follow these steps to grasp the interface better:

1. Open **`pgAdmin4`** from Windows/ macOS (in case pgAdmin is not open on your system).

2. Click on the **`Servers`** option on the left:

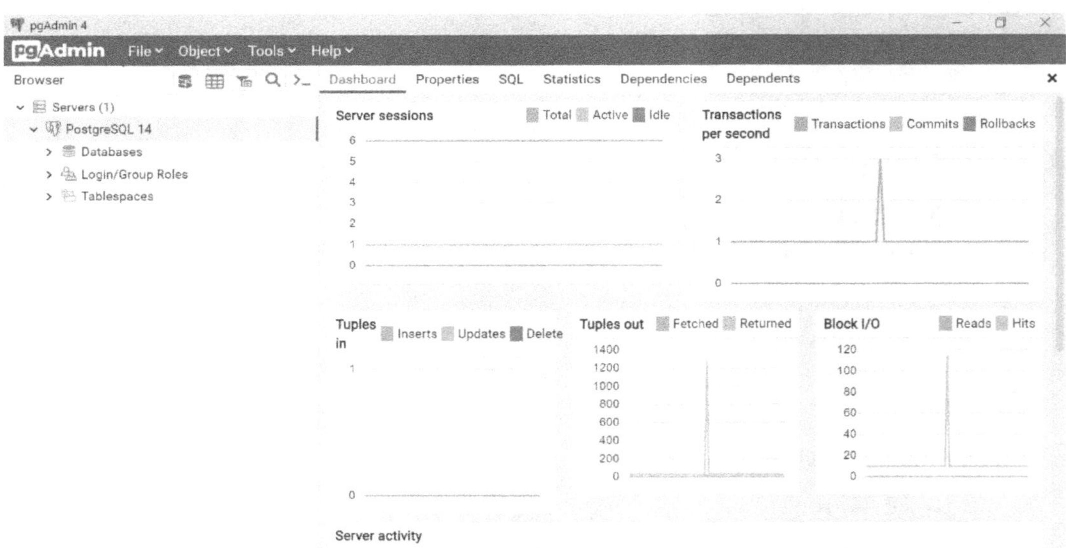

Figure 0.33: Clicking on Servers to create a database

3. Right-click on **`PostgreSQL 14`**.

4. Then click on the **`Create`** option.

5. Choose the **`Database`**... option to create a new database:

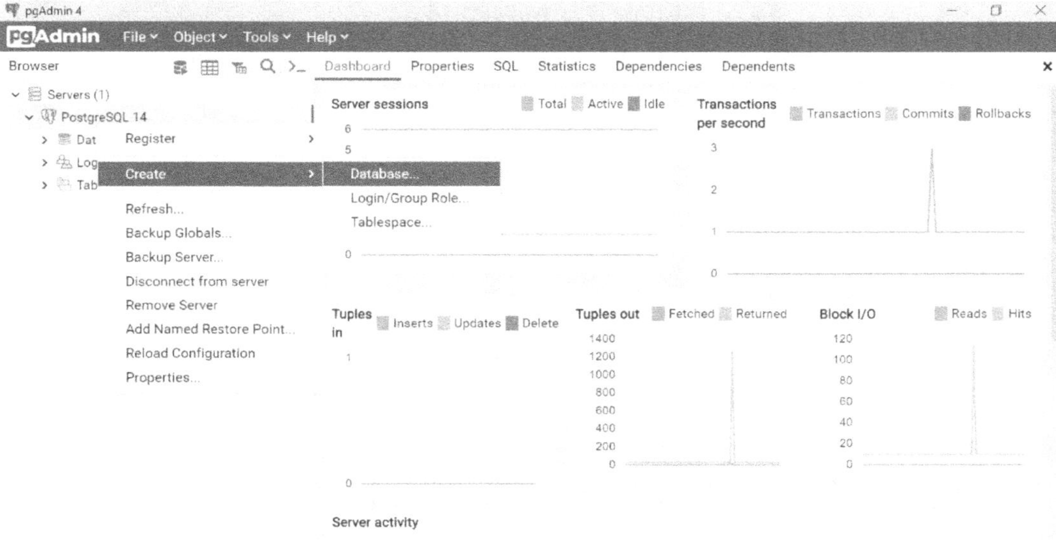

Figure 0.34: Creating a new database

This will open a Create – Database window.

6. Enter the database name, as **`TestDatabase`**.
7. Select the Owner of the database or leave it as default. For now, just use the **`Owner`** as **`postgres`**:

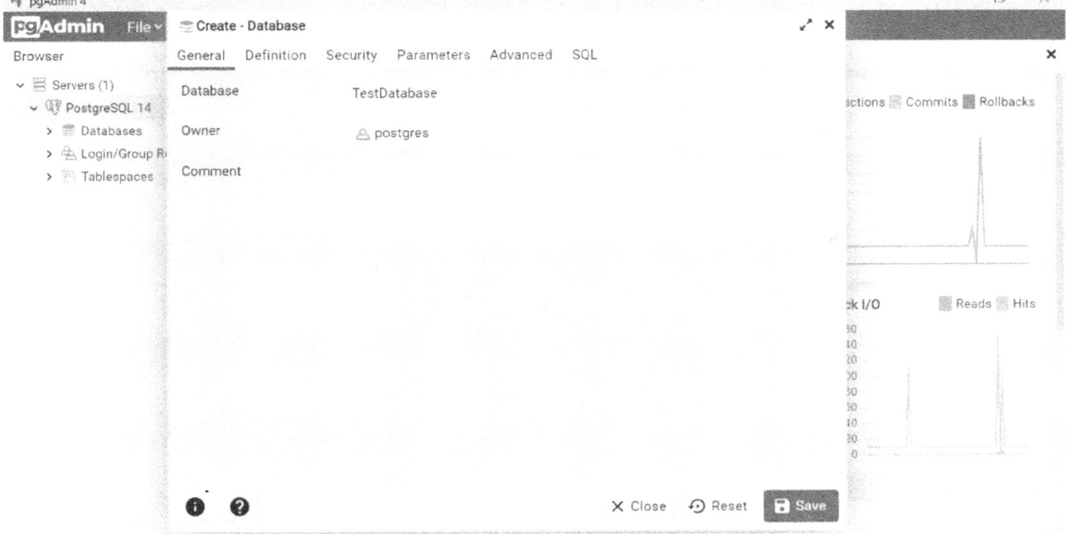

Figure 0.35: Selecting the owner of the database

8. Then click on the **Save** button. This will create a database.

9. Right-click on **Databases** and choose the **Refresh** button:

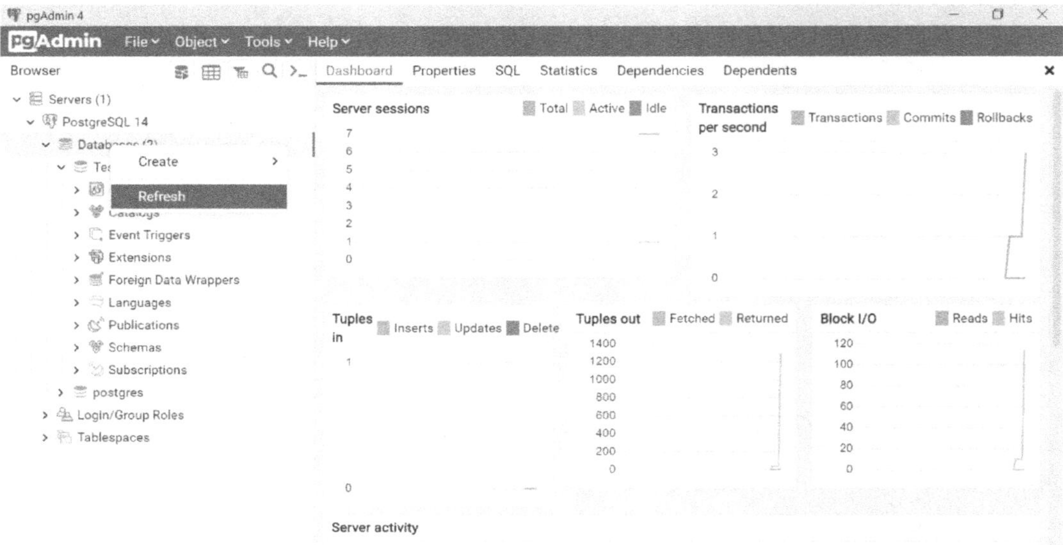

Figure 0.36: Clicking the Refresh... button after right-clicking Databases

A database with the name **TestDatabase** is now displayed within the dashboard:

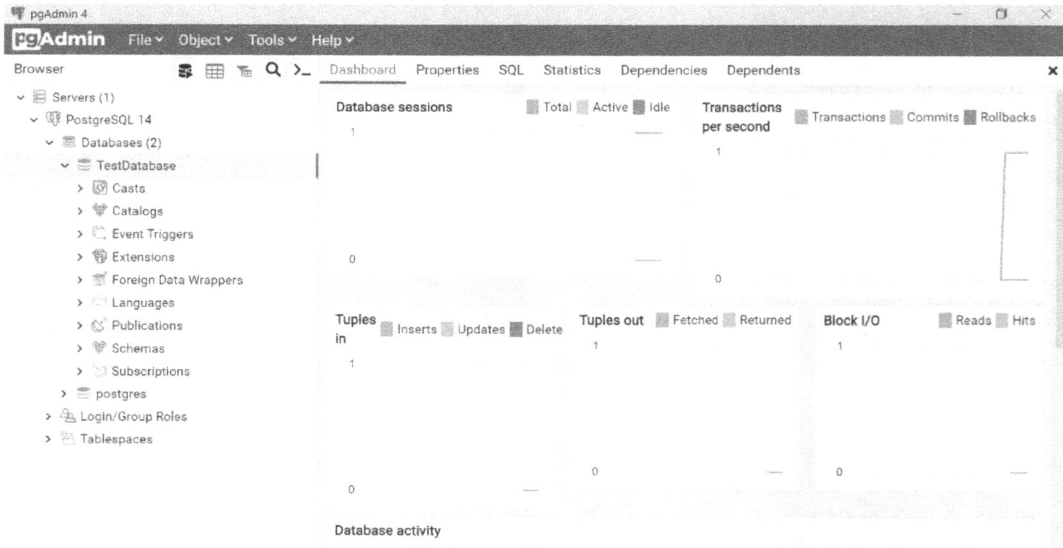

Figure 0.37: TestDatabase ready for use

Now your database is ready to be used for Windows and Mac environments.

POSTGRESQL INSTALLATION ON UBUNTU

In this example, you are using Ubuntu 20.04 for installation. Perform the following steps to do so:

1. In order to install PostgreSQL, open your Ubuntu terminal first.
2. Make sure to update your repository using the following command:

```
$ sudo apt update
```

3. Install the PostgreSQL software along with additional packages using the following command (recommended):

```
$ sudo apt install postgresql postgresql-contrib
```

> **NOTE**
>
> To install only PostgreSQL (not recommended without additional packages), use the command **`$ sudo apt install postgresql`** and then press **`Enter`**.

This installation process creates a user account called **`postgres`** that has the default **`Postgres`** role.

ACCESSING THE POSTGRES USER ACCOUNT WITH THE POSTGRES ROLE

There are two ways to start the PostgreSQL CLI using the **`postgres`** user account:

Option 1 is as follows:

1. To log in as a postgres user, use the following command:

```
$ sudo -i -u postgres
```

2. Access the CLI by using the following command:

```
$ psql
```

> **NOTE**
>
> Sometimes, while executing the preceding command, a **`psql`** error may be displayed as **`could not connect to server: No such file or directory`**. This is because of a port issue on your system. Due to this port blockage, the PostgreSQL application may not work. You can try the command again after some time.

3. To quit the CLI, use the following command:

```
$ \q
```

Option 2 is as follows:

1. To log in as a postgres user, use the following command:

```
$ sudo -u postgres psql
```

2. To quit the CLI, use the following command:

```
$ \q
```

VERIFYING THE POSTGRES USER ACCOUNT AS A POSTGRES USER ROLE

1. To verify the user account, log in and use the **`conninfo`** command:

```
$ sudo -u postgres psql
$ \conninfo
$ \q
```

Using this command, you can ensure that you are connected to the **`postgres`** database as the **`postgres`** user via port **`5432`**. If you don't want to use the default user, **`postgres`**, you can create a new user for you.

Accessing a New User and Database

1. Create a new user by using the following command and pressing **Enter**:

```
$ sudo -u postgres createuser -interactive
```

   The preceding command will ask the user to add the name of the role and its type.

2. Enter the name of the role, for example, **testUser**.
3. Next, enter **y** when prompted to set a new role to be a superuser:

```
Prompt:
Enter the name of the role to add: testUser
Shall the new role be a superuser? (y/n) y
```

   This will create a new user with the name **testUser**.

4. Create a new database with the name **testdb** using the following command:

```
$ sudo -u postgres createdb testdb
```

5. Log in to the newly created user account using the following command:

```
$ sudo -u testUser psql -d testdb
```

6. Use the following command to check the connection details:

```
$ \conninfo
```

7. To quit the CLI, use the following command:

```
$ \q
```

   Using this command, you can ensure that you are connected to the **testdb** database as the **testUser** user via port **5432**.

With these steps, you have completed the PostgreSQL installation for Ubuntu.

Downloading the Code

Download the code from GitHub at https://packt.link/sezEm. Refer to these files for the complete code.

The high-quality color images used in this book can be found at https://packt.link/5XYmX.

If you have any issues or questions about installation, please email us at **workshops@packt.com**.

1

HELLO C#

OVERVIEW

This chapter introduces you to the basics of C#. You will start by learning about the basics of the .NET Command-Line Interface (CLI) and how to use Visual Studio Code (VS Code) as a basic Integrated Development Environment (IDE). You will then learn about the various C# data types and how to declare variables for these types, before moving on to a section about arithmetic and logical operators. By the end of the chapter, you will know how to handle exceptions and errors and be able to write simple programs in C#.

INTRODUCTION

C# is a programming language created in the early 2000s by a team at Microsoft led by Anders Hejlsberg, who is also among the creators of some other popular languages, such as Delphi and Turbo Pascal, both widely used in the 1990s. Over the last 20 years, C# has grown and evolved, and today it is one of the most widely used programming languages globally, according to Stack Overflow's 2020 insights.

It has its reasons for holding such an honorable place in the tech community. C# allows you to write applications for a wide segment of markets and devices. From the banking industry, with its high-security standards, to e-commerce companies, which hold enormous volumes of transactions, it is a language trusted by companies that need both performance and reliability. Besides that, C# also makes it possible to write web, desktop, mobile, and even IoT applications, allowing you to develop for almost every kind of device.

C# was initially limited to work only on Windows; however, there have been concerted efforts by the C# team over the past few years to make it cross-platform compatible. Today, it can be used with all major OS distributions, namely, Windows, Linux, and macOS. The goal is simple: to develop, build, and run C# anywhere, letting each developer and team choose their most productive or favorite environment.

Another remarkable characteristic of C# is that it is a strongly typed programming language. You will dive into this more deeply in the upcoming sections, and you will see that strong typing enables better data security while programming.

Besides that, C# has become open source over the last few years, with Microsoft as its principal maintainer. This is highly advantageous, as it allows the language to receive continuous improvements from around the globe, with a solid backing company that both promotes and invests in it. C# is also a multi-paradigm language, meaning that you can use it to write software in many programming styles, in a beautiful, concise, and proper manner.

RUNNING AND DEVELOPING C# WITH THE .NET CLI

One term you'll hear a lot in the C# world is **.NET**. It is the foundation of C#, a framework that the language is built on top of. It has both a **Software Development Kit** (**SDK**) that allows the language to be developed and a **runtime** that allows the language to run.

That said, to start developing with C#, you only need to install the .NET SDK. This installation will provide both a **compiler** and the runtime on the development environment. In this section, you will cover the basic steps of preparing your environment for developing and running C# locally.

> **NOTE**
>
> Please refer to the *Preface* of this book for step-by-step instructions on how to download the .NET 6.0 SDK and install it on your machine.

Once the installation of the .NET 6.0 SDK is completed, you will have something called the **.NET CLI**. This **Command-Line Interface** (**CLI**) allows you to create new projects, compile them, and run them with very simple commands that you can run directly from your terminal.

After the installation, run the following command on your favorite terminal:

```
dotnet --list-sdks
```

You should see an output like this:

```
6.0.100 [/usr/local/share/dotnet/sdk]
```

This output shows that you have the 6.0.100 version of the SDK installed on your computer. That means you are ready to start developing your applications. If you type **dotnet --help**, you will notice that several commands will appear for you as options to run within the CLI. In this section, you will cover the most basic ones that you need to create and run applications: **new**, **build**, and **run**.

The **dotnet new** command allows you to create a bootstrap project to start developing. The CLI has several built-in templates, which are nothing more than basic bootstraps for various types of applications: web apps, desktop apps, and so on. You must specify two things in the **dotnet new** command:

- The template name
- The project name

The name is passed as an argument, which means you should specify it with a **-n** or **–name** flag. The command is as follows:

```
dotnet new TYPE -n NAME
```

For instance, to create a new console application named **MyConsoleApp** you can simply type:

```
dotnet new console -n MyConsoleApp
```

This will generate a new folder with a file named **MyConsoleApp.csproj**, which is the C# project file that contains all the metadata needed by the compiler to build your project, and some files needed for the application to be built and run.

Next, the **dotnet build** command allows you to build an application and make it ready to run. This command should be placed only in two locations:

- A project folder, containing a **.csproj** file.
- A folder containing a **.sln** file.

Solution (**.sln**) files are files that contain the metadata of one or more project files. They are used to organize multiple project files into single builds.

Finally, the third important command is dotnet **run**. This command allows you to properly run an application. It can be called without any arguments from the folder that contains the **.csproj** file of your .NET app, or without passing the project folder with the **--project** flag on the CLI. The **run** command also automatically builds the application prior to the run.

CREATING PROGRAMS WITH THE CLI AND VS CODE

While working through this book, you will use **Visual Studio Code** (**VS Code**) as your code editor. It works on all platforms, and you can download the version for your OS at https://code.visualstudio.com/. Although VS Code is not a complete **Integrated Development Environment** (**IDE**), it has a lot of extensions that make it a powerful tool to develop and do proper C# coding, regardless of the OS being used.

To properly develop C# code, you will primarily need to install the Microsoft C# extension. It equips VS Code with the ability to do code completion and identify errors and is available at https://marketplace.visualstudio.com/items?itemName=ms-dotnettools.csharp.

> **NOTE**
>
> Before proceeding, it is recommended that you install VS Code and the Microsoft C# extension. You can find a step-by-step breakdown of the installation process in the *Preface* of this book.

BASIC ANATOMY OF A C# PROGRAM

In order to run, every C# program needs something called an **entry point**. In C#, the standard entry point for a program is the **`Main`** method. Regardless of your program type, whether it is a web application, desktop application, or even a simple console one, the **`Main`** method will be the **entry point** for your C# program. This means that each time an application runs, the runtime searches for this method within your code and executes the code blocks inside it.

This structure is created for you by the CLI, with the **`new`** command. A **`Program.cs`** file contains a class named **`Program`**, with a method named **`Main`**, which, in turn, contains a single instruction that will be executed after the program is built and running. You will learn more about methods and classes later, but for now, just know that a class is something that usually contains a set of data and that can perform actions on this data through these **methods**.

Another important thing to note regarding basic C# concepts is **comments**. Comments allow you to place free text inside C# code files, without affecting the compiler. A comment section should always start with `//`.

EXERCISE 1.01: CREATING A CONSOLE APP THAT SAYS "HELLO WORLD"

In this exercise, you will see the CLI commands you learned about in the previous section, as you build your first ever C# program. It will be a simple console app that will print **`Hello World`** to the console.

Perform the following steps to do so:

1. Open the VS Code integrated terminal and type the following:

   ```
   dotnet new console -n Exercise1_01
   ```

   This command will create a new console application in the **`Exercise1_01`** folder.

2. On the command line, type the following:

   ```
   dotnet run --project Exercise1_01
   ```

You should see the following output:

```
> dotnet new console -n Exercise1_01
The template "Console App" was created successfully.

Processing post-creation actions...
Running 'dotnet restore' on /Users/mateusviegas/Exercise1_01/Exercise1_01.csproj...
  Determining projects to restore...
  Restored /Users/mateusviegas/Exercise1_01/Exercise1_01.csproj (in 96 ms).
Restore succeeded.

> dotnet run --project Exercise1_01
Hello, World!
```

Figure 1.1: "Hello World" output on the console

> **NOTE**
>
> You can find the code used for this exercise at https://packt.link/HErU6.

In this exercise, you created the most basic program possible with C#, a console application that prints some text to the prompt. You also learned how to use .NET CLI, which is the mechanism built within the .NET SDK to create and manage .NET projects.

Now proceed to the next section to grasp how top-level statements are written.

TOP-LEVEL STATEMENTS

You must have noticed in *Exercise 1.01* that, by default, when you create a console application, you have a **`Program.cs`** file that contains the following:

- A class named **`Program`**.
- The static void **`Main`** keywords.

You will learn about classes and methods in detail later, but for now, for the sake of simplicity, you do not need these resources to create and execute programs with C#. The latest version (.NET 6) introduced a feature that makes writing simple programs much easier and less verbose. For instance, consider the following:

```
using System;

namespace Exercise1_01
{
    class Program
    {
        static void Main(string[] args)
        {
            Console.WriteLine("Hello World!");
        }
    }
}
```

You can simply replace this snippet with two lines of code, as follows:

```
using System;
Console.WriteLine("Hello World!");
```

By using such **top-level statements**, you can write concise programs. You can simply put the statements to be executed at the top of the program. This is also useful for speeding up the learning curve with C#, as you need not worry about advanced concepts upfront. The only thing to look out for here is that the project can have only one file with top-level statements.

That is why in this chapter, you will find that all exercises will use this format, to make things as clear as possible.

DECLARING VARIABLES

You will now take your first steps in creating your own programs. This section will delve into the concept of variables—what they are and how to use them.

A **variable** is a name given to a computer memory location that holds some data that may vary. For a variable to exist, it first must be **declared** with a type and a name. It can also have a value assigned to it. The declaration of a variable can be achieved in a few different ways.

There are some basic considerations regarding naming conventions for variables in C#:

- The names must be unique, starting with a letter, and should contain only letters, digits, and the underscore character (_). The names can also begin with an underscore character.
- The names are case-sensitive; thus, **`myVariable`** and **`MyVariable`** are different names.
- Reserved keywords, such as **`int`** or **`string`**, cannot be used as names (this is a compiler restriction) unless you put an `@` symbol in front of the name, such as `@` **`int`** or `@`**`string`**.

Variables can be declared in two ways: **explicitly** and **implicitly**. Both styles of the declaration have their pros and cons, which you will explore in the next section.

DECLARING VARIABLES EXPLICITLY

A variable can be declared explicitly by writing both its type and value. Suppose you want to create two variables, **`a`** and **`b`**, both containing integers. Doing so explicitly would look like this:

```
int a = 0;
int b = 0;
```

Before a variable is used, it must have a value assigned. Otherwise, the C# compiler will give an error while building your program. The following example illustrates that:

```
int a;
int b = a; // The compiler will prompt an error on this line: Use of
unassigned local variable
```

It is also possible to declare multiple variables in the same line, like in the following snippet, where you are declaring three variables; two hold the value **`100`** and one holds the value **`10`**:

```
int a, b = 100, c = 10;
```

DECLARING VARIABLES IMPLICITLY

Remember that C# is a strongly typed programming language; this means that a variable will always have a type associated with it. It does not matter whether the type is declared implicitly or explicitly. With the **var** keyword, the C# compiler will infer the variable type based on the value that has been assigned to it.

Consider that you want to create a variable that holds some text using this method. This can be done with the following statement:

```
var name = "Elon Musk";
```

For storing text in a variable, you should start and end the text with double quotes ("). In the preceding example, by looking at the value that was assigned to **name**, C# knows that the type this variable holds is a string, even though the type is not mentioned in the statement.

EXPLICIT VERSUS IMPLICIT DECLARATION

Explicit declarations enhance readability with the type declared, and this is one of the main advantages of this technique. On the other hand, they tend to let the code become more verbose, especially when working with some data types (that you will see further ahead), such as **Collections**.

Essentially, deciding on the style of declaration depends on the personal preferences of the programmer, and may be influenced by the company's guidelines in some cases. In this journey of learning, it is recommended that you pick one that makes your learning path smoother, as there are few substantial differences from a purely technical standpoint.

In the next exercise, you will do this yourself by assigning variables to inputs that come from a user's interaction with a console application, where the user will be asked to input their name. To complete this exercise, you will make use of the following built-in methods that C# provides, which you will be using frequently in your C# journey:

- **Console.ReadLine()**: This allows you to retrieve a value that the user prompted on the console.
- **Console.WriteLine()**: This writes the value passed as an argument as an output to the console.

EXERCISE 1.02: ASSIGNING VARIABLES TO USER INPUTS

In this exercise, you will create an interactive console application. The app should ask you for your name, and once provided, it should display a greeting with your name in it.

To complete this exercise, perform the following steps:

1. Open Command Prompt and type the following:

   ```
   dotnet new console -n Exercise1_02
   ```

   This command creates a new console application in the **Exercise1_02** folder.

2. Open the **Program.cs** file. Paste the following inside the **Main** method:

   ```
   Console.WriteLine("Hi! I'm your first Program. What is your name?");
   var name = Console.ReadLine();
   Console.WriteLine($"Hi {name}, it is very nice to meet you. We have a really fun journey ahead.");
   ```

3. Save the file. On the command line, type the following:

   ```
   dotnet run --project Exercise1_02
   ```

   This outputs the following:

   ```
   Hi! I'm your first Program. What is your name?
   ```

4. Now, type your name into the console and hit **Enter** on your keyboard. For example, if you type in **Mateus**, the following will be the output:

   ```
   Hi! I'm your first Program. What is your name?
   Mateus
   Hi Mateus, it is very nice to meet you. We have a really fun journey ahead.
   ```

> **NOTE**
>
> You can find the code used for this exercise at https://packt.link/1fbVH.

You are more familiar with what variables are, how to declare them, and how to assign values to them. Now it is time to start talking about what data these variables can store and, more specifically, what types of data there are.

DATA TYPES

In this section, you will talk about the main data types within C# and their functionalities.

STRINGS

C# uses the **`string`** keyword to identify data that stores text as a sequence of characters. You can declare a string in several ways, as shown in the following snippet. However, when assigning some value to a string variable, you must place the content between a pair of double quotes, as you can see in the last two examples:

```
// Declare without initializing.
string message1;

// Initialize to null.
string message2 = null;

// Initialize as an empty string
string message3 = System.String.Empty;

// Will have the same content as the above one
string message4 = "";

// With implicit declaration
var message4 = "A random message"     ;
```

One simple but effective technique (that you used in the preceding *Exercise 1.02*) is one called **string interpolation**. With this technique, it is very simple to mix plain text values with variable values, so that the text is combined among these two. You can combine two or more strings by following these steps:

1. Before the initial quotes, insert a **$** symbol.
2. Now, inside the strings, place curly brackets and the name of the variable that you want to put into the string. In this case, this is done by putting **`{name}`** inside the initial string:

```
$"Hi {name}, it is very nice to meet you. We have a really fun
journey ahead.");
```

Another important fact to remember about strings is that they are **immutable**. This means that a string object cannot be changed after its creation. This happens because strings in C# are an array of characters. Arrays are data structures that gather objects of the same type and have a fixed length. You will cover arrays in detail in an upcoming section.

In the next exercise, you will explore string immutability.

EXERCISE 1.03: CHECKING STRING IMMUTABILITY

In this exercise, you will use two strings to demonstrate that string references are always immutable. Perform the following steps to do so:

1. Open the VS Code integrated terminal and type the following:

```
dotnet new console -n Exercise1_03
```

2. Open the **`Program.cs`** file and create a method with the **`void`** return type, which replaces part of a string like so:

```
static void FormatString(string stringToFormat)
{
stringToFormat.Replace("World", "Mars");
}
```

   In the preceding snippet, the **`Replace`** function is used to replace the first string (**`World`**, in this case) with the second one (**`Mars`**).

3. Now, create a method that does the same thing but returns the result instead:

```
static string FormatReturningString(string stringToFormat)
{
return stringToFormat.Replace("Earth", "Mars");
}
```

4. Now insert the following after the previous methods. Here, you create two string variables and observe their behavior after trying to modify them with the methods created previously:

```
var greetings = "Hello World!";
FormatString(greetings);
Console.WriteLine(greetings);

var anotherGreetings = "Good morning Earth!";
Console.WriteLine(FormatReturningString(anotherGreetings));
```

5. Finally, call **dotnet run --project Exercise1_03** from the command line. You should see the following output on the console:

```
dotnet run
Hello World!
Good morning Mars!
```

> **NOTE**
>
> You can find the code used for this exercise at https://packt.link/ZoNiw.

With this exercise, you saw the concept of string immutability in action. When you passed a string that was a reference type (**Hello World!**) as a method argument, it was not modified. That is what happens when you use the **FormatString** method, which returns **void**. Due to string immutability, a new string is created but not allocated to any variable, and the original string stays the same. With the second method, it returns a new string, and this string is then printed to the console.

COMPARING STRINGS

Even though strings are reference values, when you use the **.Equals()** method, the equality operator (==), and other operators (such as **!=**), you are actually comparing the values of the strings, as can be seen in the following example:

```
string first = "Hello.";
string second = first;
first = null;
```

Now you can compare these values and call **Console.WriteLine()** to output the result, like so:

```
Console.WriteLine(first == second);
Console.WriteLine(string.Equals(first, second));
```

Running the preceding code results in the following output:

```
False
False
```

You get this output because, even though strings are reference types, both the **==** and **.Equals** comparisons run against string values. Also, remember that strings are immutable. This means that when you assign **second** to **first** and set **first** as **null**, a new value is created for **first** and, therefore, the reference for **second** does not change.

NUMERIC TYPES

C# has its numeric types subdivided into two main categories—integral and floating-point type numbers. The integral number types are as follows:

- **sbyte**: Holds values from -128 to 127
- **short**: Holds values from -32,768 to 32,767
- **int**: Holds values from -2,147,483,648 to 2,147,483,647
- **long**: Holds values from -9,223,372,036,854,775,808 to 9,223,372,036,854,775,807

Deciding which type of integral type to use depends on the size of the values you want to store.

All these types are called **signed values**. This means that they can store both negative and positive numbers. There is also another range of types called **unsigned types**. Unsigned types are **byte**, **ushort**, **uint**, and **ulong**. The main difference between them is that signed types can store negative numbers and unsigned types can store only numbers greater than or equal to zero. You will use signed types most of the time, so do not worry about remembering this all at once.

The other category, namely, **floating-point types**, refers to the types used to store numbers with one or more decimal points. There are three floating-point types in C#:

- **float**: This occupies four bytes and can store numbers from ± 1.5 x 10−45 to ± 3.4 x 1038 with a precision range of six to nine digits. To declare a float number using **var**, you can simply append **f** to the end of the number, like so:

```
var myFloat = 10f;
```

- **double**: This occupies eight bytes and can store numbers from ± 5.0 × 10−324 to ± 1.7 × 1030 with a precision range of 15 to 17 digits. To declare a double number using var, you can append d to the end of the number, like so:

```
var myDouble = 10d;
```

- **decimal**: This occupies 16 bytes and can store numbers from ± 1.0 x 10-28 to ± 7.9228 x 1028 with a precision range from 28 to 29 digits. To declare a decimal number using var, you must simply append m to the end of the number, like so:

```
var myDecimal = 10m;
```

Choosing the floating-point type depends mainly on the degree of precision required. For instance, **decimal** is mostly used for financial applications that need a very high degree of precision and cannot rely on rounding for accurate calculations. With GPS coordinates, **double** variables might be appropriate if you want to deal with sub-meter precisions that usually have 10 digits.

Another relevant point to consider when choosing numeric types is performance. The larger the memory space allocated to a variable, the less performant the operations with these variables are. Therefore, if high precision is not a requirement, **float** variables will be better performers than **doubles**, which, in turn, will be better performers than decimals.

Here you grasped what variables are and their main types. Now you will perform some basic calculations with them, such as addition, subtraction, and multiplication. This can be done using the arithmetic operators available in C#, such as **+**, **-**, **/**, and *****. So, move on to the next exercise where you will create a basic calculator using these operators.

EXERCISE 1.04: USING THE BASIC ARITHMETIC OPERATORS

In this exercise, you will create a simple calculator that receives two inputs and shows the results between them, based on which arithmetic operation is selected.

The following steps will help you complete this exercise:

1. Open the VS Code integrated terminal and type the following:

```
dotnet new console -n Exercise1_04
```

2. Navigate to the project folder, open the **Program.cs** file, and inside the **Main** method, declare two variables that read the user input, like so:

```
Console.WriteLine("Type a value for a: ");
var a = int.Parse(Console.ReadLine());
Console.WriteLine("Now type a value for b: ");
var b = int.Parse(Console.ReadLine());
```

The preceding snippet uses the **.ReadLine** method to read the input. This method, however, gives a **string**, and you need to evaluate a number. Therefore, the **Parse** method has been used here. All the numeric types have a method called **Parse**, which receives a string and converts it into a number.

3. Next, you need to write the output of these basic operators to the console. Add the following code to the **Main** method:

```
Console.WriteLine($"The value for a is { a } and for b is { b }");
Console.WriteLine($"Sum: { a + b}");
Console.WriteLine($"Multiplication: { a * b}");
Console.WriteLine($"Subtraction: { a - b}");
Console.WriteLine($"Division: { a / b}");
```

4. Run the program using the **dotnet run** command, and you should see the following output, if you input **10** and **20**, for instance:

```
Type a value for a:
10
Now type a value for b:
20
The value for a is 10 and b is 20
Sum: 30
Multiplication: 200
Subtraction: -10
Division: 0
```

> **NOTE**
>
> You can find the code used for this exercise at https://packt.link/ldWVv.

Thus, you have built a simple calculator app in C# using the arithmetic operators. You also learned about the concept of parsing, which is used to convert strings to numbers. In the next section, you will briefly cover the topic of classes, one of the core concepts of programming in C#.

CLASSES

Classes are an integral part of coding in C# and will be covered comprehensively in *Chapter 2, Building Quality Object-Oriented Code*. This section touches upon the basics of classes so that you can begin using them in your programs.

The reserved **class** keyword within C# is used when you want to define the type of an object. An **object**, which can also be called an **instance**, is nothing more than a block of memory that has been allocated to store information. Given this definition, what a class does is act as a blueprint for an object by having some properties to describe this object and specifying the actions that this object can perform through methods.

For example, consider that you have a class named **Person**, with two properties, **Name** and **Age**, and a method that checks whether **Person** is a child. Methods are where logic can be placed to perform some action. They can return a value of a certain type or have the special **void** keyword, which indicates that they do not return anything but just execute some action. You can also have methods calling other methods:

```
public class Person
{
    public Person() { }

    public Person(string name, int age)
{
    Name = name;
    Age = age;
}

    public string Name { get; set; }

    public int Age { get; set; }

    public void GetInfo()
{
    Console.WriteLine($"Name: {Name} - IsChild? {IsChild()}");
}

    public bool IsChild()
{
    return Age < 12;
}
}
```

One question remains, though. Since classes act as blueprints (or definitions if you prefer), how do you actually allocate memory to store the information defined by a class? This is done through a process called **instantiation**. When you instantiate an object, you allocate some space in memory for it in a reserved area called the **heap**. When you assign a variable to an object, you are setting the variable to have the address of this memory space, so that each time you manipulate this variable, it points to and manipulates the data allocated at this memory space. The following is a simple example of instantiation:

```
var person = new Person();
```

Note that `Person` has properties that have two magic keywords—`get` and `set`. **Getters** define that a property value can be retrieved, and **setters** define that a property value can be set.

Another important concept here is the concept of a constructor. A constructor is a method with no return type, usually present at the top level of the class for better readability. It specifies what is needed for an object to be created. By default, a class will always have a parameter-less constructor. If another constructor with parameters is defined, the class will be constrained to only this one. In that case, if you still want to have a parameter-less constructor, you must specify one. This is quite useful, as classes can have multiple constructors.

That said, you can assign values to an object property that has a setter in the following ways:

- At the time of creation, via its constructor:

```
var person = new Person("John", 10);
```

- At the time of creation, with direct variable assignment:

```
var person = new Person() { Name = "John", Age = 10 };
```

- After the object is created, as follows:

```
var person = new Person();
person.Name = "John";
person.Age = 10;
```

There is a lot more to classes that you will see further on. For now, the main ideas are as follows:

- Classes are blueprints of objects and can have both properties and methods that describe these objects.
- Objects need to be instantiated so that you can perform operations with them.
- Classes have one parameter-less constructor by default, but can have many customized ones as required.
- **Object variables** are references that contain the memory address of a special memory space allocated to the object inside a dedicated memory section named the heap.

DATES

A date can be represented in C# using the **DateTime** value type. It is a struct with two static properties called **MinValue**, which is January 1, 0001 00:00:00, and **MaxValue**, which is December 31, 9999 11:59:59 P.M. As the names suggest, both these values represent the minimum and maximum dates according to the Gregorian calendar date format. The default value for **DateTime** objects is **MinValue**.

It is possible to construct a **DateTime** variable in various ways. Some of the most common ways are as follows:

- Assigning the current time as follows:

  ```
  var now = DateTime.Now;
  ```

  This sets the variable to the current date and time on the calling computer, expressed as the local time.

  ```
  var now = DateTime.UtcNow;
  ```

  This sets the variable to the current date and time on this computer, expressed as the **Coordinated Universal Time** (**UTC**).

- You can also use constructors for passing days, months, years, hours, minutes, and even seconds and milliseconds.

- There is also a special property available for **DateTime** objects called **Ticks**. It is a measure of the number of 100 nanoseconds elapsed since **DateTime.MinValue**. Every time you have an object of this type, you can call the **Ticks** property to get such a value.
- Another special type for dates is the **TimeSpan** struct. A **TimeSpan** object represents a time interval as days, hours, minutes, and seconds. It is useful when fetching intervals between dates. You will now see what this looks like in practice.

EXERCISE 1.05: USING DATE ARITHMETIC

In this exercise, you will use the **TimeSpan** method/struct to calculate the difference between your local time and the UTC time. To complete this exercise, perform the following steps:

1. Open the VS Code integrated terminal and type the following:

```
dotnet new console -n Exercise1_05
```

2. Open the **Program.cs** file.
3. Paste the following inside the **Main** method and save the file:

```
Console.WriteLine("Are the local and utc dates equal? {0}", DateTime.
Now.Date == DateTime.UtcNow.Date);

Console.WriteLine("\nIf the dates are equal, does it mean that
there's no TimeSpan interval between them? {0}",
(DateTime.Now.Date - DateTime.UtcNow.Date) == TimeSpan.Zero);

DateTime localTime = DateTime.Now;
DateTime utcTime = DateTime.UtcNow;
TimeSpan interval = (localTime - utcTime);

Console.WriteLine("\nDifference between the {0} Time and {1} Time:
{2}:{3} hours",
    localTime.Kind.ToString(),
    utcTime.Kind.ToString(),
```

```
    interval.Hours,
    interval.Minutes);

Console.Write("\nIf we jump two days to the future on {0} we'll be on
{1}",
    new DateTime(2020, 12, 31).ToShortDateString(),
    new DateTime(2020, 12, 31).AddDays(2).ToShortDateString());
```

In the preceding snippet, you first checked whether the current local date and UTC dates were equal. Then you checked for the interval between them, if any, using the **TimeSpan** method. Next, it printed the difference between the local and UTC time and printed the date two days ahead of the current one (**31/12/2020**, in this case).

4. Save the file. On the command line, type the following:

```
dotnet run --project Exercise1_05
```

You should see an output like the following:

```
Are the local and utc dates equal? True
If the dates are equal, does it mean there's no TimeSpan interval
between them? True

Difference between the Local Time and Utc Time: 0:0 hours


If we jump two days to the future on 31/12/2020 we'll be on
02/01/2021
```

> **NOTE**
>
> You can find the code used for this exercise at https://packt.link/WIScZ.

Note that depending on your time zone, you will likely see different output.

FORMATTING DATES

It is also possible to format **DateTime** values to localized strings. That means formatting a **DateTime** instance according to a special concept within the C# language called a **culture**, which is a representation of your local time. For instance, dates are represented differently in different countries. Now take a look at the following examples, where dates are outputted in both the format used in France and the format used in the United States:

```
var frenchDate = new DateTime(2008, 3, 1, 7, 0, 0);
Console.WriteLine(frenchDate.ToString(System.Globalization.CultureInfo.
  CreateSpecificCulture("fr-FR")));
// Displays 01/03/2008 07:00:00

var usDate = new DateTime(2008, 3, 1, 7, 0, 0);
Console.WriteLine(frenchDate.ToString(System.Globalization.CultureInfo.
CreateSpecificCulture("en-US")));

// For en-US culture, displays 3/1/2008 7:00:00 AM
```

It is also possible to explicitly define the format you want the date to be output in, as in the following example, where you pass the **yyyyMMddTHH:mm:ss** value to say that you want the date to be output as year, then month, then day, then hour, then minutes preceded by a colon, and finally, seconds, also preceded by a colon:

```
var date1 = new DateTime(2008, 3, 1, 7, 0, 0);

Console.WriteLine(date1.ToString("yyyyMMddTHH:mm:ss"));
```

The following output gets displayed:

```
    20080301T07:00:00
```

LOGICAL OPERATORS AND BOOLEAN EXPRESSIONS

You are already familiar with these. Recall that in the preceding exercise, you did the following comparison:

```
var now = DateTime.Now.Date == DateTime.UtcNow.Date;
```

This output assigns the value **true** to **now** if the dates are equal. But as you know, they might not necessarily be the same. Therefore, if the dates are different, a **false** value will be assigned. These two values are the result of such Boolean expressions and are called **Boolean values**. That is why the **now** variable has the type of **bool**.

Boolean expressions are the base for every logical comparison in every program. Based on these comparisons, a computer can execute a certain behavior in a program. Here are some other examples of Boolean expressions and variable assignments:

- Assigning the result of a comparison that checks whether **a** is greater than **b**:

```
var basicComparison = a > b;
```

- Assigning the result of a comparison that checks whether **b** is greater than or equal to **a**:

```
bool anotherBasicComparison = b >= a;
```

- Checking whether two strings are equal and assigning the result of this comparison to a variable:

```
var animal1 = "Leopard";
var animal2 = "Lion";
bool areTheseAnimalsSame = animal1 == animal2;
```

  Clearly, the result of the previous comparison would be **false** and this value will be assigned to the **areTheseAnimalsSame** variable.

Now that you have learned what Booleans are and how they work, it is time to look at some logical operators you can use to compare Boolean variables and expressions:

- The **&&** (**AND**) operator: This operator will perform an equality comparison. It will return **true** if both are equal and **false** if they are not. Consider the following example, where you check whether two strings have the length **0**:

```
bool areTheseStringsWithZeroLength = "".Length == 0 && " ".Length ==
0;
Console.WriteLine(areTheseStringsWithZeroLength);// will return false
```

- The **||** (**OR**) operator: This operator will check whether either of the values being compared is **true**. For example, here you are checking whether at least one of the strings has zero length:

```
bool isOneOfTheseStringsWithZeroLength = "".Length == 0 || " ".Length
== 0;
Console.WriteLine(isOneOfTheseStringsWithZeroLength); // will return
true
```

- The **!** (**NOT**) operator: This operator takes a Boolean expression or value and negates it; that is, it returns the opposite value. For example, consider the following example, where you negate the result of a comparison that checks whether one of the strings has zero length:

```
bool isOneOfTheseStringsWithZeroLength = "".Length == 0 || " ".Length
== 0;
bool areYouReallySure = !isOneOfTheseStringsWithZeroLength;
Console.WriteLine(areYouReallySure); // will return false
```

USING IF-ELSE STATEMENTS

Up till now, you have learned about types, variables, and operators. Now it is time to go into the mechanisms that help you to use these concepts in real-world problems—that is, decision-making statements.

In C#, **if-else** statements are some of the most popular choices for implementing branching in code, which means telling the code to follow one path if a condition is satisfied, else follow another path. They are logical statements that evaluate a Boolean expression and continue the program's execution based on this evaluation result.

For example, you can use **if-else** statements to check whether the password entered satisfies certain criteria (such as having at least six characters and one digit). In the next exercise, you will do exactly that, in a simple console application.

EXERCISE 1.06: BRANCHING WITH IF-ELSE

In this exercise, you will use **if-else** statements to write a simple credentials check program. The application should ask the user to enter their username; unless this value is at least six characters in length, the user cannot proceed. Once this condition is met, the user should be asked for a password. The password should also have a minimum of six characters containing at least one digit. Only after both these criteria are met should the program display a success message, such as **User successfully registered**.

The following steps will help you complete this exercise:

1. Inside the VS Code integrated terminal, create a new console project called **Exercise1_06**:

```
dotnet new console -n Exercise1_06
```

2. Inside the **`Main`** method, add the following code to ask the user for a username, and assign the value to a variable:

```
Console.WriteLine("Please type a username. It must have at least 6 
characters: ");
var username = Console.ReadLine();
```

3. Next, the program needs to check whether the username has more than six characters and if not, write an error message to the console:

```
if (username.Length < 6)
{
Console.WriteLine($"The username {username} is not valid.");
}
```

4. Now, within an **`else`** clause, you will continue the verification and ask the user to type a password. Once the user has entered a password, three points need to be checked. The first condition to check is whether the password has at least six characters and then whether there is at least one number. Then, if either of these conditions fails, the console should display an error message; else, it should display a success message. Add the following code for this:

```
else
{
Console.WriteLine("Now type a
password. It must have a length of at least 6 characters and also 
contain a number.");

var password = Console.ReadLine();

if (password.Length < 6)
     {
                Console.WriteLine("The password must have at least 6 
characters.");
}
     else if (!password.Any(c => char.IsDigit(c)))
     {
                Console.WriteLine("The password must contain at least 
one number.");
}
```

```
else
    {
            Console.WriteLine("User successfully registered.");
}
}
```

From the preceding snippet, you can see that if the user enters fewer than six characters, an error message is displayed as **`The password must have at least 6 characters.`**. If the password doesn't contain a single digit but satisfies the preceding condition, another error message is displayed as **`The password must contain at least one number.`**.

Notice the logical condition used for this, which is **`!password.Any(c => char.IsDigit(c))`**. You will learn more about the **=>** notation in *Chapter 2, Building Quality Object-Oriented Code*, but for now, you just need to know that this line checks every character in the password and uses the **`IsDigit`** function to check whether the character is a digit. This is done for every character, and if no digit is found, the error message is displayed. If all the conditions are met, a success message is displayed as **`User successfully registered.`**.

5. Run the program using **`dotnet run`**. You should see an output like the following:

```
Please type a username. It must have at least 6 characters:
thekingjames
Now type a password. It must have at least 6 characters and a number.
James123!"#
User successfully registered
```

> **NOTE**
>
> You can find the code used for this exercise at https://packt.link/3Q7oK.

In this exercise, you worked with **if-else** branching statements to implement a simple user registration program.

THE TERNARY OPERATOR

Another simple-to-use, yet effective, decision-making operator is the **ternary operator**. It allows you to set the value of a variable based on a Boolean comparison. For example, consider the following example:

```
var gift = person.IsChild() ? "Toy" : "Clothes";
```

Here, you are using the **?** symbol to check whether the Boolean condition placed before it is valid. The compiler runs the **IsChild** function for the **person** object. If the method returns **true**, the first value (before the **:** symbol) will be assigned to the **gift** variable. If the method returns **false**, the second value (after the **:** symbol) will be assigned to the **gift** variable.

The ternary operator is simple and makes assignments based on simple Boolean verifications even more concise. You will be using this quite often in your C# journey.

REFERENCE AND VALUE TYPES

There are two types of variables in C#, namely, **reference types** and **value types**. Variables of value types, such as structs, contain the values themselves, as the name suggests. These values are stored in a memory space called the **stack**. When a variable of such a type is declared, specific memory space is allocated to store this value, as illustrated in the following figure:

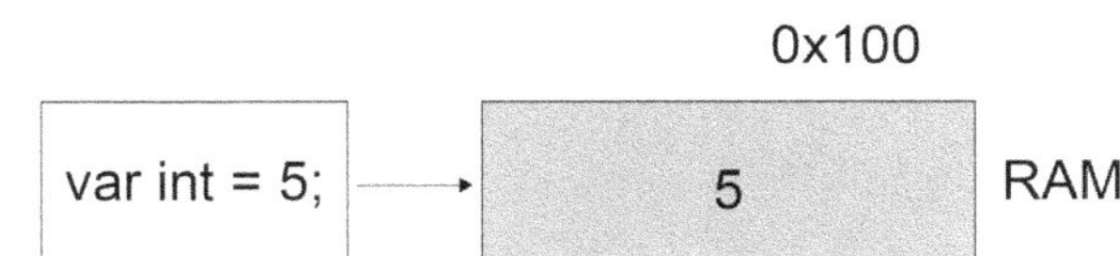

Figure 1.2: Memory allocation for a value type variable

Here, the value of the variable, which is **5**, is stored in memory at the location **0x100** in the RAM. The built-in value types for C# are **bool**, **byte**, **char**, **decimal**, **double**, **enum**, **float**, **int**, **long**, **sbyte**, **short**, **struct**, **uint**, **ulong**, and **ushort**.

The scenario for reference type variables is different, though. The three main reference types you need to know about in this chapter are **string**, array, and **class**. When a new reference type variable is assigned, what is stored in memory is not the value itself, but instead a memory address where the value gets allocated. For example, consider the following diagram:

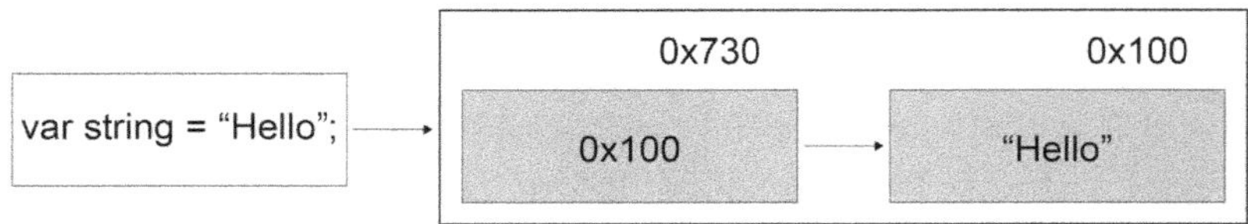

Figure 1.3: Memory allocation for a reference type variable

Here, instead of the value of the string variable (**Hello**), the address where it is allocated (**0x100**) is stored in memory. For brevity, you will not dive deep into this topic, but it is important to know the following points:

- When value type variables are passed as parameters or assigned as the value of another variable, the .NET runtime copies the value of the variable to the other object. This means that the original variable is not affected by any changes made in the newer and subsequent variables, as the values were literally copied from one place to another.
- When reference type variables are passed as parameters or assigned as the value of another variable, .NET passes the heap memory address instead of the value. This means that every subsequent change made in this variable inside a method will be reflected outside.

For instance, consider the following code, which deals with integers. Here, you declare an **int** variable named **a** and assign the value **100** to it. Later, you create another **int** variable named **b** and assign the value of **a** to it. Finally, you modify **b**, to be incremented by **100**:

```
using System;
int a = 100;
Console.WriteLine($"Original value of a: {a}");
int b = a;
Console.WriteLine($"Original value of b: {b}");
b = b + 100;
Console.WriteLine($"Value of a after modifying b: {a}");
Console.WriteLine($"Value of b after modifying b: {b}");
```

The values of **a** and **b** will be displayed in the following output:

```
Original value of a: 100
Original value of b: 100
Value of a after modifying b: 100
Value of b after modifying b: 200
```

In this example, the value from **a** was copied into **b**. From this point, any other modification you do on **b** will reflect changes only in **b** and **a** will continue to have its original value.

Now, what if you pass reference types as method arguments? Consider the following program. Here, you have a class named **Car** with two properties—**Name** and **GearType**. Inside the program is a method called **UpgradeGearType** that receives an object of the **Car** type and changes its **GearType** to **Automatic**:

```
using System;

var car = new Car();
car.Name = "Super Brand New Car";
car.GearType = "Manual";

Console.WriteLine($"This is your current configuration for the car {car.
Name}: Gea-Type - {car.GearType}");

UpgradeGearType(car);

Console.WriteLine($"You have upgraded your car {car.Name} for the
GearType {car.GearType}");

void UpgradeGearType(Car car)
{
    car.GearType = "Automatic";
}

class Car
{
    public string Name { get; set; }
    public string GearType { get; set; }
}
```

After you create a **Car instance** and call the **UpgradeGearType()** method, the output will be as the follows:

```
This is your current configuration for the car Super Brand New Car:
GearType - Manual
You have upgraded your car Super Brand New Car for the GearType Automatic
```

Thus, you see that if you pass an **object** of a reference type (**car** in this case) as an argument to a method (**UpgradeGearType** in this example), every change made inside this **object** is reflected after and outside the method call. This is because reference types refer to a specific location in memory.

EXERCISE 1.07: GRASPING VALUE AND REFERENCE EQUALITY

In this exercise, you will see how equality comparison is different for value types and reference types. Perform the following steps to do so:

1. In VS Code, open the integrated terminal and type the following:

```
dotnet new console -n Exercise1_07
```

2. Open the **Program.cs** file. In the same file, create a struct named **GoldenRetriever** with a **Name** property, as follows:

```
struct GoldenRetriever
{
    public string Name { get; set; }
}
```

3. Still in the same file, create one more class named **BorderCollie** with a similar **Name** property:

```
class BorderCollie
{
    public string Name { get; set; }
}
```

4. One final class must be created, a class named **Bernese**, also having the **Name** property, but with an extra override of the native **Equals** method:

```
class Bernese
{
    public string Name { get; set; }

    public override bool Equals(object obj)
    {
```

```
            if (obj is Bernese borderCollie && obj != null)
            {
                return this.Name == borderCollie.Name;
            }

            return false;
        }
    }
```

Here, the **this** keyword is used to refer to the current **instance** of the **borderCollie** class.

5. Finally, in the **Program.cs** file, you will create some objects for these types. Note that since you are using **top-level statements**, these declarations should be above the class and the struct declarations:

```
        var aGolden = new GoldenRetriever() { Name = "Aspen" };
        var anotherGolden = new GoldenRetriever() { Name = "Aspen" };

        var aBorder = new BorderCollie() { Name = "Aspen" };
        var anotherBorder = new BorderCollie() { Name = "Aspen" };

        var aBernese = new Bernese() { Name = "Aspen" };
        var anotherBernese = new Bernese() { Name = "Aspen" };
```

6. Now, right after the previous declarations, compare these values using the **Equals** method and assign the result to some variables:

```
var goldenComparison = aGolden.Equals(anotherGolden) ? "These Golden
Retrievers have the same name." : "These Goldens have different
names.";

var borderComparison = aBorder.Equals(anotherBorder) ? "These Border
Collies have the same name." : "These Border Collies have different
names.";

var berneseComparison = aBernese.Equals(anotherBernese) ? "These
Bernese dogs have the same name." : "These Bernese dogs have different
names.";
```

7. Finally, print the comparison results to the console with the following:

```
Console.WriteLine(goldenComparison);

Console.WriteLine(borderComparison);

Console.WriteLine(berneseComparison);
```

8. Run the program from the command line using **`dotnet run`** and you will see the following output:

```
These Golden Retrievers have the same name.
These Border Collies have different names.
These Bernese dogs have the same name.
```

> **NOTE**
>
> You can find the code used for this exercise at https://packt.link/xcWN9.

As mentioned earlier, structs are value types. Therefore, when two objects of the same struct are compared with **`Equals`**, .NET internally checks all the struct properties. If those properties have equal values, then **`true`** is returned. With **`Golden Retrievers`**, for instance, if you had a **`FamilyName`** property and this property was different between the two objects, the result of the equality comparison would be **`false`**.

For classes and all other reference types, the equality comparison is quite different. By default, **object reference** is checked on equality comparison. If the references are different (and they will be, unless the two variables are assigned to the same object), the equality comparison will return **`false`**. This explains the result you see for **`Border Collies`** in the example that the references were different for the two instances.

However, there is a method that can be implemented in reference types called **Equals**. Given two objects, the **`Equals`** method can be used for comparison following the logic placed inside the method. That is exactly what happened with the Bernese dogs example.

DEFAULT VALUE TYPES

Now that you have dealt with value and reference types, you will briefly explore the default value types. In C#, every type has a default value, as specified in the following table:

Type	Default Value
All reference types, including strings	`null`
All numerical types, either integral or floating point	`0`
`Bool`	`false`
`Char`	`'\0'`
`Enum`	`0`

Figure 1.4: Default value types table

These default values can be assigned to a variable using the **`default`** keyword. To use this word in a variable declaration, you must explicitly declare the variable type before its name. For example, consider the following snippet, where you are assigning the **`default`** value to two **`int`** variables:

```
int a = default;
int b = default;
```

Both **`a`** and **`b`** will be assigned the value **`0`** in this case. Note that it is not possible to use **`var`** in this case. This is because, for implicitly declared variables, the compiler needs a value assigned to the variable in order to infer its type. So, the following snippet will lead to an error because no type was set, either through an explicit declaration or by variable assignment:

```
var a = default;
var b = default;
```

ENHANCING DECISION MAKING WITH THE SWITCH STATEMENT

The **`switch`** statement is often used as an alternative to the if-else construct if a single expression is to be tested against three or more conditions, that is, when you want to select one of many code sections to be executed, such as the following:

```
switch (matchingExpression)
{
  case firstCondition:
    // code section
```

```
      break;
    case secondCondition:
      // code section
      break;
    case thirdCondition:
      // code section
      break;
    default:
      // code section
      break;
}
```

The matching expression should return a value that is of one of the following types: **char**, **string**, **bool**, **numbers**, **enum**, and **object**. This value will then be evaluated within one of the matching case clauses or within the default clause if it does not match any prior clause.

It is important to say that only one **switch** section in a **switch** statement will be executed. C# doesn't allow execution to continue from one **switch** section to the next. However, a **switch** statement does not know how to stop by itself. You can either use the **break** keyword if you only wish to execute something without returning or return something if that is the case.

Also, the **default** keyword on a **switch** statement is where the execution goes if none of the other options are matched. In the next exercise, you will use a **switch** statement to create a restaurant menu app.

EXERCISE 1.08: USING SWITCH TO ORDER FOOD

In this exercise, you will create a console app that lets the user select from a menu of food items available at a restaurant. The app should display an acknowledgment receipt for the order. You will use the **switch** statement to implement the logic.

Follow these steps to complete this exercise:

1. Create a new console project called **Exercise1_08**.
2. Now, create an **instance** of **System.Text.StringBuilder**. This is a class that helps build strings in many ways. Here, you are building strings line by line so that they can be properly displayed on the console:

```
var menuBuilder = new System.Text.StringBuilder();

menuBuilder.AppendLine("Welcome to the Burger Joint. ");
```

```
menuBuilder.AppendLine(string.Empty);
menuBuilder.AppendLine("1) Burgers and Fries - 5 USD");
menuBuilder.AppendLine("2) Cheeseburger - 7 USD");
menuBuilder.AppendLine("3) Double-cheeseburger - 9 USD");
menuBuilder.AppendLine("4) Coke - 2 USD");
menuBuilder.AppendLine(string.Empty);
menuBuilder.AppendLine("Note that every burger option comes with fries and ketchup!");
```

3. Display the menu on the console and ask the user to choose one of the options:

```
Console.WriteLine(menuBuilder.ToString());

Console.WriteLine("Please type one of the following options to order:");
```

4. Read the key that the user presses and assign it to a variable with the **Console.ReadKey()** method. This method works similarly to **ReadLine()**, which you have used before, with the difference that it reads the key that is immediately pressed after calling the method. Add the following code for this:

```
var option = Console.ReadKey();
```

5. Now it is time to use the **switch** statement. Use **option.KeyChar.ToString()** as the matching expression of the **switch** clause here. Keys **1**, **2**, **3**, and **4** should result in orders accepted for **burgers**, **cheeseburgers**, **double cheeseburgers**, and **Coke**, respectively:

```
switch (option.KeyChar.ToString())
{
    case "1":
        {
            Console.WriteLine("\nAlright, some burgers on the go. Please pay the cashier.");
            break;
        }
    case "2":
        {
            Console.WriteLine("\nThank you for ordering cheeseburgers. Please pay the cashier.");
            break;
        }
    case "3":
        {
            Console.WriteLine("\nThank you for ordering double cheeseburgers, hope you enjoy them. Please pay the cashier!");
```

Any other input, however, should be considered invalid and a message gets displayed, letting you know you have selected an invalid option:

```
            break;
        }
    case "4":
        {
            Console.WriteLine("\nThank you for ordering Coke. Please 
pay the cashier.");
            break;
        }
    default:
        {
            Console.WriteLine("\nSorry, you chose an invalid 
option.");
            break;
        }
}
```

6. Finally, run the program with **dotnet run --project Exercise1_08** and interact with the console to see the possible outputs. For example, if you type **1**, you should see an output like the following:

```
Welcome to the Burger Joint.

1) Burgers and Fries - 5 USD
2) Cheeseburger - 7 USD
3) Double-cheeseburger - 9 USD
4) Coke - 2 USD

Note that every burger option comes with fries and ketchup!

Please type one of the follow options to order:
1
Alright, some burgers on the go! Please pay on the following cashier!
```

> **NOTE**
>
> You can find the code used for this exercise at https://packt.link/x1Mvn.

Similarly, you should get the output for the other options as well. You have learned about branching statements in C#. There is another type of statement that you will use often while programming using C#, called iteration statements. The next section covers this topic in detail.

Iteration Statements

Iteration statements, also called **loops**, are types of statements that are useful in the real world, as you often need to continuously repeat some logical execution in your applications **while** or **until** some condition is met, such as operating with a number that must be incremented until a certain value. C# offers numerous ways of implementing such iterations, and in this section, you will examine each of these in detail.

While

The first iteration statement you will consider is the **while** statement. This statement allows a C# program to execute a set of instructions while a certain Boolean expression is evaluated to be **true**. It has one of the most basic structures. Consider the following snippet:

```
int i = 0;

while (i < 10)
{
Console.WriteLine(i);
i = i +1;
}
```

The preceding snippet shows how you can use the **while** statement. Note that the **while** keyword is followed by a pair of brackets enclosing a logical condition; in this case, the condition is that the value of **i** must be less than **10**. The code written inside the curly braces will be executed until this condition is **true**.

Thus, the preceding code will print the value of **i**, starting with **0**, up to **10**. This is fairly simplistic code; in the next exercise, you will use the **while** statement for something a little more complex, such as checking whether a number entered by you is a prime number.

EXERCISE 1.09: CHECKING WHETHER A NUMBER IS PRIME WITH A WHILE LOOP

In this exercise, you will use a **while** loop to check whether a number you enter is prime. To do so, the **while** loop will check whether the counter is less than or equal to the integer result of the division of the number by **2**. When this condition is satisfied, you check whether the remainder of the division of the number by the counter is **0**. If not, you increment the counter and continue until the loop condition is not met. If it is met, it means the number is not **false** and the loop can stop.

Perform the following steps to complete this exercise:

1. Inside the VS Code integrated terminal, create a new console project called **Exercise1_09**.
2. Inside the **Program.cs** file, create the following method, which will perform the logic you introduced at the beginning of the exercise:

```
static bool IsPrime(int number)
{

if (number ==0 || number ==1) return false;

bool isPrime = true;

int counter = 2;

while (counter <= Math.Sqrt(number))
     {
                    if (number % counter == 0)
               {
                    isPrime = false;
                    break;
}

counter++;
}

     return isPrime;
}
```

3. Now, input a number, so you can check whether it is prime:

```
Console.Write("Enter a number to check whether it is Prime: ");
var input = int.Parse(Console.ReadLine());
```

4. Now, check whether the number is prime and print the result:

```
Console.WriteLine($"{input} is prime? {IsPrime(input)}.");
```

5. Finally, on the VS Code integrated terminal, call **`dotnet run --project Exercise1_09`** and interact with the program. For example, try entering **`29`** as an input:

```
Enter a number to check whether it is Prime:
29
29 is prime? True
```

As expected, the result for **`29`** is **`true`** since it is a prime number.

> **NOTE**
>
> You can find the code used for this exercise at https://packt.link/5oNg5.

The preceding exercise aimed to show you the simple structure of a **`while`** loop with some more complex logic. It checks a number (named **`input`**) and prints whether it is a prime number. Here, you have seen the **`break`** keyword used again to stop program execution. Now proceed to learn about jump statements.

JUMP STATEMENTS

There are some other important keywords used within loops that are worth mentioning as well. These keywords are called **jump statements** and are used to transfer program executions to another part. For instance, you could rewrite the **`IsPrime`** method as follows:

```
static bool IsPrimeWithContinue(int number)
        {
        if (number == 0 || number ==1) return false;

            bool isPrime = true;
```

```
            int counter = 2;

            while (counter <= Math.Sqrt(number))
            {
                if (number % counter != 0)
                {
                    counter++;
                    continue;
                }

                isPrime = false;
                break;
            }

            return isPrime;
        }
```

Here, you have inverted the logical check. Instead of checking whether the remainder is zero and then breaking the program execution, you have checked that the remainder is not zero and, if so, have used the **continue** statement to pass the execution to the next iteration.

Now look at how you can rewrite this using another special keyword, **goto**:

```
static bool IsPrimeWithGoTo(int number)
        {
        if (number == 0 || number ==1) return false;
bool isPrime = true;

            int counter = 2;

            while (counter <= Math.Sqrt(number))
            {
                if (number % counter == 0)
                {
                    isPrime = false;
                    goto isNotAPrime;
                }
```

```
            counter++;
        }

        isNotAPrime:
        return isPrime;
    }
```

The **goto** keyword can be used to jump from one part of the code to another one defined by what is called a **label**. In this case, the label was named **isNotAPrime**. Finally, take a look at one last way of writing this logic:

```
static bool IsPrimeWithReturn(int number)
    {
    if (number == 0 || number ==1) return false;

        int counter = 2;

        while (counter <= Math.Sqrt(number))
        {
            if (number % counter == 0)
            {
                return false;
            }

            counter ++;
        }

        return true;
    }
```

Now, instead of using **break** or **continue** to stop the program execution, you simply use **return** to break the loop execution since the result that you were looking for was already found.

DO-WHILE

The **do-while** loop is like the previous one, but with one subtle difference: it executes the logic at least once, while a simple **while** statement may never be executed if the condition is not met at the first execution. It has the following structure:

```
int t = 0;
do
{
    Console.WriteLine(t);
    t++;
} while (t < 5);
```

In this example, you write the value of **t**, starting from **0**, and keep incrementing it while it is smaller than **5**. Before jumping into the next type of loop, learn about a new concept called arrays.

ARRAYS

An **array** is a data structure used to store many objects of the same type. For instance, the following example is a variable declared as an array of integer numbers:

```
int[] numbers = { 1, 2, 3, 4, 5 };
```

The first important thing to note about arrays is that they have a fixed capacity. This means that an array will have the length defined at the time of its creation and this length cannot change. The length can be determined in various ways. In the preceding example, the length is inferred by counting the number of objects in the array. However, another way of creating an array is like this:

```
var numbers = new int[5];
```

Here, you are creating an array that has the capacity of **5** integers, but you do not specify any value for the array elements. When an array of any data type is created without adding elements to it, the default values for that value type are set for each position of the array. For example, consider the following figure:

Figure 1.5: Value type array with no index assigned

The preceding figure shows that when you create an integer array of five elements, without assigning a value to any element, the array is automatically filled with the default value at every position. In this case, the default value is **0**. Now consider the following figure:

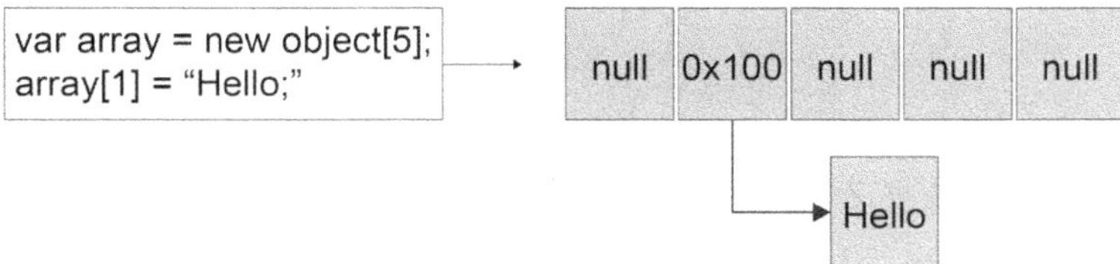

Figure 1.6: Reference type array with fixed size and only one index assigned

In the preceding example, you have created an array of five objects and assigned the **"Hello"** string value to the element at index **1**. The other positions of the array are automatically assigned the default value for objects, which is **null**.

Finally, it is worth noting that all arrays have **indexes**, which refers to the positions of the individual array elements. The first position will always have an index **0**. Thus, the positions in an array of size **n** can be specified from index **0** to **n-1**. Therefore, if you call **numbers[2]**, this means that you are trying to access the element in position **2** inside the numbers array.

FOR LOOPS

A **for** loop executes a set of instructions while a Boolean expression matches a specified condition. Just like **while** loops, jump statements can be used to stop a loop execution. It has the following structure:

```
for (initializer; condition; iterator)
{
    [statements]
}
```

The **initializer** statement is executed before the loop starts. It is used to declare and assign a local variable that will be used only inside the scope of the loop.

But in more complex scenarios, it can be used to combine other statement expressions as well. The condition specifies a Boolean condition that indicates when the loop should either continue or exit. The **iterator** is usually used to increment or decrement the variable created in the initializer section. Take the following example, where a **for** loop is used to print the elements of an integer array:

```
int[] array = { 1, 2, 3, 4, 5 };

for (int j = 0; j < array.Length - 1; j++)
{
Console.WriteLine(array[j]);
}
```

In this example, an initializer variable, **j**, has been created that is assigned **0** initially. The **for** loop will keep executing while **j** is smaller than the array length minus **1** (remember that indexes always start at **0**). After each iteration, the value of **j** is incremented by **1**. In this way, the **for** loop goes through the entire array and performs the given action, that is, printing the value of the current array element.

C# also allows the usage of **nested loops**, that is, a **loop within a loop**, as you will see in the next exercise.

EXERCISE 1.10: ORDERING AN ARRAY USING BUBBLE SORT

In this exercise, you will execute one of the simplest sorting algorithms. Bubble sort consists of going through every pair of elements inside an array and swapping them if they are unordered. In the end, the expectation is to have an array ordered in ascending order. You will use nested **for** loops to implement this algorithm.

To begin with, the array to be sorted should be passed as a parameter to this method. For each element of this array, if the current element is greater than the next, their positions should be swapped. This swap occurs by storing the value of the next element in a temporary variable, assigning the value of the current element to the next element, and finally, setting the value of the current element with the temporary value stored. Once the first element is compared to all others, a comparison starts for the second element and so on, till finally, the array is sorted.

The following steps will help you complete this exercise:

1. Create a new console project using the following command:

```
dotnet new console -n Exercise1_10
```

2. Inside the **Program.cs** file, create the method to implement the sorting algorithm. Add the following code:

```
static int[] BubbleSort(int[] array)
{
    int temp;

    // Iterate over the array
    for (int j = 0; j < array.Length - 1; j++)
    {
        // If the last j elements are already ordered, skip them
        for (int i = 0; i < array.Length - j - 1; i++)
        {
            if (array[i] > array[i + 1])
            {
                temp = array[i + 1];
                array[i + 1] = array[i];
                array[i] = temp;
            }
        }
    }

    return array;
}
```

3. Now create an **array** with some numbers, as follows:

```
int[] randomNumbers = { 123, 22, 53, 91, 787, 0, -23, 5 };
```

4. Call the **BubbleSort** method, passing the array as an argument, and assign the result to a variable, as follows:

```
int[] sortedArray = BubbleSort(randomNumbers);
```

5. Finally, you need to print the message that the array was sorted. To do so, iterate over it, printing the array elements:

```
Console.WriteLine("Sorted:");

for (int i = 0; i < sortedArray.Length; i++)
{
    Console.Write(sortedArray[i] + " ");
}
```

6. Run the program with the **`dotnet run --project Exercise1_10`** command. You should see the following output on your screen:

```
Sorted:
-23 0 5 22 53 91 123 787
```

> **NOTE**
>
> You can find the code used for this exercise at https://packt.link/cJs8y.

In this exercise, you used the two concepts learned in the last two sections: arrays and for loops. You manipulated arrays, accessing their values through indexes, and used for loops to move through these indexes.

There is another way to go through every element of an array or **group** in C#, called **`foreach`** statements. You will explore this in the following section.

FOREACH STATEMENTS

A **`foreach`** statement executes a set of instructions for each element of a collection. Just like a **`for`** loop, the **`break`**, **`continue`**, **`goto`**, and **`return`** keywords can also be used with **`foreach`** statements. Consider the following example, in which you iterate over every element of an array and write it to the console as the output:

```
var items = new int[] { 1, 2, 3, 4, 5 };

foreach (int element in items)
{
Console.WriteLine(element);
}
```

The preceding snippet prints the numbers from **1** to **5** to the console. You can use **foreach** statements with much more than arrays; they can also be used with lists, collections, and spans, which are other data structures that will be covered in later chapters.

FILE HANDLING

So far, you have been creating programs that interact mostly with CPU and memory. This section will focus on I/O operations, that is, input and output operations, on the physical disk. A great example of this type of operation is **file handling**.

C# has several classes that help you perform I/O operations. Some of these are as follows:

- **File**: This class provides methods for the manipulation of files, that is, reading, writing, creating, deleting, copying, and moving files on the disk.
- **Directory**: Like the **File** class, this class includes methods to create, move, and enumerate directories and subdirectories on the disk.
- **Path**: This provides utilities to deal with absolute and relative paths of files and directories on the disk. A **relative path** is always related to some path inside the current directory where the application is being executed, and an **absolute path** refers to an absolute location inside the hard drive.
- **DriveInfo**: This provides information about a disk drive, such as **Name**, **DriveType**, **VolumeLabel**, and **DriveFormat**.

You already know that files are mostly some sets of data located somewhere in a hard drive that can be opened for reading or writing by some program. When you open a file in a C# application, your program reads the file as a sequence of bytes through a communication channel. This communication channel is called a **stream**. Streams can be of two types:

- The **input streams** are used for reading operations.
- The **output streams** are used for writing operations.

The **Stream** class is an abstract class in C# that enables common operations regarding this byte flow. For file handling on a hard disk, you will use the **FileStream** class, designed specifically for this purpose. The following are two important properties of this class: **FileAccess** and **FileMode**.

FILEACCESS

This is an **`enum`** that provides you with options to choose a level of access when opening a specified file:

- **`Read`**: This opens a file in read-only mode.
- **`ReadWrite`**: This opens a file in read and write mode.
- **`Write`**: This opens a file in write-only mode. This is rarely used, as you usually do some reading along with the writing.

FILEMODE

This is an **`enum`** that specifies the operations that can be performed on a file. It should be used along with the access mode as some modes only work with some levels of access. Take a look at the options, as follows:

- **`Append`**: Use this when you want to add content at the end of the file. If the file does not exist, a new one will be created. For this operation, the file must have write permission; otherwise, any attempt to read fails and throws a **`NotSupportedException`** exception. Exceptions are an important concept that will be covered later in this chapter.
- **`Create`**: Use this to create a new file or overwrite an existing one. For this option, too, write permission is required. In Windows, if the file exists but is hidden, an **`UnauthorizedAccessException`** exception is thrown.
- **`CreateNew`**: This is like **`Create`** but is used to create new files and also requires write permission. However, if the file already exists, an **`IOException`** exception is thrown.
- **`Open`**: As the name suggests, this mode is used to open a file. The file must have read or read and write permissions. If the file does not exist, a **`FileNotFoundException`** exception is thrown.
- **`OpenOrCreate`**: This is like **`Open`**, except it creates a new file if it does not already exist.

EXERCISE 1.11: READING CONTENT FROM TEXT FILES

In this exercise, you will read text from a **Comma-Separated Values** (**CSV**) file. CSV files simply contain data represented by strings and separated either by colons or semicolons.

Perform the following steps to complete this exercise:

1. Open Command Prompt and type the following:

```
dotnet new console -n Exercise1_11
```

2. At the **Exercise1_11** project folder location in your computer, create a file named **products.csv** and paste the following content inside it:

```
Model;Memory;Storage;USB Ports;Screen;Condition;Price USD
Macbook Pro Mid 2012;8GB;500GB HDD;USB 2.0x2;13" 
screen;Refurbished;400
Macbook Pro Mid 2014;8GB;512GB SSD;USB 3.0x3;15" 
screen;Refurbished;750
Macbook Pro Late 2019;16GB;512GB SSD;USB 3.0x3;15" 
screen;Refurbished;1250
```

3. Open the **Program.cs** file and replace its contents with the following:

```
using System;
using System.IO;
using System.Threading.Tasks;

namespace Exercise1_11
{
    public class Program
    {
        public static async Task Main()
        {

        using (var fileStream = new FileStream("products.csv", 
FileMode.Open, FileAccess.Read))
        {
            using (var reader = new StreamReader(fileStream))
            {
                var content = await reader.ReadToEndAsync();

                var lines = content.Split(Environment.NewLine);

                foreach (var line in lines)
```

```
                {
                    Console.WriteLine(line);
                }
            }
        }
        }
    }
}
```

4. Call **dotnet run** in Command Prompt and you will get an output that is the same as the contents of the CSV file you have created.

> **NOTE**
>
> You can find the code used for this exercise at https://packt.link/5flid.

This exercise has some pretty interesting outcomes, which you are going to learn step by step. First, you opened a file using the **FileStream** class. This allows you to start streaming bytes from a file with two special properties, namely, **FileMode** and **FileAccess**. It will return a **stream** of bytes with the file contents. However, to read this content as text, you need to use the **StreamReader** class. This class enables you to read these bytes as text characters.

Notice also that your **Main** method changed from **void** to **async** Task. Additionally, the **await** keyword has been used, which is used for asynchronous operations. You will learn more about these topics in upcoming chapters. For now, you only need to know that an **async operation** is something that does not block the program execution. This means that you can output lines as they are being read; that is, you do not have to wait for all of them to be read.

In the next section, learn about the special keyword that handles files, databases, and network connections.

DISPOSABLE OBJECTS

Another special thing about the preceding exercise was the **using** keyword. It is a keyword used to clean up **unmanaged resources** from memory. These resources are special objects that handle some operational system resources, such as files, databases, and network connections. They are called **special** because they do what is called I/O operations; that is, they interact with the real resources of the machine, such as network and hard drives, not just with memory spaces.

The memory used by objects in C# is handled by something called the **garbage collector**. By default, C# handles the memory space in the stack and the heap. The only types of objects that do not perform this cleanup are called **unmanaged objects**.

Cleaning these objects from memory means that the resources will be free to be used by another process in the computer. That means a file can be handled by another one, a database connection is free to be used again by a connection pool, and so on. Those types of resources are called **disposable resources**. Every time you deal with a disposable resource, you can use the **using** keyword when creating an object. Then, the compiler knows that when the **using** statement closes, it can automatically free these resources.

EXERCISE 1.12: WRITING TO A TEXT FILE

In this exercise, you will write some text into a CSV file, again using the **FileStream** class.

Follow these steps to complete this exercise:

1. Open the VS Code integrated terminal and type the following:

```
dotnet new console -n Exercise1_12
```

2. At a preferred location on your computer, copy the **products.csv** file from the previous exercise and paste it into this exercise's folder.

3. In **Program.cs**, create a method named **ReadFile** that will receive a **FileStream** file and iterate over the file lines to output the result to the console:

```
static async Task ReadFile(FileStream fileStream)
    {
        using (var reader = new StreamReader(fileStream))
        {
            var content = await reader.ReadToEndAsync();

            var lines = content.Split(Environment.NewLine);

            foreach (var line in lines)
            {
                Console.WriteLine(line);
```

```
            }
        }
    }
```

4. Now, in your program, open the **`products.csv`** file with **`StreamWriter`** and add some more information to it, as follows:

```
            using (var file = new StreamWriter("products.
csv", append: true))
            {
                file.Write("\nOne more macbook without details.");
            }
```

5. Finally, read the contents of the file after modification:

```
using (var fileStream = new FileStream("products.csv", FileMode.Open,
                FileAccess.Read))
            {
                await ReadFile(fileStream);
            }
```

6. Call **`dotnet run --project Exercise1_12`** in the VS Code integrated terminal and you will be able to see the contents of the CSV file you just created, in addition to the line you just appended:

```
Model;Memory;Storage;USB Ports;Screen;Condition;Price USD
Macbook Pro Mid 2012;8GB;500GB HDD;USB 2.0x2;13"
screen;Refurbished;400
Macbook Pro Mid 2014;8GB;512GB SSD;USB 3.0x3;15"
screen;Refurbished;750
Macbook Pro Late 2019;16GB;512GB SSD;USB 3.0x3;15"
screen;Refurbished;1250
One more macbook without details.
```

   Note that for each run, the program will append a new line, so you will see more lines being added.

> **NOTE**
>
> You can find the code used for this exercise at https://packt.link/dUk2z.

Sometimes your program will fail to execute at some point and may not provide an output. Such an instance is called an exception error. The next section details all about such an error.

EXCEPTIONS

Exceptions indicate that a program has failed to execute at some point for some reason and can be raised by either the code itself or the .NET runtime. Usually, an exception is a severe failure and can even terminate your program's execution. Fortunately, C# provides a special way of **handling exceptions**, which is **try/ catch** blocks:

```
try
{
// some logic that might throw an exception
}
catch
{
// error handling
}
```

Inside the **try** clause, you call the code that might throw an exception, and inside the **catch** clause, you can treat the exception that was raised. For instance, consider the following example:

```
double Divide(int a, int b) => a/b;
```

This method takes two integers and returns the result of a division between them. However, what will happen if **b** is **0**? In such a case, the runtime will throw **System. DivideByZeroException**, indicating that it is not possible to execute the division. How could you handle this exception in a real-world program? You will explore this in the next exercise.

EXERCISE 1.13: HANDLING INVALID USER INPUTS WITH TRY/CATCH

In this exercise, you will create a console app that takes two inputs from you, divides the first number by the second one, and outputs the result. If you enter an invalid character, the app should throw an exception, and all of this should be handled inside the program logic.

Perform the following steps to complete this exercise:

1. Inside the VS Code integrated terminal, create a new console app called **Exercise1_13**.

2. Create the following method inside the **Program.cs** file:

```
static double Divide(int a, int b)
{
    return a / b;
}
```

3. Now, create a Boolean variable to indicate whether the division was properly executed. Assign **false** to it as its initial value:

```
bool divisionExecuted = false;
```

4. Write a **while** loop that will check whether the division happened successfully. If it did, the program should terminate. If not, the program should prompt you to input valid data and perform the division again. Add the following code to do this:

```
while (!divisionExecuted)
{
    try
    {
        Console.WriteLine("Please input a number");

        var a = int.Parse(Console.ReadLine());

        Console.WriteLine("Please input another number");

        var b = int.Parse(Console.ReadLine());

        var result = Divide(a, b);

        Console.WriteLine($"Result: {result}");

        divisionExecuted = true;
    }
    catch (System.FormatException)
    {
        Console.WriteLine("You did not input a number. Let's start 
again ... \n");
        continue;
    }
    catch (System.DivideByZeroException)
    {
```

```
            Console.WriteLine("Tried to divide by zero. Let's start again
... \n");
            continue;
        }
    }
```

5. Finally, execute the program using the **dotnet run** command and interact with the console. Try to insert strings instead of numbers and see what output you get. Look at the following output as an example:

```
Please input a number
5
Please input another number
0
Tried to divide by zero. Let's start again …

Please input a number
5
Please input another number
s
You did not input a number. Let's start again …

Please input a number
5
Please input another number
1
Result: 5
```

> **NOTE**
>
> You can find the code used for this exercise at https://packt.link/EVsrJ.

In this exercise, you handled two types of exceptions that are as follows:

- The **int.Parse(string str)** method throws **System.FormatException** if it is not possible to convert the **string** variable into an integer.
- The **double Divide(int a, int b)** method throws **System.DivideByZeroException** if b is 0.

Now that you have seen how exceptions are handled, it is important to note a rule of thumb that will help you in your C# journey, which is that *you should only catch what you can or what you need to handle*. There are only a few situations where exception handling is really needed, as follows:

- When you want to **mask** an exception, that is, catch it and pretend that nothing happened. This is known as **exception suppression**. That should take place when the exception that is thrown does not impact the flow of your program.
- When you want to control your program's execution flow to perform some alternate actions, as you did in the preceding exercise
- When you want to catch a type of exception to throw it as another type. For instance, when communicating with your web API, you might see an exception of type **HttpException** that indicates that the destination is unreachable. You could make use of a custom exception here, such as **IntegrationException**, to indicate more clearly that it happened in a part of your application that performs some integrations with external APIs.

The **throw** keyword can also be used to intentionally stop the program execution flow in certain cases. For example, consider that you are creating a **Person** object and that the **Name** property should not be **null** at the time of creation. You can enforce on this class a **contract** that says: if these parameters are not correctly provided, it cannot be used. Typically, you would do so by throwing **System.ArgumentException** or **System.ArgumentNullException**, as in the following snippet, which uses **ArgumentNullException** to do so:

```
class Person
{
Person(string name)
     {
if (string.
IsNullOrWhiteSpace(name)) throw new ArgumentNullException(nameof(name));

Name = name;
     }

    String Name { get ; set; }
}
```

Here, if the value of the **name** argument is **null** or if you only enter space characters, **ArgumentNullException** is thrown, and the program does not execute successfully. The null/white space condition is checked with the help of the **IsNullOrWhiteSpace** function, which can be used for string variables.

Now it's time to practice all that you learned in the previous sections through an activity.

ACTIVITY 1.01: CREATING A GUESSING GAME

To complete this activity, you need to create a guessing game using the concepts you have learned about and practiced so far in this chapter. In this game, first, a random number from one to 10 must be generated, not to be output to the console. The console should then prompt the user to input a number and then guess which random number has been generated, and the user should get a maximum of five chances.

Upon every incorrect input, a warning message should be displayed, letting the user know how many chances they have left, and if all five chances are exhausted with incorrect guesses, the program terminates. However, once the user guesses correctly, a success message should be displayed, before the program terminates.

The following steps will help you complete this activity:

1. Create a variable called **numberToBeGuessed** that is assigned to a random number within C#. You can use the following snippet to do so:

   ```
   new Random().Next(0, 10)
   ```

   This generates a random number for you, between **0** and **10**. You could replace **10** with a higher number if you wanted to make the game a little more difficult, or with a smaller number to make it easier, but for this activity, you will use **10** as the maximum value.

2. Create a variable called **remainingChances** that will store the remaining number of chances that the user has.

3. Create a **numberFound** variable and assign a **false** value to it.

4. Now, create a **`while`** loop that will execute while there are still some chances remaining. Within this loop, add code to output the number of chances remaining, until the correct guess is made. Then, create a variable called **`number`** that will receive the **parsed** integer for the user input. Finally, write code to check whether the **`number`** variable is the correct guess, and assign the value **`true`** to the **`numberFound`** variable if so. If not, the number of remaining chances should be reduced by **`1`**.
5. Finally, add code to inform users whether they have guessed the number correctly. You can output something such as **`Congrats! You've guessed the number with {remainingChanges} chances left!`** if they guessed correctly. If they ran out of chances, output **`You're out of chances. The number was {numberToBeGuessed}.`**.

> **NOTE**
>
> The solution to this activity can be found at https://packt.link/qclbF.

SUMMARY

This chapter gave you an overview of the fundamentals of C# and what it looks like to write programs with it. You explored everything from the variable declaration, data types, and basic arithmetic and logical operators to file and exception handling. You also explored how C# allocates memory while dealing with value and reference types.

In the exercises and activities in this chapter, you were able to solve some real-world problems and think of solutions that can be implemented with this language and its resources. You learned how to prompt for user inputs in console apps, how to handle files within a system, and finally, how to deal with unexpected inputs through exception handling.

The next chapter will cover the essentials of Object-oriented programming, diving deeper into the concept of classes and objects. You will also learn about the importance of writing clean, concise code that is easy to maintain, and the principles you can follow for writing such code.

2

BUILDING QUALITY OBJECT-ORIENTED CODE

OVERVIEW

In this chapter, you will learn how to simplify complex logic using Object-Oriented Programming (OOP). You will start by creating classes and objects, before exploring the four pillars of OOP. You will then learn about some of the best practices in coding, known as the SOLID principles, and see how you can use C# 10 features to write effective code guided by these principles. By the end of this chapter, you will be able to write clean code using object-oriented design with C#.

INTRODUCTION

How do people write software that is still maintainable even after many decades? What is the best way to model software around real-world concepts? The answer to both questions is **Object Oriented Programming** (**OOP**). OOP is a widely used paradigm in professional programming and is especially useful in enterprise settings.

OOP can be thought of as a bridge that connects real-world concepts and source code. A cat, for example, has certain defining properties, such as age, fur color, eye color, and name. The weather can be described using factors such as temperature and humidity. Both of these are real-world concepts that humans have identified and defined over time. In OOP, **classes** are what help in defining the logic of a program. When assigning concrete values to the properties of these classes, the result is an **object**. For example, using OOP, you can define a class for representing a room in a house, and then assign values to its properties (color and area) to create an object of that class.

In *Chapter 1*, *Hello C#*, you learned how to use C# to write basic programs. In this chapter, you will see how you can design your code by implementing OOP concepts and using C# at its best.

CLASSES AND OBJECTS

A class is like a blueprint that describes a concept. An object, on the other hand, is the result you get after the application of this blueprint. For example, **`weather`** can be a class, and **`25 degrees and cloudless`** could refer to an object of this class. Similarly, you can have a class named **`Dog`**, while a four-year-old **`Spaniel`** can represent an object of the **`Dog`** class.

Declaring a class in C# is simple. It starts with the **`class`** keyword, followed by the class name and a pair of curly braces. To define a class named **`Dog`**, you can write the following code:

```
class Dog
{
}
```

Right now, this class is just an empty skeleton. However, it can still be used to create objects by using the **`new`** keyword, as follows:

```
Dog dog = new Dog();
```

This creates an object named **dog**. Currently, the object is an empty shell, as it lacks properties. You will see in an upcoming section how to define properties for classes, but first, you will explore constructors.

CONSTRUCTORS

In C#, **constructor**s are functions used to create new objects. You can also use them to set the initial values of an object. Like any function, a constructor has a name, takes arguments, and can be overloaded. A class must have at least one constructor, but if needed, it can have multiple constructors with different arguments. Even if you do not explicitly define a single constructor, a class will still have a default constructor–one that does not take any arguments or perform any actions but simply assigns memory to the newly created object and its fields.

Consider the following snippet, where a constructor for the **Dog** class is being declared:

```
// Within a class named Dog
public class Dog
{
  // Constructor
  public Dog()
  {
    Console.WriteLine("A Dog object has been created");
  }
}
```

> **NOTE**
>
> You can find the code used for this example at https://packt.link/H2lUF.
> You can find the usage of the code at https://packt.link/4WoSX.

If a method has the same name as the class and does not provide a **return** type, it is a constructor. Here, the snippet of the code is within a class named **Dog**. So, the constructor is within the specified line of code. Note that by defining this constructor explicitly, you hide the default constructor. If there is one or more such custom constructors, you will no longer be able to use a default constructor. Once the new constructor is called, you should see this message printed in the console: **"A Dog object has been created"**.

FIELDS AND CLASS MEMBERS

You already know what a variable is: it has a type, a name, and a value, as you saw in *Chapter 1*, *Hello C#*. Variables can also exist in the class scope, and such a variable is called a **field**. Declaring a field is as simple as declaring a local variable. The only difference is the addition of a keyword at the start, which is the access modifier. For example, you can declare a field within the **Dog** class with the public access modifier, as follows:

```
public string Name = "unnamed";
```

This line of code states that the **Name** field, which is a string with the value **"unnamed"**, can be accessed publicly. Besides **public**, the other two main access modifiers in C# are **private** and **protected**, which you will look at them in detail later.

> **NOTE**
>
> You can find more information regarding access modifiers at https://docs.microsoft.com/en-us/dotnet/csharp/language-reference/keywords/access-modifiers.

Everything a class holds is called a **class member**. Class members can be accessed from outside of a class; however, such access needs to be granted explicitly using the **public** access modifier. By default, all members have a **private** access modifier.

You can access class members by writing the object name followed by a dot (**.**) and the member name. For example, consider the following snippet in which two objects of the **Dog** class are being created:

```
Dog sparky = new Dog();
Dog ricky = new Dog();
```

Here, you can declare two independent variables, **sparky** and **ricky**. However, you haven't explicitly assigned these names to the objects; note that these are only the variable names. To assign the names to the objects, you can write the following code using dot notation:

```
sparky.Name = "Sparky";
ricky.Name = "Ricky";
```

You can now have hands-on experience of creating classes and objects through an exercise.

EXERCISE 2.01: CREATING CLASSES AND OBJECTS

Consider that there are two books, both by an author named **New Writer**. The first one, called **First Book**, was published by **Publisher 1**. There is no description available for this book. Similarly, the second one is named **Second Book** and was published by **Publisher 2**. It has a description that simply says, **"Interesting read"**.

In this exercise, you will model these books in code. The following steps will help you complete this exercise.

1. Create a class called **Book**. Add fields for **Title**, **Author**, **Publisher**, **Description**, and the number of pages. You must print this information from outside the class, so make sure every field is **public**:

```
public class Book
{
    public string Title;
    public string Author;
    public string Publisher;
    public int Pages;
    public string Description;
}
```

2. Create a class named **Solution**, with the **Main** method. As you saw in *Chapter 1*, *Hello C#*, this class with the **Main** method is the starting point of your application:

```
public static class Solution
{
    public static void Main()
    {
    }
}
```

3. Inside the **Main** method, create an object for the first book and set the values for the fields, as follows:

```
Book book1 = new Book();
book1.Author = "New Writer";
book1.Title = "First Book";
book1.Publisher = "Publisher 1";
```

   Here, a new object named **book1** is created. Values are assigned to different fields by writing dot (**.**) followed by the field name. The first book does not have a description, so you can omit the field **book1.Description**.

4. Repeat this step for the second book. For this book, you need to set a value for the **Description** field as well:

```
Book book2 = new Book();
book2.Author = "New Writer";
book2.Title = "Second Book";
book2.Publisher = "Publisher 2";
book2.Description = "Interesting read";
```

   In practice, you will rarely see fields with public access modifiers. Data mutates easily, and you might not want to leave your program open to external changes after initialization.

5. Inside the **Solution** class, create a method named **Print**, which takes a **Book** object as an argument and prints all fields and their values. Use **string interpolation** to concatenate book information and print it to the console using **Console.WriteLine()**, as follows:

```
private static void Print(Book book)
{
    Console.WriteLine($"Author: {book.Author}, " +
                      $"Title: {book.Title}, " +
                      $"Publisher: {book.Publisher}, " +
                      $"Description: {book.Description}.");
}
```

6. Inside the **Main** method, call the **Print** method for **book1** and **book2**:

```
Print(book1);
Print(book2);
```

Upon running this code, you will see the following output on the console:

```
Author: New Writer, Title: First Book, Publisher: Publisher 1,
Description: .
Author: New Writer, Title: Second Book, Publisher: Publisher 2,
Description: Interesting read.
```

> **NOTE**
>
> You can find the code used for this exercise at https://packt.link/MGT9b.

In this exercise, you saw how to use fields and class members are used in simple programs. Now proceed to know about reference types.

REFERENCE TYPES

Suppose you have an object and the object is not created, just declared, as follows:

```
Dog speedy;
```

What would happen if you tried accessing its **Name** value? Calling **speedy.Name** would throw a **NullReferenceException** exception because **speedy** is yet to be initialized. Objects are reference types, and their default value is null until initialized. You have already worked with value types, such as **int**, **float**, and **decimal**. Now you need to grasp that there are two major differences between value and reference types.

Firstly, value types allocate memory on the stack, whereas reference types allocate memory on the **heap**. The **stack** is a temporary place in memory. As the name implies, in a stack, blocks of memory are stacked on top of each other. When you call a function, all local function variables will end up on a single block of the stack. If you call a nested function, the local variables of that function will be allocated on another block of the stack.

In the following figure, you can see which parts of code will allocate memory in the stack during execution, and which will do so in the heap. Method calls (1, 8, 10) and local variables (2, 4) will be stored in the stack. Objects (3, 5) and their members (6) will be stored on the heap. Stacks use the **Push method** to allocate data, and **Pop** to deallocate it. When memory is allocated, it comes on top of the stack. When it is deallocated, it is removed from the top as well. You deallocate memory from the stack as soon as you leave the scope of a method (8, 10, 11). Heap is much more random, and **Garbage Collector** (**GC**) automatically (unlike some other languages, where you need to do it yourself), deallocates memory.

> **NOTE**
>
> GC is a massive topic in itself. If you want to find out more, please refer to the official Microsoft documentation at https://docs.microsoft.com/en-us/dotnet/standard/garbage-collection/fundamentals.

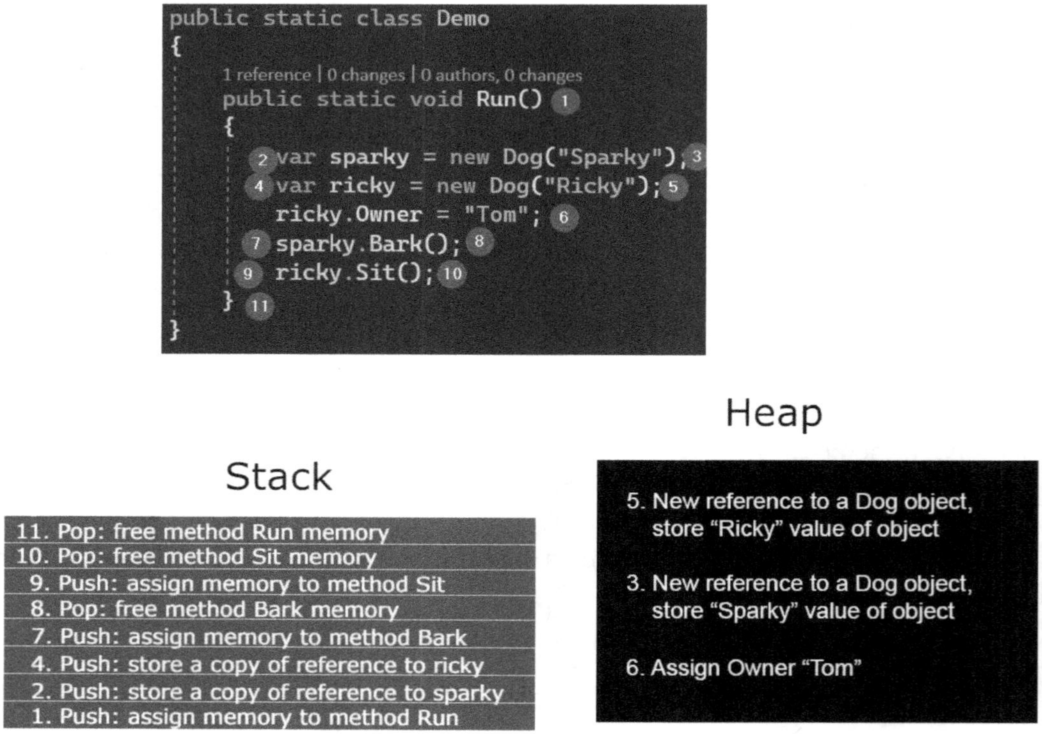

```
public static class Demo
{
    1 reference | 0 changes | 0 authors, 0 changes
    public static void Run() (1)
    {
      (2) var sparky = new Dog("Sparky"); (3)
      (4) var ricky = new Dog("Ricky"); (5)
          ricky.Owner = "Tom"; (6)
      (7) sparky.Bark(); (8)
      (9) ricky.Sit(); (10)
    } (11)
}
```

Stack

11. Pop: free method Run memory
10. Pop: free method Sit memory
9. Push: assign memory to method Sit
8. Pop: free method Bark memory
7. Push: assign memory to method Bark
4. Push: store a copy of reference to ricky
2. Push: store a copy of reference to sparky
1. Push: assign memory to method Run

Heap

5. New reference to a Dog object, store "Ricky" value of object
3. New reference to a Dog object, store "Sparky" value of object
6. Assign Owner "Tom"

Figure 2.1: Stack and heap comparison

> **NOTE**
>
> If you make too many nested calls, you will run into a **StackoverflowException** exception because the stack ran out of memory. Freeing up memory on the stack is just a matter of exiting from a function.

The second difference is that, when value types are passed to a method, their value is copied, while for reference types, only the reference is copied. This means that the reference type object's state is modifiable inside a method, unlike a value type, because a reference is simply the address of an object.

Consider the following snippet. Here, a function named **SetTo5** sets the value of the number to **5**:

```
private static void SetTo5(int number)
{
        number = 5;
}
```

Now, consider the following code:

```
int a = 2;
// a is 2
Console.WriteLine(a);
SetTo5(a);
// a is still 2
Console.WriteLine(a);
```

This should result in the following output:

```
2
2
```

If you run this code, you find that the printed value of **a** is still **2** and not **5**. This is because **a** is a value type that passed the value **2**, and therefore its value is copied. Inside a function, you never work with the original; a copy is always made.

What about reference types? Suppose you add a field named **Owner** inside the **Dog** class:

```
public class Dog
{     public string Owner;
}
```

Create a function, **ResetOwner**, that sets the value of the **Owner** field for an object to **None**:

```
private static void ResetOwner(Dog dog)
{
    dog.Owner = "None";
}
```

Now, suppose the following code is executed:

```
Dog dog = new Dog("speedy");
Console.WriteLine(dog.Owner);
ResetOwner(dog);
// Owner is "None"- changes remain
Console.WriteLine(dog.Owner);
```

This should result in the following output:

```
speedy
None
```

> **NOTE**
>
> You can find the code used for this example at https://packt.link/gj164.

If you try running this snippet of code yourself, you will first see the name **speedy** on one line and then **None** printed on another. This would change the dog's name, and the changes would remain outside the function. This is because Dog is a class, and a class is a reference type. When passed to a function, a **copy of a reference** is made. However, a **copy of a reference** points to the whole object, and therefore the changes that are made remain outside as well.

It might be confusing to hear that you pass a copy of a reference. How can you be sure you are working with a copy? To learn this, consider the following function:

```
private static void Recreate(Dog dog)
{
    dog = new Dog("Recreated");
}
```

Here, creating a new object creates a new reference. If you change the value of a reference type, you are working with a completely different object. It may be one that looks the same but is stored in a completely different place in memory. Creating an object for a passed parameter will not affect anything outside the object. Though this may sound potentially useful, you should generally avoid doing this as it can make code difficult to comprehend.

PROPERTIES

The **`Dog`** class has one flaw. Logically, you wouldn't want the name of a dog to be changed once it is assigned. However, as of now, there is nothing that prevents changing it. Think about the object from the perspective of what you can do with it. You can set the name of a dog (**`sparky.Name = "Sparky"`**) or you can get it by calling **`sparky.Name`**. However, what you want is a read-only name that can be set just once.

Most languages take care of this through setter and getter methods. If you add the **`public`** modifier to a field, this means that it can be both retrieved (read) and modified (written). It isn't possible to allow just one of these actions. However, with setters and getters, you can restrict both read and write access. In OOP, restricting what can be done with an object is key to ensuring data integrity. In C#, instead of setter and getter methods, you can use properties.

In OOP languages (for example Java), to set or get the values of a name, you would write something like this:

```
public string GetName()
{
    return Name;
}

public string SetName (string name)
{
    Name = name;
}
```

In C#, it is as simple as the following:

```
public string Name {get; set;}
```

This is a property, which is nothing but a method that reads like a field. There are two types of properties: **getters** and **setters**. You can perform both read and write operations with them. From the preceding code, if you remove **get**, it will become write-only, and if you remove **set**, it will become read-only.

Internally, the property includes a setter and a getter method with a **backing field**. A **backing field** is simply a private field that stores a value, and getter and setter methods work with that value. You can write custom getters and setters as well, as follows:

```
private string _owner;
public string Owner
{
    get
    {
        return _owner;
    }
    set
    {
        _owner = value;
    }
}
```

In the preceding snippet, the **Owner** property shows what the default getter and setter methods would look like for the **Dog** class.

Just like other members, individual parts of a property (either getter or setter) can have their own access modifier, like the following:

```
public string Name {get; private set;}
```

In this case, the getter is **public**, and the setter is **private**. All parts of the property (getter, setter, or both, as defined) take the access modifier from the property (**Name**, in this case) unless explicitly specified otherwise (as in the case of **private** set). If you do not need to set a name, you can get rid of the setter. If you need a default value, you can write the code for this as follows:

```
public string Name {get;} = "unnamed";
```

This piece of code means that the **Name** field is read-only. You can set the name only through a constructor. Note that this is not the same as a **private** set because the latter means you can still change the name within the **Dog** class itself. If no setter is provided (as is the case here), you can set the value in only one place, the constructor.

What happens internally when you create a read-only property? The following code is generated by the compiler:

```
private readonly string _name;
public string get_Name()
{
    return _name;
}
```

This shows that getter and setter properties are simply methods with a backing field. It is important to note that, if you have a property called **Name**, the **set_Name()** and **get_Name()** methods will be reserved because that's what the compiler generates internally.

You may have noticed a new keyword in the previous snippet, **readonly**. It signifies that the value of a field can only be initialized once—either during declaration or in a constructor.

Returning a backing field with a property may seem redundant sometimes. For example, consider the next snippet:

```
private string _name;

public string Name
{
    get
    {
        return "Dog's name is " + _name;
    }
}
```

This code snippet is a custom property. When a getter or a setter is more than just a basic return, you can write the property in this way to add custom logic to it. This property, without affecting the original name of a dog, will prepend **Dog's name is** before returning the name. You can make this more concise using **expression-bodied property** syntax, as follows:

```
public string Name => "Dog's name is " + _name;
```

This code does the same thing as the previous code; the **=>** operator indicates that it is a read-only property, and you return a value that is specified on the right side of the **=>** operator.

How do you set the initial value if there is no setter? The answer to that is a constructor. In OOP, a constructor serves one purpose—that is, setting the initial values of fields. Using a constructor is great for preventing the creation of objects in an invalid state.

To add some validation to the **Dog** class, you can write the following code:

```
public Dog(string name)
{
  if(string.IsNullOrWhitespace(name))
  {
    throw new ArgumentNullException("name")
  }
  Name = name;
}
```

The code you have just written will prevent an empty name from being passed when creating a **Dog** instance.

It is worth mentioning that within a class, you have access to the object itself that will be created. It might sound confusing, but it should make sense with this example:

```
private readonly string name;
public Dog(string name)
{
  this.name = name;
}
```

The **this** keyword is most often used to clear the distinction between class members and arguments. **this** refers to the object that has just been created, hence, **this.name** refers to the name of that object and **name** refers to the passed parameter.

Creating an object of the **Dog** class, and setting the initial value of a name, can now be simplified as follows:

```
Dog ricky = new Dog("Ricky");
Dog sparky = new Dog("Sparky");
```

You still have a private setter, meaning the property that you have is not entirely read-only. You can still change the value of a name within the class itself. However, fixing that is quite easy; you can simply remove the setter and it will become truly read-only.

> **NOTE**
>
> You can find the code used for this example at http://packt.link/hjHRV.

OBJECT INITIALIZATION

Often, a class has read and write properties. Usually, instead of setting the property values via a constructor, they are assigned after the creation of an object. However, in C# there is a better way—**object initialization**. This is where you create a new object and set the mutable (read and write) field values right away. If you had to create a new object of the **Dog** class and set the value of **Owner** for this object to **Tobias**, you could add the following code:

```
Dog dog = new Dog("Ricky");
dog.Owner = "Tobias";
```

This can be done using object initialization as follows:

```
Dog dog = new Dog("Ricky")
{
  Owner = "Tobias"
};
```

Setting initial properties like this when they are not a part of a constructor is generally more concise. The same applies to arrays and other collection types. Suppose you had two objects of the **Dog** class, as follows:

```
Dog ricky = new Dog("Ricky");
Dog sparky = new Dog("Sparky");
```

In such a case, one way of creating an array would be as follows:

```
Dog[] dogs = new Dog[2];
dogs[0] = ricky;
dogs[1] = sparky;
```

However, instead of this, you can just add the following code, which is more concise:

```
Dog[] dogs = {ricky, sparky};
```

In C# 10, you can simplify object initialization without providing the type, if it can be inferred from the declaration, as in the following code:

```
Dog dog = new("Dog");
```

COMPARING FUNCTIONS AND METHODS

Up until now, you might have seen the terms—function and method—used quite often, almost interchangeably. Now proceed to gain further insight into functions and methods. A **function** is a block of code that you can call using its name and some input. A **method** is a function that exists within a class.

However, in C#, you cannot have functions outside of a class. Therefore, in C#, every function is a method. Many languages, especially non-OOP languages, have only some functions that can be called methods (for example, JavaScript).

The **behavior** of a class is defined using methods. You have already defined some behavior for the **Dog** class, that is, getting its name. To finish implementing the behaviors for this class, you can implement some real-world parallels, such as sitting and barking. Both methods will be called from the outside:

```
public void Sit()
{
    // Implementation of how a dog sits
}
public void Bark()
{
    // Implementation of how a dog barks
}
```

You can call both methods like this:

```
Ricky.Sit();
Sparky.Bark();
```

In most cases, it is preferable to avoid exposing data publicly, so you should only ever expose functions publicly. Here, you might be wondering, What about properties? **Properties** are just getter and setter functions; they work with data but aren't data themselves. You should avoid **exposing** data publicly directly, for the same reason you lock your doors, or carry your phone in a case. If data were public, everyone could access it without any restrictions.

Also, data should not change when the program requires it to be constant. A **method** is a mechanism that ensures that an object is not used in invalid ways, and if it is, it's well handled.

What if you need to validate the fields consistently throughout the application? Again, properties, that is, getter and setter methods, can help with this. You can limit what you can do with data and add validation logic to it. Properties help you be in full control of how you can get and set data. Properties are handy, but it's important to use them with discretion. If you want to do something complex, something that needs extra computing, it is preferable to use a method.

For example, imagine that you have a class for an inventory made up of items, each having some weight. Here, it might make sense to have a property to return the heaviest item. If you chose to do so through a property (call it **`MaxWeight`**), you might get unexpected results; getting the heaviest item would require iterating through a collection of all items and finding the maximum by weight. This process is not as fast as you would expect. In fact, in some cases, it might even throw an error. Properties should have simple logic, otherwise working with them might yield unexpected results. Therefore, when the need for compute-heavy properties arises, consider refactoring them to a method. In this case, you would refactor the **`MaxWeight`** property into the **`GetMaxWeight`** method.

Properties should be avoided for returning results of complex calculations, as calling a property could be expensive. Getting or setting the value of a field should be straightforward. If it becomes expensive, it should no longer be treated as property.

AN EFFECTIVE CLASS

The **`Dog`** class models a **`dog`** object; therefore, it can be called a **model**. Some developers prefer to have a strict separation between data and logic. Others try to put as much logic in a model as possible, so long as it is self-contained. There is no right or wrong way here. It all depends on the context you are working with.

> **NOTE**
>
> This discussion is outside the scope of this chapter, but if you would like to know more, you can refer to the discussion on **Domain-Driven Design** (**DDD**) at https://martinfowler.com/bliki/DomainDrivenDesign.html.

It is hard to pinpoint what an effective class looks like. However, when deciding whether a method fits better in class A or class B, try asking yourself these questions:

- Would someone, who is not a programmer, know that you are talking about the class? Is it a logical representation of a real-world concept?
- How many reasons does the class have to change? Is it just one or are there more reasons?
- Is private data tightly related to public behavior?
- How often does the class change?
- How easy is it to break the code?
- Does the class do something by itself?

High cohesion is a term used to describe a class that has all its members strongly related, not only semantically, but logically as well. In contrast, a **low cohesion** class has loosely related methods and fields that probably could have a better place. Such a class is inefficient because it changes for multiple reasons and you cannot expect to look for anything inside it, as it simply has no strong logical meaning.

For example, a part of a **Computer** class could look like this:

```
class Computer
{
    private readonly Key[] keys;
}
```

However, **Computer** and **keys** are not related at the same level. There could be another class that better suits the **Key** class, that is **Keyboard**:

```
class Computer
{
    private readonly Keyboard keyboard;
}
class Keyboard
{
    private readonly Key[] keys;
}
```

> **NOTE**
>
> You can find the code used for this example at https://packt.link/FFcDa.

A keyboard is directly related to keys, just as it is directly related to a computer. Here, both **Keyboard** and the **Computer** class have high cohesion because the dependencies have a stable logical place. You can now learn more about it through an exercise.

EXERCISE 2.02: COMPARING THE AREA OCCUPIED BY DIFFERENT SHAPES

You have two sections of a backyard, one with circular tiles and the other with rectangular tiles. You would like to deconstruct one section of the backyard, but you are not sure which one it should be. Obviously, you want as little mess as possible and have decided to pick the section that occupies the least area.

Given two arrays, one for different sized rectangular tiles and the other for different-sized circular tiles, you need to find which section to deconstruct. This exercise aims to output the name of the section occupying less area, that is, **rectangular** or **circular**.

Perform the following steps to do so:

1. Create a **Rectangle** class as follows. It should have fields for **width**, **height**, and **area**:

```
public class Rectangle
{
    private readonly double _width;
    private readonly double _height;

    public double Area
    {
        get
        {
            return _width * _height;
        }
    }

    public Rectangle(double width, double height)
    {
        _width = width;
        _height = height;
    }
}
```

Here, **_width** and **_height** have been made immutable, using the **readonly** keyword. The type chosen is **double** because you will be performing **math** operations. The only property that is exposed publicly is **Area**. It will return a simple calculation: the product of width and height. The **Rectangle** is immutable, so all it needs is to be passed once through a constructor and it remains constant thereafter.

2. Similarly, create a **Circle** class as follows:

```
public class Circle
{
    private readonly double _radius;

    public Circle(double radius)
    {
        _radius = radius;
    }

    public double Area
    {
        get { return Math.PI * _radius * _radius; }
    }
}
```

The **Circle** class is similar to **Rectangle** class, except that instead of width and height, it has **radius**, and the **Area** calculation uses a different formula. The constant **PI** has been used, which can be accessed from the **Math** namespace.

3. Create a **Solution** class with a skeleton method named **Solve**:

```
public static class Solution
{
    public const string Equal = "equal";
    public const string Rectangular = "rectangular";
    public const string Circular = "circular";

    public static string Solve(Rectangle[] rectangularSection,
Circle[] circularSection)
    {
```

```
        var totalAreaOfRectangles =
CalculateTotalAreaOfRectangles(rectangularSection);
        var totalAreaOfCircles =
CalculateTotalAreaOfCircles(circularSection);

        return GetBigger(totalAreaOfRectangles, totalAreaOfCircles);
    }
}
```

Here, the **Solution** class demonstrates how the code works. For now, there are three constants based on the requirements (which section is bigger? rectangular or circular, or are they equal?). Also, the flow will be to calculate the total area of rectangles, then of circles and finally return the bigger.

Before you can implement the solution, you must first create side methods for calculating the total area of the rectangular section, calculating the total area of the circular section, and comparing the two. You will do this over the next few steps.

4. Inside **Solution** class, add a method to calculate the total area of the rectangular section:

```
private static double CalculateTotalAreaOfRectangles(Rectangle[]
rectangularSection)
{
    double totalAreaOfRectangles = 0;
    foreach (var rectangle in rectangularSection)
    {
        totalAreaOfRectangles += rectangle.Area;
    }

    return totalAreaOfRectangles;
}
```

This method goes through all the rectangles, gets the area of each, and adds it to the total sum.

5. Similarly, add a method to calculate the total area of the circular section:

```
private static double CalculateTotalAreaOfCircles(Circle[]
circularSection)
{
    double totalAreaOfCircles = 0;
    foreach (var circle in circularSection)
    {
```

```
            totalAreaOfCircles += circle.Area;
        }

        return totalAreaOfCircles;
    }
```

6. Next, add a method to get the bigger area, as follows:

```
private static string GetBigger(double totalAreaOfRectangles, double
totalAreaOfCircles)
{
    const double margin = 0.01;
    bool areAlmostEqual = Math.Abs(totalAreaOfRectangles -
totalAreaOfCircles) <= margin;
    if (areAlmostEqual)
    {
        return Equal;
    }
    else if (totalAreaOfRectangles > totalAreaOfCircles)
    {
        return Rectangular;
    }
    else
    {
        return Circular;
    }
}
```

This snippet contains the most interesting part. In most languages, numbers with a decimal point are not accurate. In fact, in most cases, if a and b are floats or doubles, it is likely that they will never be equal. Therefore, when comparing such numbers, you must consider precision.

In this code, you have defined the margin, to have an acceptable range of accuracy of your comparison for when the numbers are considered equal (for example, 0.001 and 0.0011 will be equal in this case since the margin is 0.01). After this, you can do a regular comparison and return the value for whichever section has the biggest area.

7. Now, create the **`Main`** method, as follows:

```
public static void Main()
{
    string compare1 = Solve(new Rectangle[0], new Circle[0]);
```

```
    string compare2 = Solve(new[] { new Rectangle(1, 5)}, new
Circle[0]);
    string compare3 = Solve(new Rectangle[0], new[] { new Circle(1)
});
    string compare4 = Solve(new []
    {
        new Rectangle(5.0, 2.1),
        new Rectangle(3, 3),
    }, new[]
    {
        new Circle(1),
        new Circle(10),
    });

    Console.WriteLine($"compare1 is {compare1}, " +
                      $"compare2 is {compare2}, " +
                      $"compare3 is {compare3}, " +
                      $"compare4 is {compare4}.");
}
```

Here, four sets of shapes are created for comparison. **compare1** has two empty sections, meaning they should be equal. **compare2** has a rectangle and no circles, so the rectangle is bigger. **compare3** has a circle and no rectangle, so the circles are bigger. Finally, **compare4** has both rectangles and circles, but the total area of the circles is bigger. You used string interpolation inside **Console.WriteLine** to print the results.

8. Run the code. You should see the following being printed to the console:

```
compare1 is equal, compare2 is rectangular, compare3 is circular,
compare4 is circular.
```

> **NOTE**
>
> You can find the code used for this exercise at https://packt.link/tfDCw.

What if you did not have objects? What would the section be made of in that case? For a circle, it might be viable to just pass radii, but for rectangles, you would need to pass another collinear array with widths and heights.

Object-oriented code is great for grouping similar data and logic under one shell, that is, a class, and passing those class objects around. In this way, you can simplify complex logic through simple interaction with a class.

You will now know about the four pillars of OOP.

THE FOUR PILLARS OF OOP

Efficient code should be easy to grasp and maintain, and OOP strives to achieve such simplicity. The entire concept of object-oriented design is based on four main tenets, also known as the **four pillars of OOP**.

ENCAPSULATION

The first pillar of OOP is **encapsulation**. It defines the relationship between data and behavior, placed in the same shell, that is, a class. It refers to the need to expose only what is necessary and hide everything else. When you think about encapsulation, think about the importance of security for your code: what if you leak a password, return confidential data, or make an API key public? Being reckless often leads to damage that can be hard to fix.

Security is not just limited to protection from malicious intent, but also extends to preventing manual errors. Humans tend to make mistakes. In fact, the more options there are to choose from, the more mistakes they are likely to make. Encapsulation helps in that regard because you can simply limit the number of options available to the person who will use the code.

You should prevent all access by default, and only grant explicit access when necessary. For example, consider a simplified `LoginService` class:

```
public class LoginService
{
    // Could be a dictionary, but we will use a simplified example.
    private string[] _usernames;
    private string[] _passwords;

    public bool Login(string username, string password)
    {
        // Do a password lookup based on username
        bool isLoggedIn = true;
        return isLoggedIn;
    }
}
```

This class has two **private** fields: **_usernames** and **_passwords**. The key point to note here is that neither passwords nor usernames are accessible to the public, but you can still achieve the required functionality by exposing just enough logic publicly, through the **Login** method.

> **NOTE**
>
> You can find this code used for this example at https://packt.link/6SO7a.

INHERITANCE

A police officer can arrest someone, a mailman delivers mail, and a teacher teaches one or more subjects. Each of them performs widely different duties, but what do they all have in common? In the context of the real world, they are all human. They all have a name, age, height, and weight. If you were to model each, you would need to make three classes. Each of those classes would look the same, other than one unique method for each. How could you express in code that they are all human?

The key to solving this problem is **inheritance**. It allows you to take all the properties from a parent class and transfer them to its child class. Inheritance also defines an **is-a** relationship. A police officer, a mailman, and a teacher are all humans, and so you can use inheritance. You will now write this down in code.

1. Create a **Human** class that has fields for **name**, **age**, **weight**, and **height**:

```
public class Human
{
    public string Name { get; }
    public int Age { get; }
    public float Weight { get; }
    public float Height { get; }

    public Human(string name, int age, float weight, float height)
    {
        Name = name;
        Age = age;
        Weight = weight;
        Height = height;
    }
}
```

2. A mailman is a human. Therefore, the **Mailman** class should have all that a **Human** class has, but on top of that, it should have the added functionality of being able to deliver mail. Write the code for this as follows:

```
public class Mailman : Human
{
    public Mailman(string name, int age, float weight, float height) :
base(name, age, weight, height)
    {
    }

    public void DeliverMail(Mail mail)
    {
       // Delivering Mail...
    }
}
```

   Now, look closely at the **Mailman** class. Writing **class Mailman : Human** means that **Mailman** inherits from **Human**. This means that **Mailman** takes all the properties and methods from **Human**. You can also see a new keyword, **base**. This keyword is used to tell which parent constructor is going to be used when creating **Mailman**; in this case, **Human**.

3. Next, create a class named **Mail** to represent the mail, containing a field for a message being delivered to an address:

```
public class Mail
{
   public string Message { get; }
   public string Address { get; }

   public Mail(string message, string address)
   {
       Message = message;
       Address = address;
   }
}
```

   Creating a **Mailman** object is no different than creating an object of a class that does not use inheritance.

4. Create **mailman** and **mail** variables and tell the **mailman** to deliver the mail as follows:

```
var mailman = new Mailman("Thomas", 29, 78.5f, 190.11f);
var mail = new Mail("Hello", "Somewhere far far way");
mailman.DeliverMail(mail);
```

> **NOTE**
>
> You can find the code used for this example at https://packt.link/w1bbf.

In the preceding snippet, you created **mailman** and **mail** variables. Then, you told the **mailman** to deliver the **mail**.

Generally, a base constructor must be provided when defining a child constructor. The only exception to this rule is when the parent has a parameter-less constructor. If a base constructor takes no arguments, then a child constructor using a base constructor would be redundant and therefore can be ignored. For example, consider the following snippet:

```
Public class A
{
}
Public class B : A
{
}
```

A has no custom constructors, so implementing **B** would not require a custom constructor either.

In C#, only a single class can be inherited; however, you can have a multi-level deep inheritance. For example, you could have a child class for **Mailman** named **RegionalMailman**, which would be responsible for a single region. In this way, you could go deeper and have another child class for **RegionalMailman**, called **RegionalBillingMailman**, then **EuropeanRegionalBillingMailman**, and so on.

When using inheritance, it is important to know that even if everything is inherited, not everything is visible. Just like before, **`public`** members only will be accessible from a parent class. However, in C#, there is a special modifier, named **`protected`**, that works like the **`private`** modifier. It allows child classes to access **`protected`** members (just like **`public`** members) but prevents them from being accessed from the outside of the class (just like **`private`**).

Decades ago, inheritance used to be the answer to many problems and the key to **code reuse**. However, over time, it became apparent that using inheritance comes at a price, which is **coupling**. When you apply inheritance, you couple a child class with a parent. Deep inheritance stacks class scope all the way from parent to child. The deeper the inheritance, the deeper the scope. Deep inheritance (two or more levels deep) should be avoided for the same reason you avoid global variables—it is hard to know what comes from where and hard to control the state changes. This, in turn, makes the code difficult to maintain.

Nobody wants to write duplicate code, but what is the alternative? The answer to that is **composition**. Just as a computer is composed of different parts, code should be composed of different parts as well. For example, imagine you are developing a 2D game and it has a **`Tile`** object. Some tiles contain a trap, and some tiles move. Using inheritance, you could write the code like this:

```
class Tile
{
}
class MovingTile : Tile
{
    public void Move() {}
}
class TrapTile : Tile
{
    public void Damage() {}
}
//class MovingTrapTile : ?
```

This approach works fine until you face more complex requirements. What if there are tiles that could both be a trap and move? Should you inherit from a moving tile and rewrite the **`TrapTile`** functionality there? Could you inherit both? As you have seen, you cannot inherit more than one class at a time, therefore, if you were to implement this using inheritance, you would be forced to both complicate the situation, and rewrite some code. Instead, you could think about what different tiles contain. **`TrapTile`** has a trap. **`MovingTile`** has a motor.

Both represent tiles, but the extra functionality they each have should come from different components, and not child classes. If you wanted to make this a composition-based approach, you would need to refactor quite a bit.

To solve this, keep the **Tile** class as-is:

```
class Tile
{
}
```

Now, add two components—Motor and Trap classes. Such components serve as logic providers. For now, they do nothing:

```
class Motor
{
    public void Move() { }
}
class Trap
{
    public void Damage() { }
}
```

> **NOTE**
>
> You can find the code used for this example at https://packt.link/espfn.

Next, you define a **MovingTile** class that has a single component, **_motor**. In composition, components rarely change dynamically. You should not expose class internals, so apply **private readonly** modifiers. The component itself can have a child class or change, and so should not be created from the constructor. Instead, it should be passed as an argument (see the highlighted code):

```
class MovingTile : Tile
{
    private readonly Motor _motor;

    public MovingTile(Motor motor)
    {
        _motor = motor;
    }
```

```
    public void Move()
    {
        _motor.Move();
    }
}
```

Note that the **Move** method now calls **_motor.Move()**. That is the essence of composition; the class that holds composition often does nothing by itself. It just delegates the calls of logic to its components. In fact, even though this is just an example class, a real class for a game would look quite similar to this.

You will do the same for **TrapTile**, except that instead of **Motor**, it will contain a **Trap** component:

```
class TrapTile : Tile
{
    private readonly Trap _trap;

    public TrapTile(Trap trap)
    {
        _trap = trap;
    }

    public void Damage()
    {
        _trap.Damage();
    }
}
```

Finally, it's time to create the **MovingTrapTile** class. It has two components that provide logic to the **Move** and **Damage** methods. Again, the two methods are passed as arguments to a constructor:

```
class MovingTrapTile : Tile
{
    private readonly Motor _motor;
    private readonly Trap _trap;

    public MovingTrapTile(Motor motor, Trap trap)
    {
        _motor = motor;
        _trap = trap;
    }
```

```
    public void Move()
    {
        _motor.Move();
    }
    public void Damage()
    {
        _trap.Damage();
    }
}
```

> **NOTE**
>
> You can find the code used for this example at https://packt.link/SX4qG.

It might seem that this class repeats some code from the other class, but the duplication is negligible, and the benefits are well worth it. After all, the biggest chunk of logic comes from the components themselves, and a repeated field or a call is not significant.

You may have noticed that you inherited **`Tile`**, despite not extracting it as a component for other classes. This is because **`Tile`** is the essence of all the classes that inherit it. No matter what type a tile is, it is still a tile. Inheritance is the second pillar of OOP. It is powerful and useful. However, it can be hard to get inheritance right, because in order to be maintainable, it truly needs to be very clear and logical. When choosing whether you should use inheritance, consider these factors:

- Not deep (ideally single level).
- Logical (is-a relation, as you saw in your tiles example).
- Stable and extremely unlikely for the relationship between classes to change in the future; not going to be modified often.
- Purely additive (child class should not use parent class members, except for a constructor).

If any one of these rules is broken, it is recommended to use composition instead of inheritance.

POLYMORPHISM

The third pillar of OOP is polymorphism. To grasp this pillar, it is useful to look at the meaning of the word. **Poly** means many, and **morph** means form. So, polymorphism is used to describe something that has many forms. Consider the example of a mailman, **Thomas**. **Thomas** is both a human and a mailman. **Mailman** is the specialized form and **Human** is the generalized form for Thomas. However, you can interact with **Thomas** through either of the two forms.

If you do not know the jobs for every human, you can use an **abstract** class.

An **abstract** class is a synonym for an **incomplete class**. This means that it cannot be initialized. It also means that some of its methods may not have an implementation if you mark them with the **abstract** keyword. You can implement this for the **Human** class as follows:

```
public abstract class Human
{
    public string Name { get; }

    protected Human(string name)
    {
        Name = name;
    }

    public abstract void Work();
}
```

You have created an abstract (incomplete) **Human** class here. The only difference from earlier is that you have applied the **abstract** keyword to the class and added a new **abstract** method, **public abstract void Work()**. You have also changed the constructor to protected so that it is accessible only from a child class. This is because it no longer makes sense to have it **public** if you cannot create an **abstract** class; you cannot call a **public** constructor. Logically, this means that the **Human** class, by itself, has no meaning, and it only gets meaning after you have implemented the **Work** method elsewhere (that is, in a child class).

Now, you will update the **Mailman** class. It does not change much; it just gets an additional method, that is, **Work()**. To provide an implementation for abstract methods, you must use the **override** keyword. In general, this keyword is used to change the implementation of an existing method inside a child class. You will explore this in detail later:

```
public override void Work()
{
    Console.WriteLine("A mailman is delivering mails.");
}
```

If you were to create a new object for this class and call the **Work** method, it would print **"A mailman is delivering mails."** to the console. To get a full picture of polymorphism, you will now create one more class, **Teacher**:

```
public class Teacher : Human
{
    public Teacher(string name, int age, float weight, float height) :
base(name, age, weight, height)
    {
    }

    public override void Work()
    {
        Console.WriteLine("A teacher is teaching.");
    }
}
```

This class is almost identical to **Mailman**; however, a different implementation for the **Work** method is provided. Thus, you have two classes that do the same thing in two different ways. The act of calling a method of the same name, but getting different behavior, is called **polymorphism**.

You already know about method overloading (not to be confused with overriding), which is when you have methods with the same names but different inputs. That is called **static polymorphism** and it happens during compile time. The following is an example of this:

```
public class Person
{
    public void Say()
    {
        Console.WriteLine("Hello");
```

```
    }

    public void Say(string words)
    {
        Console.WriteLine(words);
    }
}
```

The **Person** class has two methods with the same name, Say. One takes no arguments and the other takes a string as an argument. Depending on the arguments passed, different implementations of the method will be called. If nothing is passed, **"Hello"** will be printed. Otherwise, the words you pass will be printed.

In the context of OOP, polymorphism is referred to as **dynamic polymorphism**, which happens during runtime. For the rest of this chapter, polymorphism should be interpreted as dynamic polymorphism.

WHAT IS THE BENEFIT OF POLYMORPHISM?

A teacher is a human, and the way a teacher works is by teaching. This is not the same as a mailman, but a teacher also has a name, age, weight, and height, like a mailman. Polymorphism allows you to interact with both in the same way, regardless of their specialized forms. The best way to illustrate this is to store both in an array of **humans** values and make them work:

```
Mailman mailman = new Mailman("Thomas", 29, 78.5f, 190.11f);
Teacher teacher = new Teacher("Gareth", 35, 100.5f, 186.49f);
// Specialized types can be stored as their generalized forms.
Human[] humans = {mailman, teacher};
// Interacting with different human types
// as if they were the same type- polymorphism.
foreach (var human in humans)
{
    human.Work();
}
```

This code results in the following being printed in the console:

```
A mailman is delivering mails.
A teacher is teaching.
```

> **NOTE**
>
> You can find the code used for this example at https://packt.link/ovqru.

This code was **polymorphism in action**. You treated both **Mailman** and **Teacher** as **Human** and implemented the **Work** method for both. The result was different behaviors in each case. The important point to note here is that you did not have to care about the exact implementations of **Human** to implement **Work**.

How would you implement this without polymorphism? You would need to write **if** statements based on the exact type of an object to find the behavior it should use:

```
foreach (var human in humans)
{
    Type humanType = human.GetType();
    if (humanType == typeof(Mailman))
    {
        Console.WriteLine("Mailman is working...");
    }
    else
    {
        Console.WriteLine("Teaching");
    }
}
```

As you see, this is a lot more complicated and harder to grasp. Keep this example in mind when you get into a situation with many **if** statements. Polymorphism can remove the burden of all that branching code by simply moving the code for each branch into a child class and simplifying the interactions.

What if you wanted to print some information about a person? Consider the following code:

```
Human[] humans = {mailman, teacher};
foreach (var human in humans)
{
    Console.WriteLine(human);
}
```

Running this code would result in the object type names being printed to the console:

```
Chapter02.Examples.Professions.Mailman
Chapter02.Examples.Professions.Teacher
```

In C#, everything derives from the **System.Object** class, so every single type in C# has a method called **ToString()**. Each type has its own implementation of this method, which is another example of polymorphism, widely used in C#.

> **NOTE**
>
> **ToString()** is different from **Work()** in that it provides a default implementation. You can achieve that using the **virtual** keyword, which will be covered in detail later in the chapter. From the point of view of a child class, working with the **virtual** or **abstract** keyword is the same. If you want to change or provide behavior, you will override the method.

In the following snippet, a **Human** object is given a custom implementation of the **ToString()** method:

```
public override string ToString()
{
    return $"{nameof(Name)}: {Name}," +
           $"{nameof(Age)}: {Age}," +
           $"{nameof(Weight)}: {Weight}," +
           $"{nameof(Height)}: {Height}";
}
```

Trying to print information about the humans in the same foreach loop would result in the following output:

```
Name: Thomas,Age: 29,Weight: 78.5,Height: 190.11
Name: Gareth,Age: 35,Weight: 100.5,Height: 186.49
```

> **NOTE**
>
> You can find the code used for this example at https://packt.link/EGDkC.

Polymorphism is one of the best ways to use different underlying behaviors when dealing with missing type information.

ABSTRACTION

The last pillar of OOP is **abstraction**. Some say that there are only three pillars of OOP because abstraction does not really introduce much that is new. Abstraction encourages you to hide implementation details and simplify interactions between objects. Whenever you need the functionality of only a generalized form, you should not depend on its implementation.

Abstraction could be illustrated with an example of how people interact with their computers. What occurs in the internal circuitry when you turn on the computer? Most people would have no clue, and that is fine. You do not need to know about the internal workings if you only need to use some functionality. All you have to know is what you can do, and not how it works. You know you can turn a computer on and off by pressing a button, and all the complex details are hidden away. Abstraction adds little new to the other three pillars because it reflects each of them. **Abstraction is similar to encapsulation**, as it hides unnecessary details to simplify interaction. It is also similar to polymorphism because it can interact with objects without knowing their exact types. Finally, inheritance is just one of the ways to create abstractions.

You do not need to provide unnecessary details coming through implementation types when creating functions. The following example illustrates this problem. You need to make a progress bar. It should keep track of the current progress and should increment the progress up to a certain point. You could create a basic class with setters and getters, as follows:

```
public class ProgressBar
{
    public float Current { get; set; }
    public float Max { get; }

    public ProgressBar(float current, float max)
    {
        Max = max;
        Current = current;
    }
}
```

The following code demonstrates how to initialize a progress bar that starts at **0** progress and goes up to **100**. The rest of the code illustrates what happens when you want to set the new progress to 120. Progress cannot be more than **Max**, hence, if it is more than **bar.Max**, it should just remain at **bar.Max**. Otherwise, you can update the new progress with the value you set. Finally, you need to check whether the progress is complete (at **Max** value). To do so, you will compare the delta with the allowed margin of error tolerance (**0.0001**). A progress bar is complete if it is close to tolerance. So, updating progress could look like the following:

```
var bar = new ProgressBar(0, 100);
var newProgress = 120;
if (newProgress > bar.Max)
{
    bar.Current = bar.Max;
}
else
{
    bar.Current = newProgress;
}

const double tolerance = 0.0001;
var isComplete = Math.Abs(bar.Max - bar.Current) < tolerance;
```

This code does what is asked for, but it needs a lot of detail for a function. Imagine if you had to use this in other code; you would need to perform the same checks once again. In other words, it was easy to implement but complex to consume. You have so little within the class itself. A strong indicator of that is that you keep on calling the object, instead of doing something inside the class itself. Publicly, it's possible to break the object state by forgetting to check the **Max** value of progress and setting it to some high or negative value. The code that you wrote has low cohesion because to change **ProgressBar**, you would do it not within the class but somewhere outside of it. You need to create a better abstraction.

Consider the following snippet:

```
public class ProgressBar
{
    private const float Tolerance = 0.001f;

    private float _current;
    public float Current
    {
        get => _current;
        set
        {
            if (value >= Max)
            {
                _current = Max;
            }
            else if (value < 0)
            {
                _current = 0;
            }
            else
            {
                _current = value;
            }
        }
    }
```

With this code, you have hidden the nitty-gritty details. When it comes to updating progress and defining what the tolerance is, that is up to the **ProgressBar** class to decide. In the refactored code, you have a property, **Current**, with a backing field, **_current**, to store the progress. The property setter checks whether progress is more than the maximum and, if it is, it will not allow the value of **_current** to be set to a higher value, **=**. It also cannot be negative, as in those cases, the value will be adjusted to **0**. Lastly, if it is not negative and not more than the maximum, then you can set **_current** to whatever value you pass.

Clearly, this code makes it much simpler to interact with the **ProgressBar** class:

```
var bar = new ProgressBar(0, 100);
bar.Current = 120;
bool isComplete = bar.IsComplete;
```

You cannot break anything; you do not have any extra choices and all you can do is defined through minimalistic methods. When you are asked to implement a feature, it is not recommended to do more than what is asked. Try to be minimalistic and simplistic because that is key to an effective code.

Remember that well-abstracted code is full of empathy toward the reader. Just because today, it is easy to implement a class or a function, you should not forget about tomorrow. The requirements change, the implementation changes, but the structure should remain stable, otherwise, your code can break easily.

> **NOTE**
>
> You can find the code used for this example can be found at https://packt.link/U126i. The code given in GitHub is split into two contrasting examples—**ProgressBarGood** and **ProgressBarBad**. Both codes are simple **ProgressBar** but were named distinctly to avoid ambiguity.

INTERFACES

Earlier, it was mentioned that inheritance is not the proper way of designing code. However, you want to have an efficient abstraction as well as support for polymorphism, and little to no coupling. What if you wanted to have robot or ant workers? They do not have a name. Information such as height and weight are irrelevant. And inheriting from the **Human** class would make little sense. Using an **interface** solves this conundrum.

In C#, by convention, interfaces are named starting with the letter **I**, followed by their actual name. An interface is a contract that states what a class can do. It does not have any implementation. It only defines behavior for every class that implements it. You will now refactor the human example using an interface.

What can an object of the **Human** class do? It can work. Who or what can do work? A worker. Now, consider the following snippet:

```
public interface IWorker
{
    void Work();
}
```

> #### NOTE
>
> Interface **methods** will never have an access modifier. This is due to the nature of an interface. All the methods that an interface has are methods you would like to access publicly so that you can implement them. The access modifier that the **Work** method will have is the same as the interface access modifier, in this case, **public**.

An ant is not a human, but it can work. With an interface, abstracting an ant as a worker is straightforward:

```
public class Ant : IWorker
{
    public void Work()
    {
        Console.WriteLine("Ant is working hard.");
    }
}
```

Similarly, a robot is not a human, but it can work as well:

```
public class Robot : IWorker
{
    public void Work()
    {
        Console.WriteLine("Beep boop- I am working.");
    }
}
```

If you refer to the **Human** class, you can change its definition to **public abstract class Human : IWorker**. This can be read as: **Human** class implements the **IWorker** interface.

In the next snippet, **Mailman** inherits the **Human** class, which implements the **IWorker** interface:

```
public class Mailman : Human
{
    public Mailman(string name, int age, float weight, float height) : 
base(name, age, weight, height)
    {
    }

    public void DeliverMail(Mail mail)
    {
        // Delivering Mail...
    }

    public override void Work()
    {
        Console.WriteLine("Mailman is working...");
    }
}
```

If a child class inherits a parent class, which implements some interfaces, the child class will also be able to implement the same interfaces by default. However, **Human** was an abstract class and you had to provide implementation to the **abstract void Work** method.

If anyone asked what a human, an ant, and a robot have in common, you could say that they can all work. You can simulate this situation as follows:

```
IWorker human = new Mailman("Thomas", 29, 78.5f, 190.11f);
IWorker ant = new Ant();
IWorker robot = new Robot();

IWorker[] workers = {human, ant, robot};
foreach (var worker in workers)
{
    worker.Work();
}
```

This prints the following to the console:

```
Mailman is working...
Ant is working hard.
Beep boop- I am working.
```

> **NOTE**
>
> You can find the code used for the example at https://packt.link/FE2ag.

C# does not support multiple inheritance. However, it is possible to implement multiple interfaces. Implementing multiple interfaces does not count as multiple inheritance. For example, to implement a **Drone** class, you could add an **IFlyer** interface:

```
public interface IFlyer
{
    void Fly();
}
```

Drone is a flying object that can do some work; therefore it can be expressed as follows:

```
public class Drone : IFlyer, IWorker
{
    public void Fly()
    {
```

```
        Console.WriteLine("Flying");
    }

    public void Work()
    {
        Console.WriteLine("Working");
    }
}
```

Listing multiple interfaces with separating commas means the class implements each of them. You can combine any number of interfaces, but try not to overdo this. Sometimes, a combination of two interfaces makes up a logical abstraction. If every drone can fly and does some work, then you can write that in code, as follows:

```
public interface IDrone : IWorker, IFlyer
{
}
```

And the **`Drone`** class becomes simplified to **`public class Drone : IDrone`**.

It is also possible to mix interfaces with a base class (but no more than one base class). If you want to represent an ant that flies, you can write the following code:

```
public class FlyingAnt : Ant, IFlyer
{
    public void Fly()
    {
        Console.WriteLine("Flying");
    }
}
```

An interface is undoubtedly the best abstraction because depending on it does not force you to depend on any implementation details. All that is required is the logical concepts that have been defined. Implementation is prone to change, but the logic behind relations between classes is not.

If an interface defines what a class can do, is it also possible to define a contract for common data? Absolutely. An interface holds behavior, hence it can hold properties as well because they define setter and getter behavior. For example, you should be able to track the drone, and for this, it should be identifiable, that is, it needs to have an ID. This can be coded as follows:

```
public interface IIdentifiable
{
```

```
    long Id { get; }
}
public interface IDrone : IWorker, IFlyer
{
}
```

In modern software development, there are several complex low-level details that programmers use on a daily basis. However, they often do so without knowing. If you want to create a maintainable code base with lots of logic and easy-to-grasp code, you should follow these principles of abstraction:

- Keep it simple and small.
- Do not depend on details.
- Hide complexity.
- Expose only what is necessary.

With this exercise, you will grasp how OOP functions.

EXERCISE 2.03: COVERING FLOOR IN THE BACKYARD

A builder is building a mosaic with which he needs to cover an area of x square meters. You have some leftover tiles that are either rectangular or circular. In this exercise, you need to find out whether, if you shatter the tiles to perfectly fill the area they take up, can the tiles fill the mosaic completely.

You will write a program that prints **`true`**, if the mosaic can be covered with tiles, or **`false`**, if it cannot. Perform the following steps to do so:

1. Create an interface named **`IShape`**, with an **`Area`** property:

   ```
   public interface IShape
   {
       double Area { get; }
   }
   ```

   This is a **get-only** property. Note that a property is a method, so it is okay to have it in an interface.

2. Create a class called **Rectangle**, with width and height and a method for calculating area, called **Area**. Implement an **IShape** interface for this, as shown in the following code:

Rectangle.cs

```
public class Rectangle : IShape
{
    private readonly double _width;
    private readonly double _height;

    public double Area
    {
        get
        {
            return _width * _height;
        }
    }

    public Rectangle(double width, double height)
    {
```

You can find the complete code here: https://packt.link/zSquP.

The only thing required is to calculate the area. Hence, only the **Area** property is **public**. Your interface needs to implement a getter **Area** property, achieved by multiplying **width** and **height**.

3. Create a **Circle** class with a **radius** and **Area** calculation, which also implements the **IShape** interface:

```
public class Circle : IShape
{
    Private readonly double _radius;

    public Circle(double radius)
    {
        _radius = radius;
    }

    public double Area
    {
        get { return Math.PI * _radius * _radius; }
    }
}
```

4. Create a skeleton **Solution** class with a method named **IsEnough**, as follows:

```
public static class Solution
{
        public static bool IsEnough(double mosaicArea, IShape[]
tiles)
        {
    }
}
```

Both the class and the method are just placeholders for the implementation to come. The class is **static** because it will be used as a demo and it does not need to have a state. The **IsEnough** method takes the needed **mosaicArea**, an array of tiles objects, and returns whether the total area occupied by the tiles is enough to cover the mosaic.

5. Inside the **IsEnough** method, use a **for** loop to calculate the **totalArea**. Then, return whether the total area covers the mosaic area:

```
            double totalArea = 0;
            foreach (var tile in tiles)
            {
                totalArea += tile.Area;
            }
            const double tolerance = 0.0001;
            return totalArea - mosaicArea >= -tolerance;
        }
```

6. Inside the **Solution** class, create a demo. Add several sets of different shapes, as follows:

```
public static void Main()
{
    var isEnough1 = IsEnough(0, new IShape[0]);
    var isEnough2 = IsEnough(1, new[] { new Rectangle(1, 1) });
    var isEnough3 = IsEnough(100, new IShape[] { new Circle(5) });
    var isEnough4 = IsEnough(5, new IShape[]
    {
```

```
        new Rectangle(1, 1), new Circle(1), new Rectangle(1.4,1)
    });

    Console.WriteLine($"IsEnough1 = {isEnough1}, " +
                      $"IsEnough2 = {isEnough2}, " +
                      $"IsEnough3 = {isEnough3}, " +
                      $"IsEnough4 = {isEnough4}.");
}
```

Here, you use four examples. When the area to cover is **0**, then no matter what shapes you pass, it will be enough. When the area to cover is **1**, a rectangle of area **1x1** will be just enough. When it's **100**, a circle of radius **5** is not enough. Finally, for the fourth example, the area occupied by three shapes is added up, that is, a rectangle of area **1x1**, a circle of radius **1**, and the second rectangle of area **1.4x1**. The total area is **5**, which is less than the combined area of these three shapes.

7. Run the demo. You should see the following output on your screen:

```
IsEnough1 = True, IsEnough2 = True, IsEnough3 = False, IsEnough4 =
False.
```

> **NOTE**
>
> You can find the code used for this exercise at https://packt.link/EODE6.

This exercise is very similar to *Exercise 2.02*. However, even though the assignment is more complex, there is less code than in the previous assignment. By using the OOP pillars, you were able to create a simple solution for a complex problem. You were able to create functions that depend on abstraction, rather than making overloads for different types. Thus, OOP is a powerful tool, and this only scratches the surface.

Everyone can write code that works but writing code that lives for decades and is easy to grasp is hard. So, it is imperative to know about the set of best practices in OOP.

SOLID PRINCIPLES IN OOP

SOLID principles are a set of best practices for OOP. SOLID is an acronym for five principles, namely, single responsibility, open-closed, Liskov substitution, interface segregation, and dependency inversion. You will not explore each of these in detail.

SINGLE RESPONSIBILITY PRINCIPLE

Functions, classes, projects, and entire systems change over time. Every change is potentially a breaking one, so you should limit the risk of too many things changing at a time. In other words, a part of a code block should have only a single reason to change.

For a function, this means that it should do just one thing and have no side effects. In practice, this means that a function should either change, or get something, but never do both. This also means that functions responsible for high-level things should not be mixed with functions that perform low-level things. Low-level is all about implementing interactions with hardware, and working with primitives. High-level is focused on compositions of software building blocks or services. When talking about high- and low-level functions, it is usually referred to as a chain of dependencies. If function A calls function B, A is considered higher-level than B. A function should not implement multiple things; it should instead call other functions that implement doing one thing. The general guideline for this is that if you think you can split your code into different functions, then in most cases, you should do that.

For classes, it means that you should keep them small and isolated from one another. An example of an efficient class is the **`File`** class, which can read and write. If it implemented both reading and writing, it would change for two reasons (reading and writing):

```
public class File
{
    public string Read(string filePath)
    {
        // implementation how to read file contents
        // complex logic
        return "";
    }

    public void Write(string filePath, string content)
    {
        // implementation how to append content to an existing file
        // complex logic
    }
}
```

Therefore, to conform to this principle, you can split the reading code into a class called **Reader** and writing code into a class called **Writer**, as follows:

```
public class Reader
{
    public string Read(string filePath)
    {
        // implementation how to read file contents
        // complex logic
        return "";
    }
}
public class Writer
{
    public void Write(string filePath, string content)
    {
        // implementation how to append content to an existing file
        // complex logic
    }
}
```

Now, instead of implementing reading and writing by itself, the **File** class will simply be composed of a reader and writer:

```
public class File
{
    private readonly Reader _reader;
    private readonly Writer _writer;

    public File()
    {
        _reader = new Reader();
        _writer = new Writer();
    }


    public string Read(string filePath) => _reader.Read(filePath);
    public void Write(string filePath, string content) => _writer.
Write(filePath, content);
}
```

NOTE

You can find the code used for this example at https://packt.link/PBppV.

It might be confusing because what the class does essentially remains the same. However, now, it just consumes a component and is not responsible for implementing it. A high-level class (**File**) simply adds context to how lower-level classes (**Reader**, **Writer**) will be consumed.

For a module (library), it means that you should strive to not introduce dependencies, which would be more than what the consumer would want. For example, if you are using a library for logging, it should not come with some third-party logging provider-specific implementation.

For a subsystem, it means that different systems should be as isolated as possible. If two (lower level) systems need to communicate, they could call one another directly. A consideration (not mandatory) would be to have a third system (higher-level) for coordination. Systems should also be separated through a boundary (such as a contract specifying communication parameters), which hides all the details. If a subsystem is a big library collection, it should have an interface to expose what it can do. If a subsystem is a web service, it should be a collection of endpoints. In any case, a contract of a subsystem should provide only the methods that the client may want.

Sometimes, the principle is overdone and classes are split so much that making a change requires changing multiple places. It does keep true to the principle, as a class will have a single reason to change, but in such a case, multiple classes will change for the same reason. For example, suppose you have two classes: **Merchandise** and **TaxCalculator**. The **Merchandise** class has fields for **Name**, **Price**, and **Vat**:

```
public class Merchandise
{
    public string Name { get; set; }
    public decimal Price { get; set; }
    // VAT on top in %
    public decimal Vat { get; set; }
}
```

Next, you will create the **TaxCalculator** class. **vat** is measured as a percentage, so the actual price to pay will be **vat** added to the original price:

```
public static class TaxCalculator
{
    public static decimal CalculateNextPrice(decimal price, decimal vat)
    {
        return price * (1 + vat / 100);
    }
}
```

What would change if the functionality of calculating the price moved to the **Merchandise** class? You would still be able to perform the required operation. There are two key points here:

- The operation by itself is simple.
- Also, everything that the tax calculator needs come from the **Merchandise** class.

If a class can implement the logic by itself, as long as it is self-contained (does not involve extra components), it usually should. Therefore, a proper version of the code would be as follows:

```
public class Merchandise
{
    public string Name { get; set; }
    public decimal Price { get; set; }
    // VAT on top in %
    public decimal Vat { get; set; }
    public decimal NetPrice => Price * (1 + Vat / 100);
}
```

This code moves the **NetPrice** calculation to the **Merchandise** class and the **TaxCalculator** class has been removed.

> **NOTE**
>
> **Singe Responsibility Principle** (**SRP**) can be summarized in a couple of words: **split it**. You can find the code used for this example at https://packt.link/lWxNO.

OPEN-CLOSED PRINCIPLE

As mentioned previously, every change in code is potentially a breaking one. As a way around this, instead of changing existing code, it is often preferable to write new code. Every software entity should have an extension point, through which the changes should be introduced. However, after this change is done, a software entity should not be interfered with. The **Open-Closed Principle** (**OCP**) is hard to implement and takes a lot of practice, but the benefits (a minimum number of breaking changes) are well worth it.

If a multiple-step algorithm does not change, but its individual steps can change, you should split it into several functions. A change for an individual step will no longer affect the entire algorithm, but rather just that step. Such minimization of reasons for a single class or a function to change is what OCP is all about.

> **NOTE**
>
> You can find more information on OCP at https://social.technet.microsoft.com/wiki/contents/articles/18062.open-closed-principle-ocp.aspx.

Another example where you may want to implement this principle is a function working with combinations of specific values in code. This is called **hardcoding** and is generally deemed an inefficient practice. To make it work with new values, you might be tempted to create a new function, but by simply removing a hardcoded part and exposing it through function parameters, you can make it extensible. However, when you have variables that are known to be fixed and not changing, it is fine to hardcode them, but they should be flagged as constant.

Previously, you created a file class with two dependencies—**`Reader`** and **`Writer`**. Those dependencies are hardcoded, and leave you with no extension points. Fixing that will involve two things. First, add the virtual modifier for both the **`Reader`** and **`Writer`** class methods:

```
public virtual string Read(string filePath)
public virtual void Write(string filePath, string content)
```

Then, change the constructor of the **File** class so that it accepts instances of **Reader** and **Writer**, instead of hardcoding the dependencies:

```
public File(Reader reader, Writer writer)
{
    _reader = reader;
    _writer = writer;
}
```

This code enables you to override the existing reader and writer behavior and replace it with whatever behavior you want, that is, the **File** class extension point.

OCP can be summarized in a few words as **don't change it, extend it**.

LISKOV SUBSTITUTION

The **Liskov Substitution Principle** (**LSP**) is one of the most straightforward principles out there. It simply means that a child class should support all the public behavior of a parent class. If you have two classes, **Car** and **CarWreck**, where one inherits the other, then you have violated the principle:

```
class Car
{
    public object Body { get; set; }

    public virtual void Move()
    {
        // Moving
    }
}

class CarWreck : Car
{
    public override void Move()
    {
        throw new NotSupportedException("A broken car cannot start.");
    }
}
```

> **NOTE**
>
> You can find the code used for this example at https://packt.link/6nD76.

Both **Car** and **CarWreck** have a **Body** object. **Car** can move, but what about **CarWreck**? It can only stay in one place. The **Move** method is virtual because **CarWreck** intends to override it to mark it as not supported. If a child can no longer support what a parent can do, then it should no longer inherit that parent. In this case, a car wreck is not a car, it's simply a wreck.

How do you conform to this principle? All you have to do is to remove the inheritance relationship and replicate the necessary behavior and structure. In this case, **CarWreck** still has a **Body** object, but the **Move** method is unnecessary:

```
class CarWreck
{
    public object Body { get; set; }
}
```

Code changes happen quite often, and you can sometimes inadvertently use the wrong method to achieve your goals. Sometimes, you couple code in such a way that what you thought was flexible code turns out to be a complex mess. Do not use inheritance as a way of doing code reuse. Keep things small and compose them (again) instead of trying to override the existing behavior. Before things can be reusable, they should be usable. Design for simplicity and you will get flexibility for free.

LSP can be summarized in a few words: **don't fake it**.

> **NOTE**
>
> You can find more information on LSP at https://www.microsoftpressstore.com/articles/article.aspx?p=2255313.

INTERFACE SEGREGATION

The interface segregation principle is a special case of the OCP but is only applicable to contracts that will be exposed publicly. Remember, every change you make is potentially a breaking change, and this especially matters in making changes to a contract. Breaking changes are inefficient because they will often require effort to adapt to the change from multiple people.

For example, say you have an interface, **IMovableDamageable**:

```
interface IMovableDamageable
{
    void Move(Location location);
    float Hp{get;set;}
}
```

A single interface should represent a single concept. However, in this case, it does two things: move and manage **Hp** (hit points). By itself, an interface with two methods is not problematic. However, in scenarios of the implementation needing only a part of an interface, you are forced to create a workaround.

For example, score text is indestructible, but you would like it to be animated and to move it across a scene:

```
class ScoreText : IMovableDamageable
{
    public float Hp
    {
        get => throw new NotSupportedException();
        set => throw new NotSupportedException();
    }

    public void Move(Location location)
    {
        Console.WriteLine($"Moving to {location}");
    }
}

public class Location
{
}
```

> **NOTE**
>
> The point here isn't to print the location; just to give an example of where it is used. It's up to location's implementation whether it will be printed or not as such.

Taking another example, you might have a house that does not move but can be destroyed:

```
class House : IMovableDamageable
{
    public float Hp { get; set; }

    public void Move(Location location)
    {
        throw new NotSupportedException();
    }
}
```

In both scenarios, you worked around the issue by throwing **`NotSupportedException`**. However, another programmer should not be given an option to call code that never works in the first place. In order to fix the problem of representing too many concepts, you should split the **`IMoveableDamageable`** interface into **`IMoveable`** and **`IDamageable`**:

```
interface IMoveable
{
    void Move(Location location);
}
interface IDamageable
{
    float Hp{get;set;}
}
```

And the implementations can now get rid of the unnecessary parts:

```
class House : IDamageable
{
    public float Hp { get; set; }
}

class ScoreText : IMovable
{
    public void Move(Location location)
    {
        Console.WriteLine($"Moving to {location}");
    }
}
```

The **`Console.WriteLine`**, in the preceding code, would display the namespace name with the class name.

> **NOTE**
>
> **Interface segregation** can be summarized as **don't enforce it**. You can find the code used for this example at https://packt.link/32mwP.

DEPENDENCY INVERSION

Large software systems can consist of millions of classes. Each class is a small dependency, and if unmanaged, the complexity might stack into something impossible to maintain. If one low-level component breaks, it causes a ripple effect, breaking the whole chain of dependencies. The **dependency inversion principle** states that you should avoid hard dependence on underlying classes.

Dependency injection is the industry-standard way of implementing dependency inversion. Do not mix the two; one is a principle and the other refers to the implementation of this principle.

Note that you can also implement dependency inversion without dependency injection. For example, when declaring a field, instead of writing something like **`private readonly List<int> _numbers = new List<int>();`**, it is preferable to write **`private readonly IList<int> = _numbers`**, which shifts dependency to abstraction (**`IList`**) and not implementation (**`List`**).

What is dependency injection? It is the act of passing an implementation and setting it to an abstraction slot. There are three ways to implement this:

- **Constructor injection** is achieved by exposing an abstraction through the constructor argument and passing an implementation when creating an object and then assigning it to a field. Use it when you want to consistently use the same dependency in the same object (but not necessarily the same class).
- **Method injection** is done by exposing an abstraction through a method argument, and then passing an implementation when calling that method. Use it when, for a single method, a dependency might vary, and you do not plan to store the dependency throughout that object's lifetime.

- **Property injection** is implemented by exposing an abstraction through a public property, and then assigning (or not) that property to some exact implementation. Property injection is a rare way of injecting dependencies because it suggests that dependency might even be null or temporary and there are many ways in which it could break.

Given two types, **`interface IBartender { }`** and **`class Bar : Bartender { }`**, you can illustrate the three ways of dependency injection for a class called **`Bar`**.

First, prepare the **`Bar`** class for constructor injection:

```
class Bar
{
    private readonly IBartender _bartender;

    public Bar(IBartender bartender)
    {
        _bartender = bartender;
    }
}
```

The constructor injection is done as follows:

```
var bar = new Bar(new Bartender());
```

This kind of dependency injection is a dominating kind of inheritance, as it enforces stability through immutability. For example, some bars have just one bartender.

Method injection would look like this:

```
class Bar
{
    public void ServeDrinks(IBartender bartender)
    {
        // serve drinks using bartender
    }
}
```

The injection itself is as follows:

```
var bar = new Bar();
bar.ServeDrinks(new Bartender());
```

Often, this kind of dependency injection is called **interface injection** because the method often goes under an interface. The interface itself is a great idea, but that does not change the idea behind this kind of dependency injection. Use method injection when you immediately consume a dependency that you set, or when you have a complex way of setting new dependencies dynamically. For example, it makes sense to use different bartenders for serving drinks.

Finally, property injection can be done like this:

```
class Bar
{
    public IBartender Bartender { get; set; }
}
```

Bartender is now injected like this:

```
var bar = new Bar();
bar.Bartender = new Bartender();
```

For example, a bar might have bartenders changing shifts, but one bartender at a time.

> **NOTE**
>
> You can find the code used for this example at https://packt.link/JcmAT.

Property injection in other languages might have a different name: **setter injection**. In practice, components do not change that often, so this kind of dependency injection is the rarest.

For the `File` class, this should mean that instead of exposing classes (implementation), you should expose abstractions (interfaces). This means that your `Reader` and `Writer` classes should implement some contract:

```
public class Reader : IReader
public class Writer: IWriter
```

Your file class should expose reader and writer abstractions, instead of implementations, as follows:

```
private readonly IReader _reader;
private readonly IWriter _writer;
```

```
public File(IReader reader, IWriter writer)
{
    _reader = reader;
    _writer = writer;
}
```

This allows for a choice of the kind of **IReader** and **IWriter** you would like to inject. A different reader may read a different file format, or a different writer may output in a different way. You have a choice.

Dependency injection is a powerful tool that is used often, especially in an enterprise setting. It allows you to simplify complex systems by putting an interface in between and having 1:1 dependencies of implementation-abstraction-implementation.

Writing effective code that does not break can be paradoxical. It is the same as buying a tool from a shop; you can't know for sure how long it will last, or how well it will work. Code, just like those tools, might work now but break in the near future, and you will only know that it does not work if and when it breaks.

Observing and waiting, seeing how the code evolves, is the only way to know for sure if you have written an effective code. In small, personal projects, you might not even notice any changes, unless you expose the project to the public or involve other people. To most people, SOLID principles often sound like old, outdated principles, like over-engineering. But they are actually a set of best practices that have withstood the test of time, formulated by top professionals seasoned in enterprise settings. It is impossible to write perfect, SOLID code right away. In fact, in some cases, it is not even necessary (if a project is small and meant to be short-lived, for example). As someone who wants to produce quality software and work as a professional, you should practice it as early on as possible.

HOW C# HELPS WITH OBJECT-ORIENTED DESIGN

So far, the principles you have learned are not language-specific. It is time to learn how to use C# for OOP. C# is a great language because it is full of some very useful features. It is not only one of the most productive languages to work with, but it also allows you to write beautiful, hard-to-break code. With a rich selection of keywords and languages features, you can model your classes completely the way you want, making the intentions crystal clear. This section will delve deep into C# features that help with object-oriented design.

Static

Up till now in this book, you have interacted mostly with **static** code. This refers to code that does not need new classes and objects, and that can be called right away. In C#, the static modifier can be applied in five different scenarios—methods, fields, classes, constructors, and the **using** statement.

Static methods and fields are the simplest application of the **static** keyword:

```
public class DogsGenerator
{
    public static int Counter { get; private set; }

    static DogsGenerator()
    {
        // Counter will be 0 anyways if not explicitly provided,
        // this just illustrates the use of a static constructor.
        Counter = 0;
    }

    public static Dog GenerateDog()
    {
        Counter++;
        return new Dog("Dog" + Counter);
    }
}
```

> **NOTE**
>
> You can find the code used for this example at https://packt.link/748m3.

Here, you created a class called **DogsGenerator**. A **static class** cannot be initialized manually (using the **new** keyword). Internally, it is initialized, but only once. Calling the **GenerateDog** method returns a new **Dog** object with a counter next to its name, such as **Dog1**, **Dog2**, and **Dog3**. Writing a counter like this allows you to increment it from everywhere as it is **public static** and has a setter. This can be done by directly accessing the member from a class: **DogsGenerator.Counter++** will increment the counter by **1**.

Once again, note that this does not require a call through an object because a **static class** instance is the same for the entire application. However, **DogsGenerator** is not the best example of a **static class**. That's because you have just created a **global state**. Many people would say that **static** is inefficient and should be avoided because it might create unpredictable results due to being modified and accessed uncontrollably.

A public mutable state means that changes can happen from anywhere in the application. Other than being hard to grasp, such code is also prone to breaking in the context of applications with multiple threads (that is, it is not thread-safe).

> **NOTE**
>
> You will learn about threading in detail in *Chapter 5, Concurrency: Multithreading Parallel and Async Code*.

You can reduce the impact of a global state by making it publicly immutable. The benefit of doing so is that now you are in control. Instead of allowing a counter increment to happen from any place inside a program, you will change it within **DogsGenerator** only. For the **counter** property, achieving it is as simple as making the setter property **private**.

There is one valuable use case for the **static** keyword though, which is with **helper functions**. Such functions take an input and return the output without modifying any state internally. Moreover, a class that contains such functions is **static** and has no state. Another good application of the **static** keyword is creating **immutable constants**. They are defined with a different keyword (**const**). The **Math library** is probably the best example of helper functions and constants. It has constants such as **PI** and **E**, static helper methods such as **Sqrt** and **Abs**, and so on.

The **DogsGenerator** class has no members that would be applicable to an object. If all class members are **static**, then the class should be **static** as well. Therefore, you should change the class to **public static class DateGenerator**. Be aware, however, that depending on **static** is the same as depending on a concrete implementation. Although they are easy to use and straightforward, static dependencies are hard to escape and should only be used for simple code, or code that you are sure will not change and is critical in its implementation details. For that reason, the **Math** class is a **static class** as well; it has all the foundations for arithmetic calculations.

The last application of **static** is **using static**. Applying the **static** keyword before a **using** statement causes all methods and fields to be directly accessible without the need to call a **class**. For example, consider the following code:

```
using static Math;
public static class Demo
{
    public static void Run()
    {
   //No need Math.PI
        Console.WriteLine(PI);
    }
}
```

This is a static import feature in C#. By using **static Math**, all static members can be accessed directly.

SEALED

Previously, you mentioned that inheritance should be handled with great care because the complexity can quickly grow out of hand. You can carefully consider complexity when you read and write code, but can you prevent complexity by design? C# has a keyword for stopping inheritance called **sealed**. If it logically makes no sense to inherit a class, then you should mark it with the **sealed** keyword. Security-related classes should also be sealed because it is critical to keep them simple and non-overridable. Also, if performance is critical, then methods in inherited classes are slower, compared to being directly in a sealed class. This is due to how method lookup works.

PARTIAL

In .NET, it is quite popular to make desktop applications using **WinForms**. The way **WinForms** works is that you can design how your application looks, with the help of a designer. Internally, it generates UI code and all you have to do is double-click a component, which will generate event handler code. That is where the partial class comes in. All the boring, autogenerated code will be in one class and the code that you write will be in another. The key point to note is that both classes will have the same name but be in different files.

You can have as many partial classes as you want. However, the recommended number of partial classes is no more than two. The compiler will treat them as one big class, but to the user, they will seem like two separate ones. Generating code generates new class files, which will overwrite the code you write. Use **partial** when you are dealing with autogenerated code. The biggest mistake that beginners make is using **partial** to manage big complex classes. If your class is complex, it's best to split it into smaller classes, not just different files.

There is one more use case for **partial**. Imagine you have a part of code in a class that is only needed in another assembly but is unnecessary in the assembly it is originally defined in. You can have the same class in different assemblies and mark it as **partial**. That way, a part of a class that is not needed will only be used where it is needed and be hidden where it should not be seen.

VIRTUAL

Abstract methods can be overridden; however, they cannot be implemented. What if you wanted to have a method with a default behavior that could be overridden in the future? You can do this using the **virtual** keyword, as shown in the following example:

```
public class Human
{
    public virtual void SayHi()
    {
        Console.WriteLine("Hello!");
    }
}
```

Here, the **Human** class has the **SayHi** method. This method is prefixed with the virtual keyword, which means that it can change behavior in a child class, for example:

```
public class Frenchman : Human
{
    public override void SayHi()
    {
        Console.WriteLine("Bonjour!");
    }
}
```

> **NOTE**
>
> You can find the code used for this example at https://packt.link/ZpHhl.

The **Frenchman** class inherits the **Human** class and overrides the **SayHi** method. Calling **SayHi** from a **Frenchman** object will print **Bonjour**.

One of the things about C# is that its behavior is hard to override. Upon declaring a method, you need to be explicit by telling the compiler that the method can be overridden. Only **virtual** methods can be overridden. Interface methods are virtual (because they get behavior later), however, you cannot override interface methods from child classes. You can only implement an interface in a parent class.

An abstract method is the last type of virtual method and is the most similar to **virtual** in that it can be overridden as many times as you need (in child and grandchild classes).

To avoid having fragile, changing, overridable behavior, the best kind of virtual methods are the ones that come from an interface. The **abstract** and **virtual** keywords enable changing class behavior in child classes and overriding it, which can become a big issue if uncontrolled. Overriding behavior often causes both inconsistent and unexpected results, so you should be careful before using the **virtual** keyword.

INTERNAL

public, **private**, and **protected** are the three access modifiers that have been mentioned. Many beginners think that the default class modifier is **private**. However, **private** means that it cannot be called from outside a class, and in the context of a namespace, this does not make much sense. The default access modifier for a class is **internal**. This means that the class will only be visible inside the namespace it is defined in. The **internal** modifier is great for reusing classes across the same assembly, while at the same time hiding them from the outside.

CONDITIONAL OPERATORS

A null reference exception is probably the most common error in programming. For example, refer to the following code:

```
int[] numbers = null;
numbers.length;
```

This code will throw **NullReferenceException** because you are interacting with a variable that has a null value. What is the length of a null array? There is no proper answer to this question, so an exception will be thrown here.

The best way to protect against such an error is to avoid working with null values altogether. However, sometimes it is unavoidable. In those cases, there is another technique called **defensive programming**. Before using a value that might be **null**, make sure it is not **null**.

Now recall the example of the **Dog** class. If you create a new object, the value of **Owner** could be null. If you were to determine whether the owner's name starts with the letter **A**, you would need to check first whether the value of **Owner** is **null**, as follows:

```
if (dog.Owner != null)
{
    bool ownerNameStartsWithA = dog.Owner.StartsWith('A');
}
```

However, in C#, using null-conditional, this code becomes as simple as the following:

```
dog.Owner?.StartsWith('A');
```

Null-conditional (**?**) is an example of **conditional operators** in C#. It is an operator that implicitly runs an **if** statement (a specific **if** statement is based on the operator) and either returns something or continues work. The **Owner?.StartsWith('A')** part returns **true** if the condition is satisfied and **false** if it is either not satisfied or the object is **null**.

There are more conditional operators in C# that you will learn about.

TERNARY OPERATORS

There is hardly any language that does not have **if** statements. One of the most common kinds of **if** statement is **if-else**. For example, if the value of **Owner** is **null** for an instance of the **Dog** class, you can describe the instance simply as **{Name}**. Otherwise, you can better describe it as **{Name}, dog of {Owner}**, as shown in the following snippet:

```
if (dog1.Owner == null)
{
    description = dog1.Name;
}
else
```

```
{
    description = $"{dog1.Name}, dog of {dog1.Owner}";
}
```

C#, like many other languages, simplifies this by using a ternary operator:

```
description = dog1.Owner == null
    ? dog1.Name
    : $"{dog1.Name}, dog of {dog1.Owner}";
```

On the left side, you have a condition (true or false), followed by a question mark (**?**), which returns the value on the right if the condition is true, followed by a colon (**:**), which returns the value to the left if the condition is false. **$** is a string interpolation literal, which allows you to write **$"{dog1.Name}, dog of {dog1.Owner}"** over **dog1.Name + "dog of" + dog1.Owner**. You should use it when concatenating text.

Imagine there are two dogs now. You want the first dog to join the second one (that is, be owned by the owner of the second dog), but this can only happen if the second one has an owner to begin with. Normally, you would use the following code:

```
if (dog1.Owner != null)
{
    dog2.Owner = dog1.Owner;
}
```

But in C#, you can use the following code:

```
dog1.Owner = dog1.Owner ?? dog2.Owner;
```

Here, you have applied the **null-coalescing operator** (**??**), which returns the value to the right if it is **null** and the value on the left if it is not **null**. However, you can simplify this further:

```
dog1.Owner ??= dog2.Owner;
```

This means that if the value that you are trying to assign (on the left) is **null**, then the output will be the value on the right.

The last use case for the null-coalescing operator is input validation. Suppose there are two classes, **ComponentA** and **ComponentB**, and **ComponentB** must contain an initialized instance of **ComponentA**. You could write the following code:

```
public ComponentB(ComponentA componentA)
{
    if (componentA == null)
```

```
        {
            throw new ArgumentException(nameof(componentA));
        }
        else
        {
            _componentA = componentA;
        }
    }
```

However, instead of the preceding code, you can simply write the following:

```
_componentA = componentA ?? throw new
ArgumentNullException(nameof(componentA));
```

This can be read as If there is no **componentA**, then an exception must be thrown.

> **NOTE**
>
> You can find the code used for this example at https://packt.link/yHYbh.

In most cases, null operators should replace the standard **if null-else** statements. However, be careful with the way you use the **ternary operator** and limit it to simple **if-else** statements because the code can become unreadable very quickly.

OVERLOADING OPERATORS

It is fascinating how much can be abstracted away in C#. Comparing primitive numbers, multiplying, or dividing them is easy, but when it comes to objects, it is not that simple. What is one person plus another person? What is a bag of apples multiplied by another bag of apples? It is hard to say, but it can make total sense in the context of some domains.

Consider a slightly better example. Suppose you are comparing bank accounts. Finding out who has more money in a bank account is a common use case. Normally, to compare two accounts, you would have to access their members, but C# allows you to overload comparison operators so that you can compare objects. For example, imagine you had a **BankAccount** class like so:

```
public class BankAccount
{
    private decimal _balance;
```

```
    public BankAccount(decimal balance)
    {
        _balance = balance;
    }
}
```

Here, the balance amount is **private**. You do not care about the exact value of **balance**; all you want is to compare one with another. You could implement a **CompareTo** method, but instead, you will implement a comparison operator. In the **BankAccount** class, you will add the following code:

```
public static bool operator >(BankAccount account1, BankAccount account2)
    => account1?._balance > account2?._balance;
```

The preceding code is called an **operator overload**. With a custom operator overload like this, you can return true when a balance is bigger and false otherwise. In C#, operators are **public static**, followed by a return type. After that, you have the **operator** keyword followed by the actual operator that is being overloaded. The input depends on the operator being overloaded. In this case, you passed two bank accounts.

If you tried to compile the code as it is, you would get an error that something is missing. It makes sense that the comparison operators have a twin method that does the opposite. Now, add the less operator overload as follows:

```
public static bool operator <(BankAccount account1, BankAccount account2)
    => account1?._balance < account2?._balance;
```

The code compiles now. Finally, it would make sense to have an equality comparison. Remember, you will need to add a pair, equal and not equal:

```
public static bool operator ==(BankAccount account1, BankAccount
account2)
    => account1?._balance == account2?._balance;

public static bool operator !=(BankAccount account1, BankAccount
account2)
    => !(account1 == account2);
```

Next, you will create bank accounts to compare. Note that all numbers have an **m** appended, as this suffix makes those numbers **decimal**. By default, numbers with a fraction are **double**, so you need to add **m** at the end to make them **decimal**:

```
var account1 = new BankAccount(-1.01m);
var account2 = new BankAccount(1.01m);
```

```
var account3 = new BankAccount(1001.99m);
var account4 = new BankAccount(1001.99m);
```

Comparing two bank accounts becomes as simple as this now:

```
Console.WriteLine(account1 == account2);
Console.WriteLine(account1 != account2);
Console.WriteLine(account2 > account1);
Console.WriteLine(account1 < account2);
Console.WriteLine(account3 == account4);
Console.WriteLine(account3 != account4);
```

Running the code results in the following being printed to the console:

```
False
True
True
True
True
False
```

> **NOTE**
>
> You can find the code used for this example at https://packt.link/5DioJ.

Many (but not all) operators can be overloaded, but just because you can do so does not mean you should. Overloading operators can make sense in some cases, but in other cases, it might be counterintuitive. Again, remember to not abuse C# features and use them when it makes **logical** sense, and when it makes code easier to read, learn, and maintain.

NULLABLE PRIMITIVE TYPES

Have you ever wondered what to do when a primitive value is unknown? For example, say a collection of products are announced. Their names, descriptions, and some other parameters are known, but the price is revealed only before the launch. What type should you use for storing the price values?

Nullable primitive types are primitive types that might have some value or no value. In C#, to declare such a type, you have to add **?** after a primitive, as shown in the following code:

```
int? a = null;
```

Here, you declared a field that may or may not have a value. Specifically, this means that a can be unknown. Do not confuse this with a default value because, by default, the value of **int** types is **0**.

You can assign a value to a nullable field quite simply, as follows:

```
a = 1;
```

And to retrieve its value afterward, you can write the code as follows:

```
int b = a.Value;
```

GENERICS

Sometimes, you will come across situations where you do the exact same thing with different types, where the only difference is because of the type. For example, if you had to create a method that prints an **int** value, you could write the following code:

```
public static void Print(int element)
{
    Console.WriteLine(element);
}
If you need to print a float, you could add another overload:
public static void Print(float element)
{
    Console.WriteLine(element);
}
```

Similarly, if you need to print a string, you could add yet another overload:

```
public static void Print(string element)
{
    Console.WriteLine(element);
}
```

You did the same thing three times. Surely, there must be a way to reduce code duplication. Remember, in C#, all types derive from an **object** type, which has the **ToString()** method, so you can execute the following command:

```
public static void Print(object element)
{
    Console.WriteLine(element);
}
```

Even though the last implementation contains the least code, it is actually the least efficient. An object is a reference type, whereas a primitive is a value type. When you take a primitive and assign it to an object, you also create a new reference to it. This is called **boxing**. It does not come for free, because you move objects from **stack to heap**. Programmers should be conscious of this fact and avoid it wherever possible.

Earlier in the chapter, you encountered polymorphism—a way of doing different things using the same type. You can do the same things with different types as well and **generics** are what enable you to do that. In the case of the **Print** example, a generic method is what you need:

```
public static void Print<T>(T element)
{
    Console.WriteLine(element);
}
```

Using diamond brackets (**<>**), you can specify a type, **T**, with which this function works. **<T>** means that it can work with any type.

Now, suppose you want to print all elements of an array. Simply passing a collection to a **WriteLine** statement would result in printing a **reference**, instead of all the elements. Normally, you would create a method that prints all the elements passed. With the power of generics, you can have one method that prints an array of any type:

```
public static void Print<T>(T[] elements)
{
    foreach (var element in elements)
    {
        Console.WriteLine(element);
    }
}
```

Please note that the generic version is not as efficient as taking an **object** type, simply because you would still be using a **WriteLine** overload that takes an object as a parameter. When passing a generic, you cannot tell whether it needs to call an overload with an **int**, **float**, or **String**, or whether there is an exact overload in the first place. If there was no overload that takes an object for **WriteLine**, you would not be able to call the **Print** method. For that reason, the most performant code is actually the one with three overloads. It is not terribly important though because that is just one, very specific scenario where boxing happens anyway. There are so many other cases, however, where you can make it not only concise but performant as well.

Sometimes, the answer to choosing a generic or polymorphic function hides in tiny details. If you had to implement a method for comparing two elements and return **true** if the first one is bigger, you could do that in C# using an **IComparable** interface:

```
public static bool IsFirstBigger1(IComparable first, IComparable second)
{
    return first.CompareTo(second) > 0;
}
```

A generic version of this would look like this:

```
public static bool IsFirstBigger2<T>(T first, T second)
    where T : IComparable
{
    return first.CompareTo(second) > 0;
}
```

The new bit here is **where T : IComparable**. It is a generic constraint. By default, you can pass any type to a generic class or method. Constraints still allow different types to be passed, but they significantly reduce the possible options. A **generic constraint** allows only the types that conform to the constraint to be passed as a generic type. In this case, you will allow only the types that implement the **IComparable** interface. Constraints might seem like a limitation on types; however, they expose the behavior of the constrained types that you can use inside a generic method. Having constraints enables you to use the features of those types, so it is very useful. In this case, you do limit yourself to what types can be used, but at the same time, whatever you pass to the generic method will be comparable.

What if instead of returning whether the first element is bigger, you needed to return the first element itself? You could write a non-generic method as follows:

```
public static IComparable Max1(IComparable first, IComparable second)
{
    return first.CompareTo(second) > 0
        ? first
        : second;
}
```

And the generic version would look as follows:

```
public static T Max2<T>(T first, T second)
    where T : IComparable
{
    return first.CompareTo(second) > 0
        ? first
        : second;
}
```

Also, it is worth comparing how you will get a meaningful output using each version. With a non-generic method, this is what the code would look like:

```
int max1 = (int)Comparator.Max1(3, -4);
```

With a generic version, the code would be like this:

```
int max2 = Comparator.Max2(3, -4);
```

> **NOTE**
>
> You can find the code used for this example at https://packt.link/sldOp.

In this case, the winner is obvious. In the non-generic version, you have to do a cast. **Casting** in code is frowned upon because if you do get errors, you will get them during runtime and things might change and the cast will fail. Casting is also one extra action, whereas the generic version is far more fluent because it does not have a cast. Use generics when you want to work with types as-is and not through their abstractions. And returning an exact (non-polymorphic) type from a function is one of the best use cases for it.

C# generics will be covered in detail in *Chapter 4, Data Structures and LINQ*.

ENUM

The **enum** type represents a set of known values. Since it is a type, you can pass it instead of passing a primitive value to methods. **enum** holds all the possible values, hence it isn't possible to have a value that it would not contain. The following snippet shows a simple example of this:

```
public enum Gender
{
    Male,
    Female,
    Other
}
```

> **NOTE**
>
> You can find the code used for this example at https://packt.link/gP9Li.

You can now get a possible gender value as if it were in a **static class** by writing **Gender.Other**. Enums can easily be converted to an integer using casting— **(int) Gender.Male** will return 0, **(int) Gender.Female** will return 1, and so on. This is because **enum**, by default, starts numbering at 0.

Enums do not have any behavior and they are known as **constant containers**. You should use them when you want to work with constants and prevent invalid values from being passed by design.

EXTENSION METHODS

Almost always, you will be working with a part of code that does not belong to you. Sometimes, this might cause inconvenience because you have no access to change it. Is it possible to somehow extend the existing types with the functionality you want? Is it possible to do so without inheriting or creating new component classes?

You can achieve this easily through **extension methods**. They allow you to add methods on complete types and call them as if those methods were natively there.

What if you wanted to print a **string** to a console using a **Print** method, but call it from a **string** itself? **String** has no such method, but you can add it using an extension method:

```
public static class StringExtensions
{
    public static void Print(this string text)
    {
        Console.WriteLine(text);
    }
}
```

And this allows you to write the following code:

```
"Hey".Print();
```

This will print **Hey** to the console as follows:

```
Hey
```

> **NOTE**
>
> You can find the code used for this example at https://packt.link/JC5cj.

Extension methods are **static** and must be placed within a **static class**. If you look at the semantics of the method, you will notice the use of the **this** keyword. The **this** keyword should be the first argument in an extension method. After that, the function continues as normal and you can use the argument with the **this** keyword as if it was just another argument.

Use extension methods to add (extend, but not the same extensions as what happens with inheritance) new behavior to existing types, even if the type would not support having methods otherwise. With extension methods, you can even add methods to **enum** types, which is not possible otherwise.

STRUCT

A class is a reference type, but not all objects are reference types (saved on the heap). Some objects can be created on the stack, and such objects are made using structs.

A **struct** is defined like a class, but it is used for slightly different things. Now, create a **struct** named **Point**:

```
public struct Point
{
    public readonly int X;
    public readonly int Y;

    public Point(int x, int y)
    {
        X = x;
        Y = y;
    }
}
```

The only real difference here is the **struct** keyword, which indicates that this object will be saved on the stack. Also, you might have noticed that there is no use of properties. There are many people who would, instead of **Point**, type **x** and **y**. It is not a big deal, but instead of one variable, you would be working with two. This way of working with primitives is called **primitive obsession**. You should follow the principles of OOP and work with abstractions, well-encapsulated data, as well as behavior to keep things close so that they have high cohesion. When choosing where to place variables, ask yourself this question: can **x** change independently of **y**? Do you ever modify a point? Is a point a complete value on its own? The answer to all of this is **yes** and therefore putting it in a data structure makes sense. But why choose a struct over a class?

Structs are fast because they do not have any allocations on the heap. They are also fast because they are passed by value (therefore, access is direct, not through a reference). Passing them by value copies the values, so even if you could modify a struct, changes would not remain outside of a method. When something is just a simple, small composite value, you should use a struct. Finally, with structs, you get value equality.

Another effective example of a **struct** is **DateTime**. **DateTime** is just a unit of time, containing some information. It also does not change individually and supports methods such as **AddDays**, **TryParse**, and **Now**. Even though it has several different pieces of data, they can be treated as one unit, as they are date- and time-related.

Most **structs** should be immutable because they are passed by a copy of a value, so changing something inside a method will not keep those changes. You can add a **readonly** keyword to a **struct**, making all its fields **readonly**:

```
public readonly struct Point
{
    public int X { get; }
    public int Y { get; }

    public Point(int x, int y)
    {
        X = x;
        Y = y;
    }
}
```

A **readonly struct** can have either a **readonly** field or getter properties. This is useful for the future maintainers of your code base as it prevents them from doing things that you did not design for (no mutability). Structs are just tiny grouped bits of data, but they can have behavior as well. It makes sense to have a method to calculate the distance between two points:

```
public static double DistanceBetween(Point p1, Point p2)
{
    return Math.Sqrt((p1.X - p2.X) * (p1.X - p2.X) + (p1.Y - p2.Y) *
(p1.Y - p2.Y));
}
```

The preceding code has a little bit of math in it—that is, distance between two points is the square root of points x's and y's squared differences added together.

It also makes sense to calculate the distance between this and other points. You do not need to change anything because you can just reuse the existing code, passing correct arguments:

```
public double DistanceTo(Point p)
{
    return DistanceBetween(this, p);
}
```

If you wanted to measure the distance between two points, you could create them like this:

```
var p1 = new Point(3,1);
var p2 = new Point(3,4);
```

And use a member function to calculate distance:

```
var distance1 = p1.DistanceTo(p2);
```

Or a static function:

```
var distance2 = Point.DistanceBetween(p1, p2);
```

The result for each version will be as follows:

```
- 3.
```

> **NOTE**
>
> You can find the code used for this example at https://packt.link/PtQzz.

When you think about a struct, think about it as just a group of primitives. The key point to remember is that all the data members (properties or fields) in a struct must be assigned during object initialization. It needs to be done for the same reason local variables cannot be used without having a value set initially. Structs do not support inheritance; however, they do support implementing an interface.

Structs are actually a great way to have simple business logic. Structs should be kept simple and should not contain other object references within them; they should be primitive-only. However, a class can hold as many struct objects as it needs. Using structs is a great way of escaping the obsessive use of primitives and using simple logic naturally, within a tiny group of data where it belongs—that is, a **`struct`**.

RECORD

A **record** is a reference type (unlike a **`struct`**, more like a class). However, out of the box, it has methods for comparison by value (both using the **`equals`** method and the operator). Also, a record has a different default implementation of **`ToString()`**, which no longer prints a type, but instead all the properties. This is exactly what is needed in many cases, so it helps a lot. Finally, there is a lot of syntactic sugar around records, which you are about to witness.

You already know how to create custom types in C#. The only difference between different custom types is the keyword used. For record types, such a keyword is **`record`**. For example, you will now create a movie record. It has a **`Title`**, **`Director`**, **`Producer`**, **`Description`**, and a **`ReleaseDate`**:

```
public record MovieRecordV1
{
    public string Title { get; }
    public string Director { get; }
    public string Producer { get; }
    public string Description { get; set; }
    public DateTime ReleaseDate { get; }

    public MovieRecordV1(string title, string director, string producer, 
DateTime releaseDate)
    {
        Title = title;
        Director = director;
        Producer = producer;
        ReleaseDate = releaseDate;
    }
}
```

So far, you should find this very familiar, because the only difference is the keyword. Regardless of such a minor detail, you already reap major benefits.

> **NOTE**
>
> The intention of having **`MovieRecordV1`** class in chapter, as against **`MovieClass`** in GitHub code, was to have a type, similar to a class and then refactor highlighting how record helps.

Create two identical movies:

```
private static void DemoRecord()
{
    var movie1 = new MovieRecordV1(
        "Star Wars: Episode I - The Phantom Menace",
        "George Lucas",
        "Rick McCallum",
```

```
        new DateTime(1999, 5, 15));

    var movie2 = new MovieRecordV1(
        "Star Wars: Episode I - The Phantom Menace",
        "George Lucas",
        "Rick McCallum",
        new DateTime(1999, 5, 15));
}
```

So far, everything is the same. Try to print a movie to the console:

```
    Console.WriteLine(movie1);
```

The output would be as follows:

```
MovieRecordV1 { Title = Star Wars: Episode I - The Phantom Menace,
Director = George Lucas, Producer
= Rick McCallum, Description = , ReleaseDate = 5/15/1999 12:00:00 AM }
```

> **NOTE**
>
> You can find the code used for this example at https://packt.link/xylkW.

If you tried doing the same to a class or a **struct** object, you would only get a type printed. However, for a record, a default behavior is to print all of its properties and their values.

That is not the only benefit of a record. Again, a record has value-equality semantics. Comparing two movie records will compare them by their property values:

```
    Console.WriteLine(movie1.Equals(movie2));
    Console.WriteLine(movie1 == movie2);
```

This will print **true true**.

With the same amount of code, you have managed to get the most functionality by simply changing a data structure to a record. Out of the box, a record provides **Equals()**, **GetHashCode() overrides**, **== and != overrides**, and even a **ToString** override, which prints the record itself (all the members and their values). The benefits of records do not end there because, using them, you have a way to reduce a lot of boilerplate code. Take full advantage of records and rewrite your movie record:

```
public record MovieRecord(string Title, string Director, string Producer,
string Description, DateTime ReleaseDate);
```

This is a positional record, meaning all that you pass as parameters will end up in the right read-only data members as if it was a dedicated constructor. If you ran the demo again, you would notice that it no longer compiles. The major difference with this declaration is that, now, changing a description is no longer possible. Making a mutable property is not difficult, you just need to be explicit about it:

```
public record MovieRecord(string Title, string Director, string Producer,
DateTime ReleaseDate)
{
    public string Description { get; set; }
}
```

You started this paragraph with a discussion on immutability, but why is the primary focus on records? The benefits of records are actually immutability-focused. Using a **`with`** expression, you can create a copy of a record object with zero or more properties modified. So, suppose you add this to your demo:

```
var movie3 = movie2 with { Description = "Records can do that?" };
movie2.Description = "Changing original";
Console.WriteLine(movie3);
```

The code would result in this:

```
MovieRecord { Title = Star Wars: Episode I - The Phantom Menace, Director
= George Lucas, Producer
= Rick McCallum, ReleaseDate = 5/15/1999 12:00:00 AM, Description =
Records can do that? }
```

As you see, this code copies an object with just one property changed. Before records, you would need a lot of code to ensure all the members are copied, and only then would you set a value. Keep in mind that this creates a shallow copy. A **shallow copy** is an object with all the references copied. A **deep copy** is an object with all the reference-type objects recreated. Unfortunately, there is no way of overriding such behavior. Records cannot inherit classes, but they can inherit other records. They can also implement interfaces.

Other than being a reference type, records are more like structs in that they have value equality and syntactic sugar around immutability. They should not be used as a replacement for structs because structs are still preferable for small and simple objects, which have simple logic. Use records when you want immutable objects for data, which could hold other complex objects (if nested objects could have a state that changes, shallow copying might cause unexpected behavior).

INIT-ONLY SETTERS

With the introduction of records, the previous edition, C# 9, also introduced **`init`**-only setter properties. Writing **`init`** instead of **`set`** can enable object initialization for properties:

```
public class House
{
    public string Address { get; init; }
    public string Owner { get; init; }
    public DateTime? Built { get; init; }
}
```

This enables you to create a house with unknown properties:

```
var house2 = new House();
```

Or assign them:

```
var house1 = new House
{
    Address = "Kings street 4",
    Owner = "King",
    Built = DateTime.Now
};
```

Using **`init`**-only setters is especially useful when you want read-only data, which can be known or not, but not in a consistent matter.

> **NOTE**
>
> You can find the code used for this example at https://packt.link/89J99.

VALUETUPLE AND DECONSTRUCTION

You already know that a function can only return one thing. In some cases, you can use the **`out`** keyword to return a second thing. For example, converting a string to a number is often done like this:

```
var text = "123";
var isNumber = int.TryParse(text, out var number);
```

TryParse returns both the parsed number and whether the text was a number.

However, C# has a better way of returning multiple values. You can achieve this using a data structure called **ValueTuple**. It is a generic **struct** that contains from one to six public mutable fields of any (specified) type. It is just a container for holding unrelated values. For example, if you had a **dog**, a **human**, and a **Bool**, you could store all three in a **ValueTuple** struct:

```
var values1 = new ValueTuple<Dog, Human, bool>(dog, human, isDogKnown);
```

You can then access each—that is, **dog** through **values1.Item1**, **human** through **values1.Item2**, and **isDogKnown** through **values.Item3**. Another way of creating a **ValueTuple** struct is to use brackets. This does exactly the same thing as before, but using the brackets syntax:

```
var values2 = (dog, human, isDogKnown);
```

The following syntax proves extremely useful because, with it, you can declare a function that virtually returns multiple things:

```
public (Dog, Human, bool) GetDogHumanAndBool()
{
    var dog = new Dog("Sparky");
    var human = new Human("Thomas");
    bool isDogKnown = false;

    return (dog, human, isDogKnown);
}
```

> **NOTE**
>
> You can find the code used for this example at https://packt.link/OTFpm.

You can also do the opposite, using another C# feature called **deconstruction**. It takes object data members and allows you to split them apart, into separate variables. The problem with a tuple type is that it does not have a strong name. As mentioned before, every field will be called **ItemX**, where **X** is the order in which the item was returned. Working with all that, **GetDogHumanAndBool** would require the results to be assigned to three different variables:

```
var dogHumanAndBool = GetDogHumanAndBool();
var dog = dogHumanAndBool.Item1;
```

```
var human = dogHumanAndBool.Item2;
var boo = dogHumanAndBool.Item3;
```

You can simplify this and instead make use of deconstruction—assigning object properties to different variables right away:

```
var (dog, human, boo) = GetDogHumanAndBool();
```

Using deconstruction, you are able to make this a lot more readable and concise. Use **ValueTuple** when you have multiple unrelated variables and you want to return them all from a function. You do not have to always work around using the **out** keyword, nor do you have to add overhead by creating a new class. You can solve this problem by simply returning and then deconstructing a **ValueTuple** struct.

You can now have hands-on experience of using SOLID principles for writing codes incrementally through the following exercise.

EXERCISE 2.04: CREATING A COMPOSABLE TEMPERATURE UNIT CONVERTER

Temperature can be measured in different units: Celsius, Kelvin, and Fahrenheit. In the future, more units might be added. However, units do not have to be added dynamically by the user; the application either supports it or not. You need to make an application that converts temperature from any unit to another unit.

It is important to note that converting to and from that unit will be a completely different thing. Therefore, you will need two methods for every converter. As a standard unit, you will use Celsius. Therefore, every converter should have a conversion method from and to Celsius, which makes it the simplest unit of a program. When you need to convert non-Celsius to Celsius, you will need to involve two converters—one to adapt the input to the standard unit (C), and then another one to convert from C to whatever unit you want. The exercise will aid you in developing an application using the SOLID principles and C# features you have learned in this chapter, such as **record** and **enum**.

Perform the following steps to do so:

1. Create a **TemperatureUnit** that uses an **enum** type to define constants—that is, a set of known values. You do not need to add it dynamically:

```
public enum TemperatureUnit
{
    C,
    F,
    K
}
```

In this example, you will use three temperature units that are **C**, **K**, and **F**.

2. Temperature should be thought of as a simple object made of two properties: **Unit** and **Degrees**. You could either use a **record** or a **struct** because it is a very simple object with data. The best choice would be picking a **struct** here (due to the size of the object), but for the sake of practicing, you will use a **record**:

```
public record Temperature(double Degrees, TemperatureUnit Unit);
```

3. Next, add a contract defining what you want from an individual specific temperature converter:

```
public interface ITemperatureConverter
{
    public TemperatureUnit Unit { get; }
    public Temperature ToC(Temperature temperature);
    public Temperature FromC(Temperature temperature);
}
```

You defined an interface with three methods—the **Unit** property to identify which temperature the converter is for, and **ToC** and **FromC** to convert from and to standard units.

4. Now that you have a converter, add the composable converter, which has an array of converters:

```
public class ComposableTemperatureConverter
{
    private readonly ITemperatureConverter[] _converters;
```

5. It makes no sense to have duplicate temperature unit converters. So, add an error that will be thrown when a duplicate converter is detected. Also, not having any converters makes no sense. Therefore, there should be some code for validating against **null** or empty converters:

```
public class InvalidTemperatureConverterException : Exception
{
    public InvalidTemperatureConverterException(TemperatureUnit unit)
: base($"Duplicate converter for {unit}.")
    {
    }

    public InvalidTemperatureConverterException(string message) :
base(message)
```

```
        {
        }
    }
```

When creating custom exceptions, you should provide as much information as possible about the context of an error. In this case, pass the **`unit`** for which the converter was not found.

6. Add a method that requires non-empty converters:

```
private static void RequireNotEmpty(ITemperatureConverter[]
converters)
{
    if (converters?.Length > 0 == false)
    {
        throw new InvalidTemperatureConverterException("At least one
temperature conversion must be supported");
    }
}
```

Passing an array of empty converters throws an **`InvalidTemperatureConverterException`** exception.

7. Add a method that requires non-duplicate converters:

```
private static void RequireNoDuplicate(ITemperatureConverter[]
converters)
{
    for (var index1 = 0; index1 < converters.Length - 1; index1++)
    {
        var first = converters[index1];
        for (int index2 = index1 + 1; index2 < converters.Length;
index2++)
        {
            var second = converters[index2];
            if (first.Unit == second.Unit)
            {
                throw new InvalidTemperatureConverterException(first.
Unit);
            }
        }
    }
}
```

This method goes through every converter and checks that, at other indexes, the same converter is not repeated (by duplicating **TemperatureUnit**). If it finds a duplicate unit, it will throw an exception. If it does not, it will just terminate successfully.

8. Now combine it all in a constructor:

```
public ComposableTemperatureConverter(ITemperatureConverter[]
converters)
{
    RequireNotEmpty(converters);
    RequireNoDuplicate(converters);
    _converters = converters;
}
```

When creating the converter, validate against converters that are not empty and not duplicates and only then set them.

9. Next, create a **private** helper method to help you find the requisite converter, **FindConverter**, inside the composable converter:

```
private ITemperatureConverter FindConverter(TemperatureUnit unit)
{
    foreach (var converter in _converters)
    {
        if (converter.Unit == unit)
        {
            return converter;
        }
    }

    throw new InvalidTemperatureConversionException(unit);
}
```

This method returns the converter of the requisite unit and, if no converter is found, throws an exception.

10. To simplify how you search and convert from any unit to Celsius, add a **ToCelsius** method for that:

```
private Temperature ToCelsius(Temperature temperatureFrom)
{
    var converterFrom = FindConverter(temperatureFrom.Unit);
    return converterFrom.ToC(temperatureFrom);
}
```

   Here, you find the requisite converter and convert the **Temperature** to Celsius.

11. Do the same for converting from Celsius to any other unit:

```
private Temperature CelsiusToOther(Temperature celsius,
TemperatureUnit unitTo)
{
    var converterTo = FindConverter(unitTo);
    return converterTo.FromC(celsius);
}
```

12. Wrap it all up by implementing this algorithm, standardize the temperature (convert to Celsius), and then convert to any other temperature:

```
public Temperature Convert(Temperature temperatureFrom,
TemperatureUnit unitTo)
{
    var celsius = ToCelsius(temperatureFrom);
    return CelsiusToOther(celsius, unitTo);
}
```

13. Add a few converters. Start with the Kelvin converter, **KelvinConverter**:

```
public class KelvinConverter : ITemperatureConverter
{
    public const double AbsoluteZero = -273.15;

    public TemperatureUnit Unit => TemperatureUnit.K;

    public Temperature ToC(Temperature temperature)
    {
        return new(temperature.Degrees + AbsoluteZero,
TemperatureUnit.C);
    }

    public Temperature FromC(Temperature temperature)
```

```
        {
            return new(temperature.Degrees - AbsoluteZero, Unit);
        }
    }
```

The implementation of this and all the other converters is straightforward. All you had to do was implement the formula to convert to the correct unit from or to Celsius. Kelvin has a useful constant, absolute zero, so instead of having a magic number, **-273.15**, you used a named constant. Also, it is worth remembering that a temperature is not a primitive. It is both a degree value and a unit. So, when converting, you need to pass both. **ToC** will always take **TemperatureUnit.C** as a unit and **FromC** will take whatever unit the converter is identified as, in this case, **TemperatureUnit.K**.

14. Now add a Fahrenheit converter, **FahrenheitConverter**:

```
public class FahrenheitConverter : ITemperatureConverter
{
    public TemperatureUnit Unit => TemperatureUnit.F;

    public Temperature ToC(Temperature temperature)
    {
        return new(5.0/9 * (temperature.Degrees - 32), 
TemperatureUnit.C);
    }

    public Temperature FromC(Temperature temperature)
    {
        return new(9.0 / 5 * temperature.Degrees + 32, Unit);
    }
}
```

Fahrenheit is identical structure-wise; the only differences are the formulas and unit value.

15. Add a **CelsiusConverter**, which will accept a value for the temperature and return the same value, as follows:

```
    public class CelsiusConverter : ITemperatureConverter
    {
        public TemperatureUnit Unit => TemperatureUnit.C;

        public Temperature ToC(Temperature temperature)
        {
```

```
            return temperature;
        }

        public Temperature FromC(Temperature temperature)
        {
            return temperature;
        }
    }
```

CelsiusConverter is the simplest one. It does not do anything; it just returns the same temperature. The converters convert to standard temperature—Celsius to Celsius is always Celsius. Why do you need such a class at all? Without it, you would need to change the flow a bit, adding **if** statements to ignore the temperature if it was in Celsius. But with this implementation, you can incorporate it in the same flow and use it in the same way with the help of the same abstraction, **ITemperatureConverter**.

16. Finally, create a demo:

Solution.cs

```
public static class Solution
{
    public static void Main()
    {
        ITemperatureConverter[] converters = {new FahrenheitConverter(), new
KelvinConverter(), new CelsiusConverter()};
        var composableConverter = new
ComposableTemperatureConverter(converters);

        var celsius = new Temperature(20.00001, TemperatureUnit.C);

        var celsius1 = composableConverter.Convert(celsius,
TemperatureUnit.C);
        var fahrenheit = composableConverter.Convert(celsius1,
TemperatureUnit.F);
        var kelvin = composableConverter.Convert(fahrenheit,
TemperatureUnit.K);
        var celsiusBack = composableConverter.Convert(kelvin,
TemperatureUnit.C);

        Console.WriteLine($"{celsius} = {fahrenheit}");
```

You can find the complete code here: https://packt.link/ruBph.

In this example, you have created all the converters and passed them to the converters container called **composableConverter**. Then you have created a temperature in Celsius and used it to perform conversions from and to all the other temperatures.

17. Run the code and you will get the following results:

```
Temperature { Degrees = 20.00001, Unit = C } = Temperature { Degrees
= 68.000018, Unit = F }
Temperature { Degrees = 68.000018, Unit = F } = Temperature { Degrees
= -253.14998999999997, Unit = K }
Temperature { Degrees = -253.14998999999997, Unit = K } = Temperature
{ Degrees = 20.000010000000003, Unit = C }
```

> **NOTE**
>
> You can find the code used for this exercise at https://packt.link/dDRU6.

A software developer, ideally, should design code in such a way that making a change now or in the future will take the same amount of time. Using SOLID principles, you can write code incrementally and minimize the risk of breaking changes, because you never change existing code; you just add new code. As systems grow, complexity increases, and it might be difficult to learn how things work. Through well-defined contracts, SOLID enables you to have easy-to-read, and maintainable code because each piece is straightforward by itself, and they are isolated from one another.

You will now test your knowledge of creating classes and overriding operators through an activity.

ACTIVITY 2.01: MERGING TWO CIRCLES

In this activity, you will create classes and override operators to solve the following mathematics problem: A portion of pizza dough can be used to create two circular pizza bites each with a radius of three centimeters. What would be the radius of a single pizza bite made from the same amount of dough? You can assume that all the pizza bites are the same thickness. The following steps will help you complete this activity:

1. Create a **`Circle`** struct with a radius. It should be a **`struct`** because it is a simple data object, which has a tiny bit of logic, calculating area.

2. Add a property to get the area of a circle (try to use an expression-bodied member). Remember, the formula of a circle's area is **`pi*r*r`**. To use the **`PI`** constant, you will need to import the **`Math`** package.

3. Add two circles' areas together. The most natural way would be to use an overload for a plus (**+**) operator. Implement a **+** operator overload that takes two circles and returns a new one. The area of the new circle is the sum of the areas of the two old circles. However, do not create a new circle by passing the area. You need a Radius. You can calculate this by dividing the new area by **PI** and then taking the square root of the result.
4. Now create a **Solution** class that takes two circles and returns a result—the radius of the new circle.
5. Within the **main** method, create two circles with a radius of **3** cm and define a new circle, which is equal to the areas of the two other circles added together. Print the results.
6. Run the **main** method and the result should be as follows:

```
Adding circles of radius of 3 and 3 results in a new circle with a
radius 4.242640687119285
```

As you can see from this final output, the new circle will have a radius of **4.24** (rounded to the second decimal place).

> **NOTE**
>
> The solution to this activity can be found at https://packt.link/qclbF.

This activity was designed to test your knowledge of creating classes and overriding operators. Operators are not normally employed to solve this sort of problem, but in this case, it worked well.

SUMMARY

In this chapter, you learned about OOP and how it helps take complex problems and abstract them into simple concepts. C# has several useful features and, roughly every one or two years, a new language version is released. The features mentioned in this chapter are just some of the ways in which C# aids in productivity. You have seen how, by design, it allows for better, clearer code, less prone to error. C# is one of the best languages when it comes to productivity. With C#, you can make effective code, and quickly, because a lot of the boilerplate code is done for you.

Finally, you learned the SOLID principles and used them in an application. SOLID is not something you can just read and learn immediately; it takes practice, discussions with your peers, and a lot of trial and error before you get it right and start applying it consistently. However, the benefits are worth it. In modern software development, producing fast, optimal code is no longer a number one priority. Nowadays, the focus is a balance of productivity (how fast you develop) and performance (how fast your program is). C# is one of the most efficient languages out there, both in terms of performance and productivity.

In the next chapter, you will learn what functional programming is and how to work with lambdas and functional constructs such as delegates.

3

DELEGATES, EVENTS, AND LAMBDAS

OVERVIEW

In this chapter, you will learn how delegates are defined and invoked, and you will explore their wide usage across the .NET ecosystem. With this knowledge, you will move on to the inbuilt `Action` and `Func` delegates to discover how their usage reduces unnecessary boilerplate code. You will then see how multicast delegates can be harnessed to send messages to multiple parties, and how events can be incorporated into event-driven code. Along the way, you will discover some common pitfalls to avoid and best practices to follow that prevent a great application from turning into an unreliable one.

This chapter will demystify the lambda syntax style and show how it can be used effectively. By the end of the chapter, you will be able to use the lambda syntax comfortably to create code that is succinct, as well as easy to grasp and maintain.

INTRODUCTION

In the previous chapter, you learned some of the key aspects of **Object Oriented Programming** (**OOP**). In this chapter, you will build on this by looking at the common patterns used specifically in C# that enable classes to interact.

Have you found yourself working with a code that has to listen to certain signals and act on them, but you cannot be sure until runtime what those actions should be? Maybe you have a block of code that you need to reuse or pass to other methods for them to call when they are ready. Or, you may want to filter a list of objects, but need to base how you would do that on a combination of user preferences. Much of this can be achieved using interfaces, but it is often more efficient to create chunks of code that you can then pass to other classes in a type-safe way. Such blocks are referred to as **delegates** and form the backbone of many .NET libraries, allowing methods or pieces of code to be passed as parameters.

The natural extension to a delegate is the **event**, which makes it possible to offer a form of optional behavior in software. For example, you may have a component that broadcasts live news and stock prices, but unless you provide a way to opt into these services, you may limit the usability of such a component.

User Interface (**UI**) apps often provide notifications of various user actions, for example, keypresses, swiping a screen, or clicking a mouse button; such notifications follow a standard pattern in C#, which will be discussed fully in this chapter. In such scenarios, the UI element detecting such actions is referred to as a **publisher**, whereas the code that acts upon those messages is called a **subscriber**. When brought together, they form an event-driven design referred to as the **publisher-subscriber**, or **pub-sub**, pattern. You will see how this can be used in all types of C#. Remember that its usage is not just the exclusive domain of UI applications.

Finally, you will learn about lambda statements and lambda expressions, collectively known as **lambdas**. These have an unusual syntax, which can initially take a while to become comfortable with. Rather than having lots of methods and functions scattered within a class, lambdas allow for smaller blocks of code that are often self-contained and located within close proximity to where they are used in the code, thereby offering an easier way to follow and maintain code. You will learn about lambdas in detail in the latter half of this chapter. First, you will learn about delegates.

DELEGATES

The .NET delegate is similar to function pointers found in other languages, such as C++; in other words, it is like a pointer to a method to be invoked at runtime. In essence, it is a **placeholder** for a block of code, which can be something as simple as a single statement or a full-blown multiline code block, complete with complex branches of execution, that you ask other code to execute at some point in time. The term delegate hints at some form of **representative**, which is precisely what this placeholder concept relates to.

Delegates allow for minimum coupling between objects, and much less code. There is no need to create classes that are derived from specific classes or interfaces. By using a delegate, you are defining what a compatible method should look like, whether it is in a class or struct, static, or instance-based. The arguments and return type define this calling compatibility.

Furthermore, delegates can be used in a **callback** fashion, which allows multiple methods to be wired up to a single publication source. They often require much less code and provide more features than found using an interface-based design.

The following example shows how effective delegates can be. Suppose you have a class that searches for users by surname. It would probably look like this:

```
public User FindBySurname(string name)
{
    foreach(var user in _users)
       if (user.Surname == name)
          return user;
    return null;
}
```

You then need to extend this to include a search of the user's login name:

```
public User FindByLoginName(string name)
{
    foreach(var user in _users)
       if (user.LoginName == name)
          return user;
    return null;
}
```

Once again, you decide to add yet another search, this time by location:

```
public User FindByLocation(string name)
{
    foreach(var user in _users)
       if (user.Location == name)
          return user;
    return null;
}
```

You start the searches with code like this:

```
public void DoSearch()
{
  var user1 = FindBySurname("Wright");
  var user2 = FindByLoginName("JamesR");
  var user3 = FindByLocation("Scotland");
}
```

Can you see the pattern that is occurring every time? You are repeating the same code that iterates through the list of users, applying a Boolean condition (also known as a **predicate**) to find the first matching user.

The only thing that is different is that the predicate decides whether a match has been found. This is one of the common cases where delegates are used at a basic level. The **`predicate`** can be replaced with a delegate, acting as a placeholder, which is evaluated when required.

Converting this code to a delegate style, you define a delegate named **`FindUser`** (this step can be skipped as .NET contains a delegate definition that you can reuse; you will come to this later).

All you need is a single helper method, **`Find`**, which is passed a **`FindUser`** delegate instance. **Find** knows how to loop through the users, invoking the delegate passing in the user, which returns true or false for a match:

```
private delegate bool FindUser(User user);
private User Find(FindUser predicate)
{
  foreach (var user in _users)
    if (predicate(user))
      return user;
  return null;
}
```

```
public void DoSearch()
{
  var user4 = Find(user => user.Surname == "Wright");
  var user5 = Find(user => user.LoginName == "JamesR");
  var user6 = Find(user => user.Location == "Scotland");
}
```

As you can see, the code is kept together and is much more concise now. There is no need to cut and paste code that loops through the users, as that is all done in one place. For each type of search, you simply define a delegate once and pass it to **`Find`**. To add a new type of search, all you need to do is define it in a single statement line, rather than copying at least eight lines of code that repeat the looping function.

The lambda syntax is a fundamental style used to define method bodies, but its strange syntax can prove to be an obstacle at first. At first glance, lambda expressions can look odd with their **=>** style, but they do offer a cleaner way to specify a target method. The act of defining a lambda is similar to defining a method; you essentially omit the method name and use **=>** to prefix a block of code.

You will now look at another example, using interfaces this time. Consider that you are working on a graphics engine and need to calculate the position of an image onscreen each time the user rotates or zooms in. Note that this example skips any complex math calculations.

Consider that you need to transform a **`Point`** class using the **`ITransform`** interface with a single method named **`Move`**, as shown in the following code snippet:

```
public class Point
{
  public double X { get; set; }
  public double Y { get; set; }
}

public interface ITransform
{
  Point Move(double height, double width);
}
```

When the user rotates an object, you need to use **RotateTransform**, and for a zoom operation, you will use **ZoomTransform**, as follows. Both are based on the **ITransform** interface:

```
public class RotateTransform : ITransform
{
    public Point Move(double height, double width)
    {
        // do stuff
        return new Point();
    }
}

public class ZoomTransform : ITransform
{
    public Point Move(double height, double width)
    {
        // do stuff
        return new Point();
    }
}
```

So, given these two classes, a point can be transformed by creating a new **Transform** instance, which is passed to a method named **Calculate**, as shown in the following code. **Calculate** calls the corresponding **Move** method, and does some extra unspecified work on point, before returning point to the caller:

```
public class Transformer
{
    public void Transform()
    {
        var rotatePoint = Calculate(new RotateTransform(), 100, 20);
        var zoomPoint = Calculate(new ZoomTransform(), 5, 5);
    }

    private Point Calculate(ITransform transformer, double height, double
width)
    {
```

```
        var point = transformer.Move(height, width);
        //do stuff to point
        return point;
    }
}
```

This is a standard class and interface-based design, but you can see that you have made a lot of effort to create new classes with just a single numeric value from a **Move** method. It is a worthwhile idea to have the calculations broken down into an easy-to-follow implementation. After all, it could have led to a future maintenance problem if implemented in a single method with multiple if-then branches.

By re-implementing a delegate-based design, you still have maintainable code, but much less of it to look after. You can have a **TransformPoint** delegate and a new **Calculate** function that can be passed a **TransformPoint** delegate.

You can invoke a delegate by appending brackets around its name and passing in any arguments. This is similar to how you would call a standard class-level function or method. You will cover this invocation in more detail later; for now, consider the following snippet:

```
    private delegate Point TransformPoint(double height, double width);

    private Point Calculate(TransformPoint transformer, double height, 
double width)
    {
        var point = transformer(height, width);
        //do stuff to point
        return point;
    }
```

You still need the actual target **Rotate** and **Zoom** methods, but you do not have the overhead of creating unnecessary classes to do this. You can add the following code:

```
    private Point Rotate(double height, double width)
    {
        return new Point();
    }

    private Point Zoom(double height, double width)
    {
        return new Point();
    }
```

Now, calling the method delegates is as simple as the following:

```
public void Transform()
{
    var rotatePoint1 = Calculate(Rotate, 100, 20);
    var zoomPoint1 = Calculate(Zoom, 5, 5);
}
```

Notice how using delegates in this way helps eliminate a lot of unnecessary code.

> **NOTE**
>
> You can find the code used for this example at https://packt.link/AcwZA.

In addition to invoking a single placeholder method, a delegate also contains extra plumbing that allows it to be used in a **multicast** manner, that is, a way to chain multiple target methods together, each being invoked one after the other. This is often referred to as an **invocation list** or **delegate chain** and is initiated by code that acts as a publication source.

A simple example of how this multicast concept applies can be seen in UIs. Imagine you have an application that shows the map of a country. As the user moves their mouse over the map, you may want to perform various actions, such as the following:

- Changing the mouse pointer to a different shape while over a building.
- Showing a tooltip that calculates the real-world longitude and latitude coordinates.
- Showing a message in a status bar that calculates the population of the area where the mouse is hovering.

To achieve this, you would need some way to detect when the user moves the mouse over the screen. This is often referred to as the publisher. In this example, its sole purpose is to detect mouse movements and publish them to anyone who is listening.

To perform the three required UI actions, you would create a class that has a list of objects to notify when the mouse position changes, allowing each object to perform whatever activity it needs, in isolation from the others. Each of these objects is referred to as a subscriber.

When your publisher detects that the mouse has moved, you follow this pseudo code:

```
MouseEventArgs args = new MouseEventArgs(100,200)
foreach(subscription in subscriptionList)
{
   subscription.OnMouseMoved(args)
}
```

This assumes that **subscriptionList** is a list of objects, perhaps based on an interface with the **OnMouseMoved** method. It is up to you to add code that enables interested parties to subscribe to and unsubscribe from the **OnMouseMoved** notifications. It would be an unfortunate design if code that has previously subscribed has no way to unsubscribe and gets called repeatedly when there is no longer any need for it to be called.

In the preceding code, there is a fair amount of coupling between the publisher and subscribers, and you are back to using interfaces for a type-safe implementation. What if you then needed to listen for keypresses, both key down and key up? It would soon get quite frustrating having to repeatedly copy such similar code.

Fortunately, the delegate type contains all this as inbuilt behavior. You can use single or multiple target methods interchangeably; all you need to do is invoke a delegate and the delegate will handle the rest for you.

You will take an in-depth look at multicast delegates shortly, but first, you will explore the single-target method scenario.

DEFINING A CUSTOM DELEGATE

Delegates are defined in a way that is similar to that of a standard method. The compiler does not care about the code in the body of a target method, only that it can be invoked safely at some point in time.

The **delegate** keyword is used to define a delegate, using the following format:

```
public delegate void MessageReceivedHandler(string message, int size);
```

The following list describes each component of this syntax:

- Scope: An access modifier, such as **public**, **private**, or **protected**, to define the scope of the delegate. If you do not include a modifier, the compiler will default to marking it as private, but it is always better to be explicit in showing the intent of your code.
- The **delegate** keyword.

- Return type: If there is no return type, **`void`** is used.
- Delegate name: This can be anything that you like, but the name must be unique within the **namespace**. Many naming conventions (including Microsoft's) suggest adding **`Handler`** or **`EventHandler`** to your delegate's name.
- Arguments, if required.

> **NOTE**
>
> Delegates can be nested within a class or namespace; they can also be defined within the **global namespace**, although this practice is discouraged. When defining classes in C#, it is common practice to define them within a parent namespace, typically based on a hierarchical convention that starts with the company name, followed by the product name, and finally the feature. This helps to provide a more unique identity to a type.
>
> By defining a delegate without a namespace, there is a high chance that it will clash with another delegate with the same name if it is also defined in a library without the protection of a namespace. This can cause the compiler to become confused as to which delegate you are referring to.

In earlier versions of .NET, it was common practice to define custom delegates. Such code has since been replaced with various inbuilt .NET delegates, which you will look at shortly. For now, you will briefly cover the basics of defining a custom delegate. It is worthwhile know about this if you maintain any legacy C# code.

In the next exercise, you will create a custom delegate, one that is passed a **`DateTime`** parameter and returns a Boolean to indicate validity.

EXERCISE 3.01: DEFINING AND INVOKING CUSTOM DELEGATES

Say you have an application that allows users to order products. While filling in the order details, the customer can specify an order date and a delivery date, both of which must be validated before accepting the order. You need a flexible way to validate these dates. For some customers, you may allow weekend delivery dates, while for others, it must be at least seven days away. You may also allow an order to be back-dated for certain customers.

You know that delegates offer a way to vary an implementation at runtime, so that is the best way to proceed. You do not want multiple interfaces, or worse, a complex jumble of **if-then** statements, to achieve this.

Depending on the customer's profile, you can create a class named **Order**, which can be passed different date validation rules. These rules can be validated by a **Validate** method:

Perform the following steps to do so:

1. Create a new folder called **Chapter03**.
2. Change to the **Chapter03** folder and create a new console app, called **Exercise01**, using the CLI **dotnet** command, as follows:

   ```
   source\Chapter03>dotnet new console -o Exercise01
   ```

   You will see the following output:

   ```
   The template "Console Application" was created successfully.
   Processing post-creation actions...
   Running 'dotnet restore' on Exercise01\Exercise01.csproj...
     Determining projects to restore...
     Restored source\Chapter03\Exercise01\Exercise01.csproj (in 191 ms).
   Restore succeeded.
   ```

3. Open **Chapter03\Exercise01.csproj** and replace the contents with these settings:

   ```
   <Project Sdk="Microsoft.NET.Sdk">
     <PropertyGroup>
       <OutputType>Exe</OutputType>
       <TargetFramework>net6.0</TargetFramework>
     </PropertyGroup>
   </Project>
   ```

4. Open **Exercise01\Program.cs** and clear the contents.

5. The preference for using namespaces to prevent a clash with objects from other libraries was mentioned earlier, so to keep things isolated, use **Chapter03.Exercise01** as the namespace.

   To implement your date validation rules, you will define a delegate that takes a single **DateTime** argument and returns a Boolean value. You will name it **DateValidationHandler**:

```
using System;
namespace Chapter03.Exercise01
{
    public delegate bool DateValidationHandler(DateTime dateTime);
}
```

6. Next, you will create a class named **Order**, which contains details of the order and can be passed to two date validation delegates:

```
    public class Order
     {
         private readonly DateValidationHandler _orderDateValidator;
         private readonly DateValidationHandler _deliveryDateValidator;
```

   Notice how you have declared two read-only, class-level instances of **DateValidationHandler**, one to validate the order date and a second to validate the delivery date. This design assumes that the date validation rules are not going to be altered for this **Order** instance.

7. Now for the constructor, you pass the two delegates:

```
        public Order(DateValidationHandler orderDateValidator,
             DateValidationHandler deliveryDateValidator)
         {
             _orderDateValidator = orderDateValidator;
             _deliveryDateValidator = deliveryDateValidator;
         }
```

   In this design, a different class is typically responsible for deciding which delegates to use, based on the selected customer's profile.

8. You need to add the two date properties that are to be validated. These dates may be set using a UI that listens to keypresses and applies user edits directly to this class:

```
         public DateTime OrderDate { get; set; }
         public DateTime DeliveryDate { get; set; }
```

9. Now add an **IsValid** method that passes **OrderDate** to the **orderDateValidator** delegate and **DeliveryDate** to the **deliveryDateValidator** delegate:

```
        public bool IsValid() =>
            _orderDateValidator(OrderDate) &&
            _deliveryDateValidator(DeliveryDate);
    }
```

   If both are valid, then this call will return **true**. The key here is that **Order** doesn't need to know about the precise implementation of an individual customer's date validation rules, so you can easily reuse **Order** elsewhere in a program. To invoke a delegate, you simply wrap any arguments in brackets, in this case passing the correct date property to each delegate instance:

10. To create a console app to test this, add a **static** class called **Program**:

```
    public static class Program
    {
```

11. You want to create two functions that validate whether the date passed to them is valid. These functions will form the basis of your delegate target methods:

```
        private static bool IsWeekendDate(DateTime date)
        {
            Console.WriteLine("Called IsWeekendDate");
            return date.DayOfWeek == DayOfWeek.Saturday ||
                   date.DayOfWeek == DayOfWeek.Sunday;
        }

        private static bool IsPastDate(DateTime date)
        {
            Console.WriteLine("Called IsPastDate");
            return date < DateTime.Today;
        }
```

   Notice how both have the exact signature that the **DateValidationHandler** delegate is expecting. Neither is aware of the nature of the date that they are validating, as that is not their concern. They are both marked **static** as they do not interact with any variables or properties anywhere in this class.

12. Now for the **Main** entry point. Here, you create two **DateValidationHandler** delegate instances, passing **IsPastDate** to one and **IsWeekendDate** to the second. These are the target methods that will get called when each of the delegates is invoked:

```
        public static void Main()
        {
            var orderValidator = new DateValidationHandler(IsPastDate);
            var deliverValidator = new
DateValidationHandler(IsWeekendDate);
```

13. Now you can create an **Order** instance, passing in the delegates and setting the order and delivery dates:

```
            var order = new Order(orderValidator, deliverValidator)
              {
                  OrderDate = DateTime.Today.AddDays(-10),
                  DeliveryDate = new DateTime(2020, 12, 31)
              };
```

There are various ways to create delegates. Here, you have assigned them to variables first to make the code clearer (you will cover different styles later).

14. Now it's just a case of displaying the dates in the console and calling **IsValid**, which, in turn, will invoke each of your delegate methods once. Notice that a custom date format is used to make the dates more readable:

```
            Console.WriteLine($"Ordered: {order.OrderDate:dd-MMM-yy}");
            Console.WriteLine($"Delivered: {order.DeliveryDate:dd-
MMM-yy }");
            Console.WriteLine($"IsValid: {order.IsValid()}");
        }
    }
}
```

15. Running the console app produces output like this:

```
Ordered: 07-May-22
Delivered: 31-Dec-20
Called IsPastDate
Called IsWeekendDate
IsValid: False
```

This order is **not** valid as the delivery date is a Thursday, not a weekend as you require:

You have learned how to define a custom delegate and have created two instances that make use of small helper functions to validate dates. This gives you an idea of how flexible delegates can be.

> **NOTE**
>
> You can find the code used for this exercise at https://packt.link/cmL0s.

THE INBUILT ACTION AND FUNC DELEGATES

When you define a delegate, you are describing its signature, that is, the return type and a list of input parameters. With that said, consider these two delegates:

```
public delegate string DoStuff(string name, int age);
public delegate string DoMoreStuff(string name, int age);
```

They both have the same signature but vary by name alone, which is why you can declare an instance of each and have them **both** point at the **same** target method when invoked:

```
public static void Main()
{
    DoStuff stuff = new DoStuff(MyMethod);
    DoMoreStuff moreStuff = new DoMoreStuff(MyMethod);

    Console.WriteLine($"Stuff: {stuff("Louis", 2)}");
    Console.WriteLine($"MoreStuff: {moreStuff("Louis", 2)}");
}

private static string MyMethod(string name, int age)
{
    return $"{name}@{age}";
}
```

Running the console app produces the same results in both calls:

```
Stuff: Louis@2
MoreStuff: Louis@2
```

> **NOTE**
>
> You can find the code used for this example at https://packt.link/r6B8n.

It would be great if you could dispense with defining both **DoStuff** and **DoMoreStuff** delegates and use a more generalized delegate with precisely the same signature. After all, it does not matter in the preceding snippet if you create a **DoStuff** or **DoMoreStuff** delegate, since both make a call to the same target method.

.NET does, in fact, provide various inbuilt delegates that you can make use of directly, saving you the effort of defining such delegates yourself. These are the **Action** and **Func** delegates.

There are many possible combinations of **Action** and **Func** delegates, each allowing an increasing number of parameters. You can specify anywhere from zero to 16 different parameter types. With so many combinations available, it is extremely unlikely that you will ever need to define your own delegate type.

It is worth noting that **Action** and **Func** delegates were added in a later version of .NET and, as such, the use of custom delegates tends to be found in older legacy code. There is no need to create new delegates yourself.

In the following snippet, **MyMethod** is invoked using the three-argument **Func** variation; you will cover the odd-looking **<string, int, string>** syntax shortly:

```
Func<string, int, string> funcStuff = MyMethod;
Console.WriteLine($"FuncStuff: {funcStuff("Louis", 2)}");
```

This produces the same return value as the two earlier invocations:

```
FuncStuff: Louis@2
```

Before you continue exploring **Action** and **Func** delegates, it is useful to explore the **Action<string, int, string>** syntax a bit further. This syntax allows type parameters to be used to define classes and methods. These are known as **generics** and act as placeholders for a particular type. In *Chapter 4, Data Structures and LINQ*, you will cover generics in much greater detail, but it is worth summarizing their usage here with the **Action** and **Func** delegates.

The non-generic version of the **Action** delegate is predefined in .NET as follows:

```
public delegate void Action()
```

As you know from your earlier look at delegates, this is a delegate that does not take any arguments and does not have a return type; it is the simplest type of delegate available.

Contrast that with one of the generic **Action** delegates predefined in .NET:

```
public delegate void Action<T>(T obj)
```

You can see this includes a **<T>** and **T** parameter section, which means it accepts a **single-type** argument. Using this, you can declare an **Action** that is constrained to a string, which takes a single string argument and returns no value, as follows:

```
Action<string> actionA;
```

How about an **int** constrained version? This also has no return type and takes a single **int** argument:

```
Action<int> actionB;
```

Can you see the pattern here? In essence, the type that you specify can be used to declare a type at compile time. What if you wanted two arguments, or three, or four... or 16? Simple. There are **Action** and **Func** generic types that can take up to **16** different argument types. You are very unlikely to be writing code that needs more than 16 parameters.

This two-argument **Action** takes **int** and **string** as parameters:

```
Action<int, string> actionC;
```

You can spin that around. Here is another two-argument **Action**, but this takes a **string** parameter and then an **int** parameter:

```
Action<string, int> actionD;
```

These cover most argument combinations, so you can see that it is very rare to create your own delegate types.

The same rules apply to delegates that return a value; this is where the **Func** types are used. The generic **Func** type starts with a single value type parameter:

```
public delegate T Func<T>()
```

In the following example, **funcE** is a delegate that returns a Boolean value and takes no arguments:

```
Func<bool> funcE;
```

Can you guess which is the return type from this rather long **Func** declaration?

```
Func<bool, int, int, DateTime, string> funcF;
```

This gives a delegate that returns a **string** . In other words, the last argument type in a **Func** defines the return type. Notice that **funcF** takes four arguments: **bool**, **int**, **int**, and **DateTime**.

In summary, generics are a great way to define types. They save a lot of duplicate code by allowing type parameters to act as placeholders.

ASSIGNING DELEGATES

You covered creating custom delegates and briefly how to assign and invoke a delegate in *Exercise 3.01*. You then looked at using the preferred **Action** and **Func** equivalents, but what other options do you have for assigning the method (or methods) that form a delegate? Are there other ways to invoke a delegate?

Delegates can be assigned to a variable in much the same way that you might assign a class instance. You can also pass new instances or static instances around without having to use variables to do so. Once assigned, you can invoke the delegate or pass the reference to other classes so they can invoke it, and this is often done within the Framework API.

You will now look at a **Func** delegate, which takes a single **DateTime** argument and returns a **bool** value to indicate validity. You will use a **static** class containing two helper methods, which form the actual target:

```
public static class DateValidators
{
    public static bool IsWeekend(DateTime dateTime)
        => dateTime.DayOfWeek == DayOfWeek.Saturday ||
           dateTime.DayOfWeek == DayOfWeek.Sunday;
```

```
    public static bool IsFuture(DateTime dateTime)
      => dateTime.Date > DateTime.Today;
}
```

> **NOTE**
>
> You can find the code used for this example at https://packt.link/mwmxh.

Note that the **DateValidators** class is marked as **static**. You may have heard the phrase **statics are inefficient**. In other words, creating an application with many static classes is a weak practice. Static classes are instantiated the first time they are accessed by running code and remain in memory until the application is closed. This makes it difficult to control their lifetime. Defining small utility classes as static is less of an issue, provided they do indeed remain stateless. **Stateless** means they do not set any local variables. Static classes that set local states are very difficult to unit test; you can never be sure that the variable set is from one test or another test.

In the preceding snippet, **IsFuture** returns **true** if the **Date** property of the **DateTime** argument is later than the current date. You are using the static **DateTime.Today** property to retrieve the current system date. **IsWeekend** is defined using an **expression-bodied** syntax and will return **true** if the **DateTime** argument's day of the week falls on a Saturday or Sunday.

You can assign delegates the same way that you would assign regular variables (remember you do **not** have to assign a variable to pass to other classes). You will now create two validator variables, **futureValidator** and **weekendValidator**. Each constructor is passed the actual target method, the **IsFuture** or **IsWeekend** instance, respectively:

```
var futureValidator = new Func<DateTime, bool>(DateValidators.IsFuture);
var weekendValidator = new Func<DateTime, bool>(DateValidators.
IsWeekend);
```

Note that it is not valid to use the **var** keyword to assign a delegate without wrapping in the **Func** prefix:

```
var futureValidator = DateValidation.IsFuture;
```

This results in the following compiler error:

```
Cannot assign method group to an implicitly - typed variable
```

Taking this knowledge of delegates, proceed to how you can invoke a delegate.

INVOKING A DELEGATE

There are several ways to invoke a delegate. For example, consider the following definition:

```
var futureValidator = new Func<DateTime, bool>(DateValidators.IsFuture);
```

To invoke **futureValidator**, you must pass in a **DateTime** value, and it will return a **bool** value using any of these styles:

- Invoke with the null-coalescing operator:

  ```
  var isFuture1 = futureValidator?.Invoke(new DateTime(2000, 12, 31));
  ```

  This is the preferred and safest approach; you should always check for a null before calling **Invoke**. If there is a chance that a delegate does not point to an object in memory, then you must perform a null reference check before accessing methods and properties. A failure to do so will result in **NullReferenceException** being thrown. This is the runtime's way of warning you that the object is not pointing at anything.

  By using the **null-coalescing operator**, the compiler will add the null check for you. In the code, you explicitly declared **futureValidator**, so here it cannot be null. But what if you had been passed **futureValidator** from another method? How can you be sure that the caller had correctly assigned a reference?

  Delegates have additional rules that make it possible for them to throw **NullReferenceException** when invoked. In the preceding example, **futureValidator** has a single target, but as you will see later, the **multicast** feature of delegates allows multiple methods to be added and removed from a list of target methods. If all target methods are removed (which can happen), the runtime will throw a **NullReferenceException**.

- Direct Invoke

  This is the same as the previous method, but without the safety of the null check. This is not recommended for the same reason; that is, the delegate can throw a **NullReferenceException**:

  ```
  var isFuture1 = futureValidator.Invoke(new DateTime(2000, 12, 31));
  ```

- Without the **Invoke** prefix

  This looks more succinct as you simply call the delegate without the **Invoke** prefix. Again, this is not recommended due to a possible null reference:

```
var isFuture2 = futureValidator(new DateTime(2050, 1, 20));
```

Try assigning and safely invoking a delegate through an exercise by bringing them together.

EXERCISE 3.02: ASSIGNING AND INVOKING DELEGATES

In this exercise, you are going to write a console app showing how a **Func** delegate can be used to extract numeric values. You will create a **Car** class that has **Distance** and **JourneyTime** properties. You will prompt the user to enter the distance traveled yesterday and today, passing this information to a **Comparison** class that is told how to extract values and calculate their differences.

Perform the following steps to do so:

1. Change to the **Chapter03** folder and create a new console app, called **Exercise02**, using the CLI **dotnet** command:

```
source\Chapter03>dotnet new console -o Exercise02
```

2. Open **Chapter03\Exercise02.csproj** and replace the entire file with these settings:

```
<Project Sdk="Microsoft.NET.Sdk">
  <PropertyGroup>
    <OutputType>Exe</OutputType>
    <TargetFramework>net6.0</TargetFramework>
  </PropertyGroup>
</Project>
```

3. Open **Exercise02\Program.cs** and clear the contents.
4. Start by adding a record called **Car**. Include the **System.Globalization** namespace for string parsing. Use the **Chapter03.Exercise02** namespace to keep code separate from the other exercises.

5. Add two properties, **Distance** and **JourneyTime**. They will have **init**-only properties, so you will use the **init** keyword:

```
using System;
using System.Globalization;

namespace Chapter03.Exercise02
{
    public record Car
    {
        public double Distance { get; init; }
        public double JourneyTime { get; init; }
    }
```

6. Next, create a class named **Comparison** that is passed a **Func** delegate to work with. The **Comparison** class will use the delegate to extract either the **Distance** or **JourneyTime** properties and calculate the difference for two **Car** instances. By using the flexibility of delegates, **Comparison** will not know whether it is extracting **Distance** or **JourneyTime**, just that it is using a double to calculate the differences. This shows that you can reuse this class should you need to calculate other **Car** properties in the future:

```
    public class Comparison
    {
        private readonly Func<Car, double> _valueSelector;

        public Comparison(Func<Car, double> valueSelector)
        {
            _valueSelector = valueSelector;
        }
```

7. Add three properties that form the results of the calculation, as follows:

```
        public double Yesterday { get; private set; }
        public double Today { get; private set; }
        public double Difference { get; private set; }
```

8. Now for the calculation, pass two **Car** instances, one for the car journey yesterday, **yesterdayCar**, and one for today, **todayCar**:

```
        public void Compare(Car yesterdayCar, Car todayCar)
        {
```

9. To calculate a value for **Yesterday**, invoke the **valueSelector Func** delegate, passing in the **yesterdayCar** instance. Again, remember that the **Comparison** class is unaware whether it is extracting **Distance** or **JourneyTime**; it just needs to know that when the **delegate** is invoked with a **Car** argument, it will get a double number back:

```
                Yesterday = _valueSelector(yesterdayCar);
```

10. Do the same to extract the value for **Today** by using the same **Func** delegate, but passing in the **todayCar** instance instead:

```
                Today = _valueSelector(todayCar);
```

11. Now it is just a case of calculating the difference between the two extracted numbers; you don't need to use the **Func** delegate to do that:

```
                Difference = Yesterday - Today;
            }
        }
```

12. So, you have a class that knows how to invoke a **Func** delegate to extract a certain **Car** property when it is told how to. Now, you need a class to wrap up the **Comparison** instances. For this, add a class called **JourneyComparer**:

```
    public class JourneyComparer
    {
        public JourneyComparer()
        {
```

13. For the car journey, you need to calculate the difference between the **Yesterday** and **Today Distance** properties. To do so, create a **Comparison** class that is told how to extract a value from a **Car** instance. You may as well use the same name for this **Comparison** class as you will extract a car's **Distance**. Remember that the **Comparison** constructor needs a **Func** delegate that is passed a **Car** instance and returns a double value. You will add **GetCarDistance()** shortly; this will eventually be invoked by passing **Car** instances for yesterday's and today's journeys:

```
            Distance = new Comparison(GetCarDistance);
```

14. Repeat the process as described in the preceding steps for a **JourneyTime Comparison**; this one should be told to use **GetCarJourneyTime()** as follows:

```
            JourneyTime = new Comparison(GetCarJourneyTime);
```

15. Finally, add another **Comparison** property called **AverageSpeed** as follows. You will see shortly that **GetCarAverageSpeed()** is yet another function:

```
            AverageSpeed = new Comparison(GetCarAverageSpeed);
```

16. Now for the **GetCarDistance** and **GetCarJourneyTime** local functions, they are passed a **Car** instance and return either **Distance** or **JourneyTime** accordingly:

```
            static double GetCarDistance(Car car) => car.Distance;
            static double GetCarJourneyTime(Car car) => car.
JourneyTime;
```

17. **GetCarAverageSpeed**, as the name suggests, returns the average speed. Here, you have shown that the **Func** delegate just needs a compatible function; it doesn't matter what it returns as long as it is **double**. The **Comparison** class does not need to know that it is returning a calculated value such as this when it invokes the **Func** delegate:

```
            static double GetCarAverageSpeed(Car car)
              => car.Distance / car.JourneyTime;
        }
```

18. The three **Comparison** properties should be defined like this:

```
        public Comparison Distance { get; }
        public Comparison JourneyTime { get; }
        public Comparison AverageSpeed { get; }
```

19. Now for the main **Compare** method. This will be passed two **Car** instances, one for **yesterday** and one for **today**, and it simply calls **Compare** on the three **Comparison** items passing in the two **Car** instances:

```
        public void Compare(Car yesterday, Car today)
        {
            Distance.Compare(yesterday, today);
            JourneyTime.Compare(yesterday, today);
            AverageSpeed.Compare(yesterday, today);
        }
    }
```

20. You need a console app to enter the miles traveled per day, so add a class called **Program** with a static **Main** entry point:

```
    public class Program
    {
        public static void Main()
        {
```

21. You can randomly assign journey times to save some input, so add a new **Random** instance and the start of a **do-while** loop, as follows:

```
            var random = new Random();
            string input;
            do
            {
```

22. Read for yesterday's distance, as follows:

```
                Console.Write("Yesterday's distance: ");
                input = Console.ReadLine();
                double.TryParse(input, NumberStyles.Any,
                   CultureInfo.CurrentCulture, out var
distanceYesterday);
```

23. You can use the distance to create yesterday's **Car** with a random **JourneyTime**, as follows:

```
                var carYesterday = new Car
                {
                    Distance = distanceYesterday,
                    JourneyTime = random.NextDouble() * 10D
                };
```

24. Do the same for today's distance:

```
                Console.Write("    Today's distance: ");
                input = Console.ReadLine();
                double.TryParse(input, NumberStyles.Any,
                   CultureInfo.CurrentCulture, out var
distanceToday);

                var carToday = new Car
                {
```

```
                        Distance = distanceToday,
                        JourneyTime = random.NextDouble() * 10D
                    };
```

25. Now that you have two **Car** instances populated with values for yesterday and today, you can create the **JourneyComparer** instance and call **Compare**. This will then call **Compare** on your three **Comparison** instances:

```
                    var comparer = new JourneyComparer();
                    comparer.Compare(carYesterday, carToday);
```

26. Now, write the results to the console:

```
                    Console.WriteLine();
                    Console.WriteLine("Journey Details   Distance\tTime\
tAvg Speed");
                    Console.
WriteLine("-------------------------------------------------------");
```

27. Write out yesterday's results:

```
                    Console.Write($"Yesterday         {comparer.Distance.
Yesterday:N0}    \t");
                    Console.WriteLine($"{comparer.JourneyTime.
Yesterday:N0}\t {comparer.AverageSpeed.Yesterday:N0}");
```

28. Write out today's results:

```
                    Console.Write($"Today             {comparer.Distance.
Today:N0}        \t");
                    Console.WriteLine($"{comparer.JourneyTime.Today:N0}\t
{comparer.AverageSpeed.Today:N0}");
```

29. Finally, write the summary values using the **Difference** properties:

```
                    Console.WriteLine(
"=======================================================");
                    Console.Write($"Difference            {comparer.
Distance.Difference:N0}      \t");
                    Console.WriteLine($"{comparer.JourneyTime.
Difference:N0} \t{comparer.AverageSpeed.Difference:N0}");
                   Console.WriteLine(
"=======================================================");
```

30. Finish off the **do-while** loop, exiting if the user enters an empty string:

```
                }
                while (!string.IsNullOrEmpty(input));
            }
        }
    }
```

Running the console and entering distances of **1000** and **900** produces the following results:

```
Yesterday's distance: 1000
    Today's distance: 900

Journey Details   Distance       Time    Avg Speed
-------------------------------------------------
Yesterday          1,000           8       132
Today              900             4       242
=================================================
Difference        100            4       -109
```

The program will run in a loop until you enter a blank value. You will notice a different output as the **JourneyTime** is set using a random value returned by an instance of **Random** class.

> **NOTE**
>
> You can find the code used for this exercise at https://packt.link/EJTtS.

In this exercise, you have seen how a **Func<Car, double>** delegate is used to create general-purpose code that can be easily reused without the need to create extra interfaces or classes.

Now it is time to look at the second important aspect of deletes and their ability to chain multiple target methods together.

MULTICAST DELEGATES

So far, you have invoked delegates that have a single method assigned, typically in the form of a function call. Delegates offer the ability to combine a list of methods that are executed with a single invocation call, using the **multicast** feature. By using the **+=** operator, any number of additional target methods can be added to the target list. Every time the delegate is invoked, each one of the target methods gets invoked too. But what if you decide you want to remove a target method? That is where the **-=** operator is used.

In the following code snippet, you have an **Action<string>** delegate named **logger**. It starts with a single target method, **LogToConsole**. If you were to invoke this delegate, passing in a string, then the **LogToConsole** method will be called once:

```
Action<string> logger = LogToConsole;
logger("1. Calculating bill");
```

If you were to watch the call stack, you would observe these calls:

```
logger("1. Calculating bill")
--> LogToConsole("1. Calculating bill")
```

To add a new target method, you use the **+=** operator. The following statement adds **LogToFile** to the **logger** delegate's invocation list:

```
logger += LogToFile;
```

Now, every time you invoke **logger**, both **LogToConsole** and **LogToFile** will be called. Now invoke **logger** a second time:

```
logger("2. Saving order");
```

The call stack looks like this:

```
logger("2. Saving order")
--> LogToConsole("2. Saving order")
--> LogToFile("2. Saving order")
```

Again, suppose you use **+=** to add a third target method called **LogToDataBase** as follows:

```
logger += LogToDataBase
```

Now invoke it once again:

```
logger("3. Closing order");
```

The call stack looks like this:

```
logger("3. Closing order")
--> LogToConsole("3. Closing order")
--> LogToFile("3. Closing order")
--> LogToDataBase("3. Closing order")
```

However, consider that you may no longer want to include **LogToFile** in the target method list. In such a case, simply use the **-=** operator to remove it, as follows:

```
logger -= LogToFile
```

You can again invoke the delegate as follows:

```
logger("4. Closing customer");
```

And now, the call stack looks like this:

```
logger("4. Closing customer")
--> LogToConsole("4. Closing customer")
--> LogToDataBase("4. Closing customer")
```

As can be seen, this code resulted in just **two** method calls, **LogToConsole** and **LogToDataBase**.

By using delegates in this way, you can decide which target methods get called based on certain criteria at runtime. This allows you to pass this configured delegate into other methods, to be invoked as and when needed.

You have seen that **Console.WriteLine** can be used to write messages to the console window. To create a method that logs to a file (as **LogToFile** does in the preceding example), you need to use the **File** class from the **System.IO** namespace. **File** has many static methods that can be used to read and write files. You will not go into full details about **File** here, but it is worth mentioning the **File.AppendAllText** method, which can be used to create or replace a text file containing a string value, **File.Exists**, which is used to check for the existence of a file, and **File.Delete**, to delete a file.

Now it is time to practice what you have learned through an exercise.

EXERCISE 3.03: INVOKING A MULTICAST DELEGATE

In this exercise, you will use a multicast delegate to create a cash machine that logs details when a user enters their PIN and asks to see their balance. For this, you will create a **CashMachine** class that invokes a configured **logging** delegate, which you can use as a controller class to decide whether messages are sent to the file or to the console.

You will use an **Action<string>** delegate as you do not need any values to return. Using **+=**, you can control which target methods get called when your delegate is invoked by **CashMachine**.

Perform the following steps to do so:

1. Change to the **Chapter03** folder and create a new console app, called **Exercise03**, using the CLI **dotnet** command:

```
source\Chapter03>dotnet new console -o Exercise03
```

2. Open **Chapter03\Exercise03.csproj** and replace the entire file with these settings:

```
<Project Sdk="Microsoft.NET.Sdk">
  <PropertyGroup>
    <OutputType>Exe</OutputType>
    <TargetFramework>net6.0</TargetFramework>
  </PropertyGroup>
</Project>
```

3. Open **Exercise03\Program.cs** and clear the contents.

4. Add a new class called **CashMachine**.

5. Use the **Chapter03.Exercise03** namespace:

```
using System;
using System.IO;
namespace Chapter03.Exercise03
{
    public class CashMachine
    {
        private readonly Action<string> _logger;

        public CashMachine(Action<string> logger)
        {
            _logger = logger;
        }
```

The **CashMachine** constructor is passed the **Action<string>** delegate, which you can assign to a **readonly** class variable called **_logger**.

6. Add a **Log** helper function that checks whether the **_logger** delegate is null before invoking:

```
        private void Log(string message)
            => _logger?.Invoke(message);
```

7. When the **VerifyPin** and **ShowBalance** methods are called, a message should be logged with some details. Create these methods as follows:

```
        public void VerifyPin(string pin)
            => Log($"VerifyPin called: PIN={pin}");

        public void ShowBalance()
            => Log("ShowBalance called: Balance=999");
    }
```

8. Now, add a console app that configures a **logger** delegate that you can pass into a **CashMachine** object. Note that this is a common form of usage: a class that is responsible for deciding how messages are logged by other classes. Use a constant, **OutputFile**, for the filename to be used for file logging, as follows:

```
    public static class Program
    {
        private const string OutputFile = "activity.txt";

        public static void Main()
        {
```

9. Each time the program runs, it should start with a **clean text file** for logging, so use **File.Delete** to delete the output file:

```
            if (File.Exists(OutputFile))
            {
                File.Delete(OutputFile);
            }
```

10. Create a delegate instance, **logger**, that starts with a single target method, **LogToConsole**:

```
            Action<string> logger = LogToConsole;
```

11. Using the **+=** operator, add **LogToFile** as a second target method to also be called whenever the delegate is invoked by **CashMachine**:

```
            logger += LogToFile;
```

12. You will implement the two target logging methods shortly; for now, create a **cashMachine** instance and get ready to call its methods, as follows:

```
            var cashMachine = new CashMachine(logger);
```

13. Prompt for a **pin** and pass it to the **VerifyPin** method:

```
            Console.Write("Enter your PIN:");
            var pin = Console.ReadLine();
            if (string.IsNullOrEmpty(pin))
            {
                Console.WriteLine("No PIN entered");
                return;
            }
            cashMachine.VerifyPin(pin);
            Console.WriteLine();
```

    In case you enter a blank value, then it is checked and a warning is displayed. This will then close the program using a **return** statement.

14. Wait for the **Enter** key to be pressed before calling the **ShowBalance** method:

```
            Console.Write("Press Enter to show balance");
            Console.ReadLine();

            cashMachine.ShowBalance();
            Console.Write("Press Enter to quit");
            Console.ReadLine();
```

15. Now for the logging methods. They must be compatible with your **Action<string>** delegate. One writes a message to the console and the other appends it to the text file. Add these two static methods as follows:

```
            static void LogToConsole(string message)
                => Console.WriteLine(message);

            static void LogToFile(string message)
                => File.AppendAllText(OutputFile, message);
        }
    }
}
```

16. Running the console app, you see that **VerifyPin** and **ShowBalance** calls are written to the console:

```
Enter your PIN:12345
VerifyPin called: PIN=12345

Press Enter to show balance
ShowBalance called: Balance=999
```

17. For each **logger** delegate invocation, the **LogToFile** method will also be called, so when opening **activity.txt**, you should see the following line:

```
VerifyPin called: PIN=12345ShowBalance called: Balance=999
```

> **NOTE**
>
> You can find the code used for this exercise at https://packt.link/h9vic.

It is important to remember that delegates are **immutable**, so each time you use the **+=** or **-=** operators, you create a **new** delegate instance. This means that if you alter a delegate after you have passed it to a target class, you will not see any changes to the methods called from inside that target class.

You can see this in action in the following example:

MulticastDelegatesAddRemoveExample.cs

```
using System;
namespace Chapter03Examples
{
    class MulticastDelegatesAddRemoveExample
    {
        public static void Main()
        {
            Action<string> logger = LogToConsole;
            Console.WriteLine($"Logger1 #={logger.GetHashCode()}");

            logger += LogToConsole;
            Console.WriteLine($"Logger2 #={logger.GetHashCode()}");

            logger += LogToConsole;
            Console.WriteLine($"Logger3 #={logger.GetHashCode()}");
```

You can find the complete code here: https://packt.link/vqZMF.

All objects in C# have a **GetHashCode()** function that returns a unique ID. Running the code produces this output:

```
Logger1 #=46104728
Logger2 #=1567560752
Logger3 #=236001992
```

You can see that the **hashcode** is changing after each **+=** call. This shows that the object reference is changing each time.

Now look at another example using an **Action<string>** delegate. Here, you will use the **+=** operator to add target methods and then use **-=** to remove the target methods:

MulticastDelegatesExample.cs

```
using System;
namespace Chapter03Examples
{
    class MulticastDelegatesExample
    {
        public static void Main()
        {
            Action<string> logger = LogToConsole;
            logger += LogToConsole;
            logger("Console x 2");

            logger -= LogToConsole;
            logger("Console x 1");

            logger -= LogToConsole;
```

You can find the complete code here: https://packt.link/Xe0Ct.

You start with one target method, **LogToConsole**, and then add the same target method a second time. Invoking the logger delegate using **logger("Console x 2")** results in **LogToConsole** being called twice.

You then use **-=** to remove **LogToConsole twice** such that had two targets and now you do not have any at all. Running the code produces the following output:

```
Console x 2
Console x 2
Console x 1
```

However, rather than **logger("logger is now null")** running correctly, you end up with an **unhandled exception** being thrown like so:

```
System.NullReferenceException
  HResult=0x80004003
  Message=Object reference not set to an instance of an object.
  Source=Examples
  StackTrace:
   at Chapter03Examples.MulticastDelegatesExample.Main() in Chapter03\
MulticastDelegatesExample.cs:line 16
```

By removing the last target method, the **-=** operator returned a null reference, which you then assigned to the logger. As you can see, it is important to always check that a delegate is not null before trying to invoke it.

MULTICASTING WITH A FUNC DELEGATE

So far, you have used **Action<string>** delegates within **multicast** scenarios. When invoked, a string value is passed to any target method. As the target methods do not return a value, you use **Action** delegates.

You have seen that **Func** delegates are used when a return value is required from an invoked delegate. It is also perfectly legal for the C# complier to use **Func** delegates in multicast delegates.

Consider the following example where you have a **Func<string, string>** delegate. This delegate supports functions that are passed a string and return a formatted string is returned. This could be used when you need to format an email address by removing the `@` sign and dot symbols:

```
using System;
namespace Chapter03Examples
{
    class FuncExample
    {
        public static void Main()
        {
```

You start by assigning the **RemoveDots** string function to **emailFormatter** and invoke it using the **Address** constant:

```
            Func<string, string> emailFormatter = RemoveDots;
            const string Address = "admin@google.com";
            var first = emailFormatter(Address);
            Console.WriteLine($"First={first}");
```

Then you add a second target, **RemoveAtSign**, and invoke **emailFormatter** a second time:

```
            emailFormatter += RemoveAtSign;
            var second = emailFormatter(Address);
            Console.WriteLine($"Second={second}");

            Console.ReadLine();
            static string RemoveAtSign(string address)
                => address.Replace("@", "");

            static string RemoveDots(string address)
                => address.Replace(".", "");
        }
    }
}
```

Running the code produces this output:

```
First=admin@googlecom
Second=admingoogle.com
```

The first invocation returns the **admin@googlecom** string. The **dot** symbol has been removed, but the next invocation, with **RemoveAtSign** added to the target list, returns a value with only the @ symbol removed.

> **NOTE**
>
> You can find the code used for this example at https://packt.link/fshse.

Both **Func1** and **Func2** are invoked, but only the value from **Func2** is returned to both **ResultA** and **ResultB** variables, even though the correct arguments are passed in. When a **Func<>** delegate is used with multicast in this manner, all of the target **Func** instances are called, but the return value will be that of the last **Func<>** in the chain. **Func<>** is better suited in a single method scenario, although the compiler will still allow you to use it as a multicast delegate without any compilation error or warning.

WHAT HAPPENS WHEN THINGS GO WRONG?

When a delegate is invoked, all methods in the invocation list are called. In the case of single-name delegates, this will be one target method. What happens in the case of multicast delegates if one of those targets throws an exception?

Consider the following code. When the **logger** delegate is invoked, by passing in **try log this**, you may expect the methods to be called in the order that they were added: **LogToConsole**, **LogToError**, and finally **LogToDebug**:

MulticastWithErrorsExample.cs

```
using System;
using System.Diagnostics;
namespace Chapter03Examples
{
    class MulticastWithErrorsExample
    {
            public static void Main()
            {
                Action<string> logger = LogToConsole;
                logger += LogToError;
                logger += LogToDebug;

                try
                {
                    logger("try log this");
```

You can find the complete code here: https://packt.link/Ti3Nh.

If any target method throws an exception, such as the one you see in **LogToError**, then the remaining targets are **not** called.

Running the code results in the following output:

```
Console: try log this
Caught oops!
All done
```

You will see this output because the **LogToDebug** method wasn't called at all. Consider a UI with multiple targets listening to a mouse button click. The first method fires when a button is pressed and disables the button to prevent double-clicks, the second method changes the button's image to indicate success, and the third method enables the button.

If the second method fails, then the third method will not get called, and the button could remain in a disabled state with an incorrect image assigned, thereby confusing the user.

To ensure that all target methods are run regardless, you can enumerate through the invocation list and invoke each method manually. Take a look at the .NET **MulticastDelegate** type. You will find that there is a function, **GetInvocationList**, that returns an array of the delegate objects. This array contains the target methods that have been added:

```
public abstract class MulticastDelegate : Delegate
{
  public sealed override Delegate[] GetInvocationList();
}
```

You can then loop through those target methods and execute each one inside a **try**/**catch** block. Now practice what you learned through this exercise.

EXERCISE 3.04: ENSURING ALL TARGET METHODS ARE INVOKED IN A MULTICAST DELEGATE

Throughout this chapter, you have been using **Action<string>** delegates to perform various logging operations. In this exercise, you have a list of target methods for a logging delegate and you want to ensure that "all" target methods are invoked even if earlier ones fail. You may have a scenario where logging to a database or filesystem fails occasionally, maybe due to network issues. In such a situation, you will want other logging operations to at least have a chance to perform their logging activity.

Perform the following steps to do so:

1. Change to the **Chapter03** folder and create a new console app, called **Exercise04**, using the CLI **dotnet** command:

```
source\Chapter03>dotnet new console -o Exercise04
```

2. Open **Chapter03\Exercise04.csproj** and replace the entire file with these settings:

```
<Project Sdk="Microsoft.NET.Sdk">
  <PropertyGroup>
    <OutputType>Exe</OutputType>
    <TargetFramework>net6.0</TargetFramework>
  </PropertyGroup>
</Project>
```

3. Open **Exercise04\Program.cs** and clear the contents.

4. Now add a static **Program** class for your console app, including **System** and, additionally, **System.IO** as you want to create a file:

```
using System;
using System.IO;

namespace Chapter03.Exercise04
{
    public static class Program
    {
```

5. Use a **const** to name the logging file. This file is created when the program executes:

```
        private const string OutputFile = "Exercise04.txt";
```

6. Now you must define the app's **Main** entry point. Here you delete the output file if it already exists. It is best to start with an empty file here, as otherwise, the log file will keep growing every time you run the app:

```
        public static void Main()
        {
            if (File.Exists(OutputFile))
            {
                File.Delete(OutputFile);
            }
```

7. You will start with **logger** having just one target method, **LogToConsole**, which you will add shortly:

```
            Action<string> logger = LogToConsole;
```

8. You use the **InvokeAll** method to invoke the delegate, passing in **"First call"** as an argument. This will not fail as **logger** has a single valid method and you will add **InvokeAll** shortly, too:

```
InvokeAll(logger, "First call");
```

9. The aim of this exercise is to have a multicast delegate, so add some additional target methods:

```
logger += LogToConsole;
logger += LogToDatabase;
logger += LogToFile;
```

10. Try a second call using **InvokeAll** as follows:

```
InvokeAll(logger, "Second call");

Console.ReadLine();
```

11. Now for the target methods that were added to the delegate. Add the following code for this:

```
static void LogToConsole(string message)
    => Console.WriteLine($"LogToConsole: {message}");

static void LogToDatabase(string message)
    => throw new ApplicationException("bad thing 
happened!");

static void LogToFile(string message)
    => File.AppendAllText(OutputFile, message);
```

12. You can now implement the **InvokeAll** method:

```
static void InvokeAll(Action<string> logger, string arg)
{
    if (logger == null)
         return;
```

It is passed an **Action<string>** delegate that matches the **logger** delegate type, along with an **arg** string to use when invoking each target method. Before that though, it is important to check that **logger** is not already null and there is nothing you can do with a null delegate.

13. Use the delegate's **`GetInvocationList()`** method to get a list of all the target methods:

```
                var delegateList = logger.GetInvocationList();
                Console.WriteLine($"Found {delegateList.Length} items
in {logger}");
```

14. Now, loop through each item in the list as follows:

```
                foreach (var del in delegateList)
                {
```

15. After wrapping each loop element in a **`try/catch`**, cast **`del`** into an **`Action<string>`**:

```
                    try
                    {
                      var action = del as Action<string>;
```

    `GetInvocationList` returns each item as the base delegate type regardless of their actual type.

16. If it is the correct type and **not** null, then it is safe to try invoking:

```
                        if (del is Action<string> action)
                        {
                            Console.WriteLine($"Invoking '{action.
Method.Name}' with '{arg}'");
                            action(arg);
                        }
                        else
                        {
                            Console.WriteLine("Skipped null");
                        }
```

    You have added some extra details to show what is about to be invoked by using the delegate's **`Method.Name`** property.

17. Finish with a **catch** block that logs the error message if an error was caught:

```
                    }
                    catch (Exception e)
                    {
                        Console.WriteLine($"Error: {e.Message}");
                    }
                }
            }
        }
    }
}
```

18. Running the code, creates a file called **Exercise04.txt** with the following results:

```
Found 1 items in System.Action`1[System.String]
Invoking '<Main>g__LogToConsole|1_0' with 'First call'
LogToConsole: First call
Found 4 items in System.Action`1[System.String]
Invoking '<Main>g__LogToConsole|1_0' with 'Second call'
LogToConsole: Second call
Invoking '<Main>g__LogToConsole|1_0' with 'Second call'
LogToConsole: Second call
Invoking '<Main>g__LogToDatabase|1_1' with 'Second call'
Error: bad thing happened!
Invoking '<Main>g__LogToFile|1_2' with 'Second call'
```

You will see that it catches the error thrown by **LogToDatabase** and still allows **LogToFile** to be called.

> **NOTE**
>
> You can find the code used for this exercise at https://packt.link/Dp5H4.

It is now important to expand upon the multicast concept using events.

EVENTS

In the previous sections, you have created delegates and invoked them directly in the same method or passed them to another method for it to invoke when needed. By using delegates in this way, you have a simple way for code to be notified when something of interest happens. So far, this has not been a major problem, but you may have noticed that there appears to be no way to prevent an object that has access to a delegate from invoking it directly.

Consider the following scenario: you have created an application that allows other programs to register for notifications when a new email arrives by adding their target method to a delegate that you have provided. What if a program, either by mistake or for malicious reasons, decides to invoke your delegate itself? This could quite easily overwhelm all the target methods in your invocation list. Such **listener** programs should never be allowed to invoke a delegate in this way—after all, they are meant to be passive listeners.

You could add extra methods that allow listeners to add or remove their target methods from the invocation list and shield the delegate from direct access, but what if you have hundreds of such delegates available in an application? That is a great deal of code to write.

The `event` keyword instructs the C# complier to add extra code to ensure that a delegate can **only** be invoked by the class or struct that it is declared in. External code can add or remove target methods but is prevented from invoking the delegate. Attempting to do so results in a compiler error.

This pattern is commonly known as the pub-sub pattern. The object raising an event is called the event **sender** or **publisher**; the object(s) receiving the event are called **event handlers** or **subscribers**.

DEFINING AN EVENT

The `event` keyword is used to define an event and its associated delegates. Its definition looks similar to the way delegates are defined, but unlike delegates, you cannot use the global namespace to define events:

```
public event EventHandler MouseDoubleClicked
```

Events have four elements:

- Scope: An access modifier, such as **public**, **private**, or **protected**, to define the scope.
- The **event** keyword.
- Delegate type: The associated delegate, **EventHandler** in this example.
- Event name: This can be anything you like, **MouseDoubleClicked**, for example. However, the name must be unique within the namespace.

Events are typically associated with the inbuilt .NET delegates, **EventHandler**, or its generic **EventHandler<>** version. It is rare to create custom delegates for events, but you may find this in older legacy code created prior to the **Action** and generic **Action<T>** delegates.

The **EventHandler** delegate was available in early versions of .NET. It has the following signature, taking a sender **object** and an **EventArgs** parameter:

```
public delegate void EventHandler(object sender, EventArgs e);
```

The more recent generic-based **EventHandler<T>** delegate looks similar; it also takes a sender **object** and a parameter defined by the type **T**:

```
public delegate void EventHandler<T>(object sender, T e);
```

The **sender** parameter is defined as **object**, allowing any type of object to be sent to subscribers for them to identify the **sender** of the event. This can be useful in a situation where you have a centralized method that needs to work on various types of objects rather than specific instances.

For example, in a UI app, you may have one subscriber that listens for an OK button being clicked, and a second subscriber that listens for a **Cancel** button being clicked–each of these could be handled by two separate methods. In the case of multiple checkboxes used to toggle options on or off, you could use a single target method that simply needs to be told that a checkbox is the sender, and to toggle the setting accordingly. This allows you to reuse the same checkbox handler rather than creating a method for every checkbox on a screen.

It is not mandatory to include details of the sender when invoking an **EventHandler** delegate. Often, you may not want to divulge the inner workings of your code to the outside; in this case, it is common practice to pass a null reference to the delegate.

The second argument in both delegates can be used to provide extra contextual information about the event (for example, was it the left or right mouse button that was pressed?). Traditionally, this extra information was wrapped up using a class derived from **EventArgs**, but that convention has been relaxed in newer .NET versions.

There are two standard .NET delegates you should for your event definition?

- **EventHandler**: This can be used when there is no extra information to describe the event. For example, a checkbox click event may not need any extra information, it was simply clicked. In this case, it is perfectly valid to pass null or **EventArgs.Empty** as the second parameter. This delegate can often be found in legacy apps that use a class derived from **EventArgs** to describe the event further. Was it a double-click of the mouse that triggered this event? In this case, a **Clicks** property may have been added to an **EventArgs** derived class to provide such extra details.
- **EventHandler<T>**: Since the inclusion of generics in C#, this has become the more frequently used delegate for events, simply because using generics requires fewer classes to be created.

Interestingly, no matter what scope you give to your event (**public**, for example), the C# compiler will internally create a private member with that name. This is the key concept with events: only the class that defines the event may **invoke it**. Consumers are free to add or remove their interest, but they **cannot** invoke it themselves.

When an event is defined, the publisher class in which it is defined can simply invoke it as and when needed, in the same way that you invoke delegates. In the earlier examples, a point was made of always checking that the delegate is not null before invoking. The same approach should be taken with events, as you have little control over how or when a subscriber may add or remove their target methods.

When a publisher class is initially created, all events have an initial value of null. This will change to not null when any subscriber adds a target method. Conversely, as soon as a subscriber removes a target method, the event will revert to null if there are no methods left in the invocation list and all this is handled by the runtime. This is the standard behavior you saw earlier with delegates.

You can prevent an event from ever becoming null by adding an empty delegate to the end of the event definition:

```
public event EventHandler<MouseEventArgs> MouseDoubleClicked = delegate
{};
```

Rather than having the default null value, you are adding your own default delegate instance—one that does nothing. Hence the blank between the **{}** symbols.

There is a common pattern often followed when using events within a publisher class, particularly in classes that may be subclassed further. You will now see this with the help of a simple example:

1. Define a class, **MouseClickedEventArgs**, that contains additional information about the event, in this case, the number of mouse clicks that were detected:

```
using System;
namespace Chapter03Examples
{
    public class MouseClickedEventArgs
    {
        public MouseClickedEventArgs(int clicks)
        {
            Clicks = clicks;
        }

        public int Clicks { get; }
    }
```

   Observe the **MouseClickPublisher** class, This has a **MouseClicked** event defined using the generic **EventHandler<>** delegate.

2. Now add the **delegate { };** block to prevent **MouseClicked** from being null initially:

```
    public class MouseClickPublisher
    {
     public event EventHandler<MouseClickedEventArgs> MouseClicked =
delegate { };
```

3. Add an **OnMouseClicked** virtual method that gives any further subclassed **MouseClickPublisher** classes a chance to suppress or change the event notification, as follows:

```
        protected virtual void OnMouseClicked( MouseClickedEventArgs 
e)
        {
            var evt = MouseClicked;
            evt?.Invoke(this, e);
        }
```

4. Now you need a method that tracks the mouse clicks. In this example, you will not actually show how mouse clicks are detected, but you will call **OnMouseClicked**, passing in **2** to indicate a double-click.
5. Notice how you have not invoked the **MouseClicked** event directly; you always go via the **OnMouseClicked** intermediary method. This provides a way for other implementations of **MouseClickPublisher** to override the event notification if they need to:

```
        private void TrackMouseClicks()
        {
            OnMouseClicked(new MouseClickedEventArgs(2));
        }
    }
```

6. Now add a new type of publisher that is based on **MouseClickPublisher**:

```
    public class MouseSingleClickPublisher : MouseClickPublisher
    {
        protected override void OnMouseClicked(MouseClickedEventArgs 
e)
        {
            if (e.Clicks == 1)
            {
                OnMouseClicked(e);
            }
        }
    }
}
```

This **MouseSingleClickPublisher** overrides the **OnMouseClicked** method and only calls the base **OnMouseClicked** if a single click was detected. By implementing this type of pattern, you allow different types of publishers to control whether events are fired to subscribers in a customized manner.

> **NOTE**
>
> You can find the code used for this example at https://packt.link/J1EiB.

You can now practice what you learned through the following exercise.

EXERCISE 3.05: PUBLISHING AND SUBSCRIBING TO EVENTS

In this exercise, you will create an alarm clock as an example of a publisher. The alarm clock will simulate a **tick** every minute and publish a **Ticked** event. You will also add a **WakeUp** event that is published when the current time matches an alarm time. In .NET, **DateTime** is used to represent a point in time, so you will use that for the current time and alarm time properties. You will use **DateTime.Subtract** to get the difference between the current time and the alarm time and publish the **WakeUp** event when it is due.

Perform the following steps to do so:

1. Change to the **Chapter03** folder and create a new console app, called **Exercise05**, using the CLI **dotnet** command:

```
dotnet new console -o Exercise05
```

2. Open **Chapter03\Exercise05.csproj** and replace the entire file with these settings:

```
<Project Sdk="Microsoft.NET.Sdk">
  <PropertyGroup>
    <OutputType>Exe</OutputType>
    <TargetFramework>net6.0</TargetFramework>
  </PropertyGroup>
</Project>
```

3. Open **Exercise05\Program.cs** and clear the contents.

4. Add a new class called **AlarmClock**. Here you need to use a **DateTime** class, so include the **System** namespace:

```
using System;
namespace Chapter03.Exercise05
{
    public class AlarmClock
    {
```

   You will offer two events for subscribers to listen to—**WakeUp**, based on the non-generic **EventHandler** delegate (since you will not pass any extra information in this event), and **Ticked**, which uses the generic **EventHandler** delegate with a **DateTime** parameter type.

5. You will use this to pass along the current time to display in the console. Notice that both have the initial **delegate {};** safety mechanism:

```
        public event EventHandler WakeUp = delegate {};
        public event EventHandler<DateTime> Ticked = delegate {};
```

6. Include an **OnWakeUp** override as an example, but do not do the same with **Ticked**; this is to show the different invocation approaches:

```
        protected void OnWakeUp()
        {
            WakeUp.Invoke(this, EventArgs.Empty);
        }
```

7. Now add two **DateTime** properties, the alarm and clock times, as follows:

```
        public DateTime AlarmTime { get; set; }
        public DateTime ClockTime { get; set; }
```

8. A **Start** method is used to start the clock. You simulate a clock ticking once every minute for **24 hours** using a simple loop as follows:

```
        public void Start()
        {
            // Run for 24 hours
            const int MinutesInADay = 60 * 24;
```

9. For each simulated minute, increment the clock using **DateTime.AddMinute** and publish the **Ticked** event, passing in **this** (the **AlarmClock** sender instance) and the clock time:

```
            for (var i = 0; i < MinutesInADay; i++)
            {
                ClockTime = ClockTime.AddMinutes(1);
                Ticked.Invoke(this, ClockTime);
```

   ClockTime.Subtract is used to calculate the difference between the click and alarm times.

10. You pass the **timeRemaining** value to the local function, **IsTimeToWakeUp**, calling the **OnWakeUp** method and break out of the loop if it is time to wake up:

```
              var timeRemaining = ClockTime
                 .Subtract(AlarmTime)
                 .TotalMinutes;

               if (IsTimeToWakeUp(timeRemaining))
                {
                    OnWakeUp();
                    break;
                }
            }
```

11. Use the **IsTimeToWakeUp**, a relational pattern, to see whether there is less than one minute remaining. Add the following code for this:

```
            static bool IsTimeToWakeUp(double timeRemaining)
                => timeRemaining is (>= -1.0 and <= 1.0);
        }
    }
```

12. Now add a console app that subscribes to the alarm clock and its two events by starting from the static void **Main** entry point:

```
         public static class Program
    {
        public static void Main()
        {
```

13. Create the **AlarmClock** instance and use the **+=** operator to subscribe to the **Ticked** event and the **WakeUp** events. You will define **ClockTicked** and **ClockWakeUp** shortly. For now, just add the following code:

```
            var clock = new AlarmClock();
            clock.Ticked += ClockTicked;
            clock.WakeUp += ClockWakeUp;
```

14. Set the clock's current time, use **DateTime.AddMinutes** to add **120** minutes to the alarm time, and then start the clock, as follows:

```
            clock.ClockTime = DateTime.Now;
            clock.AlarmTime = DateTime.Now.AddMinutes(120);

            Console.WriteLine($"ClockTime={clock.ClockTime:t}");
            Console.WriteLine($"AlarmTime={clock.AlarmTime:t}");
            clock.Start();
```

15. Finish off **Main** by prompting for the **Enter** key to be pressed:

```
            Console.WriteLine("Press ENTER");
            Console.ReadLine();
```

16. Now you can add the event subscriber local methods:

```
            static void ClockWakeUp(object sender, EventArgs e)
            {
               Console.WriteLine();
               Console.WriteLine("Wake up");
            }
```

ClockWakeUp is passed sender and **EventArgs** arguments. You don't use either of these, but they are required for the **EventHandler** delegate. When this subscriber's method is called, you write **"Wake up"** to the console.

17. **ClockTicked** is passed the **DateTime** argument as required by the **EventHandler<DateTime>** delegate. Here, you pass the current time, so you write that to the console using **:t** to show the time in a short format:

```
             static void ClockTicked(object sender, DateTime e)
               => Console.Write($"{e:t}...");
        }
    }
}
```

18. Running the app produces this output:

```
ClockTime=14:59
AlarmTime=16:59
15:00...15:01...15:02...15:03...15:04...15:05...15:06...15:07...
15:08...15:09...15:10...15:11...15:12...15:13...15:14...15:15...
15:16...15:17...15:18...15:19...15:20...15:21...15:22...15:23...
15:24...15:25...15:26...15:27...15:28...15:29...15:30...15:31...
15:32...15:33...15:34...15:35...15:36...15:37...15:38...15:39...
15:40...15:41...15:42...15:43...15:44...15:45...15:46...15:47...
15:48...15:49...15:50...15:51...15:52...15:53...15:54...15:55...
15:56...15:57...15:58...15:59...16:00...16:01...16:02...16:03...
16:04...16:05...16:06...16:07...16:08...16:09...16:10...16:11...
16:12...16:13...16:14...16:15...16:16...16:17...16:18...16:19...
16:20...16:21...16:22...16:23...16:24...16:25...16:26...16:27...
16:28...16:29...16:30...16:31...16:32...16:33...16:34...16:35...
16:36...16:37...16:38...16:39...16:40...16:41...16:42...16:43...
16:44...16:45...16:46...16:47...16:48...16:49...16:50...16:51...
16:52...16:53...16:54...16:55...16:56...16:57...16:58...16:59...
Wake up
Press ENTER
```

In this example you see that the alarm clock simulates a tick every minute and publishes a **`Ticked`** event.

> **NOTE**
>
> You can find the code used for this exercise at https://packt.link/GPkYQ.

Now it is time to grasp the difference between events and delegates.

EVENTS OR DELEGATES?

On the face of it, events and delegates look remarkably similar:

- Events are an extended form of delegates.
- Both offer **late-bound** semantics, so rather than calling methods that are known precisely at compile-time, you can defer a list of target methods when known at runtime.
- Both are **called** using **`Invoke()`** or, more simply, the **`()`** suffix shortcut, ideally with a null check before doing so.

The key considerations are as follows:

- Optionality: Events offer an optional approach; callers can decide to opt into events or not. If your component can complete its task without needing any subscriber methods, then it is preferable to use an event-based approach.
- Return types: Do you need to handle return types? Delegates associated with events are always void.
- Lifetime: Event subscribers typically have a shorter lifetime than their publishers, leaving the publisher to continue detecting new messages even if there are no active subscribers.

STATIC EVENTS CAN CAUSE MEMORY LEAKS

Before you wrap up your look at events, it pays to be **careful** when using events, particularly those that are statically defined.

Whenever you add a subscriber's target method to a publisher's event, the publisher class will store a reference to your target method. When you have finished using a subscriber instance and it remains attached to a `static` publisher, it is possible that the memory used by your subscriber will not be cleared up.

These are often referred to as **orphaned**, **phantom**, or **ghost** events. To prevent this, always try to pair up each **+=** call with a corresponding **-=** operator.

> **NOTE**
>
> **Reactive Extensions** (**Rx**) (https://github.com/dotnet/reactive) is a great library for leveraging and taming event-based and asynchronous programming using LINQ-style operators. Rx provides a way to time-shift, for example, buffering a very chatty event into manageable streams with just a few lines of code. What's more, Rx streams are very easy to unit test, allowing you to effectively take control of time.

Now read about the interesting topic of lambda expressions.

LAMBDA EXPRESSIONS

Throughout the previous sections, you have mainly used class-level methods as targets for your delegates and events, such as the **ClockTicked** and **ClockWakeUp** methods, that were also used in *Exercise 3.05*:

```
var clock = new AlarmClock();
clock.Ticked += ClockTicked;
clock.WakeUp += ClockWakeUp;

static void ClockTicked(object sender, DateTime e)
  => Console.Write($"{e:t}...");

static void ClockWakeUp(object sender, EventArgs e)
{
    Console.WriteLine();
    Console.WriteLine("Wake up");
}
```

The **ClockWakeUp** and **ClockTicked** methods are easy to follow and step through. However, by converting them into lambda expression syntax, you can have a more succinct syntax and closer proximity to where they are in code.

Now convert the **Ticked** and **WakeUp** events to use two different lambda expressions:

```
clock.Ticked += (sender, e) =>
{
    Console.Write($"{e:t}...");
};

clock.WakeUp += (sender, e) =>
{
    Console.WriteLine();
    Console.WriteLine("Wake up");
};
```

You have used the same **+=** operator, but instead of method names, you see **(sender, e) =>** and identical blocks of code, as seen in **ClockTicked** and **ClockWakeUp**.

When defining a lambda expression, you can pass any parameters within parentheses, **()**, followed by **=>** (this is often read as **goes to**), and then by your expression/statement block:

```
(parameters) => expression-or-block
```

The code block can be as complex as you need and can return a value if it is a **Func**-based delegate.

The compiler can normally infer each of the parameter types, so you do not even need to specify their types. Moreover, you can omit the parentheses if there is only one argument and the compiler can infer its type.

Wherever a delegate (remember that **Action**, **Action<T>**, and **Func<T>** are inbuilt examples of a delegate) needs to be used as an argument, rather than creating a class or local method or function, you should consider using a lambda expression. The main reason is that this often results in less code, and that code is placed closer to the location where it is used.

Now consider another example on Lambda. Given a list of movies, you can use the **List<string>** class to store these string-based names, as shown in the following snippet:

```
using System;
using System.Collections.Generic;
namespace Chapter03Examples
{
    class LambdaExample
    {
        public static void Main()
        {
            var names = new List<string>
            {
                "The A-Team",
                "Blade Runner",
                "There's Something About Mary",
                "Batman Begins",
                "The Crow"
            };
```

You can use the **List.Sort** method to sort the names alphabetically (the final output will be shown at the end of this example):

```
names.Sort();
Console.WriteLine("Sorted names:");
foreach (var name in names)
{
    Console.WriteLine(name);
}
Console.WriteLine();
```

If you need more control over how this sort works, the **List** class has another **Sort** method that accepts a delegate of this form: **delegate int Comparison<T>(T x, T y)**. This delegate is passed two arguments of the same type (**x** and **y**) and returns an **int** value. The **int** value can be used to define the sort order of items in the list without you having to worry about the internal workings of the **Sort** method.

As an alternative, you can sort the names to exclude **"The"** from the beginning of movie titles. This is often used as an alternative way to list names. You can achieve this by passing a lambda expression, using the **( )** syntax to wrap two strings, **x, y**, that will be passed by **Sort()** when it invokes your lambda.

If **x** or **y** starts with your noise word, **"The"**, then you use the **string.Substring** function to skip the first four characters. **String.Compare** is then used to return a numeric value that compares the resulting string values, as follows:

```
const string Noise = "The ";
names.Sort( (x, y) =>
{
    if (x.StartsWith(Noise))
    {
        x = x.Substring(Noise.Length);
    }

    if (y.StartsWith(Noise))
    {
        y = x.Substring(Noise.Length);
    }

    return string.Compare(x , y);
});
```

You can then write out the sorted results to the console:

```
            Console.WriteLine($"Sorted excluding leading '{Noise}':");
            foreach (var name in names)
            {
                Console.WriteLine(name);
            }
            Console.ReadLine();
        }
    }
}
```

Running the example code produces the following output:

```
Sorted names:
Batman Begins
Blade Runner
The A-Team
The Crow
There's Something About Mary

Sorted excluding leading 'The ':
The A-Team
Batman Begins
Blade Runner
The Crow
There's Something About Mary
```

You can see that the second set of names is sorted with **"The"** is ignored.

> **NOTE**
>
> You can find the code used for this example at http://packt.link/B3NmQ.

To see these lambda statements put into practice, try your hand at the following exercise.

EXERCISE 3.06: USING A STATEMENT LAMBDA TO REVERSE WORDS IN A SENTENCE

In this exercise, you are going to create a utility class that splits the words in a sentence and returns that sentence with the words in reverse order.

Perform the following steps to do so:

1. Change to the **Chapter03** folder and create a new console app, called **Exercise06**, using the CLI **dotnet** command:

```
source\Chapter03>dotnet new console -o Exercise06
```

2. Open **Chapter03\Exercise06.csproj** and replace the entire file with these settings:

```
<Project Sdk="Microsoft.NET.Sdk">
  <PropertyGroup>
    <OutputType>Exe</OutputType>
    <TargetFramework>net6.0</TargetFramework>
  </PropertyGroup>
</Project>
```

3. Open **Exercise02\Program.cs** and clear the contents.

4. Add a new class named **WordUtilities** with a string function called **ReverseWords**. You need to include the **System.Linq** namespace to help with the string operations:

```
using System;
using System.Linq;
namespace Chapter03.Exercise06
{
    public static class WordUtilities
    {
        public static string ReverseWords(string sentence)
        {
```

5. Define a **Func<string, string>** delegate called **swapWords** that takes a string input and returns a string value:

```
            Func<string, string> swapWords =
```

6. You will accept a string input argument named **phrase**:

```
                phrase =>
```

7. Now for the lambda statement body. Use the **string.Split** function to split the **phrase** string into an array of strings using a space as the splitting character:

```
                {
                    const char Delimit = ' ';
                    var words = phrase
                        .Split(Delimit)
                        .Reverse();

                    return string.Join(Delimit, words);
                };
```

String.Reverse reverses the order of strings in the array, before finally joining the reversed words string array in a single string using **string.Join**.

8. You have defined the required **Func**, so invoke it by passing the sentence parameter and returning that as the result:

```
            return swapWords(sentence);
        }
    }
```

9. Now for a console app that prompts for a sentence to be entered, which is passed to **WordUtilities.ReverseWords**, with the result being written to the console:

```
    public static class Program
    {
        public static void Main()
        {
            do
            {
                Console.Write("Enter a sentence:");
                var input = Console.ReadLine();
                if (string.IsNullOrEmpty(input))
                {
                    break;
                }

                var result = WordUtilities.ReverseWords(input);
                Console.WriteLine($"Reversed: {result}")
```

Running the console app produces results output similar to this:

```
Enter a sentence:welcome to c#
Reversed: c# to welcome
Enter a sentence:visual studio by microsoft
Reversed: microsoft by studio visual
```

> **NOTE**
>
> You can find the code used for this exercise at https://packt.link/z12sR.

You will conclude this look at lambdas with some of the less obvious issues that you might not expect to see when running and debugging.

CAPTURES AND CLOSURES

Lambda expressions can **capture** any of the variables or parameters within the method where they are defined. The word capture is used to describe the way that a lambda expression captures or reaches up into the parent method to access any variables or parameters.

To grasp this better, consider the following example. Here you will create a **`Func<int, string>`** called **`joiner`** that joins words together using the **`Enumerable.Repeat`** method. The **`word`** variable (known as an **`Outer Variables`**) is captured inside the body of the **`joiner`** expression:

```
var word = "hello";
Func<int, string> joiner =
    input =>
    {
        return string.Join(",", Enumerable.Repeat(word, input));
    };
Console.WriteLine($"Outer Variables: {joiner(2)}");
```

Running the preceding example produces the following output:

```
Outer Variables: hello,hello
```

You invoked the **joiner** delegate by passing **2** as an argument. At that moment in time, the outer **word** variable has a value of **"hello"**, which is repeated twice.

This confirms that captured variables, from the parent method, were evaluated **only** when **Func** was invoked. Now change the value of **word** from **hello** to **goodbye** and invoke **joiner** once again, passing **3** as the argument:

```
word = "goodbye";
Console.WriteLine($"Outer Variables Part2: {joiner(3)}");
```

Running this example produces the following output:

```
Outer Variables Part2: goodbye,goodbye,goodbye
```

It is worth remembering that it does not matter where in the code you defined **joiner**. You could have changed the value of **word** to any number of strings before or after declaring **joiner**.

Taking captures one step further, if you define a variable with the same name inside a lambda, it will be scoped **locally** to the expression. This time, you have a locally defined variable, **word**, which will have no effect on the outer variable with the same name:

```
Func<int, string> joinerLocal =
    input =>
    {
        var word = "local";
        return string.Join(",", Enumerable.Repeat(word, input));
    };
Console.WriteLine($"JoinerLocal: {joinerLocal(2)}");
Console.WriteLine($"JoinerLocal: word={word}");
```

The preceding example results in the following output. Notice how the outer variable, **word**, remains unchanged from **goodbye**:

```
JoinerLocal: local,local
JoinerLocal: word=goodbye
```

Finally, you will look at the concept of **closures** that is a subtle part of the C# language and often leads to unexpected results.

In the following example, you have a variable, **actions**, that contains a **List** of **Action** delegates. You use a basic **for** loop to add five separate **Action** instances to the list. The lambda expression for each **Action** simply writes that value of **i** from the **for** loop to the console. Finally, the code simply runs through each **Action** in the **actions** list and invokes each one:

```
var actions = new List<Action>();
for (var i = 0; i < 5; i++)
{
    actions.Add( () => Console.WriteLine($"MyAction: i={i}")) ;
}

foreach (var action in actions)
{
    action();
}
```

Running the example produces the following output:

```
MyAction: i=5
MyAction: i=5
MyAction: i=5
MyAction: i=5
MyAction: i=5
```

The reason why **MyAction: i** did not start from **0** is that the value of **i**, when accessed from inside a **Action** delegate, is only evaluated once the **Action** is invoked. By the time each delegate is invoked, the outer loop has already repeated five times over.

> **NOTE**
>
> You can find the code used for this example at https://packt.link/vfOPx.

This is similar to the capture concept you observed, where the outer variables, **i** in this case, are only evaluated when invoked. You used **i** in the **for** loop to add each **Action** to the list, but by the time you invoked each action, **i** had its final value of **5**.

This can often lead to unexpected behavior, especially if you assume that an **incrementing** value for **i** is being used inside each action's loop variable. To ensure that the incrementing value of **i** is used inside each lambda expression, you need to introduce a **new** local variable inside the **for** loop, one that takes a copy of the iterator variable.

In the following code snippet, you have added the **closurei** variable. It looks very subtle, but you now have a more locally scoped variable, which you access from inside the lambda expression, rather than the iterator, **i**:

```
var actionsSafe = new List<Action>();
for (var i = 0; i < 5; i++)
{
    var closurei = i;
    actionsSafe.Add(() => Console.WriteLine($"MyAction: closurei={closurei}"));
}
foreach (var action in actionsSafe)
{
    action();
}
```

Running the example produces the following output. You can see that the incrementing value is used when each **Action** is invoked, rather than the value of **5** that you saw earlier:

```
MyAction: closurei=0
MyAction: closurei=1
MyAction: closurei=2
MyAction: closurei=3
MyAction: closurei=4
```

You have covered the key aspects of delegates and events in event-driven applications. You extended this by using the succinct coding style offered by lambdas, to be notified when events of interest occur.

You will now bring these ideas together into an activity in which you will use some of the inbuilt .NET classes with their own events. You will need to adapt these events to your own format and publish so they can be subscribed to by a console app.

Now it is time to practice all you have learned through the following activity.

Activity 3.01: Creating a Web File Downloader

You plan to investigate patterns in US storm events. To do this, you need to download storm event datasets from online sources for later analysis. The National Oceanic and Atmospheric Administration is one such source of data and can be accessed from https://www1.ncdc.noaa.gov/pub/data/swdi/stormevents/csvfiles.

You are tasked with creating a .NET Core console app that allows a web address to be entered, the contents of which are downloaded to a local disk. To be as user-friendly as possible, the application needs to use events that signal when an invalid address is entered, the progress of a download, and when it completes.

Ideally, you should try to hide the internal implementation that you use to download files, preferring to adapt any events that you use to ones that your caller can subscribe to. This form of adaption is often used to make code more maintainable by hiding internal details from callers.

For this purpose, the **`WebClient`** class in C# can be used for download requests. As with many parts of .NET, this class returns objects that implement the **`IDisposable`** interface. This is a standard interface and it indicates that the object you are using should be wrapped in a **`using`** statement to ensure that any resources or memory are cleaned away for you when you have finished using the object. **`using`** takes this format:

```
using (IDisposable) { statement_block }
```

Finally, the **`WebClient.DownloadFileAsync`** method downloads files in the background. Ideally, you should use a mechanism that allows one part of your code to **wait** for a signal to be **set** once the download has been completed. **`System.Threading.ManualResetEventSlim`** is a class that has **`Set`** and **`Wait`** methods that can help with this type of signaling.

For this activity, you will need to perform the following steps:

1. Add a progress changed **`EventArgs`** class (an example name could be **`DownloadProgressChangedEventArgs`**) that can be used when publishing progress events. This should have **`ProgressPercentage`** and **`BytesReceived`** properties.

2. The **WebClient** class from **System.Net** should be used to download a requested web file. You should create an adapter class (a suggested name is **WebClientAdapter**) that hides your internal usage of **WebClient** from your callers.
3. Your adapter class should provide three events—**DownloadCompleted**, **DownloadProgressChanged**, and **InvalidUrlRequested**—that a caller can subscribe to.
4. The adapter class will need a **DownloadFile** method that calls the **WebClient** class's **DownloadFileAsync** method to start the download request. This requires converting a string-based web address into a **Uniform Resource Identifier** (**URI**) class. The **Uri.TryCreate()** method can create an **absolute address** from the string entered via the console. If the call to **Uri.TryCreate** fails, you should publish the **InvalidUrlRequested** event to indicate this failure.
5. **WebClient** has two events—**DownloadFileCompleted** and **DownloadProgressChanged**. You should subscribe to these two events and republish them using your own similar events.
6. Create a console app that uses an instance of **WebClientAdapter** (as created in *Step 2*) and subscribe to the three events.
7. By subscribing to the **DownloadCompleted** event, you should indicate success in the console.
8. By subscribing to **DownloadProgressChanged**, you should report progress messages to the console showing the **ProgressPercentage** and **BytesReceived** values.
9. By subscribing to the **InvalidUrlRequested** event, you should show a warning on the console using a different console background color.
10. Use a **do** loop that allows the user to repeatedly enter a web address. This address and a temporary destination file path can be passed to **WebClientAdapter.DownloadFile()** until the user enters a blank address to quit.

11. Once you run the console app with various download requests, you should see an output similar to the following:

```
Enter a URL:
https://www1.ncdc.noaa.gov/pub/data/swdi/stormevents/csvfiles/
StormEvents_details-ftp_v1.0_d1950_c20170120.csv.gz
Downloading https://www1.ncdc.noaa.gov/pub/data/swdi/stormevents/
csvfiles/StormEvents_details-ftp_v1.0_d1950_c20170120.csv.gz...
Downloading...73% complete (7,758 bytes)
Downloading...77% complete (8,192 bytes)
Downloading...100% complete (10,597 bytes)
Downloaded to C:\Temp\StormEvents_details-ftp_v1.0_d1950_c20170120.
csv.gz

Enter a URL:
https://www1.ncdc.noaa.gov/pub/data/swdi/stormevents/csvfiles/
StormEvents_details-ftp_v1.0_d1954_c20160223.csv.gz
Downloading https://www1.ncdc.noaa.gov/pub/data/swdi/stormevents/
csvfiles/StormEvents_details-ftp_v1.0_d1954_c20160223.csv.gz...
Downloading...29% complete (7,758 bytes)
Downloading...31% complete (8,192 bytes)
Downloading...54% complete (14,238 bytes)
Downloading...62% complete (16,384 bytes)
Downloading...84% complete (22,238 bytes)
Downloading...93% complete (24,576 bytes)
Downloading...100% complete (26,220 bytes)
Downloaded to C:\Temp\StormEvents_details-ftp_v1.0_d1954_c20160223.
csv.gz
```

By completing this activity, you have seen how to subscribe to events from an existing .NET event-based publisher class (**`WebClient`**), adapting them to your own specification before republishing them in your adapter class (**`WebClientAdapter`**), which were ultimately subscribed to by a console app.

> **NOTE**
>
> The solution to this activity can be found at https://packt.link/qclbF.

SUMMARY

In this chapter, you took an in-depth look at delegates. You created custom delegates and saw how they could be replaced with their modern counterparts, the inbuilt **`Action`** and **`Func`** delegates. By using null reference checks, you discovered the safe way to invoke delegates and how multiple methods can be chained together to form multicast delegates. You extended delegates further to use them with the **`event`** keyword to restrict invocation and followed the preferred pattern when defining and invoking events. Finally, you covered the succinct lambda expression style and saw how bugs can be avoided by recognising the use of captures and closures.

In the next chapter, you will look at LINQ and data structures, the fundamental parts of the C# language.

4

DATA STRUCTURES AND LINQ

OVERVIEW

In this chapter, you will learn about the main collections and their primary usage in C#. You will then see how Language-Integrated Query (LINQ) can be used to query collections in memory using code that is efficient and succinct. By the end of this chapter, you will be well versed in using LINQ for operations such as sorting, filtering, and aggregating data.

INTRODUCTION

Throughout the previous chapters, you have used variables that refer to a single value, such as the **`string`** and **`double`** system types, system **`class`** instances, and your own class instances. .NET has a variety of data structures that can be used to store multiple values. These structures are generally referred to as **collections**. This chapter builds on this concept by introducing collection types from the **`System.Collections.Generic`** namespace.

You can create variables that can store **multiple object references** using collection types. Such collections include lists that resize to accommodate the number of elements and dictionaries that offer access to the elements using a unique key as an **identifier**. For example, you may need to store a list of international dialing codes using the codes as unique identifiers. In this case, you need to be certain that the same dialing code is not added to the collection twice.

These collections are instantiated like any other classes and are used extensively in most applications. Choosing the correct type of collection depends primarily on how you intend to add items and the way you would like to access such items once they are in a collection. The commonly used collection types include **`List`**, **`Set`**, and **`HashSet`**, which you will cover in detail shortly.

LINQ is a technology that offers an expressive and concise syntax for querying objects. Much of the complexities around filtering, sorting, and grouping objects can be removed using the SQL-like language, or if you prefer, a set of extension methods that can be chained together to produce collections that can be enumerated with ease.

DATA STRUCTURES

.NET provides various types of in-built data structures, such as the **`Array`**, **`List`**, and **`Dictionary`** types. At the heart of all data structures are the **`IEnumerable`** and **`ICollection`** interfaces. Classes that implement these interfaces offer a way to enumerate through the individual elements and to manipulate their items. There is rarely a need to create your own classes that derive directly from these interfaces, as all the required functionality is covered by the built-in collection types, but it is worth knowing the key properties as they are heavily used throughout .NET.

The generic version of each collection type requires a single **type parameter**, which defines the type of elements that can be added to a collection, using the standard **`<T>`** syntax of the generic types.

The **`IEnumerable`** interface has a single property, that is, **`IEnumerator<T> GetEnumerator()`**. This property returns a type that provides methods that allow the caller to iterate through the elements in the collection. You do not need to call the **`GetEnumerator()`** method directly, as the compiler will call it whenever you use a **`foreach`** statement, such as **`foreach(var book in books)`**. You will learn more about using this in the upcoming sections.

The **`ICollection`** interface has the following properties:

- **`int Count { get; }`**: Returns the number of items in the collection.
- **`bool IsReadOnly { get; }`**: Indicates if the collection is read-only. Certain collections can be marked as read-only to prevent callers from adding, deleting, or moving elements in the collection. C# will not prevent you from amending the properties of individual items in a read-only collection.
- **`void Add(T item)`**: Adds an item of type **`<T>`** to the collection.
- **`void Clear()`**: Removes all items from the collection.
- **`bool Contains(T item)`**: Returns **`true`** if the collection contains a certain value. Depending on the type of item in the collection, this can be **value-equality**, where an object is similarly based on its members, or **reference-equality**, where the object points to the same memory location.
- **`void CopyTo(T[] array, int arrayIndex)`**: Copies each element from the collection into the target array, starting with the first element at a specified index position. This can be useful if you need to skip a specific number of elements from the beginning of the collection.
- **`bool Remove(T item)`**: Removes the specified item from the collection. If there are multiple occurrences of the instance, then only the first instance is removed. This returns **`true`** if an item was successfully removed.

`IEnumerable` and **`ICollection`** are interfaces that all collections implement:

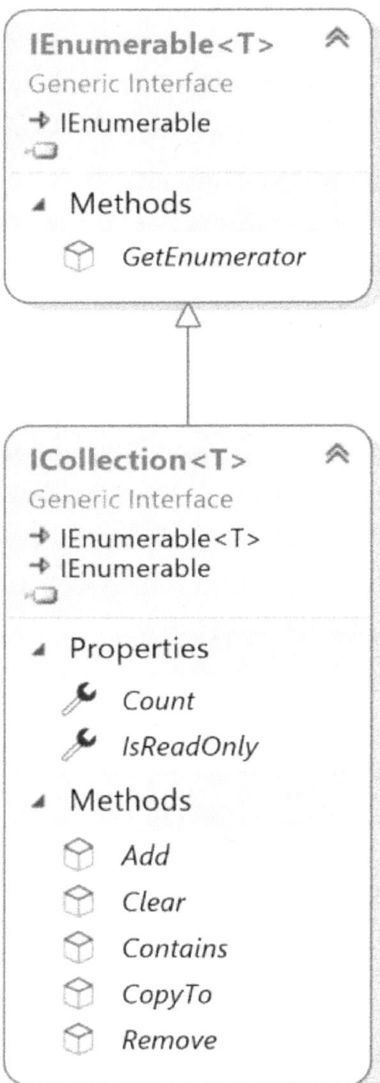

Figure 4.1: ICollection and IEnumerable class diagram

There are further interfaces that some collections implement, depending on how elements are accessed within a collection.

The **`IList`** interface is used for collections that can be accessed by index position, starting from zero. So, for a list that contains two items, **`Red`** and **`Blue`**, the element at index zero is **`Red`** and the element at index one is **`Blue`**.

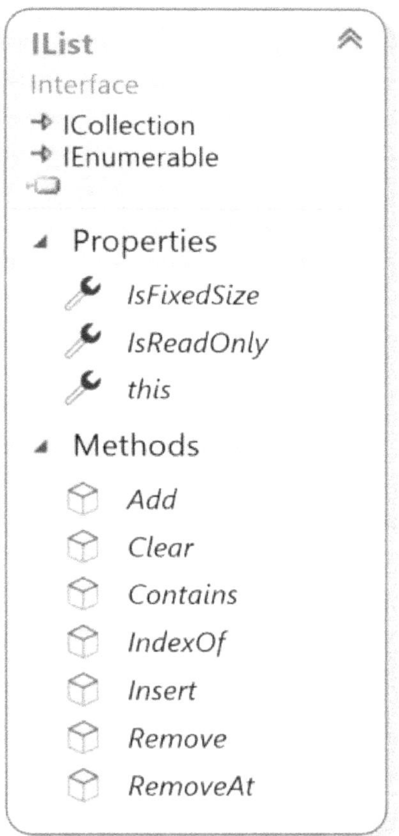

Figure 4.2: IList class diagram

The **IList** interface has the following properties:

- **T this[int index] { get; set; }**: Gets or sets the element at the specified index position.
- **int Add(T item)**: Adds the specified item and returns the index position of that item in the list.
- **void Clear()**: Removes all items from the list.
- **bool Contains(T item)**: Returns **true** if the list contains the specified item.
- **int IndexOf(T item)**: Returns the index position of the item, or **-1** if not found.
- **void Insert(int index, T item)**: Inserts the item at the index position specified.
- **void Remove(T item)**: Removes the item if it exists within the list.
- **void RemoveAt(int index)**: Removes the item at the specified index position.

You have now seen the primary interfaces common to collections. So, now you will now take a look at the main collection types that are available and how they are used.

LISTS

The **List<T>** type is one of the most extensively used collections in C#. It is used where you have a collection of items and want to control the order of items using their index position. It implements the **IList** interface, which allows items to be inserted, accessed, or removed using an index position:

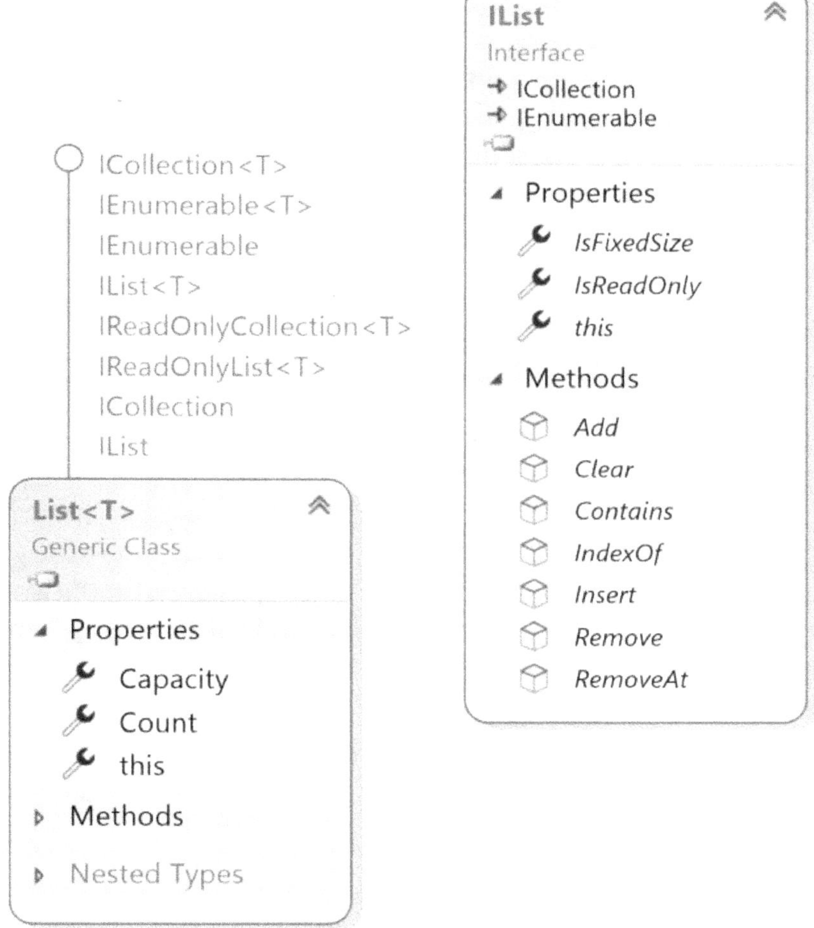

Figure 4.3: List class diagram

Lists have the following behavior:

- Items can be inserted at any position within the collection. Any trailing items will have their index position incremented.
- Items can be removed, either by index or value. This will also cause trailing items to have their index position updated.
- Items can be set using their index value.
- Items can be added to the end of the collection.
- Items can be duplicated within the collection.
- The position of items can be sorted using the various **Sort** methods.

One example of a list might be the tabs in a web browser application. Typically, a user may want to drag a browser tab amongst other tabs, open new tabs at the end, or close tabs anywhere in a list of tabs. The code to control these actions can be implemented using **List**.

Internally, **List** maintains an array to store its objects. This can be efficient when adding items to the end, but it may be inefficient when inserting items, particularly near the beginning of the list, as the index position of items will need to be recalculated.

The following example shows how the generic **List** class is used. The code uses the **List<string>** type parameter, which allows **string** types to be added to the list. Attempts to add any other type will result in a compiler error. This will show the various commonly used methods of the **List** class.

1. Create a new folder called **Chapter04** in your source code folder.
2. Change to the **Chapter04** folder and create a new console app, called **Chapter04**, using the following .NET command:

```
source\Chapter04>dotnet new console -o Chapter04
The template "Console Application" was created successfully.
```

3. Delete the **Class1.cs** file.
4. Add a new folder called **Examples**.
5. Add a new class file called **ListExamples.cs**.

6. Add the **System.Collections.Generic** namespace to access the **List<T>** class and declare a new variable called **colors**:

```
using System;
using System.Collections.Generic;

namespace Chapter04.Examples
{
    class ListExamples
    {
        public static void Main()
        {
            var colors = new List<string> {"red", "green"};
            colors.Add("orange");
```

   The code declares the new **colors** variable, which can store multiple color names as **strings**. Here, the collection initialization syntax is used so that **red** and **green** are added as part of the initialization of the variable. The **Add** method is called, adding **orange** to the list.

7. Similarly, **AddRange** adds **yellow** and **pink** to the end of the list:

```
            colors.AddRange(new [] {"yellow", "pink"});
```

8. At this point, there are five colors in the list, with **red** at index position **0** and **green** at position **1**. You can verify this using the following code:

```
            Console.WriteLine($"Colors has {colors.Count} items");
            Console.WriteLine($"Item at index 1 is {colors[1]}");
```

   Running the code produces the following output:

```
Colors has 5 items
Item at index 1 is green
```

9. Using **Insert**, **blue** can be inserted at the beginning of the list, that is, at index **0**, as shown in the following code. Note that this moves **red** from index **0** to **1** and all other colors will have their index incremented by one:

```
            Console.WriteLine("Inserting blue at 0");
            colors.Insert(0, "blue");
            Console.WriteLine($"Item at index 1 is now {colors[1]}");
```

You should see the following output on running this code:

```
Inserting blue at 0
Item at index 1 is now red
```

10. Using **foreach** you can iterate through the strings in the list, writing each string to the console, as follows:

```
            Console.WriteLine("foreach");
            foreach (var color in colors)
                Console.Write($"{color}|");
            Console.WriteLine();
```

You should get the following output:

```
foreach
blue|red|green|orange|yellow|pink|
```

11. Now, add the following code to reverse the array. Here, each **color** string is converted into an array of **char** type using **ToCharArray**:

```
            Console.WriteLine("ForEach Action:");
            colors.ForEach(color =>
            {
                var characters = color.ToCharArray();
                Array.Reverse(characters);
                var reversed = new string(characters);
                Console.Write($"{reversed}|");
            });
            Console.WriteLine();
```

This does not affect any of the values in the **colors** List, as **characters** refers to a different object. Note that **foreach** iterates through each string, whereas **ForEach** defines an Action delegate to be invoked using each string (recall that in *Chapter 3, Delegates, Events, and Lambdas*, you saw how lambda statements can be used to create **Action** delegates).

12. Running the code leads to this output:

```
ForEach Action:
eulb|der|neerg|egnaro|wolley|knip|
```

13. In the next snippet, the **List** constructor accepts a source collection. This creates a new list containing a copy of the **colors** strings in this case, which is sorted using the default **Sort** implementation:

```
            var backupColors = new List<string>(colors);
            backupColors.Sort();
```

The string type uses **value-type** semantics, which means that the **backupColors** list is populated with a **copy** of each source string value. Updating a string in one list will **not** affect the other list. Conversely, classes are defined as **reference-types** so passing a list of class instances to the constructor will still create a new list, with independent element indexes, but each element will point to the same shared reference in memory rather than an independent copy.

14. In the following snippet, prior to removing all colors (using **colors.Clear**), each value is written to the console (the list will be repopulated shortly):

```
            Console.WriteLine("Foreach before clearing:");
            foreach (var color in colors)
                Console.Write($"{color}|");
            Console.WriteLine();
            colors.Clear();
            Console.WriteLine($"Colors has {colors.Count} items");
```

Running the code produces this output:

```
Foreach before clearing:
blue|red|green|orange|yellow|pink|
Colors has 0 items
```

15. Then, **AddRange** is used again, to add the full list of colors back to the **colors** list, using the sorted **backupColors** items as a source:

```
            colors.AddRange(backupColors);
            Console.WriteLine("foreach after addrange (sorted
items):");
            foreach (var color in colors)
                Console.Write($"{color}|");
            Console.WriteLine();
```

You should see the following output:

```
foreach after addrange (sorted items):
blue|green|orange|pink|red|yellow|
```

16. The **ConvertAll** method is passed a delegate that can be used to return a new list of any type:

```
                var indexes = colors.ConvertAll(color =>
                        $"{color} is at index {colors.IndexOf(color)}");
                Console.WriteLine("ConvertAll:");
                Console.WriteLine(string.Join(Environment.NewLine, 
indexes));
```

    Here, a new **List<string>** is returned with each item being formatted using its value and the item's index in the list. As expected, running the code produces this output:

```
ConvertAll:
blue is at index 0
green is at index 1
orange is at index 2
pink is at index 3
red is at index 4
yellow is at index 5
```

17. In the next snippet, two **Contains()** methods are used to show string value-equality in action:

```
                Console.WriteLine($"Contains RED: {colors.
Contains("RED")}");
                Console.WriteLine($"Contains red: {colors.
Contains("red")}");
```

    Note that the uppercase **RED** is **not** in the list, but the lowercase **red** will be. Running the code produces this output:

```
Contains RED: False
Contains red: True
```

18. Now, add the following snippet:

```
                var existsInk = colors.Exists(color => color.
EndsWith("ink"));
                Console.WriteLine($"Exists *ink: {existsInk}");
```

    Here, the **Exists** method is passed a Predicate delegate, which returns **True** or **False** if the test condition is met. Predicate is an inbuilt delegate, which returns a boolean value. In this case, **True** will be returned if any item exists where the string value ends with the letters **ink** (**pink**, for example).

You should see the following output:

```
Exists *ink: True
```

19. You know there is already a **red** color, but it will be interesting to see what happens if you insert **red** again, twice, at the very beginning of the list:

```
            Console.WriteLine("Inserting reds");
            colors.InsertRange(0, new [] {"red", "red"});
            foreach (var color in colors)
                Console.Write($"{color}|");
            Console.WriteLine();
```

You will get the following output:

```
Inserting reds
red|red|blue|green|orange|pink|red|yellow|
```

This shows that it is possible to insert the same item more than once into a list.

20. The next snippet shows you how to use the **FindAll** method. **FindAll** is similar to the **Exists** method, in that it is passed a **Predicate** condition. All items that match that rule will be returned. Add the following code:

```
            var allReds = colors.FindAll(color => color == "red");
            Console.WriteLine($"Found {allReds.Count} red");
```

You should get an output as follows. As expected, there are three **red** items returned:

```
Found 3 red
```

21. Finishing the example, the **Remove** method is used to remove the first **red** from the list. There are still two **reds** left. You can use **FindLastIndex** to get the index of the last **red** item:

```
            colors.Remove("red");
            var lastRedIndex = colors.FindLastIndex(color => color ==
"red");
            Console.WriteLine($"Last red found at index
{lastRedIndex}");
            Console.ReadLine();
        }
    }
}
```

Running the code produces this output:

```
Last red found at index 5
```

> **NOTE**
>
> You can find the code used for this example at https://packt.link/dLbK6.

With the knowledge of how the generic **List** class is used, it is time for you to work on an exercise.

EXERCISE 4.01: MAINTAINING ORDER WITHIN A LIST

At the beginning of the chapter, web browser tabs were described as an ideal example of lists. In this exercise, you will put this idea into action, and create a class that controls the navigation of the tabs within an app that mimics a web browser.

For this, you will create a **Tab** class and a **TabController** app that allows new tabs to be opened and existing tabs to be closed or moved. The following steps will help you complete this exercise:

1. In VSCode, select your **Chapter04** project.
2. Add a new folder called **Exercises**.
3. Inside the **Exercises** folder, add a folder called **Exercise01** and add a file called **Exercise01.cs**.
4. Open **Exercise01.cs** and define a **Tab** class with a string URL constructor parameter as follows:

```
using System;
using System.Collections;
using System.Collections.Generic;

namespace Chapter04.Exercises.Exercise01
{
    public class Tab
    {
        public Tab()
        {}

        public Tab(string url) => (Url) = (url);
```

```
        public string Url { get; set; }

        public override string ToString() => Url;
    }
```

Here, the **ToString** method has been overridden to return the current URL to help when logging details to the console.

5. Create the **TabController** class as follows:

```
    public class TabController : IEnumerable<Tab>
    {
        private readonly List<Tab> _tabs = new();
```

The **TabController** class contains a List of tabs. Notice how the class inherits from the **IEnumerable** interface. This interface is used so that the class provides a way to iterate through its items, using a **foreach** statement. You will provide methods to open, move, and close tabs, which will directly control the order of items in the **_tabs** list, in the next steps. Note that you could have exposed the **_tabs** list directly to callers, but it would be preferable to limit access to the tabs through your own methods. Hence, it is defined as a **readonly** list.

6. Next, define the **OpenNew** method, which adds a new tab to the end of the list:

```
        public Tab OpenNew(string url)
        {
            var tab = new Tab(url);
            _tabs.Add(tab);
            Console.WriteLine($"OpenNew {tab}");
            return tab;
        }
```

7. Define another method, **Close**, which removes the tab from the list if it exists. Add the following code for this:

```
        public void Close(Tab tab)
        {
            if (_tabs.Remove(tab))
            {
                Console.WriteLine($"Removed {tab}");
            }
        }
```

8. To move a tab to the start of the list, add the following code:

```
        public void MoveToStart(Tab tab)
        {
            if (_tabs.Remove(tab))
            {
                _tabs.Insert(0, tab);
                Console.WriteLine($"Moved {tab} to start");
            }
```

   Here, **MoveToStart** will try to remove the tab and then insert it at index **0**.

9. Similarly, add the following code to move a tab to the end:

```
        public void MoveToEnd(Tab tab)
        {
            if (_tabs.Remove(tab))
            {
                _tabs.Add(tab);
                Console.WriteLine($"Moved {tab} to end. Index={_tabs.
IndexOf(tab)}");
            }
        }
```

   Here, calling **MoveToEnd** removes the tab first, and then adds it to the end, logging the new index position to the console.

   Finally, the **IEnumerable** interface requires that you implement two methods, **IEnumerator<Tab> GetEnumerator()** and **IEnumerable.GetEnumerator()**. These allow the caller to iterate through a collection using either a generic of type **Tab** or using the second method to iterate via an **object-based type**. The second method is a throwback to earlier versions of C# but is needed for compatibility.

10. For the actual results for both methods, you can use the **GetEnumerator** method of the **_tab** list, as that contains the tabs in list form. Add the following code to do so:

```
        public IEnumerator<Tab> GetEnumerator() => _tabs.
GetEnumerator();
        IEnumerator IEnumerable.GetEnumerator() => _tabs.
GetEnumerator();
    }
```

11. You can now create a console app that tests the controller's behavior. Start by opening three new tabs and logging the tab details via **LogTabs** (this will be defined shortly):

```
    static class Program
    {
        public static void Main()
        {
            var controller = new TabController();

            Console.WriteLine("Opening tabs...");
            var packt = controller.OpenNew("packtpub.com");
            var msoft = controller.OpenNew("microsoft.com");
            var amazon = controller.OpenNew("amazon.com");
            controller.LogTabs();
```

12. Now, move **amazon** to the start and **packt** to the end, and log the tab details:

```
            Console.WriteLine("Moving...");
            controller.MoveToStart(amazon);
            controller.MoveToEnd(packt);
            controller.LogTabs();
```

13. Close the **msoft** tab and log details once more:

```
            Console.WriteLine("Closing tab...");
            controller.Close(msoft);
            controller.LogTabs();

            Console.ReadLine();
        }
```

14. Finally, add an extension method that helps log the URL of each tab in **TabController**. Define this as an extension method for **IEnumerable<Tab>**, rather than **TabController**, as you simply need an iterator to iterate through the tabs using a **foreach** loop.

15. Use **PadRight** to left-align each URL, as follows:

```
        private static void LogTabs(this IEnumerable<Tab> tabs)
        {
            Console.Write("TABS: |");

            foreach(var tab in tabs)
```

```
                    Console.Write($"{tab.Url.PadRight(15)}|");

                Console.WriteLine();
            }
        }
    }
```

16. Running the code produces the following output:

```
Opening tabs...
OpenNew packtpub.com
OpenNew microsoft.com
OpenNew amazon.com
TABS: |packtpub.com   |microsoft.com  |amazon.com     |
Moving...
Moved amazon.com to start
Moved packtpub.com to end. Index=2
TABS: |amazon.com     |microsoft.com  |packtpub.com   |
Closing tab...
Removed microsoft.com
TABS: |amazon.com     |packtpub.com   |
```

> **NOTE**
>
> Sometimes Visual Studio might report a non-nullable property error the first time you execute the program. This is a helpful reminder that you are attempting to use a string value that may have a null value at runtime.

The three tabs are opened. **`amazon.com`** and **`packtpub.com`** are then moved before **`microsoft.com`** is finally closed and removed from the tab list.

> **NOTE**
>
> You can find the code used for this exercise at https://packt.link/iUcls.

In this exercise, you have seen how lists can be used to store multiple items of the same type while maintaining the order of items. The next section covers the **`Queue`** and **`Stack`** classes, which allow items to be added and removed in a predefined sequence.

QUEUES

The Queue class provides a **first-in, first-out** mechanism. Items are added to the end of the queue using the `Enqueue` method and are removed from the front of the queue using the `Dequeue` method. Items in the queue cannot be accessed via an index element.

Queues are typically used when you need a workflow that ensures items are processed in the order in which they are added to the queue. A typical example might be a busy online ticketing system selling a limited number of concert tickets to customers. To ensure fairness, customers are added to a queuing system as soon as they log on. The system would then dequeue each customer and process each order, in full, either until all tickets have been sold or the customer queue is empty.

The following example creates a queue containing five `CustomerOrder` records. When it is time to process the orders, each order is dequeued using the `TryDequeue` method, which will return `true` until all orders have been processed. The customer orders are processed in the order that they were added. If the number of tickets requested is more than or equal to the tickets remaining, then the customer is shown a success message. An apology message is shown if the number of tickets remaining is less than the requested amount.

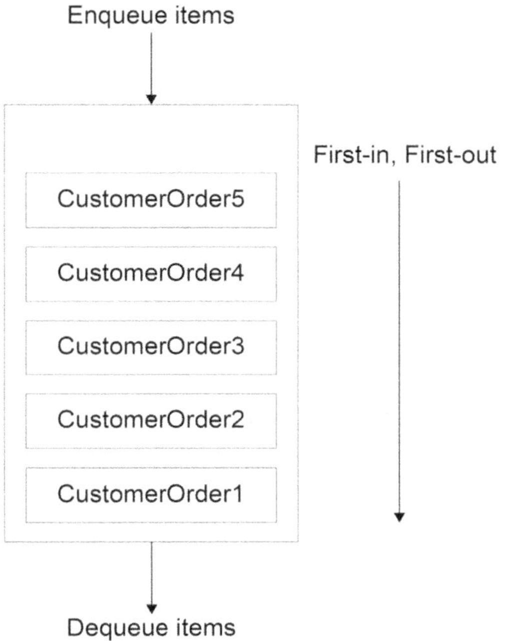

Figure 4.4: The Queue's Enqueue() and Dequeue() workflow

Perform the following steps to complete this example:

1. In the **Examples** folder of your **Chapter04** source folder, add a new class called **QueueExamples.cs** and edit it as follows:

```
using System;
using System.Collections.Generic;

namespace Chapter04.Examples
{
    class QueueExamples
    {
        record CustomerOrder (string Name, int TicketsRequested)
        {}

        public static void Main()
        {
            var ticketsAvailable = 10;
            var customers = new Queue<CustomerOrder>();
```

2. Add five orders to the queue using the **Enqueue** method as follows:

```
            customers.Enqueue(new CustomerOrder("Dave", 2));
            customers.Enqueue(new CustomerOrder("Siva", 4));
            customers.Enqueue(new CustomerOrder("Julien", 3));
            customers.Enqueue(new CustomerOrder("Kane", 2));
            customers.Enqueue(new CustomerOrder("Ann", 1));
```

3. Now, use a **while** loop that repeats until **TryDequeue** returns **false**, meaning all current orders have been processed:

```
            // Start processing orders...
            while(customers.TryDequeue(out CustomerOrder nextOrder))
            {
                if (nextOrder.TicketsRequested <= ticketsAvailable)
                {
                    ticketsAvailable -= nextOrder.TicketsRequested;
                    Console.WriteLine($"Congratulations {nextOrder.
Name}, you've purchased {nextOrder.TicketsRequested} ticket(s)");
                }
                else
                {
                    Console.WriteLine($"Sorry {nextOrder.Name},
cannot fulfil {nextOrder.TicketsRequested} ticket(s)");
```

```
                    }
                }
                Console.WriteLine($"Finished.
Available={ticketsAvailable}");
                Console.ReadLine();
            }
        }
    }
```

4. Running the example code produces the following output:

```
Congratulations Dave, you've purchased 2 ticket(s)
Congratulations Siva, you've purchased 4 ticket(s)
Congratulations Julien, you've purchased 3 ticket(s)
Sorry Kane, cannot fulfil 2 ticket(s)
Congratulations Ann, you've purchased 1 ticket(s)
Finished. Available=0
```

> **NOTE**
>
> The first time you run this program, Visual Studio might show a non-nullable type error. This error is a reminder that you are using a variable that could be a null value.

The output shows that **Dave** requested two tickets. As there are two or more tickets available, he was successful. Both **Siva** and **Julien** were also successful, but by the time **Kane** placed his order of two tickets, there was only one ticket available, so he was shown the apology message. Finally, **Ann** requested one ticket and was successful in her order.

> **NOTE**
>
> You can find the code used for this example at https://packt.link/Zb524.

STACKS

The **Stack** class provides the opposite mechanism to the **Queue** class; items are processed in **last-in, first-out** order. As with the **Queue** class, you cannot access elements via their index position. Items are added to the stack using the **Push** method and removed using the **Pop** method.

An application's **Undo** menu can be implemented using a stack. For example, in a word processor, as the user edits a document, an **Action** delegate is created, which can reverse the most recent change whenever the user presses **Ctrl** + **Z**. The most recent action is popped off the stack and the change is undone. This allows multiple steps to be undone.

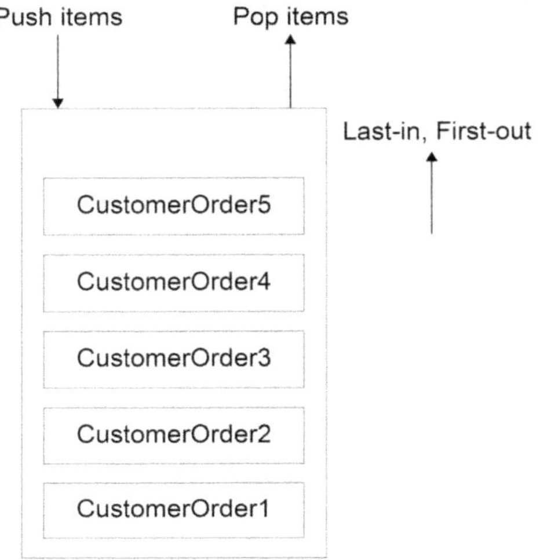

Figure 4.5: The Stack's Push() and Pop() workflow

The following example shows this in practice.

You will start by creating an **UndoStack** class that supports multiple **undo operations**. The caller decides what action should run each time the **Undo** request is called.

A typical undoable operation would be storing a copy of text prior to the user adding a word. Another undoable operation would be storing a copy of the current font prior to a new font being applied. You can start by adding the following code, where you are creating the **UndoStack** class and defining a **readonly Stack of Action** delegates, named **_undoStack**:

1. In your **Chapter04\Examples** folder, add a new class called **StackExamples.cs** and edit it as follows:

```
using System;
using System.Collections.Generic;

namespace Chapter04.Examples
{
```

```
    class UndoStack
    {
        private readonly Stack<Action> _undoStack = new
Stack<Action>();
```

2. When the user has done something, the same action can be undone. So push an **undoable Action** to the front of **_undoStack**:

```
        public void Do(Action action)
        {
            _undoStack.Push(action);
        }
```

3. The **Undo** method checks to see if there are any items to undo, then calls **Pop** to remove the most recent **Action** and invoke it, thus undoing the change that was just applied. The code for this can be added as follows:

```
        public void Undo()
        {
            if (_undoStack.Count > 0)
            {
                var undo = _undoStack.Pop();
                undo?.Invoke();
            }
        }
    }
```

4. Now, you can create a **TextEditor** class that allows edits to be added to **UndoStack**. This constructor is passed **UndoStack** as there could be multiple editors that need to add various **Action** delegates to the stack:

```
    class TextEditor
    {
        private readonly UndoStack _undoStack;

        public TextEditor(UndoStack undoStack)
        {
            _undoStack = undoStack;
        }

        public string Text {get; private set; }
```

5. Next, add the **EditText** command, which takes a copy of the **previousText** value and creates an **Action** delegate that can revert the text to its previous value, if invoked:

```
        public void EditText(string newText)
        {
            var previousText = Text;

            _undoStack.Do( () =>
            {
                Text = previousText;
                Console.Write($"Undo:'{newText}'".PadRight(40));
                Console.WriteLine($"Text='{Text}'");
            });
```

6. Now, the **newText** value should be appended to the **Text** property, using the **+=** operator. The details for this are logged to the console, using **PadRight** to improve the format:

```
            Text += newText;
            Console.Write($"Edit:'{newText}'".PadRight(40));
            Console.WriteLine($"Text='{Text}'");
        }
    }
```

7. Finally, it is time to create a console app that tests **TextEditor** and **UndoStack**. Four edits are initially made, followed by two **undo operations**, and finally two more text edits:

```
    class StackExamples
    {

        public static void Main()
        {
            var undoStack = new UndoStack();
            var editor = new TextEditor(undoStack);

            editor.EditText("One day, ");
            editor.EditText("in a ");
            editor.EditText("city ");
            editor.EditText("near by ");
```

```
            undoStack.Undo(); // remove 'near by'
            undoStack.Undo(); // remove 'city'

            editor.EditText("land ");
            editor.EditText("far far away ");

            Console.ReadLine();
        }
    }
}
```

8. Running the console app produces the following output:

```
Edit:'One day, '                        Text='One day, '
Edit:'in a '                            Text='One day, in a '
Edit:'city '                            Text='One day, in a city '
Edit:'near by '                         Text='One day, in a city near
by '
Undo:'near by '                         Text='One day, in a city '
Undo:'city '                            Text='One day, in a '
Edit:'land '                            Text='One day, in a land '
Edit:'far far away '                    Text='One day, in a land far
far away '
```

> **NOTE**
>
> Visual Studio may show non-nullable property error the first time the code is executed. This is because Visual Studio notices that the **`Text`** property can be a null value at runtime so offers a suggestion to improve the code.

The left-hand output shows the text edits and undoes operations as they are applied and the resulting **`Text`** value on the right-hand side. The two **`Undo`** calls result in **`near by`** and **`city`** being removed from the **`Text`** value, before **`land`** and **`far far away`** are finally added to the **`Text`** value.

> **NOTE**
>
> You can find the code used for this example at https://packt.link/tLVyf.

HASHSETS

The **HashSet** class provides mathematical **set operations** with collections of objects in an efficient and highly performant manner. **HashSet** does not allow duplicate elements and items are not stored in any particular order. Using the **HashSet** class is ideal for high-performance operations, such as needing to quickly find where two collections of objects overlap.

Typically, **HashSet** is used with the following operations:

- **public void UnionWith(IEnumerable<T> other)**: Produces a **set union**. This modifies **HashSet** to include the items present in the current **HashSet** instance, the other collection, or both.
- **public void IntersectWith(IEnumerable<T> other)**: Produces a **set intersect**. This modifies **HashSet** to include items present in the current **HashSet** instance and the other collection.
- **public void ExceptWith(IEnumerable<T> other)**: Produces a **set subtraction**. This removes items from the **HashSet** that are present in the current **HashSet** instance and the other collection.

HashSet is useful when you need to include or exclude certain elements from **collections**. As an example, consider that an agent manages various celebrities and has been asked to find three sets of stars:

- Those that can act **or** sing.
- Those that can act **and** sing.
- Those that can act **only** (no singers allowed).

In the following snippet, a list of actors' and singers' names is created:

1. In your **Chapter04\Examples** folder, add a new class called **HashSetExamples.cs** and edit it as follows:

```
using System;
using System.Collections.Generic;

namespace Chapter04.Examples
{
    class HashSetExamples
    {
        public static void Main()
        {
```

```
            var actors = new List<string> {"Harrison Ford", "Will 
Smith",
                                           "Sigourney Weaver"};
            var singers = new List<string> {"Will Smith", "Adele"};
```

2. Now, create a new **HashSet** instance that initially contains singers only and then use **UnionWith** to modify the set to contain a distinct set of those that can act **or** sing:

```
            var actingOrSinging = new HashSet<string>(singers);
            actingOrSinging.UnionWith(actors);
            Console.WriteLine($"Acting or Singing: {string.Join(", ",
                              actingOrSinging)}");
```

3. For those that can act **and** sing, start with a **HashSet** instance of singers, and modify the **HashSet** instance using **IntersectWith** to contain a distinct list of those that are in both collections:

```
            var actingAndSinging = new HashSet<string>(singers);
            actingAndSinging.IntersectWith(actors);
            Console.WriteLine($"Acting and Singing: {string.Join(", 
",
                              actingAndSinging)}");
```

4. Finally, for those that can **act only**, start with the actor collection, and use **ExceptWith** to remove those from the **HashSet** instance that can also sing:

```
            var actingOnly = new HashSet<string>(actors);
            actingOnly.ExceptWith(singers);
            Console.WriteLine($"Acting Only: {string.Join(", ", 
actingOnly)}");
            Console.ReadLine();
        }
    }
}
```

5. Running the console app produces the following output:

```
Acting or Singing: Will Smith, Adele, Harrison Ford, Sigourney Weaver
Acting and Singing: Will Smith
Acting Only: Harrison Ford, Sigourney Weaver
```

From the output, you can see that out of the given list of actors and singers, only **Will Smith** can act and sing.

> **NOTE**
>
> You can find the code used for this example at https://packt.link/ZdNbS.

DICTIONARIES

Another commonly used collection type is the generic **Dictionary<TK, TV>**. This allows multiple items to be added, but a unique **key** is needed to identify an item instance.

Dictionaries are commonly used to look up values using known keys. The key and value type parameters can be of any type. A value can exist in a **Dictionary** more than once, provided that its key is **unique**. Attempting to add a key that already exists will result in a runtime exception being thrown.

A common example of a **Dictionary** might be a registry of known countries that are keyed by their ISO country code. A customer service application may load customer details from a database and then use the ISO code to look up the customer's country from the country list, rather than having the extra overhead of creating a new country instance for each customer.

> **NOTE**
>
> You can find more information on standard ISO country codes at https://www.iso.org/iso-3166-country-codes.html.

The main methods used in the **Dictionary** class are as follows:

- **public TValue this[TKey key] {get; set;}**: Gets or sets a value associated with the key. An **exception** is thrown if the key does not exist.
- **Dictionary<TKey, TValue>.KeyCollection Keys { get; }**: Returns a **KeyCollection** dictionary instance that contains all keys.
- **Dictionary<TKey, TValue>.ValueCollection Values { get; }**: Returns a **ValueCollection** dictionary instance that contains all values.

- **public int Count { get; }**: Returns the number of elements in the **Dictionary**.
- **void Add(TKey key, TValue value)**: Adds the key and associated value. If the key already exists, an exception is thrown.
- **void Clear()**: Clears all keys and values from the **Dictionary**.
- **bool ContainsKey(TKey key)**: Returns **true** if the specified key exists.
- **bool ContainsValue(TValue value)**: Returns **true** if the specified value exists.
- **bool Remove(TKey key)**: Removes a value with the associated key.
- **bool TryAdd(TKey key, TValue value)**: Attempts to add the key and value. If the key already exists, an exception is "not" thrown. Returns **true** if the value was added.
- **bool TryGetValue(TKey key, out TValue value)**: Gets the value associated with the key, if it is available. Returns **true** if it was found.

The following code shows how a **Dictionary** can be used to add and navigate **Country** records:

1. In your **Chapter04\Examples** folder, add a new class called **DictionaryExamples.cs**.
2. Start by defining a **Country** record, which is passed a **Name** parameter:

```
using System;
using System.Collections.Generic;

namespace Chapter04.Examples
{
    public record Country(string Name)
    {}

    class DictionaryExamples
    {
        public static void Main()
        {
```

3. Use the **Dictionary** initialization syntax to create a **Dictionary** with five countries, as follows:

```
            var countries = new Dictionary<string, Country>
            {
                {"AFG", new Country("Afghanistan")},
                {"ALB", new Country("Albania")},
                {"DZA", new Country("Algeria")},
                {"ASM", new Country("American Samoa")},
                {"AND", new Country("Andorra")}
            };
```

4. In the next code snippet, **Dictionary** implements the **IEnumerable** interface, which allows you to retrieve a key-value pair representing the key and value items in the **Dictionary**:

```
            Console.WriteLine("Enumerate foreach KeyValuePair");
            foreach (var kvp in countries)
            {
                Console.WriteLine($"\t{kvp.Key} = {kvp.Value.Name}");
            }
```

5. Running the example code produces the following output. By iterating through each item in **countries**, you can see the five country codes and their names:

```
Enumerate foreach KeyValuePair
        AFG = Afghanistan
        ALB = Albania
        DZA = Algeria
        ASM = American Samoa
        AND = Andorra
```

6. There is an entry with the **AFG** key, so using the **set indexer** passing in **AFG** as a key allows a new **Country** record to be set that replaces the previous item with the **AGF** key. You can add the following code for this:

```
            Console.WriteLine("set indexor AFG to new value");
            countries["AFG"] = new Country("AFGHANISTAN");
            Console.WriteLine($"get indexor AFG: {countries["AFG"].
Name}");
```

7. When you run the code, adding a key for **AFG** allows you to get a value using that key:

```
set indexor AFG to new value
get indexor AFG: AFGHANISTAN

ContainsKey AGO: False
ContainsKey and: False
```

8. Key comparisons are case-sensitive with string keys, so **AGO** is present but **and** is not as the corresponding country (**Andorra**) is defined with the uppercase **AND** key. You can add the following code to check this:

```
            Console.WriteLine($"ContainsKey {"AGO"}:
                         {countries.ContainsKey("AGO")}");
            Console.WriteLine($"ContainsKey {"and"}:
                         {countries.ContainsKey("and")}"); // Case
sensitive
```

9. Using **Add** to add a new entry will throw an exception if the key already exists. This can be seen by adding the following code:

```
            var anguilla = new Country("Anguilla");
            Console.WriteLine($"Add {anguilla}...");
            countries.Add("AIA", anguilla);
            try
            {
                var anguillaCopy = new Country("Anguilla");
                Console.WriteLine($"Adding {anguillaCopy}...");
                countries.Add("AIA", anguillaCopy);
            }
            catch (Exception e)
            {
                Console.WriteLine($"Caught {e.Message}");
            }
```

10. Conversely, **TryAdd** does **not** throw an exception if you attempt to add a duplicate key. There already exists an entry with the **AIA** key, so using **TryAdd** simply returns a **false** value rather than throwing an exception:

```
            var addedAIA = countries.TryAdd("AIA", new
Country("Anguilla"));
            Console.WriteLine($"TryAdd AIA: {addedAIA}");
```

11. As the following output shows, adding **Anguilla** once using the **AIA** key is valid but attempting to add it again using the **AIA** key results in an exception being caught:

```
Add Country { Name = Anguilla }...
Adding Country { Name = Anguilla }...
Caught An item with the same key has already been added. Key: AIA

TryAdd AIA: False
```

12. **TryGetValue**, as the name suggests, allows you to try to get a value by key. You pass in a key that may be missing from the **Dictionary**. Requesting an object whose key is missing from the **Dictionary** will ensure that an exception is not thrown. This is useful if you are unsure whether a value has been added for the specified key:

```
            var tryGet = countries.TryGetValue("ALB", out Country
albania1);
            Console.WriteLine($"TryGetValue for ALB: {albania1}
                              Result={tryGet}");
            countries.TryGetValue("alb", out Country albania2);
            Console.WriteLine($"TryGetValue for ALB: {albania2}");
        }
    }
}
```

13. You should see the following output upon running this code:

```
TryGetValue for ALB: Country { Name = Albania } Result=True
TryGetValue for ALB:
```

> **NOTE**
>
> Visual Studio might report the following warning: **Warning CS8600: Converting null literal or possible null value to non-nullable type**. This is a reminder from Visual Studio that a variable may have a null value at runtime.

You have seen how the **Dictionary** class is used to ensure that only unique identities are associated with values. Even if you do not know which keys are in the **Dictionary** until runtime, you can use the **TryGetValue** and **TryAdd** methods to prevent runtime exceptions.

> **NOTE**
>
> You can find the code used for this example at https://packt.link/vzHUb.

In this example, a string key was used for the **Dictionary**. However, any type can be used as a key. You will often find that an integer value is used as a key when source data is retrieved from relational databases, as integers can often be more efficient in memory than strings. Now it is time to use this feature through an exercise.

EXERCISE 4.02: USING A DICTIONARY TO COUNT THE WORDS IN A SENTENCE

You have been asked to create a console app that asks the user to enter a sentence. The console should then split the input into individual words (using a space character as a word delimiter) and count the number of times that each word occurs. If possible, simple forms of punctuation should be removed from the output, and you are to ignore capitalized words so that, for example, **Apple** and **apple** both appear as a single word.

This is an ideal use of a **Dictionary**. The **Dictionary** will use a string as the key (a unique entry for each word) with an **int** value to count the words. You will use **string.Split()** to split a sentence into words, and **char.IsPunctuation** to remove any trailing punctuation marks.

Perform the following steps to do so:

1. In your **Chapter04\Exercises** folder, create a new folder called **Exercise02**.
2. Inside the **Exercise02** folder, add a new class called **Program.cs**.
3. Start by defining a new class called **WordCounter**. This can be marked as **static** so that it can be used without needing to create an instance:

```
using System;
using System.Collections.Generic;

namespace Chapter04.Exercises.Exercise02
{
    static class WordCounter
    {
```

4. Define a **static** method called **Process**:

```
public static IEnumerable<KeyValuePair<string, int>> Process(
    string phrase)
{
    var wordCounts = new Dictionary<string, int>();
```

   This is passed a phrase and returns **IEnumerable<KeyValuePair>**, which allows the caller to enumerate through a **Dictionary** of results. After this definition, the **Dictionary** of **wordCounts** is keyed using a **string** (each word found) and an **int** (the number of times that a word occurs).

5. You are to ignore the case of words with capital letters, so convert the string into its lowercase equivalent before using the **string.Split** method to split the phrase.

6. Then you can use the **RemoveEmptyEntries** option to remove any empty string values. Add the following code for this:

```
     var words = phrase.ToLower().Split(' ',
                 StringSplitOptions.RemoveEmptyEntries);
```

7. Use a simple **foreach** loop to iterate through the individual words found in the phrase:

```
    foreach(var word in words)
    {
        var key = word;

        if (char.IsPunctuation(key[key.Length-1]))
        {
            key = key.Remove(key.Length-1);
        }
```

   The **char.IsPunctuation** method is used to remove punctuation marks from the end of the word.

8. Use the **TryGetValue** method to check if there is a **Dictionary** entry with the current word. If so, update the **count** by one:

```
        if (wordCounts.TryGetValue(key, out var count))
        {
            wordCounts[key] = count + 1;
        }
        else
        {
            wordCounts.Add(key, 1);
```

```
                }
            }
```

If the word does not exist, add a new word key with a starting value of **1**.

9. Once all the words in the phrase have been processed, return the **wordCounts Dictionary**:

```
            return wordCounts;
        }
    }
```

10. Now, write the console app that allows the user to enter a phrase:

```
    class Program
    {
        public static void Main()
        {
            string input;
            do
            {
                Console.Write("Enter a phrase:");
                input = Console.ReadLine();
```

The **do** loop will end once the user enters an empty string; you will add the code for this in an upcoming step.

11. Call the **WordCounter.Process** method to return a key-value pair that can be enumerated through.

12. For each **key** and **value**, write the word and its count, padding each word to the right:

```
                if (!string.IsNullOrEmpty(input))
                {
                    var countsByWord = WordCounter.Process(input);
                    var i = 0;
                    foreach (var (key, value) in countsByWord)
                    {
                        Console.Write($"{key.PadLeft(20)}={value}\t");
                        i++;
                        if (i % 3 == 0)
                        {
                            Console.WriteLine();
                        }
```

```
                }
                Console.WriteLine();
```

A new line is started after every third word (using **i % 3 = 0**) for improved output formatting.

13. Finish off the **do-while** loop:

```
                }

            } while (input != string.Empty);
        }
    }
}
```

14. Running the console using the opening text from *The Gettysburg Address* of 1863 produces this output:

```
Enter a phrase:
Four score and seven years ago our fathers brought forth, upon this
continent, a new nation, conceived in liberty, and dedicated to the
proposition that all men are created equal. Now we are engaged in a
great civil war, testing whether that nation, or any nation so
conceived, and so dedicated, can long endure.
                  four=1                score=1                  and=3
                 seven=1                years=1                  ago=1
                   our=1              fathers=1              brought=1
                 forth=1                upon=1                  this=1
             continent=1                   a=2                  new=1
                nation=3            conceived=2                   in=2
               liberty=1            dedicated=2                   to=1
                   the=1          proposition=1                 that=2
                   all=1                  men=1                  are=2
               created=1                equal=1                  now=1
                    we=1              engaged=1                great=1
                 civil=1                  war=1              testing=1
               whether=1                   or=1                  any=1
                    so=2                  can=1                 long=1
                endure=1
```

NOTE

You can search online for The Gettysburg Address or visit https://rmc.library.cornell.edu/gettysburg/good_cause/transcript.htm.

From the results, you can see that each word is displayed only once and that certain words, such as **`and`** and **`that`**, appear more than once in the speech. The words are listed in the order they appear in the text, but this is not always the case with the **`Dictionary`** class. It should be assumed that the order will **not** remain fixed this way; dictionaries' values should be accessed using a key.

> **NOTE**
>
> You can find the code used for this exercise at https://packt.link/Dnw4a.

So far, you have learned about the main collections commonly used in .NET. It is now time to look at LINQ, which makes extensive use of collections based on the **`IEnumerable`** interface.

LINQ

LINQ (pronounced **link**) is short for **Language Integrated Query**. LINQ is a general-purpose language that can be used to query objects in memory by using a syntax that is similar to **Structured Query Language** (**SQL**), that is, it is used to query databases. It is an enhancement of the C# language that makes it easier to interact with objects in memory using SQL-like Query Expressions or Query Operators (implemented through a series of extension methods).

Microsoft's original idea for LINQ was to bridge the gap between .NET code and data sources, such as relational databases and XML, using LINQ providers. **LINQ providers** form a set of building blocks that can be used to query various sources of data, using a similar set of Query Operators, without the caller needing to know the intricacies of how each data source works. The following is a list of providers and how they are used:

- **LINQ to Objects**: Queries applied to objects in memory, such as those defined in a list.
- **LINQ to SQL**: Queries applied to relational databases such as SQL Server, Sybase, or Oracle.
- **LINQ to XML**: Queries applied to XML documents.

This chapter will cover LINQ to Objects. This is, by far, the most common use of LINQ providers and offers a flexible way to query collections in memory. In fact, when talking about LINQ, most people refer to LINQ to Objects, mainly due to its ubiquitous use throughout C# applications.

At the heart of LINQ is the way that collections can be converted, filtered, and aggregated into new forms using a concise and easy-to-use syntax. LINQ can be used in two interchangeable styles:

- **Query Operators**
- **Query Expressions**

Each style offers a different syntax to achieve the same result, and which one you use often comes down to personal preference. Each style can be interwoven in code easily.

QUERY OPERATORS

These are based on a series of core extension methods. The results from one method can be chained together into a programming style, which can often be easier to grasp than their expression-based counterparts.

The extension methods typically take an **`IEnumerable<T>`** or **`IQueryable<T>`** input source, such as a list, and allow a **`Func<T>`** predicate to be applied to that source. The source is generic-based, so Query Operators work with all types. It is just as easy to work with **`List<string>`** as it is with **`List<Customer>`**, for example.

In the following snippet, **`.Where`**, **`.OrderBy`**, and **`.Select`** are the extension methods being called:

```
books.Where(book => book.Price > 10)
     .OrderBy(book => book.Price)
     .Select(book => book.Name)
```

Here, you are taking the results from a **`.Where`** extension method to find all books with a unit price greater than **`10`**, which is then sorted using the **`.OrderBy`** extension method. Finally, the name of each book is extracted using the **`.Select`** method. These methods could have been declared as single lines of code, but chaining in this way provides a more intuitive syntax. This will be covered in more detail in the upcoming sections.

QUERY EXPRESSIONS

Query Expressions are an enhancement of the C# language and resemble SQL syntax. The C# compiler compiles Query Expressions into a sequence of Query Operator extension method calls. Note that not all Query Operators are available with an equivalent Query Expression implementation.

Query Expressions have the following rules:

- They start with a **from** clause.
- They can contain at least one or more optional **where**, **orderby**, **join**, **let**, and additional **from** clauses.
- They end with either a **select** or a **group** clause.

The following snippet is functionally equivalent to the Query Operator style defined in the previous section:

```
from book in books
where book.Price > 10
orderby book.Price
select book.Name
```

You will take a more in-depth look at both styles as you learn about the standard Query Operators shortly.

DEFERRED EXECUTION

Whether you choose to use Query Operators, Query Expressions, or a mixture of the two, it is important to remember that for many operators, the query that you define is not executed when it is defined, but only when it is enumerated over. This means that it is not until a **foreach** statement or a **ToList**, **ToArray**, **ToDictionary**, **ToLookup**, or **ToHashSet** method is called that the actual query is executed.

This allows queries to be constructed elsewhere in code with additional criteria included, and then used or even reused with a different collection of data. Recall that in *Chapter 3, Delegates, Lambdas, and Events*, you saw similar behavior with delegates. Delegates are not executed where they are defined, but only when they are invoked.

In the following short Query Operator example, the output will be **abz** even though **z** is added **after** the query is defined but **before** it is enumerated through. This demonstrates that LINQ queries are evaluated on demand, rather than at the point where they are declared:

```
var letters = new List<string> { "a", "b"}
var query = letters.Select(w => w.ToUpper());
letters.Add("z");
foreach(var l in query)
  Console.Write(l);
```

STANDARD QUERY OPERATORS

LINQ is driven by a core set of extension methods, referred to as standard Query Operators. These are grouped into operations based on their functionality. There are many standard Query Operators available, so for this introduction, you will explore all the main operators that you are likely to use regularly.

PROJECTION OPERATIONS

Projection operations allow you to convert an object into a new structure using only the properties that you need. You can create a new type, apply mathematical operations, or return the original object:

- **Select**: Projects each item in the source into a new form.
- **SelectMany**: Projects all items in the source, flattens the result, and optionally projects them to a new form. There is no Query Expression equivalent for **SelectMany**.

SELECT

Consider the following snippet, which iterates through a **List<string>** containing the values **Mon**, **Tues**, and **Wednes**, outputting each with the word day appended.

In your **Chapter04\Examples** folder, add a new file called **LinqSelectExamples.cs** and edit it as follows:

```
using System;
using System.Collections.Generic;
using System.Linq;

namespace Chapter04.Examples
{
```

```
    class LinqSelectExamples
    {
        public static void Main()
        {
            var days = new List<string> { "Mon", "Tues", "Wednes" };

            var query1 = days.Select(d => d + "day");
            foreach(var day in query1)
                Console.WriteLine($"Query1: {day}");
```

Looking at the Query Operator syntax first, you can see that **query1** uses the **Select** extension method and defines a **Func<T>** like this:

```
d => d + "day"
```

When executed, the variable **d** is passed to the lambda statement, which appends the word **day** to each string in the **days** list: **"Mon"**, **"Tues"**, **"Wednes"**. This returns a new **IEnumerable<string>** instance, with the original values inside the source variable, **days**, remaining unchanged.

You can now enumerate through the new **IEnumerable** instance using **foreach**, as follows:

```
            var query2 = days.Select((d, i) => $"{i} : {d}day");
            foreach (var day in query2)
                Console.WriteLine($"Query2: {day}");
```

Note that the **Select** method has another overload that allows the index position in the source and value to be accessed, rather than just the value itself. Here, **d** (the string value) and **i** (its index) are passed, using the **(d , i) =>** syntax and joined into a new string. The output will be displayed as **0 : Monday**, **1 : Tuesday**, and so on.

ANONYMOUS TYPES

Before you continue looking at **Select** projections, it is worth noting that C# does not limit you to just creating new strings from existing strings. You can project into any type.

You can also create anonymous types, which are types created by the compiler from the properties that you name and specify. For example, consider the following example, which results in a new type being created that represents the results of the **Select** method:

```
            var query3 = days.Select((d, i) => new
            {
                Index = i,
                UpperCaseName = $"{d.ToUpper()}DAY"
            });
            foreach (var day in query3)
                Console.WriteLine($"Query3: Index={day.Index},
                                  UpperCaseDay={day.
UpperCaseName}");
```

Here, **query3** results in a new type that has an Index and **UpperCaseName** property; the values are assigned using **Index = i** and **UpperCaseName = $"{d.ToUpper()}DAY"**.

These types are scoped to be available within your local method and can then be used in any local statements, such as in the previous **foreach** block. This saves you from having to create classes to temporarily store values from a **Select** method.

Running the code produces output in this format:

```
Index=0, UpperCaseDay=MONDAY
```

As an alternative, consider how the equivalent Query Expression looks. In the following example, you start with the from **day in days** expression. This assigns the name **day** to the string values in the **days** list. You then use **select** to project that to a new string, appending **"day"** to each.

This is functionally equivalent to the example in **query1**. The only difference is the code readability:

```
            var query4 = from day in days
                         select day + "day";
            foreach (var day in query4)
                Console.WriteLine($"Query4: {day}");
```

The following example snippet mixes a Query Operator and Query Expressions. The **select** Query Expression cannot be used to select a value and index, so the **Select** extension method is used to create an anonymous type with a **Name** and **Index** property:

```
                        var query5 = from dayIndex in
                          days.Select( (d, i) => new {Name = d, Index =
i})
                          select dayIndex;
            foreach (var day in query5)
                Console.WriteLine($"Query5: Index={day.Index} : {day.
Name}");

            Console.ReadLine();
        }
    }
}
```

Running the full example produces this output:

```
Query1: Monday
Query1: Tuesday
Query1: Wednesday
Query2: 0 : Monday
Query2: 1 : Tuesday
Query2: 2 : Wednesday
Query3: Index=0, UpperCaseDay=MONDAY
Query3: Index=1, UpperCaseDay=TUESDAY
Query3: Index=2, UpperCaseDay=WEDNESDAY
Query4: Monday
Query4: Tuesday
Query4: Wednesday
Query5: Index=0 : Mon
Query5: Index=1 : Tues
Query5: Index=2 : Wednes
```

Again, it largely comes down to personal choice as to which you prefer using. As queries become longer, one form may require less code than the other.

> **NOTE**
>
> You can find the code used for this example at https://packt.link/wKye0.

SELECTMANY

You have seen how **Select** can be used to project values from each item in a source collection. In the case of a source that has enumerable properties, the **SelectMany** extension method can extract the multiple items into a single list, which can then be optionally projected into a new form.

The following example creates two **City** records, each with multiple **Station** names, and uses **SelectMany** to extract all stations from both cities:

1. In your **Chapter04\Examples** folder, add a new file called **LinqSelectManyExamples.cs** and edit it as follows:

```
using System;
using System.Collections.Generic;
using System.Linq;

namespace Chapter04.Examples
{
    record City (string Name, IEnumerable<string> Stations);

    class LinqSelectManyExamples
    {
        public static void Main()
        {
            var cities = new List<City>
            {
                new City("London", new[] {"Kings Cross KGX",
                                          "Liverpool Street LVS",
                                          "Euston EUS"}),
                new City("Birmingham", new[] {"New Street NST"})
            };

            Console.WriteLine("All Stations: ");
            foreach (var station in cities.SelectMany(city => city.
Stations))
            {
                Console.WriteLine(station);
            }
```

The **Func** parameter, which is passed to **SelectMany**, requires you to specify an enumerable property, in this case, the **City** class's **Stations** property, which contains a list of string names (see the highlighted code).

Notice how a shortcut is used here, by directly integrating the query into a **foreach** statement. You are not altering or reusing the query variable, so there is no benefit in defining it separately, as done earlier.

SelectMany extracts all the station names from all of the items in the **List<City>** variable. Starting with the **City** class at element **0**, which has the name **London**, it will extract the three station names (**"Kings Cross KGX"**, **"Liverpool Street LVS"**, and **"Euston EUS"**). It will then move on to the second **City** element, named **Birmingham**, and extract the single station, named **"New Street NST"**.

2. Running the example produces the following output:

```
All Stations:
Kings Cross KGX
Liverpool Street LVS
Euston EUS
New Street NST
```

3. As an alternative, consider the following snippet. Here, you revert to using a query variable, **stations**, to make the code easier to follow:

```
            Console.Write("All Station Codes: ");
            var stations = cities
                .SelectMany(city => city.Stations.Select(s =>
s[^3..]));
            foreach (var station in stations)
            {
                Console.Write($"{station} ");
            }
            Console.WriteLine();
            Console.ReadLine();
        }
    }
}
```

Rather than just returning each **Station** string, this example uses a nested **Select** method and a **Range** operator to extract the last three characters from the station name using **s[^3..]**, where **s** is a string for each station name and **^3** indicates that the **Range** operator should extract a string that starts at the last three characters in the string.

4. Running the example produces the following output:

```
All Station Codes: KGX LVS EUS NST
```

   You can see the last three characters of each station name are shown in the output.

> **NOTE**
>
> You can find the code used for this example at https://packt.link/g8dXZ.

In the next section you will read about the filtering operations that filter a result as per a condition.

FILTERING OPERATIONS

Filtering operations allow you to filter a result to return only those items that match a condition. For example, consider the following snippet, which contains a list of orders:

1. In your **Chapter04\Examples** folder, add a new file called **LinqWhereExample.cs** and edit it as follows:

   LinqWhereExamples.cs

```
using System;
using System.Collections.Generic;
using System.Linq;

namespace Chapter04.Examples
{
    record Order (string Product, int Quantity, double Price);

    class LinqWhereExamples
    {
        public static void Main()
        {
            var orders = new List<Order>
            {
                new Order("Pen", 2, 1.99),
                new Order("Pencil", 5, 1.50),
                new Order("Note Pad", 1, 2.99),
```

   You can find the complete code here: https://packt.link/ZJpb5.

   Here, some order items are defined for various stationery products. Suppose you want to output all orders that have a quantity greater than five (this should output the **Ruler** and **USB Memory Stick** orders from the source).

2. For this, you can add the following code:

```
            Console.WriteLine("Orders with quantity over 5:");
            foreach (var order in orders.Where(o => o.Quantity > 5))
            {
                Console.WriteLine(order);
            }
```

3. Now, suppose you extend the criteria to find all products where the product is **Pen** or **Pencil**. You can chain that result into a **Select** method, which will return each order's total value; remember that **Select** can return anything from a source, even a simple extra calculation like this:

```
            Console.WriteLine("Pens or Pencils:");
            foreach (var orderValue in orders
                .Where(o => o.Product == "Pen"  || o.Product ==
"Pencil")
                .Select( o => o.Quantity * o.Price))
            {
                Console.WriteLine(orderValue);
            }
```

4. Next, the Query Expression in the following snippet uses a **where** clause to find the orders with a price less than or equal to **3.99**. This projects them into an anonymous type that has **Name** and **Value** properties, which you enumerate through using a **foreach** statement:

```
            var query = from order in orders
               where order.Price <= 3.99
               select new {Name=order.Product, Value=order.
Quantity*order.Price};
            Console.WriteLine("Cheapest Orders:");
            foreach(var order in query)
            {
                Console.WriteLine($"{order.Name}: {order.Value}");
            }

        }
    }
}
```

5. Running the full example produces this result:

```
Orders with quantity over 5:
Order { Product = Ruler, Quantity = 10, Price = 0.5 }
Order { Product = USB Memory Stick, Quantity = 6, Price = 20 }

Pens or Pencils:
3.98
7.5

Cheapest Orders:
Pen: 3.98
Pencil: 7.5
Note Pad: 2.99
Stapler: 3.99
Ruler: 5
```

Now you have seen Query Operators in action, it is worth returning to deferred execution to see how this affects a query that is enumerated multiple times over.

In this next example, you have a collection of journeys made by a vehicle, which are populated via a **TravelLog** record. The **TravelLog** class contains an **AverageSpeed** method that logs a console message each time it is executed, and, as the name suggests, returns the average speed of the vehicle during that journey:

1. In your Chapter04\Examples folder, add a new file called **LinqMultipleEnumerationExample.cs** and edit it as follows:

```
using System;
using System.Collections.Generic;
using System.Linq;

namespace Chapter04.Examples
{
    record TravelLog (string Name, int Distance, int Duration)
    {
        public double AverageSpeed()
        {
            Console.WriteLine($"AverageSpeed() called for '{Name}'");
            return Distance / Duration;
        }
    }
```

```
    class LinqMultipleEnumerationExample
    {
```

2. Next, define the console app's **Main** method, which populates a **travelLogs** list with four **TravelLog** records. You will add the following code for this:

```
        public static void Main()
        {
            var travelLogs = new List<TravelLog>
            {
                new TravelLog("London to Brighton", 50, 4),
                new TravelLog("Newcastle to London", 300, 24),
                new TravelLog("New York to Florida", 1146, 19),
                new TravelLog("Paris to Berlin", 546, 10)
            };
```

3. You will now create a **fastestJourneys** query variable, which includes a **Where** clause. This **Where** clause will call each journey's **AverageSpeed** method when enumerated.
4. Then, using a **foreach** loop, you enumerate through the items in **fastestJourneys** and write the name and distance to the console (note that you do **not** access the **AverageSpeed** method inside the **foreach** loop):

```
            var fastestJourneys = travelLogs.Where(tl =>
tl.AverageSpeed() > 50);
            Console.WriteLine("Fastest Distances:");
            foreach (var item in fastestJourneys)
            {
                Console.WriteLine($"{item.Name}: {item.Distance}
miles");
            }
            Console.WriteLine();
```

5. Running the code block will produce the following output, the **Name** and **Distance** for each journey:

```
Fastest Distances:
AverageSpeed() called for 'London to Brighton'
AverageSpeed() called for 'Newcastle to London'
AverageSpeed() called for 'New York to Florida'
New York to Florida: 1146 miles
AverageSpeed() called for 'Paris to Berlin'
Paris to Berlin: 546 miles
```

6. You can see that **AverageSpeed** was called **four** times, once for each journey as part of the **Where** condition. This is as expected so far, but now, you can reuse the same query to output the **Name** and, alternatively, the **Duration**:

```
            Console.WriteLine("Fastest Duration:");
            foreach (var item in fastestJourneys)
            {
                Console.WriteLine($"{item.Name}: {item.Duration}
hours");
            }
            Console.WriteLine();
```

7. Running this block produces the same **four** calls to the **AverageSpeed** method:

```
Fastest Duration:
AverageSpeed() called for 'London to Brighton'
AverageSpeed() called for 'Newcastle to London'
AverageSpeed() called for 'New York to Florida'
New York to Florida: 19 hours
AverageSpeed() called for 'Paris to Berlin'
Paris to Berlin: 10 hours
```

   This shows that whenever a query is enumerated, the full query is **re-evaluated every time**. This might not be a problem for a fast method such as **AverageSpeed**, but what if a method needs to access a database to extract some data? That would result in multiple database calls and, possibly, a very slow application.

8. You can use methods such as **ToList**, **ToArray**, **ToDictionary**, **ToLookup**, or **ToHashSet** to ensure that a query that could be enumerated many times is **executed once only** rather than being re-evaluated repeatedly. Continuing with this example, the following block uses the same **Where** clause but includes an extra **ToList** call to immediately execute the query and ensure it is not re-evaluated:

```
            Console.WriteLine("Fastest Duration Multiple loops:");
            var fastestJourneysList = travelLogs
                  .Where(tl => tl.AverageSpeed() > 50)
                  .ToList();
            for (var i = 0; i < 2; i++)
            {
```

```
                    Console.WriteLine($"Fastest Duration Multiple loop
iteration {i+1}:");
                    foreach (var item in fastestJourneysList)
                    {
                        Console.WriteLine($"{item.Name}: {item.Distance}
in {item.Duration} hours");
                    }
                }
            }
        }
    }
```

9. Running the block produces the following output. Notice how **`AverageSpeed`** is called **four times only** and is called prior to either of the two **`Fastest Duration Multiple loop iteration`** messages:

```
Fastest Duration Multiple loops:
AverageSpeed() called for 'London to Brighton'
AverageSpeed() called for 'Newcastle to London'
AverageSpeed() called for 'New York to Florida'
AverageSpeed() called for 'Paris to Berlin'
Fastest Duration Multiple loop iteration 1:
New York to Florida: 1146 in 19 hours
Paris to Berlin: 546 in 10 hours
Fastest Duration Multiple loop iteration 2:
New York to Florida: 1146 in 19 hours
Paris to Berlin: 546 in 10 hours
```

Notice that from the collection of journeys made by a vehicle, the code returns the average speed of the vehicle during the journeys.

> **NOTE**
>
> You can find the code used for this example at https://packt.link/CIZJE.

SORTING OPERATIONS

There are five operations to sort items in a source. Items are **primarily sorted** and that can be followed by an optional **secondary sort**, which sorts the items within their primary group. For example, you can use a primary sort to sort a list of people firstly by the **City** property and then use a secondary sort to further sort them by the **Surname** property:

- **OrderBy**: Sorts values into ascending order.
- **OrderByDescending**: Sorts values into descending order.
- **ThenBy**: Sorts values that have been primarily sorted into a secondary ascending order.
- **ThenByDescending**: Sorts values that have been primarily sorted into a secondary descending order.
- **Reverse**: Simply returns a collection where the order of elements in the source is reversed. There is no expression equivalent.

ORDERBY AND ORDERBYDESCENDING

In this example, you will use the **System.IO** namespace to query files in the host machine's **temp** folder, rather than creating small objects from lists.

The static **Directory** class offers methods that can query the filesystem. **FileInfo** retrieves details about a specific file, such as its size or creation date. The **Path.GetTempPath** method returns the system's **temp** folder. To illustrate the point, in the Windows operating system, this can typically be found at **C:\Users\username\AppData\Local\Temp**, where **username** is a specific Windows login name. This will be different for other users and other systems:

1. In your **Chapter04\Examples** folder, add a new file called **LinqOrderByExamples.cs** and edit it as follows:

```
using System;
using System.IO;
using System.Linq;
namespace Chapter04.Examples
{
    class LinqOrderByExamples
    {
        public static void Main()
        {
```

2. Use the **Directory.EnumerateFiles** method to find all filenames with the **.tmp** extension in the **temp** folder:

```
            var fileInfos = Directory.EnumerateFiles(Path.
GetTempPath(), "*.tmp")
                .Select(filename => new FileInfo(filename))
                .ToList();
```

   Here, each filename is projected into a **FileInfo** instance and chained into a populated collection using **ToList**, which allows you to further query the resulting **fileInfos** details.

3. Next, the **OrderBy** method is used to sort the earliest files by comparing the **CreationTime** property of the file:

```
            Console.WriteLine("Earliest Files");
            foreach (var fileInfo in fileInfos.OrderBy(fi =>
fi.CreationTime))
            {
                Console.WriteLine($"{fileInfo.CreationTime:dd MMM yy}:
{fileInfo.Name}");
            }
```

4. To find the largest files, re-query **fileInfos** and sort each file by its **Length** property using **OrderByDescending**:

```
            Console.WriteLine("Largest Files");
            foreach (var fileInfo in fileInfos
                                        .OrderByDescending(fi =>
fi.Length))
            {
                Console.WriteLine($"{fileInfo.Length:N0} bytes:
\t{fileInfo.Name}");
            }
```

5. Finally, use **where** and **orderby** descending expressions to find the largest files that are less than **1,000** bytes in length:

```
            Console.WriteLine("Largest smaller files");
            foreach (var fileInfo in
                from fi in fileInfos
                where fi.Length < 1000
                orderby fi.Length descending
                select fi)
            {
                Console.WriteLine($"{fileInfo.Length:N0} bytes:
\t{fileInfo.Name}");
```

```
                }
                Console.ReadLine();
            }
        }
    }
```

6. Depending on the files in your **temp** folder, you should see an output like this:

```
Earliest Files
05 Jan 21: wct63C3.tmp
05 Jan 21: wctD308.tmp
05 Jan 21: wctFE7.tmp
04 Feb 21: wctE092.tmp

Largest Files
38,997,896 bytes:       wctE092.tmp
4,824,572 bytes:        cb6dfb76-4dc9-494d-9683-ce31eab43612.tmp
4,014,036 bytes:        492f224c-c811-41d6-8c5d-371359d520db.tmp

Largest smaller files
726 bytes:      wct38BC.tmp
726 bytes:      wctE239.tmp
512 bytes:      ~DF8CE3ED20D298A9EC.TMP
416 bytes:      TFR14D8.tmp
```

With this example, you have queried files in the host machine's **temp** folder, rather than creating small objects from lists.

> **NOTE**
>
> You can find the code used for this example at https://packt.link/mWeVC.

THENBY AND THENBYDESCENDING

The following example sorts popular quotes, based on the number of words found in each.

In your **Chapter04\Examples** folder, add a new file called **LinqThenByExamples.cs** and edit it as follows:

```
using System;
using System.IO;
```

```
using System.Linq;
namespace Chapter04.Examples
{

    class LinqThenByExamples
    {
        public static void Main()
        {
```

You start by declaring a string array of quotes as follows:

```
            var quotes = new[]
            {
                "Love for all hatred for none",
                "Change the world by being yourself",
                "Every moment is a fresh beginning",
                "Never regret anything that made you smile",
                "Die with memories not dreams",
                "Aspire to inspire before we expire"
            };
```

In the next snippet, each of these string quotes is projected into a new anonymous type based on the number of words in the quote (found using **`String.Split()`**). The items are first sorted in descending order to show those with the most words and then sorted in alphabetical order:

```
            foreach (var item in quotes
                .Select(q => new {Quote = q, Words = q.Split("
").Length})
                .OrderByDescending(q => q.Words)
                .ThenBy(q => q.Quote))
            {
                Console.WriteLine($"{item.Words}: {item.Quote}");
            }

            Console.ReadLine();
        }
    }
}
```

Running the code lists the quotes in word count order as follows:

```
7: Never regret anything that made you smile
6: Aspire to inspire before we expire
6: Change the world by being yourself
6: Every moment is a fresh beginning
6: Love for all hatred for none
5: Die with memories not dreams
```

Note how the quotes with six words are shown alphabetically.

The following (highlighted code) is the equivalent Query Expression with **orderby quote.Words descending** followed by the **quote.Words** ascending clause:

```
var query = from quote in
            (quotes.Select(q => new {Quote = q, Words = q.Split("
").Length}))
            orderby quote.Words descending, quote.Words ascending
            select quote;
foreach(var item in query)
            {
                Console.WriteLine($"{item.Words}: {item.Quote}");
            }

            Console.ReadLine();
        }
    }
}
```

NOTE

You can find the code used for this example at https://packt.link/YWJRz.

Now you have sorted popular quotes based on the number of words found in each. It is time to apply the skills learnt in the next exercise.

EXERCISE 4.03: FILTERING A LIST OF COUNTRIES BY CONTINENT AND SORTING BY AREA

In the preceding examples, you have looked at code that can select, filter, and sort a collection source. You will now combine these into an exercise that filters a small list of countries for two continents (South America and Africa) and sorts the results by geographical size.

Perform the following steps to do so:

1. In your **Chapter04\Exercises** folder, create a new **Exercise03** folder.
2. Add a new class called **Program.cs** in the **Exercise03** folder.
3. Start by adding a **Country** record that will be passed the **Name** of a country, the **Continent** to which it belongs, and its **Area** in square miles:

```
using System;
using System.Linq;

namespace Chapter04.Exercises.Exercise03
{
    class Program
    {
        record Country (string Name, string Continent, int Area);

        public static void Main()
        {
```

4. Now create a small subset of country data defined in an array, as follows:

```
            var countries = new[]
            {
                new Country("Seychelles", "Africa", 176),
                new Country("India", "Asia", 1_269_219),
                new Country("Brazil", "South America",3_287_956),
                new Country("Argentina", "South America", 1_073_500),
                new Country("Mexico", "South America",750_561),
                new Country("Peru", "South America",494_209),
                new Country("Algeria", "Africa", 919_595),
                new Country("Sudan", "Africa", 668_602)
            };
```

The array contains the name of a country, the continent it belongs to, and its geographical size in square miles.

5. Your search criteria must include **South America** or **Africa**. So define them in an array rather than hardcoding the **where** clause with two specific strings:

```
            var requiredContinents = new[] {"South America",
"Africa"};
```

This offers extra code flexibility should you need to alter it.

6. Build up a query by filtering and sorting by continent, sorting by area, and using the **.Select** extension method, which returns the **Index** and **item** value:

```
            var filteredCountries = countries
                .Where(c => requiredContinents.Contains(c.Continent))
                .OrderBy(c => c.Continent)
                .ThenByDescending(c => c.Area)
                .Select( (cty, i) => new {Index = i, Country = cty});


            foreach(var item in filteredCountries)
                Console.WriteLine($"{item.Index+1}: {item.Country.
Continent}, {item.Country.Name} = {item.Country.Area:N0} sq mi");
        }
    }
}
```

You finally project each into a new anonymous type to be written to the console.

7. Running the code block produces the following result:

```
1: Africa, Algeria = 919,595 sq mi
2: Africa, Sudan = 668,602 sq mi
3: Africa, Seychelles = 176 sq mi
4: South America, Brazil = 3,287,956 sq mi
5: South America, Argentina = 1,073,500 sq mi
6: South America, Mexico = 750,561 sq mi
7: South America, Peru = 494,209 sq mi
```

Notice that **Algeria** has the largest area in **Africa**, and **Brazil** has the largest area in **South America** (based on this small subset of data). Notice how you add **1** to each **Index** for readability (since starting at zero is less user-friendly).

> **NOTE**
>
> You can find the code used for this exercise at https://packt.link/Djddw.

You have seen how LINQ extension methods can be used to access items in a data source. Now, you will learn about partitioning data, which can be used to extract subsets of items.

PARTITIONING OPERATIONS

So far, you have looked at filtering the items in a data source that match a defined condition. Partitioning is used when you need to divide a data source into two distinct sections and return either of those two sections for subsequent processing.

For example, consider that you have a list of vehicles sorted by value and want to process the five least expensive vehicles using some method. If the list is sorted in ascending order, then you could partition the data using the **Take(5)** method (defined in the following paragraphs), which will extract the first five items and discard the remaining.

There are six partitioning operations that are used to split a source, with either of the two sections being returned. There are no partitioning Query Expressions:

- **Skip**: Returns a collection that skips items up to a specified numeric position in the source sequence. Used when you need to skip the first N items in a source collection.
- **SkipLast**: Returns a collection that skips the last N items in the source sequence.
- **SkipWhile**: Returns a collection that skips items in the source sequence that match a specified condition.

- **Take**: Returns a collection that contains the first N items in the sequence.
- **TakeLast**: Returns a collection that contains the last N items in the sequence.
- **TakeWhile**: Returns a collection that contains only those items that match the condition specified.

The following example demonstrates various **Skip** and **Take** operations on an unsorted list of exam grades. Here, you use **Skip(1)** to ignore the highest grade in a sorted list.

1. In your **Chapter04\Examples** folder, add a new file called **LinqSkipTakeExamples.cs** and edit it as follows:

```
using System;
using System.Linq;

namespace Chapter04.Examples
{
    class LinqSkipTakeExamples
    {
        public static void Main()
        {
            var grades = new[] {25, 95, 75, 40, 54, 9, 99};
            Console.Write("Skip: Highest Grades (skipping first):");
            foreach (var grade in grades
                .OrderByDescending(g => g)
                .Skip(1))
            {
                Console.Write($"{grade} ");
            }
            Console.WriteLine();
```

2. Next, the relational **is** operator is used to exclude those less than **25** or greater than **75**:

```
            Console.Write("SkipWhile@ Middle Grades (excluding 25 or
75):");
            foreach (var grade in grades
                .OrderByDescending(g => g)
                .SkipWhile(g => g is <= 25 or >=75))
            {
```

```
            Console.Write($"{grade} ");
        }
        Console.WriteLine();
```

3. By using **SkipLast**, you can show the bottom half of the results. Add the code for this as follows:

```
        Console.Write("SkipLast: Bottom Half Grades:");
        foreach (var grade in grades
            .OrderBy(g => g)
            .SkipLast(grades.Length / 2))
        {
            Console.Write($"{grade} ");
        }
        Console.WriteLine();
```

4. Finally, **Take(2)** is used here to show the two highest grades:

```
        Console.Write("Take: Two Highest Grades:");
        foreach (var grade in grades
            .OrderByDescending(g => g)
            .Take(2))
        {
            Console.Write($"{grade} ");
        }
    }
  }
}
```

5. Running the example produces this output, which is as expected:

```
Skip: Highest Grades (skipping first):95 75 54 40 25 9
SkipWhile Middle Grades (excluding 25 or 75):54 40 25 9
SkipLast: Bottom Half Grades:9 25 40 54
Take: Two Highest Grades:99 95
```

This example demonstrated the various **Skip** and **Take** operations on an unsorted list of exam grades.

> **NOTE**
>
> You can find the code used for this example at https://packt.link/TsDFk.

GROUPING OPERATIONS

GroupBy groups elements that share the same attribute. It is often used to group data or provide a count of items grouped by a common attribute. The result is an enumerable **IGrouping<K, V>** type collection, where **K** is the key type and **V** is the value type specified. **IGrouping** itself is enumerable as it contains all items that match the specified key.

For example, consider the next snippet, which groups a **List** of customer orders by name. In your **Chapter04\Examples** folder, add a new file called **LinqGroupByExamples.cs** and edit it as follows:

LinqGroupByExamples.cs

```
using System;
using System.Collections.Generic;
using System.Linq;

namespace Chapter04.Examples
{
    record CustomerOrder(string Name, string Product, int Quantity);

    class LinqGroupByExamples
    {
        public static void Main()
        {
            var orders = new List<CustomerOrder>
            {
                new CustomerOrder("Mr Green", "LED TV", 4),
                new CustomerOrder("Mr Smith", "iPhone", 2),
                new CustomerOrder("Mrs Jones", "Printer", 1),
```

You can find the complete code here: https://packt.link/GbwF2.

In this example, you have a list of **CustomerOrder** objects and want to group them by the **Name** property. For this, the **GroupBy** method is passed a **Func** delegate, which selects the **Name** property from each **CustomerOrder** instance.

Each item in the **GroupBy** result contains a **Key** (in this case, the customer's **Name**). You can then sort the grouping item to show the **CustomerOrders** items sorted by **Quantity**, as follows:

```
                foreach (var item in grouping.OrderByDescending(i => 
i.Quantity))
                {
                    Console.WriteLine($"\t{item.Product} * {item.
Quantity}");
                }
            }
```

```
            Console.ReadLine();
        }
    }
}
```

Running the code produces the following output:

```
Customer Mr Green:
        LED TV * 4
        MP3 Player * 1
        Microwave Oven * 1
Customer Mr Smith:
        PC * 5
        iPhone * 2
        Printer * 2
Customer Mrs Jones:
        Printer * 1
```

You can see the data is first grouped by customer **`Name`** and then ordered by order **`Quantity`** within each customer grouping. The equivalent Query Expression is written like this:

```
            var query = from order in orders
                        group order by order.Name;
            foreach (var grouping in query)
            {
                Console.WriteLine($"Customer {grouping.Key}:");
                foreach (var item in from item in grouping
                                     orderby item.Quantity descending
                                     select item)
                {
                    Console.WriteLine($"\t{item.Product} * {item.
Quantity}");
                }
            }
```

You have now seen some of the commonly used LINQ operators. You will now bring them together in an exercise.

EXERCISE 4.04: FINDING THE MOST COMMONLY USED WORDS IN A BOOK

In *Chapter 3, Delegates, Events, and Lambdas*, you used the **WebClient** class to download data from a website. In this exercise, you will use data downloaded from *Project Gutenberg*.

> **NOTE**
>
> Project Gutenberg is a library of 60,000 free eBooks. You can search online for Project Gutenberg or visit https://www.gutenberg.org/.

You will create a console app that allows the user to enter a URL. Then, you will download the book's text from the Project Gutenberg URL and use various LINQ statements to find the most frequent words in the book's text.

Additionally, you want to exclude some common **stop-words**; these are words such as **and**, **or**, and **the** that appear regularly in English, but add little to the meaning of a sentence. You will use the **Regex.Split** method to help split words more accurately than a simple space delimiter. Perform the following steps to do so:

> **NOTE**
>
> You can find more information on Regex can be found at https://packt.link/v4hGN.

1. In your **Chapter04\Exercises** folder, create a new **Exercise04** folder.
2. Add a new class called **Program.cs** in the **Exercise04** folder.
3. First, define the **TextCounter** class. This will be passed the path to a file, which you will add shortly. This should contain common English stop-words:

```
using System;
using System.Collections.Generic;
using System.IO;
using System.Linq;
using System.Net;
using System.Text;
using System.Text.RegularExpressions;

namespace Chapter04.Exercises.Exercise04
```

```
{
    class TextCounter
    {
        private readonly HashSet<string> _stopWords;

        public TextCounter(string stopWordPath)
        {
            Console.WriteLine($"Reading stop word file:
{stopWordPath}");
```

4. Using **File.ReadAllLines**, add each word into the **_stopWords HashSet**.

```
            _stopWords = new HashSet<string>(File.
ReadAllLines(stopWordPath));
        }
```

   You have used a **HashSet**, as each stop-word is unique.

5. Next, the **Process** method is passed a string that contains the book's text and the maximum number of words to show.

6. Return the result as a **Tuple<string, int>** collection, which saves you from having to create a class or record to hold the results:

```
        public IEnumerable<Tuple<string, int>> Process(string text,
                                                       int
maximumWords)
        {
```

7. Now perform the query part. Use **Regex.Split** with the pattern **@"\s+"** to split all the words.

   In its simplest form, this pattern splits a string into a list of words, typically using a space or punctuation marks to identify word boundaries. For example, the string **Hello Goodbye** would be split into an array that contains two elements, **Hello** and **Goodbye**. The returned string items are filtered via **where** to ensure all stop-words are ignored using the **Contains** method. The words are then grouped by value, **GroupBy(t=>t)**, projected to a **Tuple** using the word as a **Key**, and the number of times it occurs using **grp.Count**.

8. Finally, you sort by **Item2**, which for this **Tuple** is the word count, and then take only the required number of words:

```
            var words = Regex.Split(text.ToLower(), @"\s+")
                .Where(t => !_stopWords.Contains(t))
                .GroupBy(t => t)
                .Select(grp => Tuple.Create(grp.Key, grp.Count()))
```

```
                .OrderByDescending(tup => tup.Item2) //int
                .Take(maximumWords);

            return words;
        }
    }
```

9. Now start creating the main console app:

```
    class Program
    {
        public static void Main()
        {
```

10. Include a text file called **StopWords.txt** in the **Chapter04** source folder:

```
            const string StopWordFile = "StopWords.txt";
            var counter = new TextCounter(StopWordFile);
```

> **NOTE**
>
> You can find **StopWords.txt** on GitHub at https://packt.link/Vi8JH, or you can download any standard stop-word file, such as NLTK's https://packt.link/ZF1Tf. This file should be saved in the **Chapter04\Exercises** folder.

11. Once **TextCounter** has been created, prompt the user for a URL:

```
            string address;
            do
            {
                //https://www.gutenberg.org/files/64333/64333-0.txt
                Console.Write("Enter a Gutenberg book URL: ");
                address = Console.ReadLine();

                if (string.IsNullOrEmpty(address))
                    continue;
```

12. Enter a valid address and create a new **WebClient** instance and download the data file into a temporary file.

13. Perform extra processing to the text file before passing its contents to **TextCounter**:

```
using var client = new WebClient();
var tempFile = Path.GetTempFileName();
Console.WriteLine("Downloading...");
client.DownloadFile(address, tempFile);
```

The Gutenberg text files contain extra details such as the author and title. These can be read by reading each line in the file. The actual text of the book doesn't begin until finding a line that starts ***** START OF THE PROJECT GUTENBERG EBOOK**, so you need to read each line looking for this start message too:

```
Console.WriteLine($"Processing file {tempFile}");
const string StartIndicator = "*** START OF THE PROJECT GUTENBERG EBOOK";
//Title: The Little Review, October 1914(Vol. 1, No. 7)
//Author: Various
var title = string.Empty;
var author = string.Empty;
```

14. Next, append each line read into a **StringBuilder** instance, which is efficient for such string operations:

```
var bookText = new StringBuilder();
var isReadingBookText = false;
var bookTextLineCount = 0;
```

15. Now parse each line inside **tempFile**, looking for the **Author**, **Title**, or the **StartIndicator**:

```
foreach (var line in File.ReadAllLines(tempFile))
{
    if (line.StartsWith("Title"))
    {
        title = line;
    }
    else if (line.StartsWith("Author"))
    {
        author = line;
    }
    else if (line.StartsWith(StartIndicator))
```

```
            {
                isReadingBookText = true;
            }
            else if (isReadingBookText)
            {
                bookText.Append(line);
                bookTextLineCount++;
            }
        }
```

16. If the book text is found, provide a summary of lines and characters read before calling the **`counter.Process`** method. Here, you want the top **50** words:

```
        if (bookTextLineCount > 0)
        {
            Console.WriteLine($"Processing
{bookTextLineCount:N0} lines ({bookText.Length:N0} characters)..");
          var wordCounts = counter.Process(bookText.
ToString(), 50);
          Console.WriteLine(title);
          Console.WriteLine(author);
```

17. Once you have the results, use a **`foreach`** loop to output the word count details, adding a blank line to the output after every third word:

```
            var i = 0;
            //deconstruction
            foreach (var (word, count) in wordCounts)
            {
                Console.Write($"'{word}'={count}\t\t");
                i++;
                if (i % 3 == 0)
                {
                    Console.WriteLine();
                }
            }
            Console.WriteLine();
        }
        else
        {
```

18. Running the console app, using **`https://www.gutenberg.org/files/64333/64333-0.txt`** as an example URL produces the following output:

```
Reading stop word file: StopWords.txt
Enter a Gutenberg book URL: https://www.gutenberg.org/
files/64333/64333-0.txt
Downloading...
Processing file C:\Temp\tmpB0A3.tmp
Processing 4,063 lines (201,216 characters)..
Title: The Little Review, October 1914 (Vol. 1, No. 7)
Author: Various
'one'=108           'new'=95            'project'=62
'man'=56            'little'=54         'life'=52
'would'=51          'work'=50           'book'=42
'must'=42           'people'=39         'great'=37
'love'=37           'like'=36           'gutenberg-tm'=36
'may'=35            'men'=35            'us'=32
'could'=30          'every'=30          'first'=29
'full'=29           'world'=28          'mr.'=28
'old'=27            'never'=26          'without'=26
'make'=26           'young'=24          'among'=24
'modern'=23         'good'=23           'it.'=23
'even'=22           'war'=22            'might'=22
'long'=22           'cannot'=22         '_the'=22
'many'=21           'works'=21          'electronic'=21
'always'=20         'way'=20            'thing'=20
'day'=20            'upon'=20           'art'=20
'terms'=20          'made'=19
```

> **NOTE**
>
> Visual Studio might show the following when the code is run for the first time: **`warning SYSLIB0014: 'WebClient.WebClient()' is obsolete: 'WebRequest, HttpWebRequest, ServicePoint, and WebClient are obsolete. Use HttpClient instead.`**
>
> This is a recommendation to use the newer **`HttpClient`** class instead of the **`WebClient`** class. Both are, however, functionally equivalent.

The output shows a list of words found amongst the **4,063** lines of text downloaded. The counter shows that **one**, **new**, and **project** are the most popular words. Notice how **mr.**, **gutenberg-tm**, **it.**, and **_the** appear as words. This shows that the Regex expression used is not completely accurate when splitting words.

> **NOTE**
>
> You can find the code used for this exercise at https://packt.link/Q7Pf8.

An interesting enhancement to this exercise would be to sort the words by count, include a count of the stop words found, or find the average word length.

AGGREGATION OPERATIONS

Aggregation operations are used to compute a single value from a collection of values in a data source. An example could be the maximum, minimum, and average rainfall from data collected over a month:

- **Average**: Calculates the average value in a collection.
- **Count**: Counts the items that match a predicate.
- **Max**: Calculates the maximum value.
- **Min**: Calculates the minimum value.
- **Sum**: Calculates the sum of values.

The following example uses the **Process.GetProcess** method from the **System.Diagnostics** namespace to retrieve a list of processes currently running on the system:

In your **Chapter04\Examples** folder, add a new file called **LinqAggregationExamples.cs** and edit it as follows:

```
using System;
using System.Diagnostics;
using System.Linq;
namespace Chapter04.Examples
{
    class LinqAggregationExamples
    {
```

```
        public static void Main()
        {
```

First, **Process.GetProcesses().ToList()** is called to retrieve a list of the active processes running on the system:

```
            var processes = Process.GetProcesses().ToList();
```

Then, the **Count** extension method obtains a count of the items returned. Count has an additional overload, which accepts a **Func** delegate used to filter each of the items to be counted. The **Process** class has a **PrivateMemorySize64** property, which returns the number of bytes of memory the process is currently consuming, so you can use that to count the **small** processes, that is, those using less than **1,000,000** bytes of memory:

```
            var allProcesses = processes.Count;
            var smallProcesses = processes.Count(proc =>
                                          proc.PrivateMemorySize64 <
1_000_000);
```

Next, the **Average** extension method returns the overall average of a specific value for all items in the **processes** list. In this case, you use it to calculate the average memory consumption, using the **PrivateMemorySize64** property again:

```
            var average = processes.Average(p => p.PrivateMemorySize64);
```

The **PrivateMemorySize64** property is also used to calculate the maximum and minimum memory used for all processes, along with the total memory, as follows:

```
            var max = processes.Max(p => p.PrivateMemorySize64);
            var min = processes.Min(p => p.PrivateMemorySize64);
            var sum = processes.Sum(p => p.PrivateMemorySize64);
```

Once you have calculated the statistics, each value is written to the console:

```
            Console.WriteLine("Process Memory Details");
            Console.WriteLine($"  All Count: {allProcesses}");
            Console.WriteLine($"Small Count: {smallProcesses}");
            Console.WriteLine($"    Average: {FormatBytes(average)}");
            Console.WriteLine($"    Maximum: {FormatBytes(max)}");
            Console.WriteLine($"    Minimum: {FormatBytes(min)}");
            Console.WriteLine($"      Total: {FormatBytes(sum)}");
        }
```

In the preceding snippet, the **Count** method returns the number of all processes and, using the **Predicate** overload, you **Count** those where the memory is less than 1,000,000 bytes (by examining the **process.PrivateMemorySize64** property). You can also see that **Average**, **Max**, **Min**, and **Sum** are used to calculate statistics for process memory usage on the system.

> **NOTE**
>
> The aggregate operators will throw **InvalidOperationException** with the error **Sequence contains no elements** if you attempt to calculate using a source collection that contains no elements. You should check the **Count** or **Any** methods prior to calling any aggregate operators.

Finally, **FormatBytes** formats the amounts of memory into their megabyte equivalents:

```
        private static string FormatBytes(double bytes)
        {
            return $"{bytes / Math.Pow(1024, 2):N2} MB";
        }
    }
}
```

Running the example produces results similar to this:

```
Process Memory Details
  All Count: 305
Small Count: 5
    Average: 38.10 MB
    Maximum: 1,320.16 MB
    Minimum: 0.06 MB
      Total: 11,620.03 MB
```

From the output you will observe how the program retrieves a list of processes currently running on the system.

> **NOTE**
>
> You can find the code used for this example at https://packt.link/HI2eV.

QUANTIFIER OPERATIONS

Quantifier operations return a **bool** that indicates whether **all** or **some** elements in a sequence match a **Predicate** condition. This is often used to verify any elements in a collection match some criteria, rather than relying on **Count**, which enumerates **all** items in the collection, even if you need just one result.

Quantifier operations are accessed using the following extension methods:

- **All**: Returns **true** if **all** elements in the source sequence match a condition.
- **Any**: Returns **true** if **any** element in the source sequence matches a condition.
- **Contains**: Returns **true** if the source sequence contains the specified item.

The following card-dealing example selects three cards at random and returns a summary of those selected. The summary uses the **All** and **Any** extension methods to determine whether any of the cards were clubs or red and whether all cards were diamonds or an even number:

1. In your **Chapter04\Examples** folder, add a new file called **LinqAllAnyExamples.cs**.
2. Start by declaring an **enum** that represents each of the four suits in a pack of playing cards and a **record** class that defines a playing card:

```
using System;
using System.Collections.Generic;
using System.Linq;
namespace Chapter04.Examples
{
    enum PlayingCardSuit
    {
        Hearts,
        Clubs,
        Spades,
        Diamonds
    }

    record PlayingCard (int Number, PlayingCardSuit Suit)
    {
```

3. It is common practice to override the **ToString** method to provide a user-friendly way to describe an object's state at runtime. Here, the card's number and suit are returned as a string:

```
        public override string ToString()
        {
            return $"{Number} of {Suit}";
        }
    }
```

4. Now create a class to represent a deck of cards (for ease, only create cards numbered one to 10). The deck's constructor will populate the **_cards** collection with **10** cards for each of the suits:

```
    class Deck
    {
        private readonly List<PlayingCard> _cards = new();
        private readonly Random _random = new();

        public Deck()
        {
            for (var i = 1; i <= 10; i++)
            {
                _cards.Add(new PlayingCard(i, PlayingCardSuit.
Hearts));
                _cards.Add(new PlayingCard(i, PlayingCardSuit.
Clubs));
                _cards.Add(new PlayingCard(i, PlayingCardSuit.
Spades));
                _cards.Add(new PlayingCard(i, PlayingCardSuit.
Diamonds));
            }
        }
```

5. Next, the **Draw** method randomly selects a card from the **_cards** List, which it removes before returning to the caller:

```
        public PlayingCard Draw()
        {
            var index = _random.Next(_cards.Count);
            var drawnCard = _cards.ElementAt(index);
            _cards.Remove(drawnCard);
```

```
            return drawnCard;
        }
    }
```

6. The console app selects three cards using the deck's **Draw** method. Add the code for this as follows:

```
class LinqAllAnyExamples
{

    public static void Main()
    {
        var deck = new Deck();
        var hand = new List<PlayingCard>();

        for (var i = 0; i < 3; i++)
        {
            hand.Add(deck.Draw());
        }
```

7. To show a summary, use the **OrderByDescending** and **Select** operations to extract the user-friendly **ToString** description for each **PlayingCard**. This is then joined into a single delimited string as follows:

```
        var summary = string.Join(" | ",
            hand.OrderByDescending(c => c.Number)
                .Select(c => c.ToString()));
        Console.WriteLine($"Hand: {summary}");
```

8. Using **All** or **Any**, you can provide an overview of the cards and their score using the **Sum** of the card numbers. By using **Any**, you determine whether **any** of the cards in the hand are a club (the suit is equal to **PlayingCardSuit.Clubs**):

```
        Console.WriteLine($"Any Clubs: {hand.Any(card => card.
Suit == PlayingCardSuit.Clubs)}");
```

9. Similarly, **Any** is used to see if **any** of the cards belong to the **Hearts** or **Diamonds** suits, and therefore, are **Red**:

```
        Console.WriteLine($"Any Red: {hand.Any(card => card.Suit
==
PlayingCardSuit.Hearts || card.Suit == PlayingCardSuit.Diamonds)}");
```

10. In the next snippet, the **All** extension looks at every item in the collection and returns **true**, in this case, if **all** cards are **Diamonds**:

```
            Console.WriteLine($"All Diamonds: {hand.All(card => card.
Suit == PlayingCardSuit.Diamonds)}");
```

11. All is used again to see if all card numbers can be divided by two without a remainder, that is, whether they are even:

```
            Console.WriteLine($"All Even: {hand.All(card => card.
Number % 2 == 0)}");
```

12. Conclude by using the **Sum** aggregation method to calculate the value of the cards in the hand:

```
            Console.WriteLine($"Score :{hand.Sum(card => card.
Number)}");
        }
    }
}
```

13. Running the console app produces output like this:

```
Hand: 8 of Spades | 7 of Diamonds | 6 of Diamonds
Any Clubs: False
Any Red: True
All Diamonds: False
All Even: False
Score :21
```

The cards are randomly selected so you will have different hands each time you run the program. In this example, the score was **21**, which is often a winning hand in card games.

> **NOTE**
>
> You can find the code used for this example at https://packt.link/xPuTc.

JOIN OPERATIONS

Join operations are used to join two sources based on the association of objects in one data source with those that share a common attribute in a second data source. If you are familiar with database design, this can be thought of as a primary and foreign key relationship between tables.

A common example of a join is one where you have a one-way relationship, such as **Orders**, which has a property of type **Products**, but the **Products** class does not have a collection property that represents a backward relationship to a collection of **Orders**. By using a **Join** operator, you can create a backward relationship to show **Orders** for **Products**.

The two join extension methods are the following:

- **Join**: Joins two sequences using a key selector to extract pairs of values.
- **GroupJoin**: Joins two sequences using a key selector and groups the resulting items.

The following example contains three **Manufacturer** records, each with a unique **ManufacturerId**. These numeric IDs are used to define various **Car** records, but to save memory, you will not have a direct memory reference from **Manufacturer** back to **Car**. You will use the **Join** method to create an association between the **Manufacturer** and **Car** instances:

1. In your **Chapter04\Examples** folder, add a new file called **LinqJoinExamples.cs**.
2. First, declare the **Manufacturer** and **Car** records as follows:

```
using System;
using System.Collections.Generic;
using System.Linq;

namespace Chapter04.Examples
{
    record Manufacturer(int ManufacturerId, string Name);
    record Car (string Name, int ManufacturerId);
```

3. Inside the **Main** entry point, create two lists, one for the manufacturers and the other to represent the **cars**:

LinqJoinExamples.cs

```
    class LinqJoinExamples
    {

        public static void Main()
        {

            var manufacturers = new List<Manufacturer>
            {
                new(1, "Ford"),
                new(2, "BMW"),
                new(3, "VW")
            };

            var cars = new List<Car>
            {
                new("Focus", 1),
                new("Galaxy", 1),
                new("GT40", 1),
```

You can find the complete code here: https://packt.link/Ue7Fj.

4. At this point, there is no direct reference, but as you know, you can use **ManufacturerId** to link the two together using the **int** IDs. You can add the following code for this:

```
            var joinedQuery = manufacturers.Join(
                cars,
                manufacturer => manufacturer.ManufacturerId,
                car => car.ManufacturerId,
                (manufacturer, car) => new
                        {ManufacturerName = manufacturer.Name,
                         CarName = car.Name});


            foreach (var item in joinedQuery)
            {
                Console.WriteLine($"{item}");
            }
        }
    }
}
```

In the preceding snippet, the **`Join`** operation has various parameters. You pass in the **`cars`** list and define which properties in the **`manufacturer`** and **`car`** classes should be used to create the join. In this case, **`manufacturer.ManufacturerId = car.ManufacturerId`** determines the correct join.

Finally, the **`manufacturer`** and **`car`** arguments return a new anonymous type that contains the **`manufacturer.Name`** and **`car.Name`** properties.

5. Running the console app produces the following output:

```
{ ManufacturerName = Ford, CarName = Focus }
{ ManufacturerName = Ford, CarName = Galaxy }
{ ManufacturerName = Ford, CarName = GT40 }
{ ManufacturerName = BMW, CarName = 1 Series }
{ ManufacturerName = BMW, CarName = 2 Series }
{ ManufacturerName = VW, CarName = Golf }
{ ManufacturerName = VW, CarName = Polo }
```

As you can see, each of the **`Car`** and **`Manufacturer`** instances has been joined correctly using **`ManufacturerId`**.

6. The equivalent Query Expression would be as follows (note that in this case, it is a more concise format than the Query Operator syntax):

```
var query = from manufacturer in manufacturers
            join car in cars
              on manufacturer.ManufacturerId equals car.
ManufacturerId
              select new
              {
                ManufacturerName = manufacturer.Name, CarName = car.
Name
              };
foreach (var item in query)
{
  Console.WriteLine($"{item}");
}
```

> **NOTE**
>
> You can find the code used for this example at http://packt.link/Wh8jK.

Before you finish exploring LINQ, there is one more area related to LINQ Query Expressions—the **let** clause.

USING A LET CLAUSE IN QUERY EXPRESSIONS

In earlier Query Expressions, you are often required to repeat similar-looking code in various clauses. Using a **let** clause, you can introduce new variables inside an Expression Query and reuse the variable's value throughout the rest of the query. For example, consider the following query:

```
var stations = new List<string>
{
    "Kings Cross KGX",
    "Liverpool Street LVS",
    "Euston EUS",
    "New Street NST"
};

var query1 = from station in stations
             where station[^3..] == "LVS" || station[^3..] == "EUS" ||
                   station[0..^3].Trim().ToUpper().EndsWith("CROSS")
             select new { code= station[^3..],
                          name= station[0..^3].Trim().ToUpper()};
```

Here, you are searching for a station with the **LVS** or **EUS** code or a name ending in **CROSS**. To do this, you must extract the last three characters using a range, **station[^3..]**, but you have duplicated that in two **where** clauses and the final projection.

The station code and station names could both be converted into local variables using the **let** clause:

```
var query2 = from station in stations
             let code = station[^3..]
             let name = station[0..^3].Trim().ToUpper()
             where code == "LVS" || code == "EUS" ||
                   name.EndsWith("CROSS")
             select new {code, name};
```

Here, you have defined **code** and **name** using a **let** clause and reused them throughout the query. This code looks much neater and is also easier to follow and maintain.

Running the code produces the following output:

```
Station Codes:
KGX : KINGS CROSS
LVS : LIVERPOOL STREET
EUS : EUSTON
Station Codes (2):
KGX : KINGS CROSS
LVS : LIVERPOOL STREET
EUS : EUSTON
```

> **NOTE**
>
> You can find the code used for this example at https://packt.link/b2KiG.

By now you have seen the main parts of LINQ. Now you will now bring these together into an activity that filters a set of flight records based on a user's criteria and provides various statistics on the subset of flights found.

ACTIVITY 4.01: TREASURY FLIGHT DATA ANALYSIS

You have been asked to create a console app that allows the user to download publicly available flight data files and apply statistical analysis to the files. This analysis should be used to calculate a count of the total records found, along with the average, minimum, and maximum fare paid within that subset.

The user should be able to enter a number of commands and each command should add a specific filter based on the flight's class, origin, or destination properties. Once the user has entered the required criteria, the **go** command must be entered, and the console should run a query and output the results.

The data file you will use for this activity contains details of flights made by the UK's HM Treasury department between January 1 to December 31, 2011 (there are 714 records.) You will need to use **`WebClient.DownloadFile`** to download the data from the following URL: https://www.gov.uk/government/uploads/system/uploads/attachment_data/file/245855/HMT_-_2011_Air_Data.csv

> **NOTE**
>
> The website might open differently for Internet Explorer or Google Chrome. This depends on how IE or Chrome are configured on your machine. Using **`WebClient.DownloadFile`**, you can download the data as suggested.

Ideally, the program should download data once and then reread it from the local filesystem each time it is started.

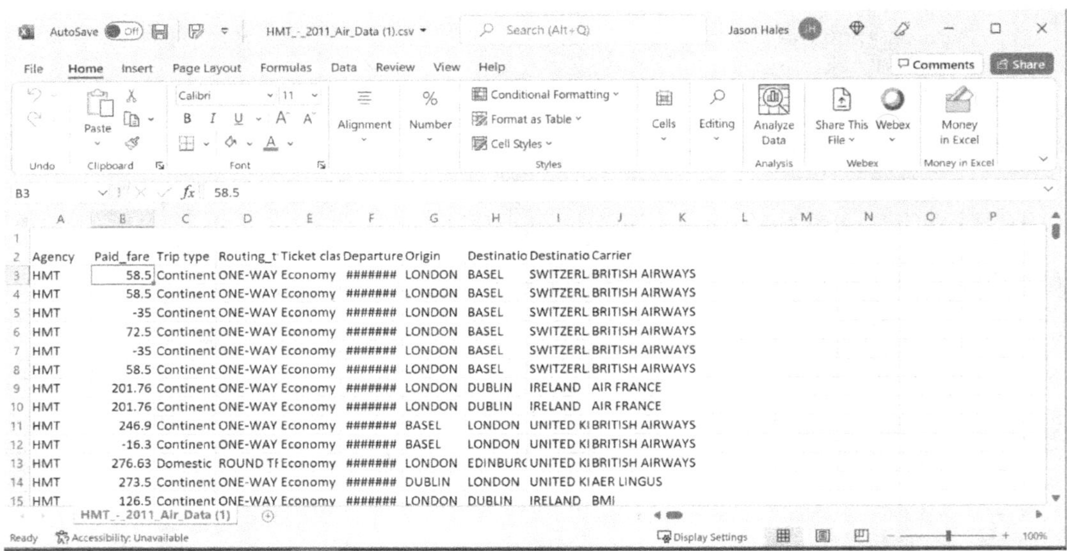

Figure 4.6: Preview of HM Treasury traffic data in Excel

Once downloaded, the data should then be read into a suitable record structure before being added to a collection, which allows various queries to be applied. The output should show the following aggregate values for all rows that match the user's criteria:

- Record count
- Average fare
- Minimum fare
- Maximum fare

The user should be able to enter the following console commands:

- **`Class c`**: Adds a class filter, where **`c`** is a flight class to search for, such as **`economy`** or **`Business class`**.
- **`Origin o`**: Adds an **`origin`** filter, where o is the flight origin, such as **`dublin`**, **`london`**, or **`basel`**.
- **`Destination d`**: Adds a destination filter, where **`d`** is the flight destination, such as **`delhi`**.
- **`Clear`**: Clears all filters.
- **`go`**: Applies the current filters.

If a user enters multiple filters of the same type, then these should be treated as an **`OR`** filter.

An **`enum`** can be used to identify the filter criteria type entered, as shown in the following line of code:

```
enum FilterCriteriaType {Class, Origin, Destination}
```

Similarly, a record can be used to store each filter type and comparison operand, as follows:

```
record FilterCriteria(FilterCriteriaType Filter, string Operand)
```

Each filter specified should be added to a **`List<FilterCriteria>`** instance. For example, if the user enters two origin filters, one for **`dublin`** and another for **`london`**, then the list should contain two objects, each representing an origin type filter.

When the user enters the **go** command, a query should be built that performs the following steps:

- Extracts all **class** filter values into a list of strings (**List<string>**).
- Extracts all **origin** filter values into **List<string>**.
- Extracts all **destination** filter values into **List<string>**.
- Uses a **where** extension method to filter the fight records for each criteria type specified using the **List<string>**. It contains a method to perform a case-insensitive search.

The following steps will help you complete this activity:

1. Create a new folder called **Activities** in the **Chapter04** folder.
2. Add a new folder called **Activity01** to that new folder.
3. Add a new class file called **Flight.cs**. This will be a **Record** class with fields that match those in the flight data. A **Record** class should be used as it offers a simple type purely to hold data rather than any form of behavior.
4. Add a new class file called **FlightLoader.cs**. This class will be used for downloading or importing data. **FlightLoader** should include a list of the field index positions within the data file, to be used when reading each line of data and splitting the contents into a string array, for example:

```
public const int Agency = 0;
public const int PaidFare = 1;
```

5. Now for the **FlightLoader** implementation, use a **static** class to define the index of known field positions in the data file. This will make it easier to handle any future changes in the layout of the data.
6. Next, a **Download** method should be passed a URL and destination file. Use **WebClient.DownloadFile** to download the data file and then defer to **Import** to process the downloaded file.
7. An **Import** method is to be added. This is passed the name of the local file to import (downloaded using the **Import** method) and will return a list of **Flight** records.
8. Add a class file called **FilterCriteria.cs**. This should contain a **FilterCriteriaType enum** definition. You will offer filters based on the flight's class, origin, and destination properties, so **FilterCriteriaType** should represent each of these.

9. Now, for the main filtering class, add a new class file called **FlightQuery.cs**. The constructor will be passed a **FlightLoader** instance. Within it, create a list named **_flights** to contain the data imported via **FlightLoader**. Create a **List<FilterCriteria>** instance named **_filters** that represent each of the criteria items that are added, each time the user specifies a new filter condition.
10. The **Import** and **Download** methods of **FlightLoader** should be called by the console at startup, allowing previously downloaded data to be processed, via the **_loader** instance.
11. Create a **Count** variable that returns the number of flight records that have been imported.
12. When the user specifies a filter to add, the console will call **AddFilter**, passing an **enum** to define the criteria type and the string value being filtered for.
13. **RunQuery** is the main method that returns those flights that match the user's criteria. You need to use the built-in **StringComparer.InvariantCultureIgnoreCase** comparer to ensure string comparison ignores any case differences. You define a query variable that calls **Select** on the flights; at the moment, this would result in a filtered result set.
14. Each of the types of filter available is string-based, so you need to extract all the string items. If there are any items to filter, you add an extra **Where** call to the query for each type (**Class**, **Destination**, or **Origin**). Each **Where** clause uses a **Contains** predicate, which examines the associated property.
15. Next, add the two helper methods used by **RunQuery**. **GetFiltersByType** is passed each of the **FilterCriteriaType** enums that represent a known type of criteria type and finds any of these in the list of filters using the **.Where** method. For example, if the user added two **Destination** criteria such as India and Germany, this would result in the two strings **India** and **Germany** being returned.
16. **FormatFilters** simply joins a list of **filterValues** strings into a user-friendly string with the word **OR** between each item, such as **London OR Dublin**.
17. Now create the main console app. Add a new class called **Program.cs**, which will allow the user to input requests and process their commands.
18. Hardcode the download URL and destination filename.

19. Create the main **FlightQuery** class, passing in a **FlightLoader** instance. If the app has been run before, you can **Import** the local flight data, or use **Download** if not.
20. Show a summary of the records imported and the available commands.
21. When the user enters a command, there might also be an argument, such as **destination united kingdom**, where **destination** is the command and **united kingdom** is the argument. To determine this, use the **IndexOf** method to find the location of the first space character in the input, if any.
22. For the **go** command, call **RunQuery** and use various aggregation operators on the results returned.
23. For the remaining commands, clear or add filters as requested. If the **Clear** command is specified, call the query's **ClearFilters** method, which will clear the list of criteria items.
24. If a **class** filter command is specified, call **AddFilter** specifying the **FilterCriteriaType.Class enum** and the string **Argument**.
25. The same pattern should be used for **Origin** and **Destination** commands. Call **AddFilter**, passing in the required **enum** value and the argument.

    The console output should be similar to the following, here listing the commands available to the user:

```
Commands: go | clear | class value | origin value | destination value
```

26. The user should be able to add two class filters, for **economy** or **Business Class** (all string comparisons should be case-insensitive), as shown in the following snippet:

```
Enter a command:class economy
Added filter: Class=economy
Enter a command:class Business Class
Added filter: Class=business class
```

27. Similarly, the user should be able to add an origin filter as follows (this example is for **london**):

```
Enter a command:origin london
Added filter: Origin=london
```

28. Adding the destination filter should look like this (this example is for **zurich**):

```
Enter a command:destination zurich
Added filter: Destination=zurich
```

29. Entering **go** should show a summary of all filters specified, followed by the results for flights that match the filters:

```
Enter a command:go
Classes: economy OR business class
Destinations: zurich
Origins: london
Results: Count=16, Avg=266.92, Min=-74.71, Max=443.49
```

> **NOTE**
>
> The solution to this activity can be found at https://packt.link/qclbF.

SUMMARY

In this chapter, you saw how the **IEnumerable** and **ICollection** interfaces form the basis of .NET data structures, and how they can be used to store multiple items. You created different types of collections depending on how each collection is meant to be used. You learned that the **List** collection is most extensively used to store collections of items, particularly if the number of elements is not known at compile time. You saw that the **Stack** and **Queue** types allow the order of items to be handled in a controlled manner, and how the **HashSet** offers set-based processing, while the **Dictionary** stores unique values using a key identifier.

You then further explored data structures by using LINQ Query Expressions and Query Operators to apply queries to data, showing how queries can be altered at runtime depending on filtering requirements. You sorted and partitioned data and saw how similar operations can be achieved using both Query Operators and Query Expressions, each offering a preference and flexibility based on context.

In the next chapter, you will see how parallel and asynchronous code can be used to run complex or long-running operations together.

5

CONCURRENCY: MULTITHREADING PARALLEL AND ASYNC CODE

OVERVIEW

C# and .NET provide a highly effective way to run concurrent code, making it easy to perform complex and often time-consuming actions. In this chapter, you will explore the various patterns that are available, from creating tasks using the **Task** factory methods to continuations to link tasks together, before moving on to the **async**/**await** keywords, which vastly simplify such code. By the end of this chapter, you will see how C# can be used to execute code that runs concurrently and often produces results far quicker than a single-threaded application.

INTRODUCTION

Concurrency is a generalized term that describes the ability of software to do more than one thing at the same time. By harnessing the power of concurrency, you can provide a more responsive user interface by offloading CPU-intensive activities from the main UI thread. On the server side, taking advantage of modern processing power through multi-processor and multi-core architectures, scalability can be achieved by processing operations in parallel.

Multithreading is a form of concurrency whereby multiple threads are used to perform operations. This is typically achieved by creating many **`Thread`** instances and coordinating operations between them. It is regarded as a legacy implementation, having largely been replaced by parallel and async programming; you may well find it used in older projects.

Parallel programming is a class of multithreading where similar operations are run independently of each other. Typically, the same operation is repeated using multiple loops, where the parameters or target of the operation themselves vary by iteration. .NET provides libraries that shield developers from the low-level complexities of thread creation. The phrase **embarrassingly parallel** is often used to describe an activity that requires little extra effort to be broken down into a set of tasks that can be run in parallel, often where there are few interactions between sub-tasks. One such example of parallel programming could be counting the number of words found in each text file within a folder. The job of opening a file and scanning through the words can be split into parallel tasks. Each task executes the same lines of code but is given a different text file to process.

Asynchronous programming is a more recent form of concurrency where an operation, once started, will complete at some point in the future, and the calling code is able to continue with other operations. This completion is often known as a **promise** (or a future in other languages) and is implemented through the task and its generic **`Task<>`** equivalent. In C# and .NET, async programming has become the preferred means to achieve concurrent operations.

A common application of asynchronous programming is where multiple slow-running or **expensive dependencies** need to be initialized and marshaled prior to calling a final step that should be called only when all or some of the dependencies are ready to be used. For example, a mobile hiking application may need to wait for a reliable GPS satellite signal, a planned navigation route, and a heart-rate monitoring service to be ready before the user can start hiking safely. Each of these distinct steps would be initialized using a dedicated task.

Another very common use case for asynchronous programming occurs in UI applications where, for example, saving a customer's order to a database could take 5-10 seconds to complete. This may involve validating the order, opening a connection to a remote server or database, packaging and sending the order in a format that can be transmitted over the wire, and then finally waiting for confirmation that the customer's order has been successfully stored in a database. In a single-threaded application, this would take much longer, and this delay would soon be noticed by the user. The application would become unresponsive until the operation was completed. In this scenario, the user may rightly think the application has crashed and may try to close it. That is not an ideal user experience.

Such issues can be mitigated by using async code that performs any of the slow operations using a dedicated task for each. These tasks may choose to provide feedback as they progress, which the UI's main thread can use to notify the user. Overall, the operation should complete sooner, thus freeing the user to continue interacting with the app. In modern applications, users have come to expect this method of operation. In fact, many UI guidelines suggest that if an operation may take more than a few seconds to complete, then it should be performed using async code.

Note that when code is executing, whether it's synchronous or asynchronous code, it is run within the context of a **`Thread`** instance. In the case of asynchronous code, this **`Thread`** instance is chosen by the .NET scheduler from a pool of available threads.

The **`Thread`** class has various properties but one of the most useful is **`ManagedThreadId`**, which will be used extensively throughout this chapter. This integer value serves to uniquely identify a thread within your process. By examining **`Thread.ManagedThreadId`**, you can determine that multiple thread instances are being used. This can be done by accessing the **`Thread`** instance from within your code using the static **`Thread.CurrentThread`** method.

For example, if you started five long-running tasks and examined the **`Thread.ManagedThreadId`** for each, you would observe five unique IDs, possibly numbered as two, three, four, five, and six. In most cases, the thread with ID number one is the process's **main thread**, created when the process first starts.

Keeping track of thread IDs can be quite useful, especially when you have time-consuming operations to perform. As you have seen, using concurrent programming, multiple operations can be executed at the same time, rather than using a traditional single-threaded approach, where one operation must complete before a subsequent operation can start.

In the physical world, consider the case of building a train tunnel through a mountain. Starting at one side of a mountain and tunneling through to the other side could be made considerably faster if two teams started on opposite sides of the mountain, both tunneling toward each other. The two teams could be left to work independently; any issues experienced by a team on one side should not have an adverse effect on the other side's team. Once both sides have completed their tunneling, there should be one single tunnel, and the construction could then continue with the next task, such as laying the train line.

The next section will look at using the C# **Task** class, which allows you to execute blocks of code at the same time and independently of each other. Consider again the example of the UI app, where the customer's order needs to be saved to a database. For this, you would have two options:

Option 1 is to create a C# **Task** that performs each step one after another:

- Validate the order.
- Connect to the server.
- Send the request.
- Wait for a response.

Option 2 is to create a C# **Task** for each of the steps, executing each in parallel where possible.

Both options achieve the same end result, freeing the UI's main thread to respond to user interactions. Option one may well be slower to finish, but the upside is that this would require simpler code. However, Option two would be the preferred choice as you are offloading multiple steps, so it should complete sooner. Although, this could involve additional complexity as you may need to coordinate each of the individual tasks as they are complete.

In the upcoming sections, you will first get a look at how Option one could be approached, that is, using a single **Task** to run blocks of code, before moving on to the complexity of Option two where multiple tasks are used and coordinated.

RUNNING ASYNCHRONOUS CODE USING TASKS

The **Task** class is used to execute blocks of code asynchronously. Its usage has been somewhat superseded by the newer **async** and **await** keywords, but this section will cover the basics of creating tasks as they tend to be pervasive in larger or mature C# applications and form the backbone of the **async/await** keywords.

In C#, there are three ways to schedule asynchronous code to run using the **Task** class and its generic equivalent **Task<T>**.

CREATING A NEW TASK

You'll start off with the simplest form, one that performs an operation but does not return a result back to the caller. You can declare a **Task** instance by calling any of the **Task** constructors and passing in an **Action** based delegate. This delegate contains the actual code to be executed at some point in the future. Many of the constructor overloads allow **cancellation tokens** and **creation options** to further control how the **Task** runs.

Some of the commonly used constructors are as follows:

- **public Task(Action action)**: The **Action** delegate represents the body of code to be run.
- **public Task(Action action, CancellationToken cancellationToken)**: The **CancellationToken** parameter can be used as a way to interrupt the code that is running. Typically, this is used where the caller has been provided with a means to request that an operation be stopped, such as adding a **Cancel** button that a user can press.
- **public Task(Action action, TaskCreationOptions creationOptions)**: **TaskCreationOptions** offers a way to control how the **Task** is run, allowing you to provide hints to the scheduler that a certain **Task** might take extra time to complete. This can help when running related tasks together.

The following are the most often used **Task** properties:

- **public bool IsCompleted { get; }**: Returns **true** if the **Task** completed (completion does not indicate success).
- **public bool IsCompletedSuccessfully { get; }**: Returns **true** if the **Task** completed successfully.
- **public bool IsCanceled { get; }**: Returns **true** if the **Task** was canceled prior to completion.
- **public bool IsFaulted { get; }**: Returns **true** if the **Task** has thrown an unhandled exception prior to completion.
- **public TaskStatus Status { get; }**: Returns an indicator of the task's current status, such as **Canceled**, **Running**, or **WaitingToRun**.
- **public AggregateException Exception { get; }**: Returns the exception, if any, that caused the **Task** to end prematurely.

Note that the code within the **Action** delegate is not executed until sometime after the **Start()** method is called. This may well be some milliseconds after and is determined by the .NET scheduler.

Start here by creating a new VS Code console app, adding a utility class named **Logger**, which you will use in the exercises and examples going forward. It will be used to log a message to the console along with the current time and current thread's **ManagedThreadId**.

The steps for this are as follows:

1. Change to your source folder.
2. Create a new console app project called **Chapter05** by running the following command:

```
source>dotnet new console -o Chapter05
```

3. Rename the **Class1.cs** file to **Logger.cs** and remove all the template code.

4. Be sure to include the **System** and **System.Threading** namespaces. **System.Threading** contains the **Threading** based classes:

```
using System;
using System.Threading;

namespace Chapter05
{
```

5. Mark the **Logger** class as static so that it can be used without having to create an instance to use:

```
    public static class Logger
    {
```

> **NOTE**
>
> If you use the **Chapter05** namespace, then the **Logger** class will be accessible to code in examples and activities, provided they also use the **Chapter05** namespace. If you prefer to create a folder for each example and exercise, then you should copy the file **Logger.cs** into each folder that you create.

6. Now declare a **static** method called **Log** that is passed a **string message** parameter:

```
        public static void Log(string message)
        {
            Console.WriteLine($"{DateTime.Now:T} [{Thread.
CurrentThread.ManagedThreadId:00}] {message}");
        }
    }
}
```

When invoked, this will log a message to the console window using the **WriteLine** method. In the preceding snippet, the string interpolation feature in C# is used to define a string using the **$** symbol; here, **:T** will format the current time (**DateTime.Now**) into a time-formatted string and **:00** is used to include **Thread.ManagedThreadId** with a leading 0.

Thus, you have created the static Logger class that will be used throughout the rest of this chapter.

> **NOTE**
>
> You can find the code used for this example at https://packt.link/cg6c5.

In the next example, you will use the **`Logger`** class to log details when a thread is about to start and finish.

7. Start by adding a new class file called **`TaskExamples.cs`**:

```
using System;
using System.Threading;
using System.Threading.Tasks;

namespace Chapter05.Examples
{
    class TaskExamples
    {
```

8. The **`Main`** entry point will log that **`taskA`** is being created:

```
        public static void Main()
        {
            Logger.Log("Creating taskA");
```

9. Next, add the following code:

```
            var taskA = new Task(() =>
            {
                Logger.Log("Inside taskA");
                Thread.Sleep(TimeSpan.FromSeconds(5D));
                Logger.Log("Leaving taskA");
            });
```

Here, the simplest **`Task`** constructor is passed an **`Action`** lambda statement, which is the actual target code that you want to execute. The target code writes the message **`Inside taskA`** to the console. It pauses for five seconds using **`Thread.Sleep`** to block the current thread, thus simulating a long-running activity, before finally writing **`Leaving taskA`** to the console.

10. Now that you have created **taskA**, confirm that it will only invoke its target code when the **Start()** method is called. You will do this by logging a message immediately before and after the method is called:

```
            Logger.Log($"Starting taskA. Status={taskA.Status}");
            taskA.Start();
            Logger.Log($"Started taskA. Status={taskA.Status}");
            Console.ReadLine();
        }
    }
}
```

11. Copy the contents of **Logger.cs** file to same folder as the **TaskExamples.cs** example.

12. Next run the console app to produce the following output:

```
10:47:34 [01] Creating taskA
10:47:34 [01] Starting taskA. Status=Created
10:47:34 [01] Started taskA. Status=WaitingToRun
10:47:34 [03] Inside taskA
10:47:39 [03] Leaving taskA
```

Note that the task's status is **WaitingToRun** even after you've called **Start**. This is because you are asking the .NET scheduler to schedule the code to run—that is, to add it to its queue of pending actions. Depending on how busy your application is with other tasks, it may not run immediately after you've called **Start**.

> **NOTE**
>
> You can find the code used for this example at https://packt.link/DHxt3.

In earlier versions of C#, this was the main way to create and start **Task** objects directly. It is no longer recommended and is only included here as you may find it used in older code. Its usage has been replaced by the **Task.Run** or **Task.Factory.StartNew** static factory methods, which offer a simpler interface for the most common usage scenarios.

USING TASK.FACTORY.STARTNEW

The static method **Task.Factory.StartNew** contains various overloads that make it easier to create and configure a **Task**. Notice how the method is named **StartNew**. It creates a **Task** and automatically starts the method for you. The .NET team recognized that there is little value in creating a **Task** that is not immediately started after it is first created. Typically, you would want the **Task** to start performing its operation right away.

The first parameter is the familiar **Action** delegate to be executed, followed by optional cancelation tokens, creation options, and a **TaskScheduler** instance.

The following are some of the common overloads:

- **Task.Factory.StartNew(Action action)**: The **Action** delegate contains the code to execute, as you have seen previously.
- **Task.Factory.StartNew(Action action, CancellationToken cancellationToken)**: Here, **CancellationToken** coordinates the cancellation of the task.
- **Task.Factory.StartNew(Action<object> action, object state, CancellationToken cancellationToken, TaskCreationOptions creationOptions, TaskScheduler scheduler)**: The **TaskScheduler** parameter allows you to specify a type of low-level scheduler responsible for queuing tasks. This option is rarely used.

Consider the following code, which uses the first and simplest overload:

```
var taskB = Task.Factory.StartNew((() =>
{
  Logger.Log("Inside taskB");
  Thread.Sleep(TimeSpan.FromSeconds(3D));
  Logger.Log("Leaving taskB");
}));
Logger.Log($"Started taskB. Status={taskB.Status}");
Console.ReadLine();
```

Running this code produces the following output:

```
21:37:42 [01] Started taskB. Status=WaitingToRun
21:37:42 [03] Inside taskB
21:37:45 [03] Leaving taskB
```

From the output, you can see that this code achieves the same result as creating a **`Task`** but is more concise. The main point to consider is that **`Task.Factory.StartNew`** was added to C# to make it easier to create tasks that are started for you. It was preferable to use **`StartNew`** rather than creating tasks directly.

> **NOTE**
>
> The term **Factory** is often used in software development to represent methods that help create objects.

`Task.Factory.StartNew` provides a highly configurable way to start tasks, but in reality, many of the overloads are rarely used and need a lot of extra parameters to be passed to them. As such, **`Task.Factory.StartNew`** itself has also become somewhat obsolete in favor of the newer **`Task.Run`** static method. Still, the **`Task.Factory.StartNew`** is briefly covered as you may see it used in legacy C# applications.

USING TASK.RUN

The alternative and preferred **`static`** factory method, **`Task.Run`**, has various overloads and was added later to .NET to simplify and shortcut the most common task scenarios. It is preferable for newer code to use **`Task.Run`** to create started tasks, as far fewer parameters are needed to achieve common threading operations.

Some of the common overloads are as follows:

- **`public static Task Run(Action action)`**: Contains the **`Action`** delegate code to execute.
- **`public static Task Run(Action action, CancellationToken cancellationToken)`**: Additionally contains a cancelation token used to coordinate the cancellation of a task.

For example, consider the following code:

```
var taskC = Task.Run(() =>
{
  Logger.Log("Inside taskC");
  Thread.Sleep(TimeSpan.FromSeconds(1D));
  Logger.Log("Leaving taskC");
  });
Logger.Log($"Started taskC. Status={taskC.Status}");
Console.ReadLine();
```

Running this code will produce the following output:

```
21:40:27 [03] Inside taskC
21:40:27 [01] Started taskC. Status=WaitingToRun
21:40:28 [03] Leaving taskC
```

As you can see, the output is pretty similar to the outputs of the previous two code snippets. Each wait for a shorter time than its predecessor before the associated **Action** delegate completes.

The main difference is that creating a **Task** instance directly is an obsolete practice but will allow you to add an extra logging call before you explicitly call the **Start** method. That is the only benefit in creating a **Task** directly, which is not a particularly compelling reason to do so.

Running all three examples together produces this:

```
21:45:52 [01] Creating taskA
21:45:52 [01] Starting taskA. Status=Created
21:45:52 [01] Started taskA. Status=WaitingToRun
21:45:52 [01] Started taskB. Status=WaitingToRun
21:45:52 [01] Started taskC. Status=WaitingToRun
21:45:52 [04] Inside taskB
21:45:52 [03] Inside taskA
21:45:52 [05] Inside taskC
21:45:53 [05] Leaving taskC
21:45:55 [04] Leaving taskB
21:45:57 [03] Leaving taskA
```

You can see various **ManagedThreadIds** being logged and that **taskC** completes before **taskB**, which completes before **taskA**, due to the decreasing number of seconds specified in the **Thread.Sleep** calls in each case.

It is preferable to favor either of the two static methods, but which should you use when scheduling a new task? **Task.Run** should be used for the **majority** of cases where you need to simply offload some work onto the thread pool. Internally, **Task.Run** defers down to **Task.Factory.StartNew**.

Task.Factory.StartNew should be used where you have more advanced requirements, such as defining where tasks are queued, by using any of the overloads that accept a **TaskScheduler** instance, but in practice, this is seldom the requirement.

> **NOTE**
>
> You can find more information on **Task.Run** and **Task.Factory.StartNew** at https://devblogs.microsoft.com/pfxteam/task-run-vs-task-factory-startnew/ and https://blog.stephencleary.com/2013/08/startnew-is-dangerous.html.

So far, you have seen how small tasks can be started, each with a small delay before completion. Such delays can simulate the effect caused by code accessing slow network connections or running complex calculations. In the following exercise, you'll extend your **Task.Run** knowledge by starting multiple tasks that run increasingly longer numeric calculations.

This serves as an example to show how potentially complex tasks can be started and allowed to run to completion in isolation from one another. Note that in a traditional synchronous implementation, the throughput of such calculations would be severely restricted, owing to the need to wait for one operation to complete before the next one can commence. It is now time to practice what you have learned through an exercise.

EXERCISE 5.01: USING TASKS TO PERFORM MULTIPLE SLOW-RUNNING CALCULATIONS

In this exercise, you will create a recursive function, **Fibonacci**, which calls itself twice to calculate a cumulative value. This is an example of potentially slow-running code rather than using **Thread.Sleep** to simulate a slow call. You will create a console app that repeatedly prompts for a number to be entered. The larger this number, the longer each task will take to calculate and output its result. The following steps will help you complete this exercise:

1. In the **Chapter05** folder, add a new folder called **Exercises**. Inside that folder, add a new folder called **Exercise01**. You should have the folder structure as **Chapter05\Exercises\Exercise01**.
2. Create a new file called **Program.cs**.
3. Add the recursive **Fibonacci** function as follows. You can save a little processing time by returning **1** if the requested iteration is less than or equal to **2**:

```
using System;
using System.Globalization;
using System.Threading;
using System.Threading.Tasks;

namespace Chapter05.Exercises.Exercise01
{
  class Program
  {
        private static long Fibonacci(int n)
        {
            if (n <= 2L)
                return 1L;

            return Fibonacci(n - 1) + Fibonacci(n - 2);
        }
```

4. Add the **static Main** entry point to the console app and use a **do**-loop to prompt for a number to be entered.

5. Use **`int.TryParse`** to convert the string into an integer if the user enters a string:

```
            public static void Main()
            {
                string input;
                do
                {
                    Console.WriteLine("Enter number:");
                    input = Console.ReadLine();
                    if (!string.IsNullOrEmpty(input) &&
                        int.TryParse(input, NumberStyles.Any, 
CultureInfo.CurrentCulture, out var number))
```

6. Define a lambda statement that captures the current time using **`DateTime.Now`**, calls the slow-running **`Fibonacci`** function, and logs the time taken to run:

```
                    {
                        Task.Run(() =>
                        {
                            var now = DateTime.Now;
                            var fib = Fibonacci(number);
                            var duration = DateTime.Now.Subtract(now);
                            Logger.Log($"Fibonacci {number:N0} = {fib:N0} 
(elapsed time: {duration.TotalSeconds:N0} secs)");
                        });
                    }
```

The lambda is passed to **`Task.Run`** and will be started by **`Task.Run`** shortly, freeing the **`do-while`** loop to prompt for another number.

7. The program shall exit the loop when an empty value is entered:

```
                } while (input != string.Empty);
            }
        }
    }
```

8. For running the console app, start by entering the numbers **1** and then **2**. As these are very quick calculations, they both return in under one second.

> **NOTE**
>
> The first time you run this program, Visual Studio will show a warning similar to **"Converting null literal or possible null value to non-nullable type"**. This is a reminder that you are using a variable that could be a null value.

```
Enter number:1
Enter number:2
11:25:11 [04] Fibonacci 1 = 1 (elapsed time: 0 secs)
Enter number:45
11:25:12 [04] Fibonacci 2 = 1 (elapsed time: 0 secs)
Enter number:44
Enter number:43
Enter number:42
Enter number:41
Enter number:40
Enter number:10
11:25:35 [08] Fibonacci 41 = 165,580,141 (elapsed time: 4 secs)
11:25:35 [09] Fibonacci 40 = 102,334,155 (elapsed time: 2 secs)
11:25:36 [07] Fibonacci 42 = 267,914,296 (elapsed time: 6 secs)
Enter number: 39
11:25:36 [09] Fibonacci 10 = 55 (elapsed time: 0 secs)
11:25:37 [05] Fibonacci 43 = 433,494,437 (elapsed time: 9 secs)
11:25:38 [06] Fibonacci 44 = 701,408,733 (elapsed time: 16 secs)
Enter number:38
11:25:44 [06] Fibonacci 38 = 39,088,169 (elapsed time: 1 secs)
11:25:44 [05] Fibonacci 39 = 63,245,986 (elapsed time: 2 secs)
11:25:48 [04] Fibonacci 45 = 1,134,903,170 (elapsed time: 27 secs)
```

Notice how the **ThreadId** is **[04]** for both **1** and **2**. This shows that the same thread was used by **Task.Run** for both iterations. By the time **2** was entered, the previous calculation had already been completed. So .NET decided to reuse thread **04** again. The same occurs for the value **45**, which took **27** seconds to complete even though it was the third requested.

You can see that entering values above **40** causes the elapsed time to increase quite dramatically (for each increase by one, the time taken almost doubles). Starting with higher numbers and descending downward, you can see that the calculations for **41**, **40**, and **42** were all completed before **44** and **43**, even though they were started at similar times. In a few instances, the same thread appears twice. Again, this is .NET re-using idle threads to run the task's action.

> **NOTE**
>
> You can find the code used for this exercise at https://packt.link/YLYd4.

COORDINATING TASKS

In the previous *Exercise 5.01*, you saw how multiple tasks can be started and left to run to completion without any interaction between the individual tasks. One such scenario is a process that needs to search a folder looking for image files, adding a copyright watermark to each image file found. The process can use multiple tasks, each working on a distinct file. There would be no need to coordinate each task and its resulting image.

Conversely, it is quite common to start various long-running tasks and only continue when some or all of the tasks have completed; maybe you have a collection of complex calculations that need to be started and can only perform a final calculation once the others have completed.

In the *Introduction* section, it was mentioned that a hiking application needed a GPS satellite signal, navigation route, and a heart rate monitor before it could be used safely. Each of these dependencies can be created using a **Task** and only when all of them have signaled that they are ready to be used should the application then allow the user to start with their route.

Over the next sections, you will cover various ways offered by C# to coordinate tasks. For example, you may have a requirement to start many independent tasks running, each running a complex calculation, and need to calculate a final value once all the previous tasks have completed. You may either like to start downloading data from multiple websites but want to cancel the downloads that are taking too long to complete. The next section will cover this scenario.

WAITING FOR TASKS TO COMPLETE

`Task.Wait` can be used to wait for an individual task to complete. If you are working with multiple tasks, then the static **`Task.WaitAll`** method will wait for all tasks to complete. The **`WaitAll`** overloads allow cancellation and timeout options to be passed in, with most returning a Boolean value to indicate success or failure, as you can see in the following list:

- **`public static bool WaitAll(Task[] tasks, TimeSpan timeout)`**: This is passed an array of **`Task`** items to wait for. It returns **`true`** if **all** of the tasks complete within the maximum time period specified (**`TimeSpan`** allows specific units such as hours, minutes, and seconds to be expressed).
- **`public static void WaitAll(Task[] tasks, CancellationToken cancellationToken)`**: This is passed an array of **`Task`** items to wait for, and a cancellation token that can be used to coordinate the cancellation of the tasks.
- **`public static bool WaitAll(Task[] tasks, int millisecondsTimeout, CancellationToken cancellationToken)`**: This is passed an array of **`Task`** items to wait for and a cancellation token that can be used to coordinate the cancellation of the tasks. **`millisecondsTimeout`** specifies the number of milliseconds to wait for all tasks to complete by.
- **`public static void WaitAll(params Task[] tasks)`**: This allows an array of **`Task`** items to wait for.

  If you need to wait for any task to complete from a list of tasks, then you can use **`Task.WaitAny`**. All of the **`WaitAny`** overloads return either the index number of the first completed task or **`-1`** if a timeout occurred (the maximum amount of time to wait for).

  For example, if you pass an array of five Task items and the last Task in that array completes, then you will be returned the value four (array indexes always start counting at zero).

- **`public static int WaitAny(Task[] tasks, int millisecondsTimeout, CancellationToken cancellationToken)`**: This is passed an array of **`Task`** items to wait for, the number of milliseconds to wait for any **`Task`** to complete by, and a cancellation token that can be used to coordinate the cancellation of the tasks.

- **`public static int WaitAny(params Task[] tasks)`**: This is passed an array of **`Task`** items to wait for any **`Task`** to be completed.
- **`public static int WaitAny(Task[] tasks, int millisecondsTimeout)`**: Here, you pass the number of milliseconds to wait for any tasks to complete.
- **`public static int WaitAny(Task[] tasks, CancellationToken cancellationToken) CancellationToken`**: This is passed a cancellation token that can be used to coordinate the cancellation of the tasks.
- **`public static int WaitAny(Task[] tasks, TimeSpan timeout)`**: This is passed the maximum time period to wait for.

Calling **`Wait`**, **`WaitAll`**, or **`WaitAny`** will block the current thread, which can negate the benefits of using a task in the first place. For this reason, it is preferable to call these from within an **awaitable task**, such as via **`Task.Run`** as the following example shows.

The code creates **`outerTask`** with a lambda statement, which itself then creates two inner tasks, **`inner1`**, and **`inner2`**. **`WaitAny`** is used to get the index of the **first** inner task to complete. In this example, **`inner2`** will complete first as it pauses for a shorter time, so the resulting index value will be **`1`**:

`TaskWaitAnyExample.cs`

```
var outerTask = Task.Run( () =>
{
    Logger.Log("Inside outerTask");
    var inner1 = Task.Run(() =>
    {
        Logger.Log("Inside inner1");
        Thread.Sleep(TimeSpan.FromSeconds(3D));
    });
    var inner2 = Task.Run(() =>
     {
         Logger.Log("Inside inner2");
         Thread.Sleep(TimeSpan.FromSeconds(2D));
     });

     Logger.Log("Calling WaitAny on outerTask");
```

You can find the complete code here: http://packt.link/CicWk.

When the code runs, it produces the following output:

```
15:47:43 [04] Inside outerTask
15:47:43 [01] Press ENTER
15:47:44 [04] Calling WaitAny on outerTask
15:47:44 [05] Inside inner1
15:47:44 [06] Inside inner2
15:47:46 [04] Waitany index=1
```

The application remains responsive because you called **WaitAny** from inside a **Task**. You have not blocked the application's main thread. As you can see, thread ID **01** has logged this message: **15:47:43 [01] Press ENTER**.

This type of pattern can be used in cases where you need to fire and forget a task. For example, you may want to log an informational message to a database or a log file, but it is not essential that the flow of the program is altered if either task fails to complete.

A common progression from fire-and-forget tasks is those cases where you need to wait for several tasks to complete within a certain time limit. The next exercise will cover this scenario.

EXERCISE 5.02: WAITING FOR MULTIPLE TASKS TO COMPLETE WITHIN A TIME PERIOD

In this exercise, you will start three long-running tasks and decide your next course of action if they all completed within a certain randomly selected time span.

Here, you will see the generic **Task<T>** class being used. The **Task<T>** class includes a **Value** property that can be used to access the result of **Task** (in this exercise, it is a string-based generic, so **Value** will be a string type). You won't use the **Value** property here as the purpose of this exercise is to show that void and generic tasks can be waited for together. Perform the following steps to complete this exercise:

1. Add the main entry point to the console app:

```
using System;
using System.Threading;
using System.Threading.Tasks;

namespace Chapter05.Exercises.Exercise02
{
    class Program
    {
        public static void Main()
        {
            Logger.Log("Starting");
```

2. Declare a variable named **taskA**, passing **Task.Run** a lambda that pauses the current thread for **5** seconds:

```
            var taskA = Task.Run( () =>
            {
                Logger.Log("Inside TaskA");
                Thread.Sleep(TimeSpan.FromSeconds(5));
                Logger.Log("Leaving TaskA");
                return "All done A";
            });
```

3. Create two more tasks using the **method group syntax**:

```
            var taskB = Task.Run(TaskBActivity);
            var taskC = Task.Run(TaskCActivity);
```

   As you may recall, this shorter syntax can be used if the compiler can determine the type of argument required for a zero- or single-parameter method.

4. Now pick a random maximum timeout in seconds. This means that either of the two tasks may **not** complete before the timeout period has elapsed:

```
            var timeout = TimeSpan.FromSeconds(new Random().Next(1, 
10));
            Logger.Log($"Waiting max {timeout.TotalSeconds} 
seconds...");
```

   Note that each of the tasks will still run to completion as you have not added a mechanism to stop executing the code inside the body of the **Task.Run Action** lambda.

5. Call **WaitAll**, passing in the three tasks and the **timeout** period:

```
            var allDone = Task.WaitAll(new[] {taskA, taskB, taskC}, 
timeout);
            Logger.Log($"AllDone={allDone}: TaskA={taskA.Status}, 
TaskB={taskB.Status}, TaskC={taskC.Status}");

            Console.WriteLine("Press ENTER to quit");
            Console.ReadLine();
        }
```

   This will return **true** if all tasks complete in time. You will then log the status of all tasks and wait for **Enter** to be pressed to exit the application.

6. Finish off by adding two slow-running **Action** methods:

```
        private static string TaskBActivity()
        {
            Logger.Log($"Inside {nameof(TaskBActivity)}");
            Thread.Sleep(TimeSpan.FromSeconds(2));
            Logger.Log($"Leaving {nameof(TaskBActivity)}");
            return "";
        }

        private static void TaskCActivity()
        {
            Logger.Log($"Inside {nameof(TaskCActivity)}");
            Thread.Sleep(TimeSpan.FromSeconds(1));
            Logger.Log($"Leaving {nameof(TaskCActivity)}");
        }
    }
}
```

   Each will log a message when starting and leaving a task, after a few seconds. The useful **nameof** statement is used to include the name of the method for extra logging information. Often, it is useful to examine log files to see the name of a method that has been accessed rather than hardcoding its name as a literal string.

7. Upon running the code, you will see the following output:

```
14:46:28 [01] Starting
14:46:28 [04] Inside TaskBActivity
14:46:28 [05] Inside TaskCActivity
14:46:28 [06] Inside TaskA
14:46:28 [01] Waiting max 7 seconds...
14:46:29 [05] Leaving TaskCActivity
14:46:30 [04] Leaving TaskBActivity
14:46:33 [06] Leaving TaskA
14:46:33 [01] AllDone=True: TaskA=RanToCompletion,
TaskB=RanToCompletion, TaskC=RanToCompletion
Press ENTER to quit
```

While running the code, a seven-second timeout was randomly picked by the runtime. This allowed all tasks to complete in time, so **`true`** was returned by **`WaitAll`** and all tasks had a **`RanToCompletion`** status at that point. Notice that the thread ID, in square brackets, is different for all three tasks.

8. Run the code again:

```
14:48:20 [01] Starting
14:48:20 [01] Waiting max 2 seconds...
14:48:20 [05] Inside TaskCActivity
14:48:20 [06] Inside TaskA
14:48:20 [04] Inside TaskBActivity
14:48:21 [05] Leaving TaskCActivity
14:48:22 [04] Leaving TaskBActivity
14:48:22 [01] AllDone=False: TaskA=Running, TaskB=Running,
TaskC=RanToCompletion
Press ENTER to quit
14:48:25 [06] Leaving TaskA
```

This time the runtime picked a two-second maximum wait time, so the **`WaitAll`** call times out with **`false`** being returned.

You may have noticed from the output that **`Inside TaskBActivity`** can sometimes appear before **`Inside TaskCActivity`**. This demonstrates the .NET scheduler's queuing mechanism. When you call **`Task.Run`**, you are asking the scheduler to add this to its queue. There may only be a matter of milliseconds between the time that you call **`Task.Run`** and when it invokes your lambda, but this can depend on how many other tasks you have recently added to the queue; a greater number of pending tasks could increase that time period.

Interestingly, the output shows **Leaving TaskBActivity**, but the **taskB** status was still **Running** just after **WaitAll** finished waiting. This indicates that there can sometimes be a very slight delay when a timed-out task's status is changed.

Some three seconds after the **Enter** key is pressed, **Leaving TaskA** is logged. This shows that the **Action** within any timed-out tasks will continue to run, and .NET will not stop it for you.

> **NOTE**
>
> You can find the code used for this exercise at https://packt.link/5IH0o.

CONTINUATION TASKS

So far, you have created tasks that are independent of one another, but what if you need to continue a task with the results of the previous task? Rather than blocking the current thread, by calling **Wait** or accessing the **Result** property, this can be achieved using the **Task ContinueWith** methods.

These methods return a new task, referred to as a **continuation** task, or more simply, a continuation, which can consume the previous task's or the antecedent's results.

As with standard tasks, they do not block the caller thread. There are several **ContinueWith** overloads available, many allowing extensive customization. A few of the more commonly used overloads are as follows:

- **public Task ContinueWith(Action<Task<TResult>> continuationAction)**: This defines a generic **Action<T>** based **Task** to run when the previous task completes.
- **public Task ContinueWith(Action<Task<TResult>> continuationAction, CancellationToken cancellationToken)**: This has a task to run and a cancellation token that can be used to coordinate the cancellation of the task.
- **public Task ContinueWith(Action<Task<TResult>> continuationAction, TaskScheduler scheduler)**: This also has a task to run and a low-level **TaskScheduler** that be used to queue the task.

- **public Task ContinueWith(Action<Task<TResult>> continuationAction, TaskContinuationOptions continuationOptions)**: A task to run, with the behavior for the task specified with **TaskContinuationOptions**. For example, specifying **NotOnCanceled** indicates that you do **not** want the continuation to be called if the previous task is canceled.

Continuations have an initial **WaitingForActivation** status. The .NET Framework will execute this task once the antecedent task or tasks have completed. It is important to note that you do not need to start a continuation and attempting to do so will result in an exception.

The following example simulates calling a long-running function, **GetStockPrice** (this may be some sort of web service or database call that takes a few seconds to return):

ContinuationExamples.cs

```
    class ContinuationExamples
    {
        public static void Main()
        {
            Logger.Log("Start...");
            Task.Run(GetStockPrice)
                .ContinueWith(prev =>
                {
                    Logger.Log($"GetPrice returned {prev.Result:N2}, 
status={prev.Status}");
                 });

            Console.ReadLine();
        }

```

You can find the complete code here: http://packt.link/rpNcx.

The call to **GetStockPrice** returns a **double**, which results in the generic **Task<double>** being passed to as a continuation (see the highlighted part). The **prev** parameter is a generic **Action** of type **Task<double>**, allowing you to access the antecedent task and its **Result** to retrieve the value returned from **GetStockPrice**.

If you hover your mouse over the **`ContinueWith`** method, you will see the IntelliSense description for it as follows:

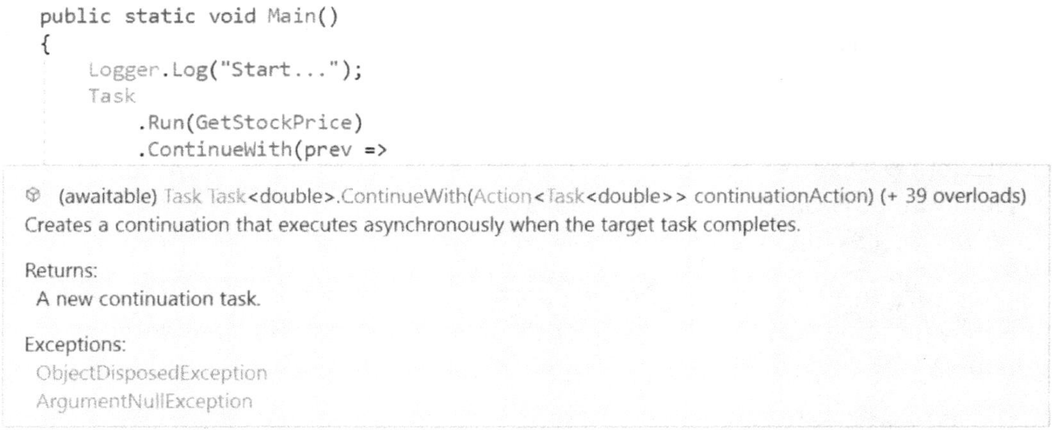

```
public static void Main()
{
    Logger.Log("Start...");
    Task
        .Run(GetStockPrice)
        .ContinueWith(prev =>
```

(awaitable) Task Task<double>.ContinueWith(Action<Task<double>> continuationAction) (+ 39 overloads)
Creates a continuation that executes asynchronously when the target task completes.

Returns:
A new continuation task.

Exceptions:
ObjectDisposedException
ArgumentNullException

Figure 5.1: ContinueWith method signature

> **NOTE**
>
> The **`ContinueWith`** method has various options that can be used to fine-tune behavior, and you can get more details about them from https://docs.microsoft.com/en-us/dotnet/api/system.threading.tasks.taskcontinuationoptions.

Running the example produces an output similar to the following:

```
09:30:45 [01] Start...
09:30:45 [03] Inside GetStockPrice
09:30:50 [04] GetPrice returned 76.44, status=RanToCompletion
```

In the output, thread **`[01]`** represents the console's main thread. The task that called **`GetStockPrice`** was executed by thread ID **`[03]`**, yet the continuation was executed using a different thread, thread (**`[04]`**).

> **NOTE**
>
> You can find the code used for this example at https://packt.link/rpNcx.

The continuation running on a different thread may not be a problem, but it certainly will be an issue if you are working on UWP, WPF, or WinForms UI apps where it's essential that UI elements are updated using the main UI thread (unless you are using binding semantics).

It is worth noting that the **TaskContinuationOptions.OnlyOnRanToCompletion** option can be used to ensure the continuation only runs if the antecedent task has run to completion first. For example, you may create a **Task** that fetches customers' orders from a database and then use a continuation task to calculate the average order value. If the previous task fails or is canceled by the user, then there is no point in wasting processing power to calculate the average if the user no longer cares about the result.

> **NOTE**
>
> The **ContinueWith** method has various options that can be used to fine-tune behavior, and you can see https://docs.microsoft.com/en-us/dotnet/api/system.threading.tasks.taskcontinuationoptions for more details.

If you access the **Task<T> Result** property on a **failed** or **canceled** antecedent task, this will result in an **AggregateException** being thrown. This will be covered in more detail later.

USING TASK.WHENALL AND TASK.WHENANY WITH MULTIPLE TASKS

You have seen how a single task can be used to create a continuation task, but what if you have multiple tasks and need to continue with a final operation when any or all of the previous tasks have completed?

Earlier, the **Task.WaitAny** and **Task.WaitAll** methods were used to wait for tasks to complete, but these block the current thread. This is where **Task.WhenAny** and **Task.WhenAll** can be used. They return a new **Task** whose **Action** delegate is called **when** any, or all, of the preceding tasks have completed.

There are four **WhenAll** overloads, two that return a **Task** and two that return a generic **Task<T>** allowing the task's result to be accessed:

1. **public static Task WhenAll(IEnumerable<Task> tasks)**: This continues when the collection of tasks completes.

2. **`public static Task WhenAll(params Task[] tasks)`**: This continues when the array of tasks completes.

3. **`public static Task<TResult[]> WhenAll<TResult>(params Task<TResult>[] tasks)`**: This continues when the array of generic **`Task<T>`** items complete.

4. **`public static Task<TResult[]> WhenAll<TResult>(IEnumerable<Task<TResult>> tasks)`**: This continues when the collection of generic **`Task<T>`** items complete.

`WhenAny` has a similar set of overloads but returns the **`Task`** or **`Task<T>`** that is the **first** task to complete. You'll next perform a few exercises showing **`WhenAll`** and **`WhenAny`** in practice.

EXERCISE 5.03: WAITING FOR ALL TASKS TO COMPLETE

Say you have been asked by a car dealer to create a console application that calculates the average sales value for cars sold across different regions. A dealership is a busy place, but they know it may take a while to fetch and calculate the average. For this reason, they want to enter a maximum number of seconds that they are prepared to wait for the average calculation. Any longer and they will leave the app and ignore the result.

The dealership has 10 regional sales hubs. To calculate the average, you need to first invoke a method called **`FetchSales`**, which returns a list of **`CarSale`** items for each of these regions.

Each call to **`FetchSales`** could be to a potentially slow-running service (you will implement random pauses to simulate such a delay) so you need to use a **`Task`** for each as you can't know for sure how long each call will take to complete. You also do not want slow-running tasks to affect other tasks, but to calculate a valid average, it's important to have **all** results returned before continuing.

Create a **`SalesLoader`** class that implements **`IEnumerable<CarSale> FetchSales()`** to return the car sales details. Then, a **`SalesAggregator`** class should be passed a list of **`SalesLoader`** (in this exercise, there will be 10 loader instances, one for each region). The aggregator will wait for all loaders to finish using **`Task.WhenAll`** before continuing with a task that calculates the average across all regions.

Perform the following steps to do so:

1. First, create a **CarSale** record. The constructor accepts two values, the name of the car and its sale price (**name** and **salePrice**):

```
using System;
using System.Collections.Generic;
using System.Globalization;
using System.Linq;
using System.Threading;
using System.Threading.Tasks;

namespace Chapter05.Exercises.Exercise03
{
    public record CarSale
    {
        public CarSale(string name, double salePrice)
            => (Name, SalePrice) = (name, salePrice);

        public string Name { get; }
        public double SalePrice { get; }
    }
```

2. Now create an interface, **ISalesLoader**, that represents the sales data loading service:

```
    public interface ISalesLoader
    {
        public IEnumerable<CarSale> FetchSales();
    }
```

   It has just one call, **FetchSales**, returning an enumerable of type **CarSale**. For now, it's not important to know how the loader works; just that it returns a list of car sales when called. Using an interface here allows using various types of loader as needed.

3. User the aggregator class to call an **ISalesLoader** implementation:

```
    public static class SalesAggregator
    {
       public static Task<double> Average(IEnumerable<ISalesLoader>
loaders)
       {
```

It is declared as **static** as there is no state between calls. Define an **Average** function that is passed an enumerable of **ISalesLoader** items and returns a generic **Task<Double>** for the final average calculation.

4. For each of the loader parameters, use a LINQ projection to pass a **loader.FetchSales** method to **Task.Run**:

```
            var loaderTasks = loaders.Select(ldr => Task.Run(ldr.
FetchSales));
            return Task
                .WhenAll(loaderTasks)
                .ContinueWith(tasks =>
```

Each of these will return a **Task<IEnumerable<CarSale>>** instance. **WhenAll** is used to create a single task that continues when **all** of the loader tasks have completed via a **ContinueWith** call.

5. Use the LINQ **SelectMany** to grab all of the **CarSale** items from every loader call result, before calling the Linq **Average** on the **SalePrice** field of each **CarSale** item:

```
                {
                    var average = tasks.Result
                        .SelectMany(t => t)
                        .Average(car => car.SalePrice);

                    return average;
                });
        }
    }
}
```

6. Implement the **ISalesLoader** interface from a class called **SalesLoader**:

```
    public class SalesLoader : ISalesLoader
    {
        private readonly Random _random;
        private readonly string _name;

        public SalesLoader(int id, Random rand)
        {
            _name = $"Loader#{id}";
            _random = rand;
        }
```

The constructor will be passed an **int** variable used for logging and a **Random** instance to help create a random number of **CarSale** items.

7. Your **ISalesLoader** implementation requires a **FetchSales** function. Include a random delay of between **1** and **3** seconds to simulate a less reliable service:

```
        public IEnumerable<CarSale> FetchSales()
        {
            var delay = _random.Next(1, 3);
            Logger.Log($"FetchSales {_name} sleeping for {delay}
seconds ...");
            Thread.Sleep(TimeSpan.FromSeconds(delay));
```

   You are trying to test that your application behaves with various time delays. Hence, the random class use.

8. Use **Enumerable.Range** and **random.Next** to pick a random number from one to five:

```
            var sales = Enumerable
                .Range(1, _random.Next(1, 5))
                .Select(n => GetRandomCar())
                .ToList();

            foreach (var car in sales)
                Logger.Log($"FetchSales {_name} found: {car.Name} @
{car.SalePrice:N0}");

            return sales;
        }
```

   This is the total number of **CarSale** items to return using your **GetRandomCar** function.

9. Use the **GetRandomCar** to generate a **CarSale** item with a random manufacturer's name from a hardcoded list.

10. Use the **carNames.length** property to pick a random index number between zero and four for the car's name:

```
        private readonly string[] _carNames = { "Ford", "BMW",
"Fiat", "Mercedes", "Porsche" };
        private CarSale GetRandomCar()
        {
            var nameIndex = _random.Next(_carNames.Length);
            return new CarSale(
                _carNames[nameIndex], _random.NextDouble() * 1000);
        }
    }
```

11. Now, create your console app to test this out:

```
    public class Program
    {
        public static void Main()
        {
            var random = new Random();
            const int MaxSalesHubs = 10;

            string input;
            do
            {
                Console.WriteLine("Max wait time (in seconds):");
                input = Console.ReadLine();

                if (string.IsNullOrEmpty(input))
                    continue;
```

Your app will repeatedly ask for a maximum time that the user is prepared to wait while data is downloaded. Once all the data has been downloaded, the app will use this to calculate an average price. Pressing **Enter** alone will result in the program loop ending. **MaxSalesHubs** is the maximum number of sales hubs to request data for.

12. Convert the entered value into an **int** type, then use **Enumerable.Range** again to create a random number of new **SalesLoader** instances (you have up to 10 different sales hubs):

```
                if (int.TryParse(input, NumberStyles.Any,
CultureInfo.CurrentCulture, out var maxDelay))
                {
                    var loaders = Enumerable.Range(1,
                                      random.Next(1,
MaxSalesHubs))
                     .Select(n => new SalesLoader(n, random))
                     .ToList();
```

13. Pass loaders to the static **SalesAggregator.Average** method to receive a **Task<Double>**.

14. Call **Wait**, passing in the maximum wait time:

```
                    var averageTask = SalesAggregator.
Average(loaders);
                    var hasCompleted = averageTask.Wait(
                           TimeSpan.FromSeconds(maxDelay));
                    var average = averageTask.Result;
```

If the **Wait** call does return in time, then you will see a **true** value for has completed.

15. Finish off by checking **hasCompleted** and log a message accordingly:

```
                    if (hasCompleted)
                    {
                        Logger.Log($"Average={average:N0}");
                    }
                    else
                    {
                        Logger.Log("Timeout!");
                    }
                }
            } while (input != string.Empty);
        }
    }
}
```

16. When running the console app and entering a short maximum wait of **1** second, you see three loader instances randomly created:

```
Max wait time (in seconds):1
10:52:49 [04] FetchSales Loader#1 sleeping for 1 seconds ...
10:52:49 [06] FetchSales Loader#3 sleeping for 1 seconds ...
10:52:49 [05] FetchSales Loader#2 sleeping for 1 seconds ...
10:52:50 [04] FetchSales Loader#1 found: Mercedes @ 362
10:52:50 [04] FetchSales Loader#1 found: Ford @ 993
10:52:50 [06] FetchSales Loader#3 found: Fiat @ 645
10:52:50 [05] FetchSales Loader#2 found: Mercedes @ 922
10:52:50 [06] FetchSales Loader#3 found: Ford @ 9
10:52:50 [05] FetchSales Loader#2 found: Porsche @ 859
10:52:50 [05] FetchSales Loader#2 found: Mercedes @ 612
10:52:50 [01] Timeout!
```

Each loader sleeps for **1** second (you can see various thread IDs are logged) before returning a random list of **CarSale** records. You soon reach the maximum timeout value, hence the message **Timeout!** with no average value displayed.

17. Enter a larger timeout period of **10** seconds:

```
Max wait time (in seconds):10
20:08:41 [05] FetchSales Loader#1 sleeping for 2 seconds ...
20:08:41 [12] FetchSales Loader#4 sleeping for 1 seconds ...
20:08:41 [08] FetchSales Loader#2 sleeping for 1 seconds ...
20:08:41 [11] FetchSales Loader#3 sleeping for 1 seconds ...
20:08:41 [15] FetchSales Loader#5 sleeping for 2 seconds ...
20:08:41 [13] FetchSales Loader#6 sleeping for 2 seconds ...
20:08:41 [14] FetchSales Loader#7 sleeping for 1 seconds ...
20:08:42 [08] FetchSales Loader#2 found: Porsche @ 735
20:08:42 [08] FetchSales Loader#2 found: Fiat @ 930
20:08:42 [11] FetchSales Loader#3 found: Porsche @ 735
20:08:42 [12] FetchSales Loader#4 found: Porsche @ 735
20:08:42 [08] FetchSales Loader#2 found: Porsche @ 777
20:08:42 [11] FetchSales Loader#3 found: Ford @ 500
20:08:42 [12] FetchSales Loader#4 found: Ford @ 500
20:08:42 [12] FetchSales Loader#4 found: Porsche @ 710
20:08:42 [14] FetchSales Loader#7 found: Ford @ 144
```

```
20:08:43 [05] FetchSales Loader#1 found: Fiat @ 649
20:08:43 [15] FetchSales Loader#5 found: Ford @ 779
20:08:43 [13] FetchSales Loader#6 found: Porsche @ 763
20:08:43 [15] FetchSales Loader#5 found: Fiat @ 137
20:08:43 [13] FetchSales Loader#6 found: BMW @ 415
20:08:43 [15] FetchSales Loader#5 found: Fiat @ 853
20:08:43 [15] FetchSales Loader#5 found: Porsche @ 857
20:08:43 [01] Average=639
```

Entering a value of **10** seconds allow **7** random loaders to complete in time and to finally create the average value of **639**.

> **NOTE**
>
> You can find the code used for this exercise at https://packt.link/kbToQ.

So far, this chapter has considered the various ways that individual tasks can be created and how static **Task** methods are used to create tasks that are started for us. You saw how **Task.Factory.StartNew** is used to create configured tasks, albeit with a longer set of configuration parameters. The **Task.Run** methods, which were more recently added to C#, are preferable by using their more concise signatures for most regular scenarios.

Using continuations, single and multiple tasks can be left to run in isolation, only continuing with a final task when all or any of the preceding tasks have run to completion.

Now it is time to look at the **async** and **wait** keywords to run asynchronous code. These keywords are a relatively new addition to the C# language. The **Task.Factory.StartNew** and **Task.Run** methods can be found in older C# applications, but hopefully, you will see that **async/await** provides a much clearer syntax.

ASYNCHRONOUS PROGRAMMING

So far, you have created tasks and used the static **`Task`** factory methods to run and coordinate such tasks. In earlier versions of C#, these were the only ways to create tasks.

The C# language now provides the **`async`** and **`await`** keywords to **mark** a method as asynchronous. This is the preferred way to run **asynchronous code**. Using the **`async/await`** style results in less code and the code that is created is generally easier to grasp and therefore easier to maintain.

> **NOTE**
>
> You may often find that legacy concurrent-enabled applications were originally created using **`Task.Factory.StartNew`** methods are subsequently updated to use the equivalent **`Task.Run`** methods or are updated directly to the **`async/await`** style.

The **`async`** keyword indicates that the method will return to the caller before it has had a chance to complete its operations, therefore the caller should wait for it to complete at some point in time.

Adding the **`async`** keyword to a method instructs the compiler that it may need to generate additional code to create a **state machine**. In essence, a state machine extracts the logic from your original method into a series of delegates and local variables that allows code to continue onto the next statement following an **`await`** expression. The compiler generates delegates that can jump back to the same location in the method once they have completed.

> **NOTE**
>
> You don't normally see this extra complied code, but if you are interested in learning more about state machines in C#, visit https://devblogs.microsoft.com/premier-developer/dissecting-the-async-methods-in-c.

Adding the **async** keyword does not mean that **all** or **any** part of the method will actually run in an asynchronous manner. When an **async** method is executed, it starts off running synchronously until it comes to a section of code with the **await** keyword. At this point, the **awaitable** block of code (in the following example, the **BuildGreetings** call is awaitable due to the preceding **async** keyword) is checked to see if it has already been completed. If so, it continues executing **synchronously**. If not, the asynchronous method is paused and returns an incomplete **Task** to the caller. This will be complete once the **async** code has been completed.

In the following console app, the entry point, **static Main**, has been marked as **async** and the **Task** return type added. You cannot mark a **Main** entry point, which returns either **int** or **void**, as **async** because the runtime must be able to return a **Task** result to the calling environment when the console app closes:

AsyncExamples.cs

```
using System;
using System.Threading;
using System.Threading.Tasks;

namespace Chapter05.Examples
{
    public class AsyncExamples
    {
        public static async Task Main()
         {
             Logger.Log("Starting");
             await BuildGreetings();

             Logger.Log("Press Enter");
             Console.ReadLine();
```

You can find the complete code here: http://packt.link/CsCek.

Running the example produces an output like this:

```
18:20:31 [01] Starting
18:20:31 [01] Morning
18:20:41 [04] Morning...Afternoon
18:20:42 [04] Morning...Afternoon...Evening
18:20:42 [04] Press Enter
```

As soon as **Main** runs, it logs **Starting**. Notice how the **ThreadId** is **[01]**. As you saw earlier, the console app's main thread is numbered as **1** (because the **Logger.Log** method uses the **00** format string, which adds a leading **0** to numbers in the range zero to nine).

Then the asynchronous method **BuildGreetings** is called. It sets the string **message** variable to **"Morning"** and logs the message. The **ThreadId** is still **[01]**; this is currently running synchronously.

So far, you have been using **Thread.Sleep** to block the calling thread in order or simulate long-running operations, but **async/await** makes it easier to simulate slow actions using the static **Task.Delay** method and awaiting that call. **Task.Delay** returns a task so it can also be used in continuation tasks.

Using **Task.Delay**, you will make two distinct awaitable calls (one that waits for 10 seconds and the second for two seconds), before continuing and appending to your local **message** string. The two **Task.Delay** calls could have been any method in your code that returns a **Task**.

The great thing here is that each awaited section gets its correct state in the order that it was declared in the code, irrespective of waiting 10 (or two) seconds prior. The thread IDs have all changed from **[01]** to **[04]**. This tells you that a different thread is running these statements. Even the very last **Press Enter** message has a different thread to the original thread.

Async/await makes it easier to run a series of task-based codes using the familiar **WhenAll**, **WhenAny**, and **ContinueWith** methods interchangeably.

The following example shows how multiple **async/await** calls can be applied at various stages in a program using a mixture of various awaitable calls. This simulates an application that makes a call to a database (**FetchPendingAccounts**) to fetch a list of user accounts. Each user in the pending accounts list is given a unique ID (using a task for each user).

Based on the user's region, an account is then created in the **northern** region or the other region, again, using a task for each. Finally, an awaitable **Task.WhenAll** call signals that everything has been completed.

```
using System;
using System.Collections.Generic;
using System.Linq;
using System.Threading.Tasks;

namespace Chapter05.Examples
{
```

Use an **enum** to define a **RegionName**:

```
    public enum RegionName { North, East, South, West };
```

A **User** record constructor is passed a **userName** and the user's **region**:

```
    public record User
    {
        public User(string userName, RegionName region)
            => (UserName, Region) = (userName, region);

        public string UserName { get; }
        public RegionName Region { get; }

        public string ID { get; set; }
    }
```

AccountGenerator is the main controlling class. It contains an **async CreateAccounts** method that can be awaited by a console app (this is implemented at the end of the example):

```
    public class AccountGenerator
    {
        public async Task CreateAccounts()
        {
```

Using the **await** keyword, you define an awaitable call to **FetchPendingAccounts**:

```
            var users = await FetchPendingAccounts();
```

For each one of the users returned by **FetchPendingAccounts**, you make an awaitable call to **GenerateId**. This shows that a loop can contain multiple awaitable calls. The runtime will set the user ID for the correct user instance:

```
            foreach (var user in users)
            {
                var id = await GenerateId();
                user.ID = id;
            }
```

Using a Linq **Select** function, you create a list of tasks. For each user, a Northern or Other account is created based on the user's region (each one of the calls is a **Task** per user):

```
            var accountCreationTasks = users.Select(
                user => user.Region == RegionName.North
                    ? Task.Run(() => CreateNorthernAccount(user))
                    : Task.Run(() => CreateOtherAccount(user)))
                .ToList();
```

The list of account creation tasks is awaited using the **static WhenAll** call. Once this completes, **UpdatePendindAccounts** will be called passing in the updated user list. This shows that you can pass lists of tasks between **async** statements:

```
            Logger.Log($"Creating {accountCreationTasks.Count}
accounts");
            await Task.WhenAll(accountCreationTasks);

            var updatedAccountTask = UpdatePendingAccounts(users);
            await updatedAccountTask;

            Logger.Log($"Updated {updatedAccountTask.Result} pending
accounts");
        }
```

The **FetchPendingAccounts** method returns a **Task** containing a list of users (here you simulate a delay of **3** seconds using **Task.Delay**):

```
        private async Task<List<User>> FetchPendingAccounts()
        {
            Logger.Log("Fetching pending accounts...");
            await Task.Delay(TimeSpan.FromSeconds(3D));

            var users = new List<User>
            {
                new User("AnnH", RegionName.North),
                new User("EmmaJ", RegionName.North),
                new User("SophieA", RegionName.South),
                new User("LucyG", RegionName.West),
            };
```

```
            Logger.Log($"Found {users.Count} pending accounts");

            return users;
        }
```

GenerateId uses **Task.FromResult** to generate a globally unique ID using the **Guid** class. **Task.FromResult** is used when you want to return a result but do not need to create a running task as you would with **Task.Run**:

```
        private static Task<string> GenerateId()
        {
            return Task.FromResult(Guid.NewGuid().ToString());
        }
```

The two **bool** task methods create either a northern account or other account. Here, you return **true** to indicate that each account creation call was successful, regardless:

```
        private static async Task<bool> CreateNorthernAccount(User user)
        {
            await Task.Delay(TimeSpan.FromSeconds(2D));
            Logger.Log($"Created northern account for {user.UserName}");
            return true;
        }

        private static async Task<bool> CreateOtherAccount(User user)
        {
            await Task.Delay(TimeSpan.FromSeconds(1D));
            Logger.Log($"Created other account for {user.UserName}");
            return true;
        }
```

Next, **UpdatePendingAccounts** is passed a list of users. For each user, you create a task that simulates a slow-running call to update each user and returning a count of the number of users subsequently updated:

```
        private static async Task<int>
UpdatePendingAccounts(IEnumerable<User> users)
        {
            var updateAccountTasks = users.Select(usr => Task.Run(
                async () =>
                {
                    await Task.Delay(TimeSpan.FromSeconds(2D));
                    return true;
                }))
                .ToList();

            await Task.WhenAll(updateAccountTasks);

            return updateAccountTasks.Count(t => t.Result);
        }
    }
```

Finally, the console app creates an **AccountGenerator** instance and waits for **CreateAccounts** to finish before writing an **All done** message:

```
    public static class AsyncUsersExampleProgram
    {
        public static async Task Main()
        {
            Logger.Log("Starting");

            await new AccountGenerator().CreateAccounts();

            Logger.Log("All done");
            Console.ReadLine();
        }
    }

}
```

Running the console app produces this output:

```
20:12:38 [01] Starting
20:12:38 [01] Fetching pending accounts...
20:12:41 [04] Found 4 pending accounts
20:12:41 [04] Creating 4 accounts
20:12:42 [04] Created other account for SophieA
20:12:42 [07] Created other account for LucyG
20:12:43 [04] Created northern account for EmmaJ
20:12:43 [05] Created northern account for AnnH
20:12:45 [05] Updated 4 pending accounts
20:12:45 [05] All done
```

Here, you can see that thread **[01]** writes the **Starting** message. This is the application's main thread. Note, too, that the main thread also writes **Fetching pending accounts...** from the **FetchPendingAccounts** method. This is still running synchronously as the awaitable block (**Task.Delay**) has not yet been reached.

Threads **[4]**, **[5]**, and **[7]** create each of the four user accounts. You used **Task.Run** to call the **CreateNorthernAccount** or **CreateOtherAccount** methods. Thread **[5]** runs the last statement in **CreateAccounts: Updated 4 pending accounts**. The thread numbers might differ in your system because .NET uses an internal pool of threads which vary based on how busy each thread is.

> **NOTE**
>
> You can find the code used for this example at https://packt.link/ZIK8k.

ASYNC LAMBDA EXPRESSIONS

Chapter 3, *Delegates, Events, and Lambdas*, looked at lambda expressions and how they can be used to create succinct code. You can also use the **async** keyword with lambda expressions to create code for an event handler that contains various **async** code.

The following example uses the **WebClient** class to show two different ways to download data from a website (this will be covered in great detail in *Chapter 8, Creating and Using Web API Clients* and *Chapter 9, Creating API Services*).

```
using System;
using System.Net;
using System.Net.Http
using System.Threading.Tasks;

namespace Chapter05.Examples
{
    public class AsyncLambdaExamples
    {
        public static async Task Main()
        {
            const string Url = "https://www.packtpub.com/";

            using var client = new WebClient();
```

Here, you add your own event handler to the **WebClient** class **DownloadDataCompleted** event using a lambda statement that is prefixed with the **async** keyword. The compiler will allow you to add awaitable calls inside the body of the lambda.

This event will be fired after **DownloadData** is called and the data requested has been downloaded for us. The code uses an awaitable block **Task.Delay** to simulate some extra processing on a different thread:

```
            client.DownloadDataCompleted += async (sender, args) =>
            {
                Logger.Log("Inside DownloadDataCompleted...looking 
busy");
                await Task.Delay(500);
                Logger.Log("Inside DownloadDataCompleted..all done now");
            };
```

You invoke the **DownloadData** method, passing in your URL and then logging the length of the web data received. This particular call itself will block the main thread until data is downloaded. **WebClient** offers a task-based asynchronous version of the **DownloadData** method called **DownloadDataTaskAsync**. So it's recommended to use the more modern **DownloadDataTaskAsync** method as follows:

```
            Logger.Log($"DownloadData: {Url}");
            var data = client.DownloadData(Url);
            Logger.Log($"DownloadData: Length={data.Length:N0}");
```

Once again, you request the same URL but can simply use an **await** statement, which will be run once the data download has been completed. As you can see, this requires less code and has a cleaner syntax:

```
            Logger.Log($"DownloadDataTaskAsync: {Url}");
            var downloadTask = client.DownloadDataTaskAsync(Url);
            var downloadBytes =  await downloadTask;
            Logger.Log($"DownloadDataTaskAsync: Length={downloadBytes.
Length:N0}");

            Console.ReadLine();
        }
    }
}
```

Running the code produces this output:

```
19:22:44 [01] DownloadData: https://www.packtpub.com/
19:22:45 [01] DownloadData: Length=278,047
19:22:45 [01] DownloadDataTaskAsync: https://www.packtpub.com/
19:22:45 [06] Inside DownloadDataCompleted...looking busy
19:22:45 [06] DownloadDataTaskAsync: Length=278,046
19:22:46 [04] Inside DownloadDataCompleted..all done now
```

> **NOTE**
>
> When running the program, you may see the following warning: **"Warning SYSLIB0014: 'WebClient.WebClient()' is obsolete: 'WebRequest, HttpWebRequest, ServicePoint, and WebClient are obsolete. Use HttpClient instead.'"**. Here, Visual Studio has suggested that the **HttpClient** class be used, as **WebClient** has been marked as obsolete.

DownloadData is logged by thread **[01]**, the main thread, which is blocked for around one second until the download completes. The size of the downloaded file is then logged using the **downloadBytes.Length** property.

The **DownloadDataTaskAsync** request is handled by thread **06**. Finally, the delayed code inside the **DownloadDataCompleted** event handler completes via thread **04**.

> **NOTE**
>
> You can find the code used for this example at https://packt.link/IJEaU.

CANCELING TASKS

Task cancelation is a two-step approach:

- You need to add a way to request a cancelation.
- Any cancelable code needs to support this.

You cannot provide cancelation without both mechanisms in place.

Typically, you will start a long-running task that supports cancelation and provide the user with the ability to cancel the operation by pressing a button on a UI. There are many real-world examples where such cancellation is needed, such as image processing where multiple images need to be altered allowing a user to cancel the remainder of the task if they run out of time. Another common scenario is sending multiple data requests to different web servers and allowing slow-running or pending requests to be canceled as soon as the first response is received.

In C#, **CancellationTokenSource** acts as a top-level object to initiate a cancelation request with its **Token** property, **CancellationToken**, being passed to concurrent/slow running code that can periodically check and act upon this cancellation status. Ideally, you would not want low-level methods to arbitrarily cancel high-level operations, hence the separation between the source and the token.

There are various **CancellationTokenSource** constructors, including one that will initiate a cancel request after a specified time has elapsed. Here are a few of the **CancellationTokenSource** methods, offering various ways to initiate a cancellation request:

- **public bool IsCancellationRequested { get; }**: This returns **true** if a cancellation has been requested for this token source (a caller has called the **Cancel** method). This can be inspected at intervals in the target code.
- **public CancellationToken Token { get; }**: The **CancellationToken** that is linked to this source object is often passed to **Task.Run** overloads, allowing .NET to check the status of pending tasks or for your own code to check while running.
- **public void Cancel()**: Initiates a request for cancellation.
- **public void Cancel(bool throwOnFirstException)**: Initiates a request for cancellation and determines whether further operations are to be processed should an exception occur.
- **public void CancelAfter(int millisecondsDelay)**: Schedules a cancel request after a specified number of milliseconds.

CancellationTokenSource has a **Token** property. **CancellationToken** contains various methods and properties that can be used for code to detect a cancellation request:

- **public bool IsCancellationRequested { get; }**: This returns **true** if a cancellation has been requested for this token.
- **public CancellationTokenRegistration Register(Action callback)**: Allows code to register a delegate that will be executed by the system if this token is canceled.
- **public void ThrowIfCancellationRequested()**: Calling this method will result in **OperationCanceledException** being thrown if a cancellation has been requested. This is typically used to break out of loops.

Throughout the previous examples, you may have spotted that **CancellationToken** can be passed to many of the static **Task** methods. For example, **Task.Run**, **Task.Factory.StartNew**, and **Task.ContinueWith** all contain overrides that accept **CancellationToken**.

.NET will not try to interrupt or stop any of your code once it is running, no matter how many times you call **Cancel** on a **CancellationToken**. Essentially, you pass these tokens into target code, but it is up to that code to periodically check the cancellation status whenever it can, such as within a loop, and then decide how it should act upon it. This makes logical sense; how would .NET know at what point it was safe to interrupt a method, maybe one that has hundreds of lines of code?

Passing **CancellationToken** to **Task.Run** only provides a hint to the queue scheduler that it may not need to start a task's action, but once started, .NET will not stop that running code for you. The running code itself must subsequently observe the cancelation status.

This is analogous to a pedestrian waiting to cross a road at a set of traffic lights. Motor vehicles can be thought of as tasks that have been started elsewhere. When the pedestrian arrives at the crossing and they press a button (calling **Cancel** on **CancellationTokenSource**), the traffic lights should eventually change to red so that the moving vehicles are requested to stop. It is up to each individual driver to observe that the red light has changed (**IsCancellationRequested**) and then decide to stop their vehicle. The traffic light does not forcibly stop each vehicle (.NET runtime). If a driver notices that the vehicle behind is too close and stopping soon may result in a collision, they may decide to not stop immediately. A driver that is not observing the traffic light status at all may fail to stop.

The next sections will continue with exercises that show **async/await** in action, some of the commonly used options for canceling tasks, in which you will need to control whether pending tasks should be allowed to run to completion or interrupted, and when you should aim to catch exceptions.

EXERCISE 5.04: CANCELING LONG-RUNNING TASKS

You will create this exercise in two parts:

- One that uses a **Task** that returns a double-based result.
- Second that provides a fine-grained level of control by inspecting the **Token.IsCancellationRequested** property.

Perform the following steps to complete this exercise:

1. Create a class called **SlowRunningService**. As the name suggests, the methods inside the service have been designed to be slow to complete:

```
using System;
using System.Globalization;
using System.Threading;
using System.Threading.Tasks;

namespace Chapter05.Exercises.Exercise04
{
    public class SlowRunningService
    {
```

2. Add the first slow-running operation, **Fetch**, which is passed a delay time (implemented with a simple **Thread.Sleep** call), and the cancellation token, which you pass to **Task.Run**:

```
        public Task<double> Fetch(TimeSpan delay, CancellationToken 
token)
        {
            return Task.Run(() =>
                {
                    var now = DateTime.Now;

                    Logger.Log("Fetch: Sleeping");
                    Thread.Sleep(delay);
                    Logger.Log("Fetch: Awake");

                    return DateTime.Now.Subtract(now).TotalSeconds;
                },
                token);
        }
```

When **Fetch** is called, the token may get canceled before the sleeping thread awakes.

3. To test whether **Fetch** will just stop running or return a number, add a console app to test this. Here, use a default delay (**DelayTime**) of **3** seconds:

```
    public class Program
    {
        private static readonly TimeSpan DelayTime=TimeSpan.
FromSeconds(3);
```

4. Add a helper function to prompt for a maximum number of seconds that you are prepared to wait. If a valid number is entered, convert the value entered into a **TimeSpan**:

```
        private static TimeSpan? ReadConsoleMaxTime(string message)
        {
            Console.Write($"{message} Max Waiting Time (seconds):");

            var input = Console.ReadLine();

            if (int.TryParse(input, NumberStyles.Any, CultureInfo.
CurrentCulture, out var intResult))
            {
                return TimeSpan.FromSeconds(intResult);
            }

            return null;
        }
```

5. Add a standard **Main** entry point for the console app. This is marked async and returns a **Task**:

```
public static async Task Main()
        {
```

6. Create an instance of the service. You will use the same instance in a loop, shortly:

```
            var service = new SlowRunningService();
```

7. Now add a **do**-loop that repeatedly asks for a maximum delay time:

```
            Console.WriteLine($"ETA: {DelayTime.TotalSeconds:N}
seconds");

            TimeSpan? maxWaitingTime;
              while (true)
              {
                  maxWaitingTime = ReadConsoleMaxTime("Fetch");
                  if (maxWaitingTime == null)
                      break;
```

   This allows you to try various values to see how that affects the cancel token and the results you receive back. In the case of a **null** value, you will **break** out of the **do**-loop.

8. Create **CancellationTokenSource**, passing in the maximum waiting time:

```
                  using var tokenSource = new CancellationTokenSource(
maxWaitingTime.Value);
                  var token = tokenSource.Token;
```

   This will trigger a cancellation without having to call the **Cancel** method yourself.

9. Using the **CancellationToken.Register** method, pass an **Action** delegate to be invoked when the token gets signaled for cancellation. Here, simply log a message when that occurs:

```
                  token.Register(() => Logger.Log($"Fetch: Cancelled
token={token.GetHashCode()}"));
```

10. Now for the main activity, call the service's **Fetch** method, passing in the default **DelayTime** and the token:

```
                  var resultTask = service.Fetch(DelayTime, token);
```

11. Before you await **resultTask**, add a **try-catch** block to catch any **TaskCanceledException**:

```
                try
                {
                    await resultTask;

                    if (resultTask.IsCompletedSuccessfully)
                        Logger.Log($"Fetch: Result={resultTask.
Result:N0}");
                    else
                        Logger.Log($"Fetch: Status={resultTask.
Status}");
                }
                catch (TaskCanceledException ex)
                {
                    Logger.Log($"Fetch: TaskCanceledException {ex.
Message}");
                }
            }
        }
    }
}
```

When using cancelable tasks, there is a possibility that they will throw **TaskCanceledException**. In this case, that is okay as you do expect that to happen. Notice that you only access the **resultTask.Result** if the task is marked as **IsCompletedSuccessfully**. If you attempt to access the **Result** property of a faulted task, then **AggregateException** instance is thrown. In some older projects, you may see non-async/await code that catches **AggregateException**.

12. Run the app and enter a waiting time greater than the ETA of three seconds, **5** in this case:

```
ETA: 3.00 seconds
Fetch Max Waiting Time (seconds):5
16:48:11 [04] Fetch: Sleeping
16:48:14 [04] Fetch: Awake
16:48:14 [04] Fetch: Result=3
```

As expected, the token was not canceled prior to completion, so you see **Result=3** (the elapsed time in seconds).

13. Try this again. For the cancellation to be triggered and detected, enter **2** for the number of seconds:

```
Fetch Max Waiting Time (seconds):2
16:49:51 [04] Fetch: Sleeping
16:49:53 [08] Fetch: Cancelled token=28589617
16:49:54 [04] Fetch: Awake
16:49:54 [04] Fetch: Result=3
```

Notice that the **Cancelled token** message is logged **before** the **Fetch** awakes, but you still end up receiving a result of **3** seconds with no **TaskCanceledException** message. This emphasizes the point that passing a cancellation token to **Start.Run** does not stop the task's action from starting, and more importantly, it did not interrupt it either.

14. Finally, use **0** as the maximum waiting time, which will effectively trigger the cancellation immediately:

```
Fetch Max Waiting Time (seconds):
0
16:53:32 [04] Fetch: Cancelled token=48717705
16:53:32 [04] Fetch: TaskCanceledException A task was canceled.
```

You will see the canceled token message and **TaskCanceledException** being caught, but there are no **Sleeping** or **Awake** messages logged at all. This shows that the **Action** passed to **Task.Run** was not actually started by the runtime. When you pass a **CancelationToken** to **Start.Run**, the task's **Action** gets queued but **TaskScheduler** will not run the action if it notices that the token has been canceled prior to starting; it just throws **TaskCanceledException**.

Now for an alternative slow-running method, one that allows you to support cancellable actions via a loop that polls for a change in the cancellation status.

15. In the **SlowRunningService** class, add a **FetchLoop** function:

```
        public Task<double?> FetchLoop(TimeSpan delay, 
CancellationToken token)
        {
            return Task.Run(() =>
            {
                const int TimeSlice = 500;
                var iterations = (int)(delay.TotalMilliseconds / 
TimeSlice);
```

```
                Logger.Log($"FetchLoop: Iterations={iterations}
token={token.GetHashCode()}");

                var now = DateTime.Now;
```

This produces a result similar to the earlier **Fetch** function but its purpose is to show how a function can be broken into a repeating loop that offers the ability to examine **CancellationToken** as each loop iteration runs.

16. Define the body of a **for...next** loop, which checks, for each iteration, if the **IsCancellationRequested** property is **true** and simply returns a nullable double if it detects that a cancellation has been requested:

```
                for (var i = 0; i < iterations; i++)
                {
                    if (token.IsCancellationRequested)
                    {
                        Logger.Log($"FetchLoop: Iteration {i + 1}
detected cancellation token={token.GetHashCode()}");
                        return (double?)null;
                    }

                    Logger.Log($"FetchLoop: Iteration {i + 1}
Sleeping");
                    Thread.Sleep(TimeSlice);
                    Logger.Log($"FetchLoop: Iteration {i + 1}
Awake");
                }

                Logger.Log("FetchLoop: done");

                return DateTime.Now.Subtract(now).TotalSeconds;
            }, token);
        }
```

This is a rather firm way to exit a loop, but as far as this code is concerned, nothing else needs to be done.

> **NOTE**
>
> You could have also used a **continue** statement and cleaned up before returning. Another option is to call **token.ThrowIfCancellationRequested()** rather than checking **token.IsCancellationRequested**, which will force you to exit the **for** loop.

17. In the **Main** console app, add a similar **while** loop that calls the **FetchLoop** method this time. The code is similar to the previous looping code:

```
            while (true)
                {
                    maxWaitingTime = ReadConsoleMaxTime("FetchLoop");
                    if (maxWaitingTime == null)
                        break;

                    using var tokenSource = new
CancellationTokenSource(maxWaitingTime.Value);
                    var token = tokenSource.Token;
                    token.Register(() => Logger.Log($"FetchLoop:
Cancelled token={token.GetHashCode()}"));
```

18. Now call the **FetchLoop** and await the result:

```
                    var resultTask = service.FetchLoop(DelayTime, token);

                    try
                    {
                        await resultTask;
                        if (resultTask.IsCompletedSuccessfully)
                            Logger.Log($"FetchLoop: Result={resultTask.
Result:N0}");
                        else
                            Logger.Log($"FetchLoop: Status={resultTask.
Status}");
                    }
                    catch (TaskCanceledException ex)
                    {
                        Logger.Log($"FetchLoop: TaskCanceledException
{ex.Message}");
                    }
                }
```

19. Running the console app and using a **5**-second maximum allows all the iterations to run through with none detecting a cancellation request. The result is **3** as expected:

```
FetchLoop Max Waiting Time (seconds):5
17:33:38 [04] FetchLoop: Iterations=6 token=6044116
17:33:38 [04] FetchLoop: Iteration 1 Sleeping
17:33:38 [04] FetchLoop: Iteration 1 Awake
17:33:38 [04] FetchLoop: Iteration 2 Sleeping
```

```
17:33:39 [04] FetchLoop: Iteration 2 Awake
17:33:39 [04] FetchLoop: Iteration 3 Sleeping
17:33:39 [04] FetchLoop: Iteration 3 Awake
17:33:39 [04] FetchLoop: Iteration 4 Sleeping
17:33:40 [04] FetchLoop: Iteration 4 Awake
17:33:40 [04] FetchLoop: Iteration 5 Sleeping
17:33:40 [04] FetchLoop: Iteration 5 Awake
17:33:40 [04] FetchLoop: Iteration 6 Sleeping
17:33:41 [04] FetchLoop: Iteration 6 Awake
17:33:41 [04] FetchLoop: done
17:33:41 [04] FetchLoop: Result=3
```

20. Use **2** as the maximum. This time the token is auto-triggered during iteration **4** and spotted by iteration **5**, so you are returned a null result:

```
FetchLoop Max Waiting Time (seconds):
2
17:48:47 [04] FetchLoop: Iterations=6 token=59817589
17:48:47 [04] FetchLoop: Iteration 1 Sleeping
17:48:48 [04] FetchLoop: Iteration 1 Awake
17:48:48 [04] FetchLoop: Iteration 2 Sleeping
17:48:48 [04] FetchLoop: Iteration 2 Awake
17:48:48 [04] FetchLoop: Iteration 3 Sleeping
17:48:49 [04] FetchLoop: Iteration 3 Awake
17:48:49 [04] FetchLoop: Iteration 4 Sleeping
17:48:49 [06] FetchLoop: Cancelled token=59817589
17:48:49 [04] FetchLoop: Iteration 4 Awake
17:48:49 [04] FetchLoop: Iteration 5 detected cancellation
token=59817589
17:48:49 [04] FetchLoop: Result=
```

21. By using **0**, you see the same output as the earlier **`Fetch`** example:

```
FetchLoop Max Waiting Time (seconds):
0
17:53:29 [04] FetchLoop: Cancelled token=48209832
17:53:29 [08] FetchLoop: TaskCanceledException A task was canceled.
```

The action doesn't get a chance to run. You can see a **`Cancelled token`** message and **`TaskCanceledException`** being logged.

By running this exercise, you have seen how long-running tasks can be automatically marked for cancellation by the .NET runtime if they do not complete within a specified time. By using a **for** loop, a task was broken down into small iterative steps, which provided a frequent opportunity to detect if a cancellation was requested.

> **NOTE**
>
> You can find the code used for this exercise at https://packt.link/xa1Yf.

EXCEPTION HANDLING IN ASYNC/AWAIT CODE

You have seen that canceling a task can result in **TaskCanceledException** being thrown. Exception handling for asynchronous code can be implemented in the same way you would for standard synchronous code, but there are a few things you need to be aware of.

When code in an **async** method causes an exception to be thrown, the task's status is set to **Faulted**. However, an exception will not be rethrown until the awaited expression gets rescheduled. What this mean is that if you do not await a call, then it's possible for exceptions to be thrown and to go completely unobserved in code.

Unless you absolutely cannot help it, you should not create **async void** methods. Doing so makes it difficult for the caller to await your code. This means they cannot catch any exceptions raised, which by default, will terminate a program. If the caller is not given a **Task** reference to await, then there is no way for them to tell if the called method ran to completion or not.

The general exception to this guideline is in the case of fire-and-forget methods as mentioned at the start of the chapter. A method that asynchronously logs the usage of the application may not be of such critical importance, so you may not care if such calls are successful or not.

It is possible to detect and handle unobserved task exceptions. If you attach an event delegate to the static **TaskScheduler.UnobservedTaskException** event, you can receive a notification that a task exception has gone unobserved. You can attach a delegate to this event as follows:

```
TaskScheduler.UnobservedTaskException += (sender, args) =>
{
  Logger.Log($"Caught UnobservedTaskException\n{args.Exception}");
};
```

The runtime considers a task exception to be **unobserved** once the task object is finalized.

> **NOTE**
>
> You can find the code used for this example at https://packt.link/OkH7r.

Continuing with some more exception handling examples, see how you can catch a specific type of exception as you would with synchronous code.

In the following example, the **CustomerOperations** class provides the **AverageDiscount** function, which returns **Task<int>**. However, there is a chance that it may throw **DivideByZeroException**, so you will need to catch that; otherwise, the program will crash.

```
using System;
using System.Threading.Tasks;

namespace Chapter05.Examples
{
    class ErrorExamplesProgram
    {
        public static async Task Main()
        {
            try
            {
```

Create a **CustomerOperations** instance and wait for the **AverageDiscount** method to return a value:

```
                var operations = new CustomerOperations();
                var discount = await operations.AverageDiscount();
                Logger.Log($"Discount: {discount}");
            }
            catch (DivideByZeroException)
            {
                Console.WriteLine("Caught a divide by zero");
            }

            Console.ReadLine();
```

```
        }

        class CustomerOperations
        {
            public async Task<int> AverageDiscount()
            {
                Logger.Log("Loading orders...");
                await Task.Delay(TimeSpan.FromSeconds(1));
```

Choose a random value for **ordercount** between **0** and **2**. An attempt to divide by zero will result in an exception being thrown by the .NET runtime:

```
                var orderCount = new Random().Next(0, 2);
                var orderValue = 1200;
                return orderValue / orderCount;
            }
        }
    }
}
```

The results show that when **orderCount** was zero, you did catch **DivideByZeroException** as expected:

```
15:47:21 [01] Loading orders...
Caught a divide by zero
```

Running a second time, there was no error caught:

```
17:55:54 [01] Loading orders...
17:55:55 [04] Discount: 1200
```

On your system you may find that the program needs to be run multiple times before the **DivideByZeroException** is raised. This is due to the use of a random instance to assign a value to **orderCount**.

> **NOTE**
>
> You can find the code used for this example at https://packt.link/18kOK.

So far, you have created single tasks that may throw exceptions. The following exercise will look at a more complex variant.

EXERCISE 5.05: HANDLING ASYNC EXCEPTIONS

Imagine you have a **CustomerOperations** class that can be used to fetch a list of customers via a **Task**. For each customer, you need to run an extra **async** task, which goes off to a service to calculate the total value of that customer's orders.

Once you have your customer list, the customers need to be sorted in descending order of sales, but due to some security restrictions, you are not allowed to read a customer's **TotalOrders** property if their region name is **West**. In this exercise you will create a copy of the **RegionName** enum that was used in the earlier example.

Perform the following steps to complete this exercise:

1. Start by adding the **Customer** class:

```
using System;
using System.Collections.Generic;
using System.Linq;
using System.Threading.Tasks;

namespace Chapter05.Exercises.Exercise05
{
    public enum RegionName { North, East, South, West };

     public class Customer
     {
         private readonly RegionName _protectedRegion;

         public Customer(string name, RegionName region, RegionName protectedRegion)
         {
```

The constructor is passed the customer **name** and their **region**, along with a second region that identifies the **protectedRegion** name. If the customer's **region** is the same as this **protectedRegion**, then throw an access violation exception on any attempt to read the **TotalOrders** property.

2. Then add a **CustomerOperations** class:

```
public class CustomerOperations
{
    public const RegionName ProtectedRegion = RegionName.West;
```

   This knows how to load a customer's name and populate their total order value. The requirement here is that customers from the **West** region need to have a restriction hardcoded, so add a constant called **ProtectedRegion** that has **RegionName.West** as a value.

3. Add a **FetchTopCustomers** function:

```
        public async Task<IEnumerable<Customer>> FetchTopCustomers()
        {
            await Task.Delay(TimeSpan.FromSeconds(2));

            Logger.Log("Loading customers...");
            var customers = new List<Customer>
            {
            new Customer("Rick Deckard", RegionName.North,
ProtectedRegion),
            new Customer("Taffey Lewis", RegionName.North,
ProtectedRegion),
            new Customer("Rachael", RegionName.North,
ProtectedRegion),
            new Customer("Roy Batty", RegionName.West,
ProtectedRegion),
            new Customer("Eldon Tyrell", RegionName.East,
ProtectedRegion)
            };
```

   This returns a **Task** enumeration of **Customer** and is marked as **async** as you will make further **async** calls to populate each customer's order details inside the function. Await using **Task.Delay** to simulate a slow-running operation. Here, a sample list of customers is hardcoded. Create each **Customer** instance, passing their name, actual region, and the protected region constant, **ProtectedRegion**.

4. Add an **await** call to **FetchOrders** (which will be declared shortly):

```
            await FetchOrders(customers);
```

5. Now, iterate through the list of customers, but be sure to wrap each call to **TotalOrders** with a **try-catch** block that explicitly checks for the access violation exception that will be thrown if you attempt to view a protected customer:

```
            var filteredCustomers = new List<Customer>();
            foreach (var customer in customers)
            {
                try
                {
                    if (customer.TotalOrders > 0)
                        filteredCustomers.Add(customer);
                }
                catch (AccessViolationException e)
                {
                    Logger.Log($"Error {e.Message}");
                }
            }
```

6. Now that the **filteredCustomers** list has been populated with a filtered list of customers, use the Linq **OrderByDescending** extension method to return the items sorted by each customer's **TotalOrders** value:

```
            return filteredCustomers.OrderByDescending(c =>
c.TotalOrders);
        }
```

7. Finish off **CustomerOperations** with the **FetchOrders** implementation.

8. For each customer in the list, use an **async** lambda that pauses for **500** milliseconds before assigning a random value to **TotalOrders**:

```
        private async Task FetchOrders(IEnumerable<Customer>
customers)
        {
            var rand = new Random();

            Logger.Log("Loading orders...");
            var orderUpdateTasks = customers.Select(
              cust => Task.Run(async () =>
              {
                    await Task.Delay(500);
```

```
                cust.TotalOrders = rand.Next(1, 100);
            }))
            .ToList();
```

The delay could represent another slow-running service.

9. Wait for **orderUpdateTasks** to complete using **Task.WhenAll**:

```
            await Task.WhenAll(orderUpdateTasks);
        }
    }
```

10. Now create a console app to run the operation:

```
    public class Program
    {
        public static async Task Main()
        {
            var ops = new CustomerOperations();
            var resultTask = ops.FetchTopCustomers();

            var customers = await resultTask;

            foreach (var customer in customers)
            {
                Logger.Log($"{customer.Name} ({customer.Region}):
{customer.TotalOrders:N0}");
            }
            Console.ReadLine();
        }
    }
}
```

11. On running the console, there are no errors as **Roy Batty** from the **West** region was skipped safely:

```
20:00:15 [05] Loading customers...
20:00:16 [05] Loading orders...
20:00:16 [04] Error Cannot access orders for Roy Batty
20:00:16 [04] Rachael (North): 56
20:00:16 [04] Taffey Lewis (North): 19
20:00:16 [04] Rick Deckard (North): 10
20:00:16 [04] Eldon Tyrell (East): 6
```

In this exercise, you saw how exceptions can be handled gracefully with asynchronous code. You placed a **try-catch** block at the required location, rather than over-complicating and adding too many unnecessary levels of nested **try-catch** blocks. When the code was run, an exception was caught that did not crash the application.

> **NOTE**
>
> You can find the code used for this exercise at https://packt.link/4ozac.

THE AGGREGATEEXCEPTION CLASS

At the beginning of the chapter, you saw that the **Task** class has an **Exception** property of type **AggregateException**. This class contains details about one or more errors that occur during an asynchronous call.

AggregateException has various properties, but the main ones are as follows:

- **public ReadOnlyCollection<Exception> InnerExceptions { get; }**: A collection of exceptions that caused the current exception. A single asynchronous call can result in multiple exceptions being raised and collected here.
- **public AggregateException Flatten()**: Flattens all of the **AggregateException** instances in the **InnerExeceptions** property into a single new instance. This saves you from having to iterate over **AggregateException** nested with the exceptions list.
- **public void Handle(Func<Exception, bool> predicate)**: Invokes the specified Func handler on every exception in this aggregate exception. This allows the handler to return **true** or **false** to indicate whether each exception was handled. Any remaining unhandled exceptions will be thrown for the caller to catch as required.

When something goes wrong and this exception is caught by a caller, **InnerExceptions** contains a list of the exceptions that caused the current exception. These can be from multiple tasks, so each individual exception is added to the resulting task's **InnerExceptions** collection.

You may often find **async** code with a **try-catch** block that catches **AggregateException** and logs each of **InnerExceptions** details. In this example, **BadTask** returns an **int** based task, but it can be the cause of an exception when run. Perform the following steps to complete this example:

```
using System;
using System.Collections.Generic;
using System.Linq;
using System.Threading.Tasks;

namespace Chapter05.Examples
{
    class WhenAllErrorExamples
    {+
```

It sleeps for **1,000** milliseconds before throwing the **InvalidOperationException** in case the number passed in is an even number (using the % operator to see if the number can be divided by **2** with no remainder):

```
        private static async Task<int> BadTask(string info, int n)
        {
            await Task.Delay(1000);

            Logger.Log($"{info} number {n} awake");

            if (n % 2 == 0)
            {
                Logger.Log($"About to throw one {info} number {n}"…");
                throw new InvalidOperationException"($"Oh dear from 
{info} number "n}");
            }

            return n;
        }
```

Add a helper function, **CreateBadTasks**, that creates a collection of five bad tasks. When started, each of the tasks will eventually throw an exception of type **InvalidOperationException**:

```
        private static IEnumerable<Task<int>> CreateBadTasks(string info)
        {
            return Enumerable.Range(0, 5)
                .Select(i => BadTask(info, i))
                .ToList();
        }
```

Now, create the console app's **Main** entry point. You pass the results of **CreateBadTasks** to **WhenAll**, passing in the string **[WhenAll]** to make it easier to see what is happening in the output:

```
        public static async Task Main()
        {
            var whenAllCompletedTask = Task.
WhenAll(CreateBadTasks("[WhenAll]"));
```

Before you attempt to await the **whenAllCompletedTask** task, you need to wrap it in **try-catch**, which catches the base **Exception** type (or a more specific one if you are expecting that).

You cannot catch **AggregateException** here as it's the first exception inside the **Task** that you receive, but you can still use the **Exception** property of **whenAllCompletedTask** to get at the **AggregateException** itself:

```
            try
            {
                await whenAllCompletedTask;
            }
            catch (Exception ex)
            {
```

You've caught an exception, so log its type (this will be **InvalidOperationException** instance that you threw) and the message:

```
                Console.WriteLine($"WhenAll Caught {ex.GetType().Name},
Message={ex.Message}");
```

Now you can examine **whenAllCompletedTask**, iterating though this task's **AggregateException** to see its **InnerExceptions** list:

```
                Console.WriteLine($"WhenAll Task.
Status={whenAllCompletedTask.Status}");
               foreach (var ie in whenAllCompletedTask.Exception.
InnerExceptions)
               {
                   Console.WriteLine($"WhenAll Caught Inner Exception:
{ie.Message}");
               }
            }
            Console.ReadLine();
        }
    }
}
```

Running the code, you'll see five tasks that sleep, and eventually, numbers **0**, **2**, and **4** each throw **InvalidOperationException**, which you will catch:

```
17:30:36 [05] [WhenAll] number 3 awake
17:30:36 [09] [WhenAll] number 1 awake
17:30:36 [07] [WhenAll] number 0 awake
17:30:36 [06] [WhenAll] number 2 awake
17:30:36 [04] [WhenAll] number 4 awake
17:30:36 [06] About to throw one [WhenAll] number 2...
17:30:36 [04] About to throw one [WhenAll] number 4...
17:30:36 [07] About to throw one [WhenAll] number 0...
WhenAll Caught InvalidOperationException, Message=Oh dear from [WhenAll]
number 0
WhenAll Task.Status=Faulted
WhenAll Caught Inner Exception: Oh dear from [WhenAll] number 0
WhenAll Caught Inner Exception: Oh dear from [WhenAll] number 2
WhenAll Caught Inner Exception: Oh dear from [WhenAll] number 4
```

Notice how **number 0** appears to be the only error that was caught (**(Message=Oh** dear from **[WhenAll] number 0**). However, by logging each entry in the **InnerExceptions** list, you see all **three** erroneous tasks with **number 0** appearing once again.

You can try the same code, but this time use **WhenAny**. Remember that **WhenAny** will complete when the first task in the list completes, so notice the complete lack of **error handling** in this case:

```
            var whenAnyCompletedTask = Task.
WhenAny(CreateBadTasks("[WhenAny]"));
            var result = await whenAnyCompletedTask;
            Logger.Log($"WhenAny result: {result.Result}");
```

Unless you wait for all tasks to complete, you may miss an exception raised by a task when using **WhenAny**. Running this code results in not a single error being caught and the app does **not** break. The result is **3** as that completed first:

```
18:08:46 [08] [WhenAny] number 2 awake
18:08:46 [10] [WhenAny] number 0 awake
18:08:46 [10] About to throw one [WhenAny] number 0...
18:08:46 [07] [WhenAny] number 3 awake
18:08:46 [09] [WhenAny] number 1 awake
18:08:46 [07] WhenAny result: 3
18:08:46 [08] About to throw one [WhenAny] number 2...
18:08:46 [06] [WhenAny] number 4 awake
18:08:46 [06] About to throw one [WhenAny] number 4...
```

You will finish this look at **async/await** code by looking at some of the newer options in C# around handling streams of **async** results. This provides a way to efficiently iterate through the items of a collection without the calling code having to wait for the entire collection to be populated and returned before it can start processing the items in the list.

> **NOTE**
>
> You can find the code used for this example at https://packt.link/SuCXK.

IASYNCENUMERABLE STREAMS

If your application targets .NET 5, .NET6, .NET Core 3.0, .NET Standard 2.1, or any of the later versions, then you can use **`IAsyncEnumerable`** streams to create awaitable code that combines the **`yield`** keyword into an enumerator to iterate asynchronously through a collection of objects.

> **NOTE**
>
> Microsoft's documentation provides this definition of the **`yield`** keyword: When a **`yield`** return statement is reached in the iterator method, expression is returned, and the current location in code is retained. Execution is restarted from that location the next time that the iterator function is called.

Using the **`yield`** statement, you can create methods that return an enumeration of items to the caller. Additionally, the caller does not need to wait for the **entire list** of items to be returned before they can start traversing each item in the list. Instead, the caller can access each item as soon as it becomes available.

In this example, you will create a console app that replicates an insurance quoting system. You will make five requests for an insurance quote, once again using **`Task.Delay`** to simulate a 1-second delay in receiving each delay.

For the list-based approach, you can only log each quote once all five results have been received back to the **`Main`** method. Using **`IAsyncEnumerable`** and the **`yield`** keyword, the same one second exists between quotes being received, but as soon as each quote is received, the **`yield`** statement allows the calling **`Main`** method to receive and process the value quoted. This is ideal if you want to start processing items right away or potentially do not want the overhead of having thousands of items in a list for longer than is needed to process them individually:

```
using System;
using System.Collections.Generic;
using System.Threading.Tasks;
namespace Chapter05.Examples
{
    class AsyncEnumerableExamplesProgram
    {
        public static async Task Main()
        {
```

Start by **awaiting** for **GetInsuranceQuotesAsTask** to return a list of strings and iterate through each, logging the details of each quote. This code will wait for all quotes to be received before logging each item:

```
            Logger.Log("Fetching Task quotes...");
            var taskQuotes = await GetInsuranceQuotesAsTask();
            foreach(var quote in taskQuotes)
            {
                Logger.Log($"Received Task: {quote}");
            }
```

Now for the **async** stream version. If you compare the following code to the preceeding code block, you'll see that there are fewer lines of code needed to iterate through the items returned. This code does not wait for all quote items to be received but instead writes out each quote as soon as it is received from **GetInsuranceQuotesAsync**:

```
            Logger.Log("Fetching Stream quotes...");
            await foreach (var quote in GetInsuranceQuotesAsync())
            {
                Logger.Log($"Received Stream: {quote}");
            }

            Logger.Log("All done...");

            Console.ReadLine();
        }
```

The **GetInsuranceQuotesAsTask** method returns a **Task** of strings. Between each of the five quotes, you wait for one second to simulate a delay, before adding the result to the list and finally returning the entire list back to the caller:

```
        private static async Task<IEnumerable<string>>
GetInsuranceQuotesAsTask()
        {
            var rand = new Random();
            var quotes = new List<string>();

            for (var i = 0; i < 5; i++)
            {
                await Task.Delay(1000);
```

```
                quotes.Add($"Provider{i}'s quote is {rand.Next(5, 10)}");
            }

            return quotes;
        }
```

The **GetInsuranceQuotesAsync** method contains the same delay between each quote, but rather than populating a list to return back to the caller, the **yield** statement is used to allow the **Main** method to process each quote item immediately:

```
        private static async IAsyncEnumerable<string>
GetInsuranceQuotesAsync()
        {
            var rand = new Random();

            for (var i = 0; i < 5; i++)
            {
                await Task.Delay(1000);
                yield return $"Provider{i}'s quote is {rand.Next(5,
10)}";
            }
        }
    }
}
```

Running the console app produces the following output:

```
09:17:57 [01] Fetching Task quotes...
09:18:02 [04] Received Task: Provider0's quote is 7
09:18:02 [04] Received Task: Provider1's quote is 9
09:18:02 [04] Received Task: Provider2's quote is 9
09:18:02 [04] Received Task: Provider3's quote is 8
09:18:02 [04] Received Task: Provider4's quote is 8
09:18:02 [04] Fetching Stream quotes...
09:18:03 [04] Received Stream: Provider0's quote is 7
09:18:04 [04] Received Stream: Provider1's quote is 8
09:18:05 [05] Received Stream: Provider2's quote is 9
09:18:06 [05] Received Stream: Provider3's quote is 8
09:18:07 [04] Received Stream: Provider4's quote is 7
09:18:07 [04] All done...
```

Thread **[04]** logged all five task-based quote details five seconds after the app started. Here, it waited for all quotes to be returned before logging each quote. However, notice that each of the stream-based quotes was logged as soon as it was yielded by threads **4** and **5** with 1 second between them.

The overall time taken for both calls is the same (5 seconds in total), but **yield** is preferrable when you want to start processing each result as soon as it is ready. This is often useful in UI apps where you can provide early results to the user.

> **NOTE**
>
> You can find the code used for this example at https://packt.link/KarKW.

PARALLEL PROGRAMMING

So far, this chapter has covered async programming using the **Task** class and **async/await** keywords. You have seen how tasks and **async** blocks of code can be defined and the flow of a program can be finely controlled as these structures complete.

The **Parallel Framework** (**PFX**) offers further ways to utilize multicore processors to efficiently run concurrent operations. The phrase **TPL** (**Task Parallel Library**) is generally used to refer to the **Parallel** class in C#.

Using the Parallel Framework, you do not need to worry about the complexity of creating and reusing threads or coordinating multiple tasks. The framework manages this for you, even adjusting the number of threads that are used, in order to maximize throughput.

For parallel programming to be effective, the order in which each task executes must be irrelevant and all tasks should be independent of each other, as you cannot be certain when one task completes and the next one begins. Coordinating negates any benefits. Parallel programming can be broken down into two distinct concepts:

- Data parallelism
- Task parallelism

DATA PARALLELISM

Data parallelism is used when you have multiple data values, and the same operation is to be applied concurrently to each of those values. In this scenario, processing each of the values is partitioned across different threads.

A typical example might be calculating the prime numbers from one to 1,000,000. For each number in the range, the same function needs to be applied to determine whether the value is a prime. Rather than iterating through each number one at a time, an asynchronous approach would be to split numbers across multiple threads.

TASK PARALLELISM

Conversely, **task parallelism** is used where a collection of threads all performs a different action, such as calling different functions or sections of code, concurrently. One such example is a program that analyzes the words found in a book, by downloading the book's text and defining separate tasks to do the following:

- Count the number of words.
- Find the longest word.
- Calculate the average word length.
- Count the number of noise words (the, and, of, for example).

Each of these tasks can be run concurrently and they do not depend on each other.

For the **`Parallel`** class, the Parallel Framework provides various layers that offer parallelism, including **Parallel Language Integrated Query** (**PLINQ**). PLINQ is a collection of extension methods that add the power of parallel programming to the LINQ syntax. The PLINQ won't be covered here in detail, but the **`Parallel`** class will be covered in more detail.

> **NOTE**
>
> If you're interested in learning more about PLINQ, you can refer to the online documentation at https://docs.microsoft.com/en-us/dotnet/standard/parallel-programming/introduction-to-plinq.

THE PARALLEL CLASS

The **Parallel** class contains just three **static** methods but there are numerous overloads providing options to control and influence how actions are performed. Each of the methods **block** the current thread, and if an exception occurs, whilst an iterator is working, the trailing iterators are stopped and an exception is thrown to the caller. Due to this blocking behavior, the **Parallel** class is often called from within an awaitable block such as **Task.Run**.

It is worth remembering that the runtime may run the required operations in parallel only if it thinks that is warranted. In the case of individual steps completing sooner than others, the runtime may decide that the overhead of running the remaining operations in parallel is not justified.

Some of the commonly used **Parallel** method overloads are as follows:

- **public static ParallelLoopResult For(int from, int to, Action<int> body)**: This data parallelism call executes a loop by invoking the body **Action** delegate, passing in an **int** value across the from and to numeric range. It returns **ParallelLoopResult**, which contains details of the loop once completed.
- **public static ParallelLoopResult For(int from, int to, ParallelOptions options, Action<int, ParallelLoopState> body)**: A data parallelism call that executes a loop across the numeric range. **ParallelOptions** allows loop options to be configured and **ParallelLoopState** is used to monitor or manipulate the state of the loop as it runs. It returns **ParallelLoopResult**.
- **public static ParallelLoopResult ForEach<TSource>(IEnumerable<TSource> source, Action<TSource, ParallelLoopState> body)**: A data parallelism call that invokes the **Action** body on each item in the **IEnumerable** source. It returns **ParallelLoopResult**.
- **public static ParallelLoopResult ForEach<TSource>(Partitioner<TSource> source, Action<TSource> body)**: An advanced data parallelism call that invokes the **Action** body and allows you to specify **Partitioner** to provide partitioning strategies optimized for specific data structures to improve performance. It returns **ParallelLoopResult**.
- **public static void Invoke(params Action[] actions)**: A task parallelism call that executes each of the actions passed.

- **public static void Invoke(ParallelOptions parallelOptions, params Action[] actions)**: A task parallelism call that executes each of the actions and allows **ParallelOptions** to be specified to configure method calls.

The **ParallelOptions** class can be used to configure how the **Parallel** methods operate:

- **public CancellationToken CancellationToken { get; set; }**: The familiar cancelation token that can be used to detect within loops if cancellation has been requested by a caller.
- **public int MaxDegreeOfParallelism { get; set; }**: An advanced setting that determines the maximum number of concurrent tasks that can be enabled at a time.
- **public TaskScheduler? TaskScheduler { get; set; }**: An advanced setting that allows a certain type of task queue scheduler to be set.

ParallelLoopState can be passed into the body of an **Action** for that action to then determine or monitor flow through the loop. The most commonly used properties are as follows:

- **public bool IsExceptional { get; }**: Returns **true** if an iteration has thrown an unhandled exception.
- **public bool IsStopped { get; }**: Returns **true** if an iteration has stopped the loop by calling the **Stop** method.
- **public void Break()**: The **Action** loop can call this to indicate execution should cease beyond the current iteration.
- **public void Stop()**: Requests that the loop should cease execution at the current iteration.
- **ParallelLoopResult**, as returned by the **For** and **ForEach** methods, contains a completion status for the **Parallel** loop.
- **public bool IsCompleted { get; }**: Indicates that the loop ran to completion and did not receive a request to end before completion.
- **public long? LowestBreakIteration { get; }**: If **Break** is called **while** the loop runs. This returns the index of the lowest iteration the loop arrived at.

Using the **Parallel** class does not automatically mean that a particular bulk operation will complete any faster. There is an overhead in scheduling tasks, so care should be taken when running tasks that are too short or too long. Sadly, there is no simple metric that determines an optimal figure here. It is often a case of profiling to see if operations do indeed complete faster using the **Parallel** class.

> **NOTE**
>
> You can find more information on data and task parallelism at https://docs.microsoft.com/en-us/dotnet/standard/parallel-programming/potential-pitfalls-in-data-and-task-parallelism.

PARALLEL.FOR AND PARALLEL.FOREACH

These two methods offer data parallelism. The same operation is applied to a collection of data objects or numbers. To benefit from these, each operation should be CPU-bound, that is it should require CPU cycles to execute rather than being IO-bound (accessing a file, for example).

With these two methods, you define an **Action** to be applied, which is passed an object instance or number to work with. In the case of **Parallel.ForEach**, the **Action** is passed an object reference parameter. A numeric parameter is passed to **Parallel.For**.

As you saw in *Chapter 3, Delegates, Events, and Lambdas*, the **Action** delegate code can be as simple or complex as you need:

```
using System;
using System.Threading.Tasks;
using System.Globalization;
using System.Threading;

namespace Chapter05.Examples
{
    class ParallelForExamples
    {
        public static async Task Main()
        {
```

In this example, calling **Parallel.For**, you pass an inclusive **int** value to start from (**99**) and an exclusive end value (**105**). The third argument is a lambda statement, **Action**, that you want invoked over each iteration. This overload uses **Action<int>**, passing an integer via the **i** argument:

```
            var loopResult = Parallel.For(99, 105, i =>
            {
                Logger.Log($"Sleep iteration {i}");
                Thread.Sleep(i * 10);
                Logger.Log($"Awake iteration {i}");
            });
```

Examine the **ParallelLoopResult IsCompleted** property:

```
            Console.WriteLine($"Completed: {loopResult.IsCompleted}");
            Console.ReadLine();
        }
    }
}
```

Running the code, you'll see that it stops at **104**. Each iteration is executed by a set of different threads and the order appears somewhat random with certain iterations awaking before others. You have used a relatively short delay (using **Thread.Sleep**) so the parallel task scheduler may take a few additional milliseconds to activate each iteration. This is the reason why the orders in which iterations are executed should be independent of each other:

```
18:39:37 [10] Sleep iteration 104
18:39:37 [03] Sleep iteration 100
18:39:37 [06] Sleep iteration 102
18:39:37 [04] Sleep iteration 101
18:39:37 [01] Sleep iteration 99
18:39:37 [07] Sleep iteration 103
18:39:38 [03] Awake iteration 100
18:39:38 [01] Awake iteration 99
18:39:38 [06] Awake iteration 102
18:39:38 [04] Awake iteration 101
18:39:38 [07] Awake iteration 103
18:39:38 [10] Awake iteration 104
Completed: True
```

Using the **ParallelLoopState** override, you can control the iterations from with the **Action** code. In the following example, the code checks to see if it is at iteration number **15**:

```
            var loopResult1 = Parallel.For(10, 20,
              (i, loopState) =>
              {
                Logger.Log($"Inside iteration {i}");
                if (i == 15)
                {
                    Logger.Log($"At {i}…break when you're ready");
```

Calling **Break** on **loopState** communicates that the **Parallel** loop should cease further iterations as soon as it can:

```
                    loopState.Break();
                }
              });
            Console.WriteLine($"Completed: {loopResult1.IsCompleted},
LowestBreakIteration={loopResult1.LowestBreakIteration}");
            Console.ReadLine();
```

From the results, you can see you got to item **17** before things actually stopped, despite asking to break at iteration **15**, as can be seen from the following snippet:

```
19:04:48 [03] Inside iteration 11
19:04:48 [03] Inside iteration 13
19:04:48 [03] Inside iteration 15
19:04:48 [03] At 15...break when you're ready
19:04:48 [01] Inside iteration 10
19:04:48 [05] Inside iteration 14
19:04:48 [07] Inside iteration 17
19:04:48 [06] Inside iteration 16
19:04:48 [04] Inside iteration 12
Completed: False, LowestBreakIteration=15
```

The code used **ParallelLoopState.Break**; this indicates the loop **should** cease at the next iteration if possible. In this case, you actually arrived at iteration **17** despite requesting a stop at iteration **15**. This generally occurs when the runtime has already started a subsequent iteration and then detects a **Break** request just after. These are requests to stop; the runtime may run extra iterations before it can stop.

Alternatively, the **ParallelLoopState.Stop** method can be used for a more abrupt stop. An alternative **Parallel.For** overload allows state to be passed into each loop and return a single aggregate value.

To better learn about these overloads, you will calculate the value of **pi** in the next example. This is an ideal task for **Parallel.For** as it means repeatedly calculating a value, which is aggregated before being passed to the next iteration. The higher the number of iterations, the more accurate the final number.

> **NOTE**
>
> You can find more information on the formula at https://www.mathscareers.org.uk/article/calculating-pi/.

You use a loop to prompt the user to enter the number of series (the number of decimal places to be shown) as a multiple of a million (to save typing many zeroes):

```
            double series;
            do
            {
                Console.Write("Pi Series (in millions):");
                var input = Console.ReadLine();
```

Try to parse the input:

```
                if (!double.TryParse(input, NumberStyles.Any,
CultureInfo.CurrentCulture, out series))
                {
                    break;
                }
```

Multiply the entered value by one million and pass it to the awaitable **CalcPi** function (which will be defined shortly):

```
                var actualSeries = series * 1000000;
                Console.WriteLine($"Calculating PI {actualSeries:N0}");
                var pi = await CalcPi((int)(actualSeries));
```

You eventually receive the value of **`pi`**, so use the string interpolation feature to write **`pi`** to **`18`** decimal places using the **`:N18`** numeric format style:

```
            Console.WriteLine($"PI={pi:N18}");

        }
```

Repeat the loop until **0** is entered:

```
        while (series != 0D);
        Console.ReadLine();
```

Now for the **`CalcPi`** function. You know that the **`Parallel`** methods all block the calling thread, so you need to use **`Task.Run`** which will eventually return the final calculated value.

The concept of **thread synchronization** will be covered briefly. There is a danger when using multiple threads and shared variables that one thread may read a value from memory and attempt to write a new value at the same time a different thread is trying to do the same operation, with its own value and what it thinks is the correct current value, when it may have read an already out-of-date shared value.

To prevent such issues, a **mutual-exclusion lock** can be used so that a given thread can execute its statements while it holds a lock and then releases that lock when finished. All other threads are blocked from acquiring the lock and are forced to wait until the lock is released.

This can be achieved using the **`lock`** statement. All of the complexities are handled by the runtime when the **`lock`** statement is used to achieve thread synchronization. The **`lock`** statement has the following form:

```
lock (obj){ //your thread safe code here }.
```

Conceptually, you can think of the **`lock`** statement as a narrow gate that has enough room to allow just one person to pass through at a time. No matter how long a person takes to pass through the gate and what they do while they are there, everyone else must wait to get through the gate until the person with the key has left (releasing the lock).

Returning to the **CalcPi** function:

```
        private static Task<double> CalcPi(int steps)
        {
            return Task.Run(() =>
            {
                const int StartIndex = 0;
                var sum = 0.0D;
                var step = 1.0D / (double)steps;
```

The **gate** variable is of type **object** and used with the **lock** statement inside the lambda to protect the **sum** variable from unsafe access:

```
                var gate = new object();
```

This is where things get a little more complex, as you use the **Parallel.For** overload, which additionally allows you to pass in extra parameters and delegates:

- **fromInclusive**: The start index (**0** in this case).
- **toExclusive**: The end index (steps).
- **localInit**: A **Func** delegate that returns the **initial state** of data local to each iteration.
- **body**: The actual **Func** delegate that calculates a value of Pi.
- **localFinal**: A **Func** delegate that performs the final action on the local state of each iteration.

```
                Parallel.For(
                    StartIndex,
                    steps,

                    () => 0.0D,                  // localInit

                    (i, state, localFinal) =>    // body
                    {
                        var x = (i + 0.5D) * step;
                        return localFinal + 4.0D / (1.0D + x * x);
                    },

                    localFinal =>                //localFinally
                    {
```

Here, you now use the **lock** statement to ensure that only one thread at a time can increment the value of **sum** with its own correct value:

```
                    lock (gate)
                        sum += localFinal;
                });

            return step * sum;
        });
    }
```

By using the **lock(obj)** statement, you have provided a minimum level of thread safety, and running the program produces the following output:

```
Pi Series (in millions):1
Calculating PI 1,000,000
PI=3.141592653589890000

Pi Series (in millions):20
Calculating PI 20,000,000
PI=3.141592653589810000

Pi Series (in millions):30
Calculating PI 30,000,000
PI=3.141592653589750000
```

Parallel.ForEach follows similar semantics; rather than a range of numbers being passed to the **Action** delegate, you pass a collection of objects to work with.

> **NOTE**
>
> You can find the code used for this example at https://packt.link/1yZu2.

The following example shows **Parallel.ForEach** using **ParallelOptions** along with a cancelation token. In this example, you have a console app that creates 10 customers. Each customer has a list containing the value of all orders placed. You want to simulate a slow-running service that fetches a customer's order on demand. Whenever any code accesses the **Customer.Orders** property, the list is populated only once though. Here, you will use another **lock** statement per customer instance to ensure the list is safely populated.

An **Aggregator** class will iterate through the list of customers and calculate the total and average order costs using a **Parallel.ForEach** call. Allow the user to enter a maximum time period that they are prepared to wait for all of the aggregations to complete and then show the top five customers.

Start by creating a **Customer** class whose constructor is passed a **name** argument:

```
using System;
using System.Collections.Generic;
using System.Globalization;
using System.Linq;
using System.Threading;
using System.Threading.Tasks;

namespace Chapter05.Examples
{
    public class Customer
    {
        public Customer(string name)
        {
            Name = name;
            Logger.Log($"Created {Name}");
        }

        public string Name { get; }
```

You want to populate the **Orders** list on demand and once only per customer, so use another **lock** example that ensures the list of orders is safely populated once. You simply use the **Orders get** accessor to check for a null reference on the **_orders** variable, before creating a random number of order values using the **Enumerable.Range** LINQ method to generate a range of numbers.

Note, you also simulate a slow request by adding **Thread.Sleep** to block the thread that is accessing this customer's orders for the first time (as you're using the **Parallel** class, this will be a background thread rather than the main thread):

ParallelForEachExample.cs

```
        private readonly object _orderGate = new object();
        private IList<double> _orders;
        public IList<double> Orders
        {
            get
            {
                lock (_orderGate)
                {
                    if (_orders != null)
                         return _orders;

                     var random = new Random();
                     var orderCount = random.Next(1000, 10000);

```

You can find the complete code here: https://packt.link/Nmx3X.

The **Total** and **Average** properties that will be calculated by your **Aggregator** class are as follows:

```
        public double? Total { get; set; }
        public double? Average { get; set; }
    }
```

Looking at the **Aggregator** class, note that its **Aggregate** method is passed a list of customers to work with and **CancellationToken**, which will automatically raise a cancellation request based on the console user's preferred timespan. The method returns a bool-based **Task**. The result will indicate whether the operation was canceled partway through processing the customers:

```
    public static class Aggregator
    {
        public static Task<bool> Aggregate(IEnumerable<Customer> 
customers, CancellationToken token)
        {
            var wasCancelled = false;
```

The main **`Parallel.ForEach`** method is configured by creating a **`ParallelOptions`** class, passing in the cancellation token. When invoked by the **`Parallel`** class, the **`Action`** delegate is passed a **`Customer`** instance (**`customer =>`**) that simply sums the order values and calculates the average which is assigned to the customer's properties.

Notice how the **`Parallel.ForEach`** call is wrapped in a **`try-catch`** block that catches any exceptions of type **`OperationCanceledException`**. If the maximum time period is exceeded, then the runtime will throw an exception to stop processing. You must catch this; otherwise, the application will crash with an unhandled exception error:

`ParallelForEachExample.cs`

```
            return Task.Run(() =>
            {
                var options = new ParallelOptions { CancellationToken = token };

                try
                {
                    Parallel.ForEach(customers, options,
                        customer =>
                        {
                             customer.Total = customer.Orders.Sum();
                             customer.Average = customer.Total /
                                                customer.Orders.Count;
                             Logger.Log($"Processed {customer.Name}");
                        });
                 }
```

You can find the complete code here: https://packt.link/FfVNA.

The main console app prompts for a maximum waiting time, **`maxWait`**:

```
    class ParallelForEachExampleProgram
    {
        public static async Task Main()
        {
            Console.Write("Max waiting time (seconds):");
            var input = Console.ReadLine();
            var maxWait = TimeSpan.FromSeconds(int.TryParse(input,
NumberStyles.Any, CultureInfo.CurrentCulture, out var inputSeconds)
                ? inputSeconds
                : 5);
```

Create **100** customers that can be passed to the aggregator:

```
            var customers = Enumerable.Range(1, 10)
                .Select(n => new Customer($"Customer#{n}"))
                .ToList();
```

Create **CancellationTokenSource** instance, passing in the maximum wait time. As you saw earlier, any code that uses this token will be interrupted with a cancellation exception should the time limit be exceeded:

```
            var tokenSource = new CancellationTokenSource(maxWait);
            var aggregated = await Task.Run(() => Aggregator.
Aggregate(customers,
                                       tokenSource.Token));
```

Once the task completes, you simply take the top five customers ordered by total. The **PadRight** method is used to align the customer's name in the output:

```
            var topCustomers = customers
                .OrderByDescending(c => c.Total)
                .Take(5);

            Console.WriteLine($"Cancelled: {aggregated }");
            Console.WriteLine("Customer      \tTotal         \tAverage  \
tOrders");

            foreach (var c in topCustomers)
            {
                Console.WriteLine($"{c.Name.PadRight(10)}\t{c.Total:N0}\
t{c.Average:N0}\t\t{c.Orders.Count:N0}");
            }

            Console.ReadLine();

        }
    }
}
```

Running the console app with a short time of **1** second produces this output:

```
Max waiting time (seconds):1
21:35:56 [01] Created Customer#1
21:35:56 [01] Created Customer#2
21:35:56 [01] Created Customer#3
21:35:56 [01] Created Customer#4
21:35:56 [01] Created Customer#5
21:35:56 [01] Created Customer#6
21:35:56 [01] Created Customer#7
21:35:56 [01] Created Customer#8
21:35:56 [01] Created Customer#9
21:35:56 [01] Created Customer#10
21:35:59 [07] Processed Customer#5
21:35:59 [04] Processed Customer#3
21:35:59 [10] Processed Customer#7
21:35:59 [06] Processed Customer#2
21:35:59 [05] Processed Customer#1
21:35:59 [11] Processed Customer#8
21:35:59 [08] Processed Customer#6
21:35:59 [09] Processed Customer#4
21:35:59 [05] Caught The operation was canceled.
Cancelled: True
Customer        Total           Average         Orders
Customer#1      23,097,348      2,395           9,645
Customer#4      19,029,182      2,179           8,733
Customer#8      15,322,674      1,958           7,827
Customer#6      9,763,247       1,568           6,226
Customer#2      6,189,978       1,250           4,952
```

The operation of creating **10** customers ran using Thread **01** as this was intentionally synchronous.

> **NOTE**
>
> Visual Studio may show the following warning the first time you run the program: **Non-nullable field '_orders' must contain a non-null value when exiting constructor. Consider declaring the field as nullable.** This is a suggestion to check the code for the possibility of a null reference.

Aggregator then starts processing each of the customers. Notice how different threads are used and processing does not start with the first customer either. This is the task scheduler deciding which task is next in the queue. You only managed to process eight of the customers before the token raised the cancelation exception.

> **NOTE**
>
> You can find the code used for this example at https://packt.link/1LDxI.

You have looked at some of the features available in the **Parallel** class. You can see that it provides a simple yet effective way to run code across multiple tasks or pieces of data.

The phrase **embarrassingly parallel** was covered under *Parallel Programming* section at the beginning of the chapter and refers to cases in which a series of tasks can be broken down into small independent chunks. Using the **Parallel** class is an example of this and can be a great utility.

The next section will bring these concurrency concepts into an activity that uses multiple tasks to generate a sequence of images. As each of the images can take a few seconds to create, you will need to offer the user a way to cancel any remaining tasks if the user so chooses.

ACTIVITY 5.01: CREATING IMAGES FROM A FIBONACCI SEQUENCE

In *Exercise 5.01*, you looked at a recursive function to create a value called a Fibonacci number. These numbers can be joined into what is known as a Fibonacci sequence and used to create interesting spiral-shaped images.

For this activity, you need to create a console application that allows various inputs to be passed to a sequence calculator. Once the user has entered their parameters, the app will start the time-consuming task of creating 1,000 images.

Each image in the sequence may take a few seconds to compute and create so you will need to provide a way to cancel the operation midway using **`TaskCancellationSource`**. If the user cancels the task, they should still be able to access the images that were created prior to the cancellation request. Essentially, you are allowing the user to try different parameters to see how this affects output images.

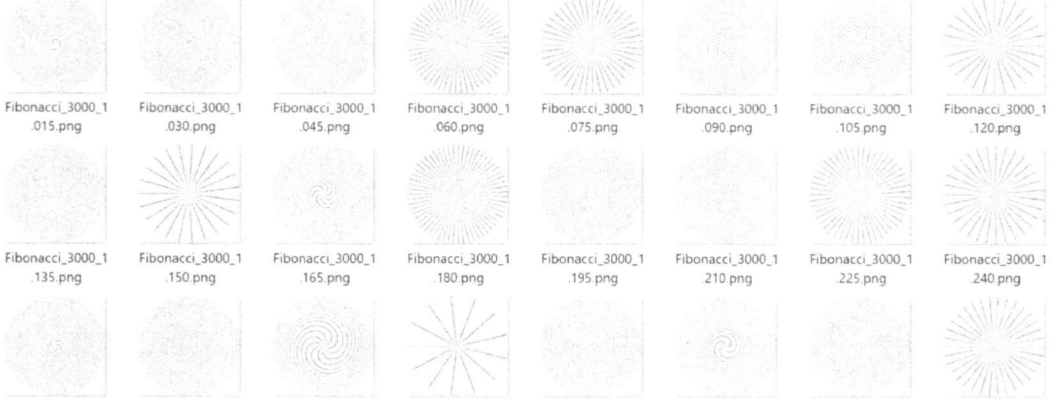

Figure 5.2: Fibonacci sequence image files

This is an ideal example for the **Parallel** class or **async/await** tasks if you prefer. The following inputs will be needed from the user:

- Input the value for **phi** (values between **1.0** and **6.0** provide ideal images).
- Input the number of images to create (the suggestion is **1,000** per cycle).
- Input the optional number of points per image (a default of **3,000** is recommended).
- Input the optional image size (defaults to **800** pixels).
- Input the optional point size (defaults to **5**).
- Next input the optional file format (defaults to **.png** format).
- The console app should use a loop that prompts for the preceding parameters and allows the user to enter new criteria while images are created for previous criteria.
- If the user presses **Enter** whilst a previous set of images is still being created, then that task should be canceled.
- Pressing **x** should close the application.

As this activity is aimed at testing your asynchronous skills, rather than math or image processing, you have the following classes to help with calculations and image creation:

- The **Fibonacci** class defined here calculates **X** and **Y** coordinates for successive sequence items. For each image loop, return a list of **Fibonacci** classes.
- Create the first element by calling **CreateSeed**. The remainder of the list should use **CreateNext**, passing in the previous item:

FibonacciSequence.cs

```
public class Fibonacci
{
    public static Fibonacci CreateSeed()
    {
        return new Fibonacci(1, 0D, 1D);
    }

    public static Fibonacci CreateNext(Fibonacci previous, double angle)
    {
         return new Fibonacci(previous, angle);
     }

     private Fibonacci(int index, double theta, double x)
     {
         Index = index;
```

You can find the complete code here: http://packt.link/I7C6A.

- Create a list of Fibonacci items using the following **`FibonacciSequence.Calculate`** method. This will be passed the number of points to be drawn and the value of **`phi`** (both as specified by the user):

FibonacciSequence.cs

```
public static class FibonacciSequence
{
    public static IList<Fibonacci> Calculate(int indices, double phi)
    {
        var angle = phi.GoldenAngle();

        var items = new List<Fibonacci>(indices)
        {
            Fibonacci.CreateSeed()
         };

         for (var i = 1; i < indices; i++)
         {
             var previous = items.ElementAt(i - 1);
             var next = Fibonacci.CreateNext(previous, angle);
```

You can find the complete code here: https://packt.link/gYK4N.

- Export the generated data to **.png** format image files using the **dotnet add package** command to add a reference to the **System.Drawing.Common** namespace. Within your project's source folder, run this command:

```
source\Chapter05>dotnet add package System.Drawing.Common
```

- This image creation class **ImageGenerator** can be used to create each of the final image files:

ImageGenerator.cs

```
using System.Collections.Generic;
using System.Drawing;
using System.Drawing.Drawing2D;
using System.Drawing.Imaging;
using System.IO;

namespace Chapter05.Activities.Activity01
{
    public static class ImageGenerator
     {
          public static void ExportSequence(IList<Fibonacci> sequence,
              string path, ImageFormat format,
              int width, int height, double pointSize)
          {
              double minX = 0;
```

You can find the complete code here: http://packt.link/a8Bu7.

To complete this activity, perform the following steps:

1. Create a new console app project.
2. The generated images should be saved in a folder within the system's **Temp** folder, so use **Path.GetTempPath()** to get the **Temp** path and create a subfolder called **Fibonacci** using **Directory.CreateDirectory**.
3. Declare a **do**-loop that repeats the following *Step 4* to *Step 7*.
4. Prompt the user to enter a value for **phi** (this typically ranges from **1.0** to **6.00**). You will need to read the user's input as a string and use **double.TryParse** to attempt to convert their input into a valid double variable.

5. Next, prompt the user to enter a value for the number of image files to create (**1,000** is an acceptable example value). Store the parsed input in an **int** variable called **imageCount**.
6. If either of the entered values is empty, this will indicate that the user pressed the **Enter** key alone, so break out of the **do**-loop. Ideally, **CancellationTokenSource** can also be defined and used to cancel any pending calculations.
7. The value of **phi** and **imageCount** should be passed to a new method called **GenerateImageSequences**, which returns a **Task**.
8. The **GenerateImageSequences** method needs to use a loop that iterates for each of the image counts requested. Each iteration should increment **phi**, and a constant value (a suggestion is **0.015**) before awaiting a **Task.Run** method that calls **FibonacciSequence.Calculate**, passing in **phi** and a constant for the number of points (**3,000** provides an acceptable example value). This will return a list of Fibonacci items.
9. **GenerateImageSequences** should then pass the generated Fibonacci list to the image creator **ImageGenerator.ExportSequence**, awaiting using a **Task.Run** call. An image size of **800** and a point size of **5** are recommended constants for the call to **ExportSequence**.
10. Running the console app should produce the following console output:

```
Using temp folder: C:Temp\Fibonacci\
Phi (eg 1.0 to 6.0) (x=quit, enter=cancel):1
Image Count (eg 1000):1000
Creating 1000 images...
20:36:19 [04] Saved Fibonacci_3000_1.015.png
20:36:19 [06] Saved Fibonacci_3000_1.030.png
20:36:20 [06] Saved Fibonacci_3000_1.090.png
```

You will find that various image files have been generated in the Fibonacci folder in the system's **Temp** folder:

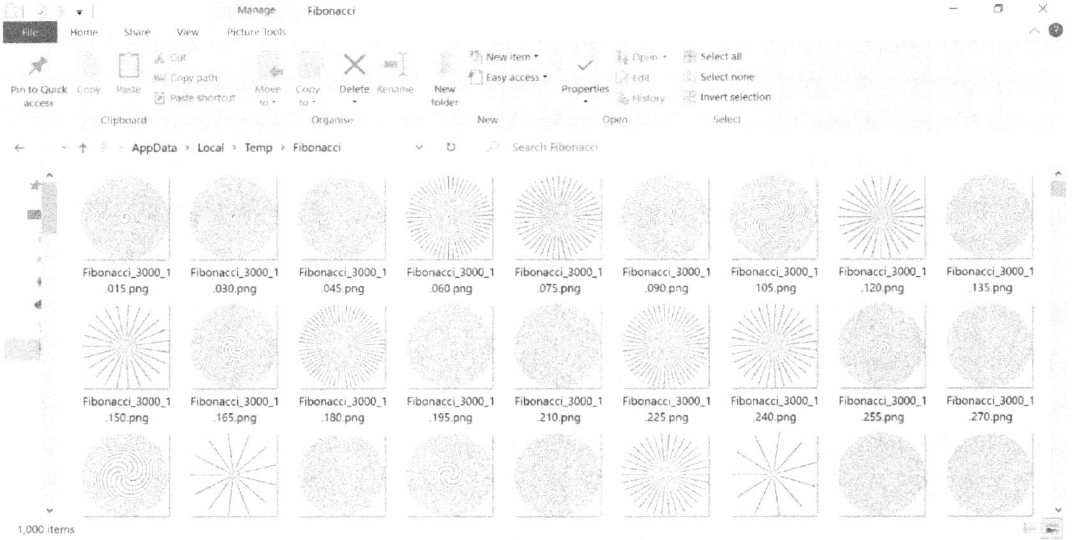

Figure 5.3: Windows 10 Explorer image folder contents (a subset of images produced)

By completing this activity, you have seen how multiple long-running operations can be started and then coordinated to produce a single result, with each step running in isolation, allowing other operations to continue as necessary.

> **NOTE**
>
> The solution to this activity can be found at https://packt.link/qclbF.

SUMMARY

In this chapter, you considered some of the power and flexibility that concurrency provides. You started by passing target actions to tasks that you created and then looked at the static **`Task`** factory helper methods. By using continuation tasks, you saw that single tasks and collections of tasks can be coordinated to perform aggregate actions.

Next, you studied the **`async/await`** keywords that can help you write simpler and more concise code that is, hopefully, easier to maintain.

This chapter looked at how C# provides, with relative ease, concurrency patterns that make it possible to leverage the power of multicore processors. This is great for offloading time-consuming calculations, but it does come at a price. You saw how the **`lock`** statement can be used to safely prevent multiple threads from reading or writing to a value simultaneously.

In the next chapter, you will look at how Entity Framework and SQL Server can be used to interact with relational data in C# applications. This chapter is about working with databases. If you are unfamiliar with database structure or would like a refresher on the basics of PostgreSQL, please refer to the bonus chapter available in the GitHub repository for this book.

6

ENTITY FRAMEWORK WITH SQL SERVER

OVERVIEW

This chapter introduces you to the basics of database design, storage, and processing using SQL and C#. You will learn about the Entity Framework (EF), and Object-Relational Mapper (ORM) and use them to convert database results into C# objects. You will then learn about the main performance pitfalls of SQL and EF and how to find and fix them.

Finally, you will delve into enterprise practices of working with databases by looking at Repository and Command Query Responsibility Segregation (CQRS) patterns and also by setting up a local database for development and testing. By the end of this chapter, you will be able to create and design your own database using PostgreSQL Server and use EF to hook a C# backend to it.

INTRODUCTION

There are multiple types of databases, but the most common one is relational, and the language for managing relational databases is SQL. SQL is optimized for data persistence. However, executing business rules in it is inefficient. Therefore, before consumption, data is often fetched in application memory and transformed into objects. This transformation is called **object-relational mapping**.

There is a lot of complexity in mapping database records to objects. However, this complexity is mitigated by **Object-Relational Mapper** (**ORM**). Some ORMs only do mapping (called **micro-ORMs**), but many popular ORMs also abstract away database language and allow you to use the same language to execute business rules and process data:

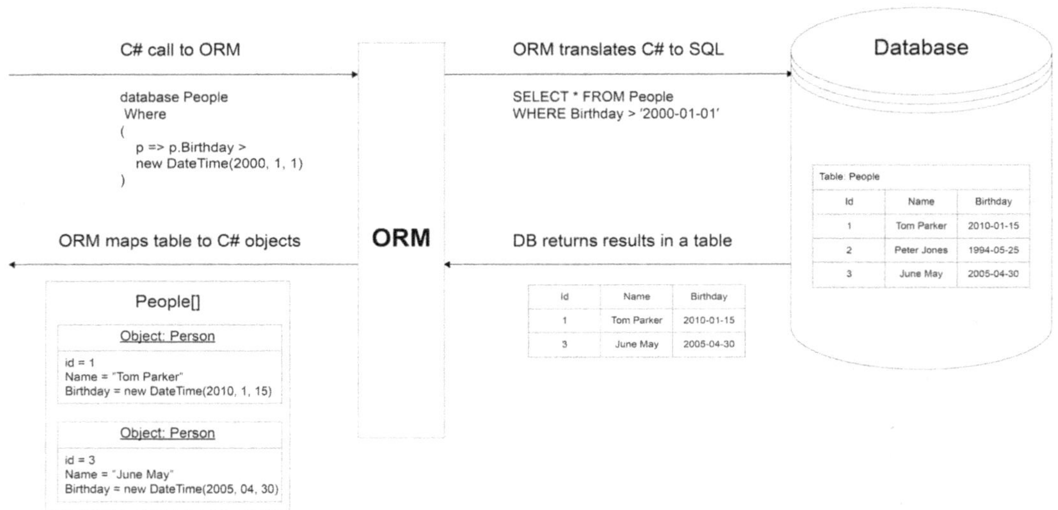

Figure 6.1: How an ORM works in translating C# to SQL and back

The focus of this chapter will be on **Entity Framework** (**EF**)—the most popular ORM in .NET. In the practical sections of this chapter, you will use it to rapidly prototype relational databases, and then make queries against them. It's worth mentioning that internally, whenever databases are involved, you are interacting with the **ADO.NET** part of .NET.

Before proceeding, however, it's recommended that you install the latest version of PostgreSQL with PostgreSQL Server found here: https://www.enterprisedb.com/downloads/postgres-postgresql-downloads. You can find the installation instructions for this in the *Preface*.

This chapter will use the **AdventureWorks** database, which is an adaptation of a popular example database that Microsoft often uses; it will be defined in detail in the following section.

> **NOTE**
>
> For those who are interested in learning the basics of databases and how to work with PostgreSQL, a reference chapter has been included in the GitHub repository of this book. You can access it at https://packt.link/sezEm.

CREATING A DEMO DATABASE BEFORE YOU START

You will use **Adventureworks** as an example because it is a common database used by Microsoft and has just enough complexity to learn about databases topic.

Perform the following steps to do so:

1. Open the command line and make a directory where you will call **AdventureWorks** database and move to that directory:

```
C:\<change-with-your-download-path-to-The-C-Sharp-Workshop>\
Chapter06\AdventureWorks\>
```

> **NOTE**
>
> Replace **<change-with-your-download-path-to-The-C-Sharp-Workshop>** with a directory where you downloaded the The-C-Sharp-Workshop repository.

2. Create an empty **Adventureworks** database by running the following command in the console:

```
psql -U postgres -c "CREATE DATABASE \"Adventureworks\";"
```

3. Create tables and populate them with data using the installation script.

> **NOTE**
>
> The installation script is found at https://packt.link/0SHd5.

4. Run the following command pointing to the installation script:

```
psql -d Adventureworks -f install.sql -U postgres
```

MODELING DATABASES USING EF

Working with a database from any other language comes with an interesting problem and that is, how do you convert table rows into C# objects? In C#, communicating with a database requires a database connection and SQL statements. Executing the statements will bring up a **results reader**, which is very similar to a table. Using the results reader dictionary, you can go through the results and map them into a new object.

The code for this would look like the following:

```
using var connection = new NpgsqlConnection(Program.
GlobalFactoryConnectionString);
connection.Open();
NpgsqlCommand command = new NpgsqlCommand("SELECT * FROM factory.
product", connection);
var reader = command.ExecuteReader();
var products = new List<Product>();
while (reader.Read())
{
    products.Add(new Product
    {
        Id = (int)reader["id"],
        //ManufacturerId = (int)reader["ManufacturerId"],
        Name = (string)reader["name"],
        Price = (decimal)reader["price"]
    });
}

return products;
```

Don't worry about the details of this code yet; it will be broken down soon. For now, it is enough to know that the preceding snippet returns all rows from the `factory.product` table and maps the results to a list named `products`. Using this approach may be okay when working with a single table, but when joins are involved, it becomes tricky. Mapping from one type to another, as has been done here, is very granular and can become tedious. In order to run this example, go to https://packt.link/2oxXn and comment all lines within `static void Main(string[] args)` body except `Examples.TalkingWithDb.Raw.Demo.Run();`.

NOTE

You can find the code used for this example at https://packt.link/7uIJq.

Another factor to consider is that when you deal with SQL from the client side, you should be careful. You should not assume that a user will use your program as intended. So, you should therefore add validation on both the client and server sides. For example, if a textbox requires a user ID to be entered, the client could enter **105** and get the details of the user of that ID. The query for this would be as follows:

```
SELECT * FROM Users WHERE UserId = 105
```

A user could also enter **105 or 1 = 1**, which is always true and thus this query returns all users:

```
SELECT * FROM Users WHERE UserId = 105 or 1 = 1
```

At best, this breaks your application. At worst, it leaks all the data. This kind of exploit is called **SQL injection**.

A simple yet effective way to solve the problem of accepting any kind of user input is to use an ORM as it allows you to convert database tables into C# objects and vice versa. In the .NET ecosystem, the three ORMs most commonly used are EF, Dapper, and NHibernate. Dapper is effective when top performance is needed because working with it involves executing raw SQL statements. Such ORMs are called micro-ORMs because they just do the mapping and nothing else.

NHibernate originated with the Java ecosystem and was one of the first ORMs in .NET. NHibernate, just like EF, solves a bigger problem than micro-ORMs by trying to abstract away SQL and database-related low-level details. Using a full-fledged ORM, such as EF or Nhibernate, often means that you don't need to write SQL to communicate with a database. In fact, the two ORMs allow you to generate complex databases out of the objects you have. The opposite is also possible (that is, you can generate objects out of databases you already have).

In the next sections, the focus will be on EF. Why not Dapper? Because Dapper requires knowledge of SQL and you want to make use of a simplified syntax. Why not NHibernate? Because NHibernate is old, it has too many configuration options, none of which are useful for getting started with ORMs.

Before delving into EF, you first need to connect to a database. So, proceed to learn about connection string and security.

CONNECTION STRING AND SECURITY

No matter what language you use, connecting to a database will always involve using a **connection string**. It contains three important parts:

- IP or a server name.
- The name of the database you would like to connect to.
- Some sort of security credentials (or none, if using a trusted connection only used for databases on the same network).

To connect to the local database you were previously working on in the *Modeling Databases Using EF* section (**`new NpgsqlConnection(ConnectionString))`**, you could use the following connection string (the password has been obfuscated for security reasons):

```
"Host=localhost;Username=postgres;Password=*****;Database=
globalfactory2021"
```

The connection string will be used when you will add the environment variables in your OS. This is detailed ahead. Different databases use different connections. For example, the following databases use these connections:

- SQL Server: **`SqlConnection`**
- PostgreSQL: **`NpgsqlConnection`**
- MySql: **`MySqlConnection`**
- SQLite: **`SqliteConnection`**

The connection object is the touching point between .NET and SQL database because it is only through it that you can communicate with a database.

Hardcoding a **connection string** comes with a few problems:

- To change a connection string, the program must be recompiled.
- It's not secure. The connection string can be viewed by everyone who knows how to decompile code (or worse, is publicly visible if it's an open-source project).

Therefore, a connection string is usually stored in a configuration file. This does not solve the problem of sensitive parts of a connection string being stored. To fix that, often, either the whole string or a part of it is replaced during the application's deployment. There are three main ways to securely store and retrieve application secrets:

- Environment variables: These are variables unique to a system and can be accessed by any application on the same machine. This is the simplest secure approach and might not be safe in a production environment.
- Secret Manager tool (available in both .NET and .NET Core applications): Similar to environment variables but more .NET specific, it will store all secrets on the local machine as well but in a file called **secrets.json**. This option, too, might not be safe in a production environment.
- Key vault: This is the most secure approach because, unlike the other two, it is not coupled with a specific environment. Key vaults store secrets in one centralized place; usually remotely. This approach is most commonly used for enterprise applications. In the context of Azure, Azure Key Vault is the best choice and is perfect for a production environment.

In the following example, you'll try to securely store the connection string you made previously. You will use the simplest secure approach that is suitable for a development environment—that is, environment variables. This approach fits local development the best because the other two require third-party tools to set up and take much longer.

> **NOTE**
>
> Before you continue, make sure to go through *Exercise 1* of the *Reference Chapter*, *A Primer for Simple Databases and SQL*. It has the steps needed to create a new database with the needed tables.

Adding an environment variable in your OS is just a matter of performing some simple steps. Perform the following steps in Windows to set the environment variables:

1. Go to **Control Panel**.
2. Click **System & Security** and choose **System**.
3. Type **Environmental Variables** in the search box.
4. Then choose **Edit Environment Variables for your account** from the list displayed.
5. Inside the **Environment Variables** window, click **New** under the **System Variables** window.

6. Inside the New System variable window, type **GlobalFactory** beside the **Variable name**.
7. Beside **Variable value**, paste the following:

```
Host=localhost;Username=postgres;Password=*****;Database=
globalfactory2021
```

8. Next click **OK** on all windows to set your environment variables.

> **NOTE**
>
> Here the password would carry your database superuser password which you entered while creating the **globalfactory2021** database in PostgreSQL.

- Mac: From the command line, find **bash-profile: ~/.bash-profile f**. Open it using any text editor, then at the end of the file, add **export GlobalFactory='Host=localhost;Username=postgres;Password=*****;Database=globalfactory2021'**. Lastly, run **source ~/.bash-profile**, which will update the environment variables.
- Linux: From the command line, run this: **export GlobalFactory='Host=localhost;Username=postgres;Password=*****;Database=globalfactory2021'**.

Getting the environment variable instead of an in-memory one can now be done by placing a property in **Program.cs**, at the top of the class, as follows:

```
public static string ConnectionString { get; } = Environment.
GetEnvironmentVariable("GlobalFactory", EnvironmentVariableTarget.User);
```

This line returns the value of the **GlobalFactory** environment variable, configured for the local user. In the preceding snippet, you have added this line to **Program.cs** and made it static because that makes it easily accessible throughout the application. While in big applications, it is not a practice you would want to go for; however, for your purposes here, this is fine.

Before you grasp models—the centerpiece of a program—you need to know about the major versions of EF.

WHICH ONE TO CHOOSE—EF OR EF CORE?

There are two major versions of EF—EF and EF Core. Both are widely used, but you should be aware of some factors before making the choice that fits your project's requirements the best. EF was first released in 2008. At that time, there was no .NET Core and C# was for **Windows only** and strictly required .NET Framework. Currently, the latest major version of EF is 6 and it's likely that there won't be any other major version, because in 2016, along with .NET Core 1.0 came EF Core 1 (a rework of EF 6).

EF Core was initially named EF 7. However, it was a complete rewrite of EF 6 and therefore was soon renamed EF Core 1.0. EF works only on .NET and is for Windows only, whereas .NET Core works only on .NET Core and is multi-platform.

Feature-wise, both frameworks are similar and are still being developed. However, the focus these days is on EF Core because the future of C# is associated with .NET 6, which is a **multi-platform** framework. At the time of writing this book, EF 6 has a richer set of features. However, EF Core is quickly catching up and is likely to soon be ahead. If your project's specifications do not require working with .NET Framework, it is preferable to stick with EF Core.

> **NOTE**
>
> For the latest list of differences between the two, please refer to a comparison by Microsoft here: https://docs.microsoft.com/en-us/ef/efcore-and-ef6/.

Before you proceed, install the EF Core NuGet package so that you get access to the EF Core API. With the project open in **Visual Studio Code** (**VS Code**), run the following line in the terminal:

```
dotnet add package Microsoft.EntityFrameworkCore
```

By itself, **`EntityFrameworkCore`** is just a tool to abstract away database structures. To connect it with a specific database provider, you will need another package. Here you are using PostgreSQL. Therefore, the package you will install is **`Npgsql.EntityFrameworkCore.PostgreSQL`**. In order to install it, from the VS Code console, run the following:

```
dotnet add package Npgsql.EntityFrameworkCore.PostgreSQL
```

You are now aware of the two versions of EF and how they work with .NET Framework and .NET. The next section will delve into the models which are the heart of a program.

MODEL

A class designed to represent a business object is called a **model**. It always has data managed by properties or methods. Models are the centerpiece of a program. They don't depend on anything; other parts of a program point to them.

An object to which an ORM maps data tables is called an **entity**. In simple applications, an entity and a model are the same class. In complex applications, a change to a database is a common thing. That means that entities change often, and if you do not have a separate class for a model, your model would be impacted as well. Business logic should be isolated from database changes, and it is therefore recommended to have two classes—one for an entity and one for a model.

Before you continue with the next section, have a quick look at the **`factory.product`** and **`factory.manufacturer`** tables. One manufacturer makes many products. The following **Entity Relationship** (**ER**) diagram illustrates this relationship in *Figure 6.2*.

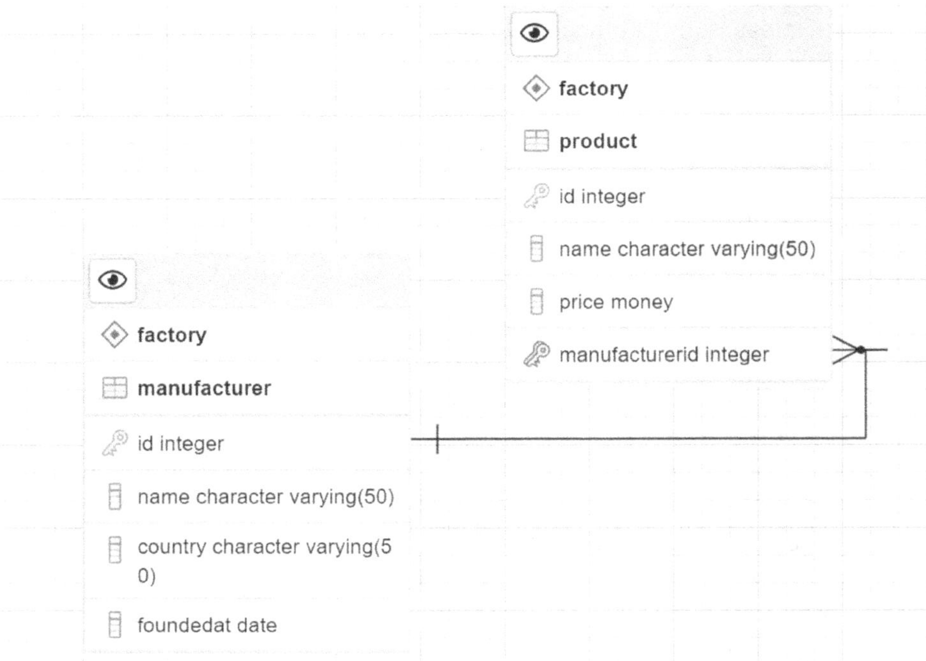

Figure 6.2: ER diagram of products and manufacturers

An entity, ideally, should mirror table columns. You can mirror columns through properties. For example, a **factory.product** table has **id**, **name**, **price**, and **manufacturerId**. An object that maps to that would look like this:

```
public class Product
{
    public int id { get; set; }
    public string name { get; set; }
    public decimal price { get; set; }
    public int manufacturerId { get; set; }
}
```

You know that only the price of a product can change; the rest of the properties would not. However, in the preceding snippet, a setter has still been written for every property. This is because entities created through an ORM always need to have all properties with setters, or else it might not set the value.

An entity should be designed to match a table structure, but it does not always have to be that way. For example, if the **Id** property were renamed to **PrimaryKey**, you could still use EF the same way by using a special data annotation **[Key]**:

```
public class Product
{
    [Key]
    public int PrimaryKey { get; set; }
    public string name { get; set; }
    public decimal price { get; set; }
    public int manufacturerId { get; set; }
}
```

Data Annotation is an attribute that adds metadata to a property. You can use it to provide a different name, have a constraint column as a key, add the minimum and maximum lengths for fields, add precision, declare a field as mandatory, and more. On their own, data annotations don't do anything. They don't add logic to a model. Some other components will consume annotated objects, which will involve reading their attributes and performing actions based on that.

Your model (illustrating the ER diagram from *Figure 6.2*) is almost complete, but there are a few problems to be addressed:

- First, the table-model mapping is missing a schema (**factory**, in this case), and so you need to specify it explicitly using a **Table** attribute.
- Second, by default if you wanted to also retrieve a **manufacturer**, you would need another query. You can fix this by adding a **navigational property** that refers to the manufacturer. But why should you use a navigational property? If there were only an ID, you would need a separate query to get the related entity. However, using navigational properties, you can use **eager loading** and get two or more entities at once.

The following code snippet will show you how to create a **Manufacturer** class and fix these issues for both models:

```
[Table("manufacturer", Schema = "factory")]
public class Manufacturer
{
    public int id { get; set; }
    public string name { get; set; }
    public string country { get; set; }
    public virtual ICollection<Product> Products { get; set; } = new
List<Product>();
}
```

Note the new **List<Product>();** part. It is needed so that if a table does not yet have products, the code would still function when you try to add a new product without throwing **NullReferenceException**.

In the following snippet of code, a model is created for a product table:

```
[Table("product", Schema = "factory")]
public class Product
{
    public int id { get; set; }
    public string name { get; set; }
    public decimal price { get; set; }
    public int manufacturerId { get; set; }
    public virtual Manufacturer Manufacturer { get; set; }
}
```

The two models are complete for mapping to tables from your database. You did not replace an ID property with a navigational property; both are present. If you did not do this, it would require the parent entity (**Manufacturer**) to be fetched before you could do anything with a product. With this approach, you can work with a product in isolation from the manufacturer. All you need is an ID link.

On top of the mentioned fixes, you made your navigational properties (**Manufacturer** and **Products**) virtual. This is needed to enable lazy loading for EF. **Lazy loading** means that there is no data loaded in a property until that property is referenced.

Finally, it is worth mentioning that for manufacturer products, you used **ICollection** and not **IEnumerable** or other collections. This makes sense because EF will need to populate the collection when it retrieves and map the items. **List** or even **Set** could work, but when designing object-oriented code, you should focus on the highest abstraction you can depend on, in this case it is **ICollection**.

> **NOTE**
>
> You can find the code used for this example at https://packt.link/gfgB1.

In order to run this example, go to https://packt.link/2oxXn and comment all lines within the **static voidMain(string[] args)** body, except **Examples.TalkingWithDb.Orm.Demo.Run();**.

You are now clear about the entity, models, entity relationship, data annotation, eager loading, and lazy loading. The next section will show you how to combine everything and communicate with a database through EF Core.

DBCONTEXT AND DBSET

DbContext is what EF uses as an abstraction to a database. A new database abstraction must be derived from the **DbContext** class and provide a way of connecting to a database. Just like a database contains one or more tables, **DbContext** contains one or more **DbSet** entities. For example, consider the following code:

```
public class FactoryDbContext : DbContext
{
    public DbSet<Product> Products { get; set; }
```

```
    public DbSet<Manufacturer> Manufacturers { get; set; }

    protected override void OnConfiguring(DbContextOptionsBuilder
optionsBuilder)
    {
        if (!optionsBuilder.IsConfigured)
        {
            optionsBuilder.UseNpgsql(Program.
GlobalFactoryConnectionString);
        }
    }
}
```

Here, **FactoryDbContext** is an abstraction of the database that you created before, with two tables: **Products** and **Manufacturers**. The **OnConfiguring** method takes **DbContextOptionsBuilder**, which allows you to specify what database you want to connect to and how that connection is made. In this case, you are using PostgreSQL Server and specifying a database to connect to. Please note that in case there is an already configured database provider then you will not use **Npgsql** in the **if** statement i.e., the **if (!optionsBuilder.IsConfigured)** statement.

It is important to note that you should not completely depend on a specific database provider for two reasons:

- Firstly, changing a database provider is easy; it is just a matter of using a different extension method on a builder.
- Secondly, EF has an in-memory database provider, which is effective for testing. Alternatively, you could use SQLite as well as a lightweight database meant just for testing.

Currently, your database abstraction needs improvement because it only lets you communicate with the SQL Server database. Instead of hardcoding the options, you will inject them. **Injecting** allows you to configure an existing class differently, without modifying it. You do not need to change the models to be able to choose the database you want to connect to. You can specify which database you want to connect to by passing an **options** object through the **FactoryDbContext** constructor:

```
public FactoryDbContext(DbContextOptions<FactoryDbContext> options)
    : base(options)
{
}
```

The default constructor is for the default provider, which will be used when no options are supplied. In this case, the context was designed to use PostgreSQL; therefore, you would add the following code:

```
public FactoryDbContext()
    : base(UsePostgreSqlServerOptions())
{
}
```

DbContext can be configured using **DbContextOptions**. In this example, you need to configure a database provider (PostgreSQL) and a connection string. Choose the provider using **DbContextOptionsBuilder**. The **UseNpgsql** is how you hook the PostgreSQL provider with your database context, as shown here:

```
protected static DbContextOptions UsePostgreSqlServerOptions()
{
    return new DbContextOptionsBuilder()
        .UseNpgsql(Program.ConnectionString)
        .Options;
}
```

The full **DbContext** now looks like this:

FactoryDbContext.cs

```
public class FactoryDbContext : DbContext
{
    public DbSet<Product> Products { get; set; }
    public DbSet<Manufacturer> Manufacturers { get; set; }


    public FactoryDbContext(DbContextOptions<FactoryDbContext> options)
        : base(options)
    {
    }

    public FactoryDbContext()
        : base(UsePostgreSqlServerOptions())
    {
    }
```

The complete code can be found here: https://packt.link/0uVPP.

In order to run this example, go to https://packt.link/2oxXn and comment all lines within the **static void Main(string[] args)** body, except **Examples.TalkingWithDb.Orm.Demo.Run();**.

To get the products from the database you have made, you first connect to a database by initializing an instance of your **DbContext**. You then call a wanted **DbSet** from that context and send a call to a database by calling **ToList()**:

```
using var context = new FactoryDbContext();
var products = context.Products.ToList();
```

In this case, you create a **FactoryDbContext** (which creates a connection to the **GlobalFactory** database) and the **context.Products.ToList()** equates to a **SELECT * FROM Products** SQL statement.

> **NOTE**
>
> The two lines mentioned are not included within GitHub. They are trivial and are here only for illustrative purposes.

When you initialize a **DbContext**, you almost always create a connection to a database, and if not managed, you might eventually run out of connections inside a connection pool (a collection of available connections). **DbContext** is an unmanaged resource; it implements the **IDisposable** interface, and so it needs explicit cleanup. Here, you applied a C# feature—**inline using**—which disposes of the object after it leaves the scope it is at:

```
using var context = new FactoryDbContext()
```

When you have a **DbContext**, getting data from it is trivial:

- Access a **DbSet**.
- Convert it into a list.

Why do you need to make any conversions, though? That is because **DbSet**, much like **IEnumerable**, is lazy-loaded. It encapsulates the SQL needed to execute. So, unless you explicitly demand it (for example, by calling **ToList**), there won't be any data queried. Calling **ToList** does the actual call to a database and retrieves all the products.

You now know all about databases. The next section will touch on the **AdventureWorks** database which is a common database for teaching SQL to beginners.

ADVENTUREWORKS DATABASE

AdventureWorks is a database used for learning purposes. It contains dozens of tables and has hundreds of records in each table. The tables are focused on wholesale, which is a common scenario in enterprise applications. In other words, the **AdventureWorks** database provides examples for learning with closeness to real-world problems.

> **NOTE**
>
> You must first create the **AdventureWorks** database in PostgreSQL. You can find the steps to create this database in the reference chapter placed on GitHub.

The previous sections covered entity, models, and how to combine everything and communicate with a database. You also learned about **DbContext** and **DbSet**. This concludes the theoretical portion of this section. In the following section, you will put this into practice with an exercise.

EXERCISE 6.01: READING STOCK LOCATIONS FROM ADVENTUREWORKS DATABASE

The simplest use case of EF is to read data tables into C# objects. This exercise will teach you how to create a data entity class and add correct attributes to it. For this, you will create an inventory **location** table within the example **AdventureWorks** database. Perform the following steps to do so:

1. Create a **Location** entity. It should have **LocationId**, **Name**, **Costrate**, **Availability**, and **ModifiedDate** properties, as follows:

```
[Table("location", Schema = "production")]
public class Location
{
    [Column("locationid")]
    public int LocationId { get; set; }
    [Column("name")]
    public string Name { get; set; }
    [Column("costrate")]
    public double Costrate { get; set; }
    [Column("availability")]
```

```
    public double Availability { get; set; }
    [Column("modifieddate")]
    public DateTime ModifiedDate { get; set; }
}
```

A **[Table]** attribute has been applied because you need to specify a schema as well as a properly capitalized table name. On top of that, every column name needs to be explicitly specified using the **[Column]** attribute since the capitalization does not match.

2. Create a class named **AdventureWorksContext**, which inherits the **DbContext**, as follows:

```
public class AdventureWorksContext : DbContext
{
    public DbSet<Location> Locations { get; set; }

    public AdventureWorksContext()
        : base(UsePostgreSqlServerOptions())
    {
    }

    protected static DbContextOptions UsePostgreSqlServerOptions()
    {
        return new DbContextOptionsBuilder()
            .UseNpgsql(Program.AdventureWorksConnectionString)
            .Options;
    }
```

Inheriting **DbContext** is necessary if you want to reuse the base functionality of database abstraction such as connecting to a database. The use of base functionality is visible in the two base constructors. In the parameterized constructor, you use PostgreSQL; in non-parameterized you can supply whatever database provider you choose.

3. Now use the **Program.AdventureWorksConnectionString** connection string as follows:

```
Host=localhost;Username=postgres;Password=****;Database=
Adventureworks. DbSet<Location>Locations
```

This represents the needed **location** table.

> **NOTE**
>
> Please keep your PostgreSQL passwords safe. Don't write them in code in plaintext, instead use environment variables or secrets.

4. Connect to a database:

```
var db = new AdventureWorksContext();
```

This is as simple as creating a new **DbContext**.

5. Get all products by adding the following code:

```
var locations = db.Locations.ToList();
```

6. Now that you have queried the locations and no longer need to keep the connection open, it is better to disconnect from the database. In order to disconnect from the database, call the **Dispose** method as follows:

```
db.Dispose();
```

7. Print the results by adding the following code:

```
foreach (var location in locations)
{
    Console.WriteLine($"{location.LocationId} {location.Name}
{location.Costrate} {location.Availability} {location.ModifiedDate}");
}
```

The code itself is run from https://packt.link/2oxXn. Make sure to comment all lines within **static void Main(string[] args)** body, except **Exercises.Exercise03.Demo.Run()**. When you run the code, the following output gets displayed:

```
1 Tool Crib 0 0 2008-04-30 00:00:00
2 Sheet Metal Racks 0 0 2008-04-30 00:00:00
3 Paint Shop 0 0 2008-04-30 00:00:00
4 Paint Storage 0 0 2008-04-30 00:00:00
5 Metal Storage 0 0 2008-04-30 00:00:00
6 Miscellaneous Storage 0 0 2008-04-30 00:00:00
7 Finished Goods Storage 0 0 2008-04-30 00:00:00
10 Frame Forming 22,5 96 2008-04-30 00:00:00
20 Frame Welding 25 108 2008-04-30 00:00:00
```

```
30 Debur and Polish 14,5 120 2008-04-30 00:00:00
40 Paint 15,75 120 2008-04-30 00:00:00
45 Specialized Paint 18 80 2008-04-30 00:00:00
50 Subassembly 12,25 120 2008-04-30 00:00:00
60 Final Assembly 12,25 120 2008-04-30 00:00:00
```

Working with EF is simple. As you can see from this exercise, it is intuitive and feels like a natural extension to C#.

> **NOTE**
>
> You can find the code used for this exercise at https://packt.link/9Weup.

QUERYING A DATABASE—LINQ TO SQL

One of the more interesting features of EF is that running SQL statements is very much like working with a collection. For example, say you want to retrieve a product by its name. You can get a product by name the same way you would be using LINQ:

```
public Product GetByName(string name)
{
    var product = db.Products.FirstOrDefault(p => p.Name == name);
    return product;
}
```

Here, **`FirstOrDefault`** returns the first matching product by its name. If no product by that name exists, then it returns a **`null`**.

What about finding a unique element by its ID? In that case, you would use a special method (**`Find`**), which either gets an entity from a database or, if one with the same ID has been retrieved recently, returns it from memory:

```
public Product GetById(int id)
{
    var product = db.Products.Find(id);
    return product;
}
```

When using a primary key, it is better to use **Find** instead of **Where** because it has a slightly different meaning in the context of EF. Instead of trying to create a SQL query and execute it, **Find** will check whether this item has already been accessed and will retrieve it from a cache, rather than going through a database. This makes for more efficient operations.

What about finding all products by the related manufacturer ID? You can create a method that returns an **IEnumerable<Product>** for this purpose, named **GetByManufacturer**, as follows:

```
public IEnumerable<Product> GetByManufacturer(int manufacturerId)
{     var products = db
          .Products
          .Where(p => p.Manufacturer.Id == manufacturerId)
          .ToList();

     return products;
}
```

You might be wondering why you should choose to use **Where** instead of **Find** here. That is because you are getting many products by their foreign key **manufacturerId**. Be careful not to mix foreign and primary keys; **Find** is used only for primary keys.

In order to run this example, go to https://packt.link/2oxXn and comment all lines within **static void Main(string[] args)** body except **Examples.Crud.Demo.Run();**.

> **NOTE**
>
> You can find the code used for this example at https://packt.link/pwcwx.

Now, how about retrieving related entities? If you simply call **db.Manufacturers.ToList()**, you will have null products. This is because the products will not be retrieved automatically unless explicitly specified. If you didn't call **ToList()**, you could make use of **lazy-loading** (that is, loading the required entities on demand), but that would result in a very suboptimal solution as you would always be querying child entities for every parent.

A proper solution is to call **`Include(parent => parent.ChildToInclude)`**:

```
db.Manufacturers
.Include(m => m.Products)
.ToList();
```

This approach is called **eager loading**. With this approach, you specify which child entities should be retrieved immediately. There will be scenarios where child entities will have their child entities; there, you could call **`ThenInclude`**. In order to run this example, comment all lines within **`static void Main(string[] args)`** body except **`Examples.Crud.Demo.Run();`** in **`Program.cs`**.

> **NOTE**
>
> You can find the code used for this example at https://packt.link/c82nA.

Remember when it was established that trying to get everything from a table is not the right thing to do in most cases? Eager loading has the same problem. So, what should you do if you only want some properties? It's time to learn about the other side of LINQ.

QUERY SYNTAX

Query syntax is an alternative syntax to LINQ lambdas. It is very similar to SQL. The main advantage of query syntax over lambdas is that it feels more natural to write queries when you have complex joins and want only some of the data back. Imagine you wanted to get all product-manufacturer name pairs. You cannot simply get manufacturers and include products; you only want two products. If you tried using LINQ, the code would like the following:

```
db.Products
.Join(db.Manufacturers,
    p => p.ManufacturerId, m => m.Id,
    (p, m) => new {Product = p.Name, Manufacturer = m.Name})
.ToList();
```

The same operation using query syntax looks like this:

```
(from p in db.Products
join m in db.Manufacturers
```

```
    on p.ManufacturerId equals m.Id
select new {Product = p.Name, Manufacturer = m.Name}
).ToList();
```

Break the code down:

```
from p in db.Products
```

Now select all products and their columns:

```
join m in db.Manufacturers
```

For every product, add manufacturer columns like this:

```
on p.ManufacturerId equals m.Id
```

Manufacturer columns are added only for products which have **ManufacturerId** equal to the **Id** of the manufacturer (**INNER JOIN**).

> **NOTE**
>
> Why can't you write **==** instead of **equals**? That is because, in LINQ query syntax, **equals** completes a join; it is not just a comparison of two values.

The **select** part is the same in both lambda and query syntax; however, it's worth mentioning what you selected. The **select new {...}** means that you create a new anonymous object to have all the things you want to select. The idea is to later use this to return a strongly typed object that you need. Therefore, after a **ToList** method, you are likely to perform another **select** operation to map the results for the final return. You cannot do the mapping right away because before you do **ToList**, you are still working with an expression that is yet to be converted into SQL. Only after **ToList** is called can you be sure that you are working with C# objects.

Finally, you may be wondering why the join is surrounded by brackets before calling **ToList**. That's because you were still in query syntax mode and the only way to escape it and go back to normal LINQ is by surrounding it with brackets.

If you struggle to remember LINQ query syntax, remember a **foreach** loop:

```
foreach(var product in db.Products)
```

The query syntax of **from** is as follows:

```
from product in db.Products
```

The highlighted part in the preceding code snippet is the syntax parts that overlap on both. This also applies to join. The two make the most of query syntax.

Both lambda and query syntax has the same performance metrics because, in the end, the query syntax will be compiled into the lambda equivalent. When making complex joins, it might make more sense to use a query syntax because it will look closer to SQL and therefore might be easier to grasp.

Now run the code. In order to run this example, comment all lines within **`static void Main(string[] args)`** body except **`Examples.Crud.Demo.Run();`** in **`Program.cs`**:

> **NOTE**
>
> You can find the code used for this example at https://packt.link/c82nA.

You now know that query syntax is an alternative syntax to LINQ lambdas. But how you can perform the remaining operations with rows that are, create, update, and delete, using query syntax? The next section details how that can be done.

THE REST OF CRUD

Adding, updating, and removing data using query syntax is also similar to basic LINQ. However, similar to executing the queries by calling **`ToList`**, it involves one extra step that is, committing the changes. Consider the following code where you are creating a new product:

```
var product = new Product
{
    Name = "Teddy Bear",
    Price = 10,
    ManufacturerId = 1
};
db.Products.Add(product);
db.SaveChanges();
```

This code should look almost completely familiar, except for the last line. The **`SaveChanges`** method is used to run the actual SQL. If you don't call it, nothing will happen, and the changes will be gone after disconnecting from the database. Also, when adding a child entity (**`product`**), you don't have to get a parent entity (**`manufacturer`**). All you have to do is to provide a link between the two via the foreign key (**`ManufacturerId`**).

Why do you need an extra method to apply your changes? Wouldn't it be simpler to call **Add** and immediately have a new product row created? In practice, it is not that simple. What happens if multiple **Add** methods of different entities need to be performed, and what if one of them fails? Should you allow some of them to succeed, while others fail? The worst thing you can do is to put your database in an invalid state, or, in other words, break data integrity. You need a mechanism to either complete fully or fail without affecting anything.

In the SQL context, such commands that are run together are called a **transaction**. You can do two things with transactions—either commit or roll them back. In EF, every action, other than a query, results in a transaction. The **SaveChanges** completes the transaction, whereas a command failing rolls the transaction back.

If you were to call plain SQL commands in C#, you would need to create a parameterized SQL command, provide each argument separately, and concatenate SQL for multi-query updates. For a small entity it may be easy; however, as the size grows the complexity increases as well. Using EF, you don't need to care about low-level details, such as passing arguments to a command. For example, with EF, adding a **manufacturer** with a few products is as simple as adding a **manufacturer** to a **Manufacturers** list:

```
var manufacturer = new Manufacturer
{
    Country = "Lithuania",
    Name = "Toy Lasers",
    Products = new List<Product>
    {
        new()
        {
            Name = "Laser S",
            Price = 4.01m
        },
        new()
        {
            Name = "Laser M",
            Price = 7.99m
        }
    }
};
db.Manufacturers.Add(manufacturer);
db.SaveChanges();
```

As you can see, creating manufacturers is nearly the same as adding an element to a list. The major difference is the need to complete the changes using **db.SaveChanges()** method.

What about updating an existing product? Set the price of a product to **45.99**:

```
var productUpdate = new Product
{
    Id = existingProduct.Id,
    Price = 45.99m,
    ManufacturerId = existingProduct.ManufacturerId,
    Name = existingProduct.Name
};
db.Products.Update(productUpdate);
 db.SaveChanges();
```

If you look carefully at this code, you are provided with not only the updated **Price** and an existing item **Id** but also all other fields. This is because there is no way for EF to know whether you want to set existing values to null or only set the new values. But don't worry; logically, there is rarely a case when you update something out of nowhere. You should have a set of items loaded somewhere. Therefore, updating an existing object would simply be a matter of setting a new value of a property of that object.

Of course, there are exceptions when you want to update just one thing. In that case, you can have a dedicated method and be completely in control. In the following snippet, you will update product values, but only when they are not null:

```
var productToUpdate = db.Products.Find(productUpdate.Id);
var anyProductToUpdate = productToUpdate != null;
if (anyProductToUpdate)
{
    productToUpdate.Name = productUpdate.Name ?? productToUpdate.Name;

    productToUpdate.ManufacturerId = (productUpdate.ManufacturerId !=
default)
        ? productUpdate.ManufacturerId
        : productToUpdate.ManufacturerId;

    productToUpdate.Price = (productUpdate.Price != default)
        ? productUpdate.Price
```

```
        : productToUpdate.Price;

    db.SaveChanges();
}
```

Here, you would only update the values if they were not the default ones. Ideally, when working in situations like this (in which you only want to update some of the fields), you should have a dedicated model for the updated fields, send those fields, and map them using libraries such as **AutoMapper**.

> **NOTE**
>
> To learn more about AutoMapper, refer to their official documentation at https://docs.automapper.org/en/stable/Getting-started.html.

What about deleting existing rows from a database? This involves first getting the object you want to remove and only then remove it. For example, say you want to remove a product with a particular ID:

```
var productToDelete = db.Products.Find(productId);
if (productToDelete != null)
{
    db.Products.Remove(productToDelete);
    db.SaveChanges();
}
```

Once again, removing something from a database is nearly the same as removing an element from a list with a small difference that **db.SaveChanges()** is used to confirm the changes. In order to run this example, comment all lines within **static void Main(string[] args)** body except **Examples.Crud.Demo.Run();** in **Program.cs**.

> **NOTE**
>
> You can find the code used for this example at https://packt.link/bH5c4.

You have grasped that the basic concept of CRUD is a combination of four functions—create, read, update, and delete. Now it is time to put this into practice in the following exercise.

EXERCISE 6.02: UPDATING PRODUCTS AND MANUFACTURERS TABLE

You have already created a **GlobalFactory** database with **Products** and **Manufacturers** tables, and you now have enough components to perform full **Create, Read, Update and Delete** (**CRUD**) on the database. In this exercise, you will use **FactoryDbContext** to create methods inside a new class called **GlobalFactoryService**, which can accomplish the following tasks:

- Add a list of manufacturers in the US.
- Add a list of products to all manufacturers in the US.
- Update any one product in the US with a given discount price.
- Remove any one product from the US region.
- Get all manufacturers from the US and their products.

Perform the following steps to complete this exercise:

1. First, create a **GlobalFactoryService** class.
2. Create **FactoryDbContext** inside a constructor and inject the context. Injecting the context means that you have a choice of setting it up in any way you want (for example, using different providers).
3. Create a constructor that accepts **FactoryDbContext** as an argument, as follows:

```
public class GlobalFactoryService : IDisposable
{
    private readonly FactoryDbContext _context;

    public GlobalFactoryService(FactoryDbContext context)
    {
        _context = context;
    }
```

4. Create a **public void CreateManufacturersInUsa(IEnumerable<string> names)** method, as follows:

```
public void CreateManufacturersInUsa(IEnumerable<string> names)
{
    var manufacturers = names
        .Select(name => new Manufacturer()
        {
            Name = name,
            Country = "USA"
        });

    _context.Manufacturers.AddRange(manufacturers);
    _context.SaveChanges();
}
```

   A manufacturer has only two custom fields—**Name** and **Country**. In this case, the value of the **Country** is known to be **"USA"**. All you have to do is to pass a list of manufacturer **names** and build **Manufacturers** by combining the value of the **Country** with their name.

5. To create the products, create a **public void CreateUsaProducts(IEnumerable<Product> products)** method.
6. Then get all the manufacturers in the US.
7. Finally, iterate each manufacturer and add all the products to each of them:

```
public void CreateUsaProducts(IEnumerable<Product> products)
{
    var manufacturersInUsa = _context
        .Manufacturers
        .Where(m => m.Country == "USA")
        .ToList();

    foreach (var product in products)
    {
```

```
        manufacturersInUsa.ForEach(m => m.Products.Add(
            new Product {Name = product.Name, Price = product.Price}
            ));
    }

    _context.SaveChanges();
}
```

Note that in this example, you have recreated a new product every time that you add the same product to a manufacturer. This is done because even though the product has the same properties, it belongs to a different manufacturer. In order for that distinction to be set, you need to pass different objects. If you do not do that, the products will be assigned to the same (last referenced) manufacturer.

8. Create a **public void SetAnyUsaProductOnDiscount(decimal discountedPrice)** method.
9. To set any USA product on discount, first get all the products from the US region and then select the first of them (order doesn't matter).
10. Next set a new **Price** for that product, and call **SaveChanges()** to confirm it:

```
public void SetAnyUsaProductOnDiscount(decimal discountedPrice)
{
    var anyProductInUsa = _context
        .Products
        .FirstOrDefault(p => p.Manufacturer.Country == "USA");

    anyProductInUsa.Price = discountedPrice;

    _context.SaveChanges();
}
```

11. Create a **public void RemoveAnyProductInUsa()** method.
12. To delete an item, simply select the first product in the **"USA"** group and remove it:

```
public void RemoveAnyProductInUsa()
{
    var anyProductInUsa = _context
        .Products
        .FirstOrDefault(p => p.Manufacturer.Country == "USA");
```

```
        _context.Remove(anyProductInUsa);
        _context.SaveChanges();
    }
```

> **NOTE**
>
> Observe that the **SaveChanges** has been called after every step.

13. In order to get a manufacturers from USA, create a **public IEnumerable<Manufacturer> GetManufacturersInUsa()** method.
14. Call the **ToList()** at the end of a query so that the SQL gets executed:

```
      public IEnumerable<Manufacturer> GetManufacturersInUsa()
      {
          var manufacturersFromUsa = _context
              .Manufacturers
              .Include(m => m.Products)
              .Where(m => m.Country == "USA")
              .ToList();

          return manufacturersFromUsa;
      }
    }
```

15. Create a **Demo** class where you call all functions:

Demo.cs

```
public static class Demo
{
    public static void Run()
    {
        var service = new GlobalFactoryService(new FactoryDbContext());
        service.CreateManufacturersInUsa(new []{"Best Buy", "Iron Retail"});
        service.CreateUsaProducts(new []
        {
            new Product
            {
                Name = "Toy computer",
                Price = 20.99m
            },
            new Product
            {
```

The complete code can be found here: https://packt.link/qMYbi.

In order to run this exercise, comment all lines within **static void Main(string[] args)** body except **Exercises.Exercise02.Demo. Run();** in **Program.cs**. The output of the preceding code will be as follows:

```
Best Buy:
Loli microphone 5
Iron Retail:
Toy computer 20,99
Loli microphone 7,51
```

This output shows exactly what you wanted to achieve. You created two manufacturers: **Best Buy** and **Iron Retail**. Each of them had two products, but from the first manufacturer, **Best Buy**, you removed one. Therefore, only a single product appears under it, as opposed to two products under **Iron Retail**.

> **NOTE**
>
> You can find the code used for this exercise at https://packt.link/uq97N.

At this point, you know how to interact with an existing database. However, what you have done so far is manually written models to fit the **GlobalFactory** database you have created. Using EF, you only need one side—either a database or a **DbContext** schema. In the next sections, you will learn how to work with either approach.

DATABASE FIRST

In some cases, you won't have to design a database yourself. Often, an architect will do that for you and then a database administrator will handle further changes. In other cases, you may get to work with some really old projects and a legacy database. Both scenarios are perfect for a **database first approach** because you can generate a **DbContext** schema with all the needed models using an existing database.

The project selected must be an executable project. For example, **`WebApi`** and **`ConsoleApp`** are okay; however, a class library is not (you cannot run a class library; you can only reference it from other applications). So, install EF tools by running this in the console:

```
dotnet add package Microsoft.EntityFrameworkCore.tools
```

Finally, run the following:

```
dotnet ef dbcontext scaffold
"Host=localhost;Username=postgres;Password=****;Database=Adventureworks"
Npgsql.EntityFrameworkCore.PostgreSQL -o your/models/path --schema
"production"
```

This command reads the database schema (you specified to generate the database from all the schemas rather than just one production schema) and generates models out of it. You used the **`AdventureWorks`** database. Using the **`-o`** flag, you select the output directory, and using the **`–schema`** flag, you specify the schemas you would like to generate the database from.

> **NOTE**
>
> The generated models from an existing database can be found at https://packt.link/8KIOK.

The models generated are quite interesting. They reveal two things that have not yet been talked about. When you created a **`Manufacturer`** class (read the *Modeling Databases Using EF* section), you did not initialize a collection of products from a constructor. This is not a big issue, but instead of not returning data, you get a null reference exception, which might not be what you want. None of the models, no matter how simple or complex they are, have attributes.

You are almost done with the db-first approach. The next section will revisit **`DbContext`** and inspect how EF does it so you can then apply what you learned in a code-first approach.

REVISITING DBCONTEXT

By logically grasping the following snippet, **AdventureWorksContext**, you will notice that the default configuration is passed slightly differently from the one created in the *DbContext and DbSet* section. Instead of directly using a connection string for SQL Server, the generated context uses the **OnConfiguring** method to double-check the given context options and if they are unconfigured, set one. This is a cleaner approach because you don't have to manually initialize the builder yourself and prevent unconfigured options:

```
public globalfactory2021Context()
        {
        }
        public
globalfactory2021Context(DbContextOptions<globalfactory2021Context>
options)
            : base(options)
        {
        }
        protected override void OnConfiguring(DbContextOptionsBuilder
optionsBuilder)
        {
            if (!optionsBuilder.IsConfigured)
            {
                optionsBuilder.UseNpgsql(Program.
GlobalFactoryConnectionString);
            }
        }
```

Next, there is a method named **OnModelCreating**. It is a method that takes **ModelBuilder** that is used to dynamically build models for your database. **ModelBuilder** directly replaces the attribute-based approach because it allows you to keep the models attribute-free and add whatever constraints or mappings are needed when the context is initialized. It includes column names, constraints, keys, and indexes.

ModelBuilder allows you to use Fluent API (that is, **method chaining**), which in turn allows you to add extra configurations to models. Consider the following single, fully configured model:

globalfactory2021Context.cs

```
protected override void OnModelCreating(ModelBuilder modelBuilder)
{
modelBuilder.Entity<Manufacturer>(entity =>
{
              entity.ToTable("manufacturer", "factory");

              entity.Property(e => e.Id)
                      .HasColumnName("id")
                      .UseIdentityAlwaysColumn();

              entity.Property(e => e.Country)
                      .IsRequired()
                      .HasMaxLength(50)
                      .HasColumnName("country");
```

The complete code can be found here: https://packt.link/S5s6d.

Looking at this part of **ModelBuilder** will give you a full picture of how the model maps to a table and its columns, keys, indexes, and relations. The generated code is broken down for you. To begin configuring an entity, you need to call the following:

```
modelBuilder.Entity< Manufacturer >(entity =>
```

Mapping to the table and schema looks like this:

```
entity.ToTable("manufacturer", "factory");
```

You can also add constraints (for example, to make sure that a field is not null) and set the character limit and name of a column the property maps to. In the following code, you're doing so for **Name**:

```
entity.Property(e => e.Name)
          .IsRequired()
          .HasMaxLength(50)
    .HasColumnName("name");
```

Lastly, some entities have multiple navigational properties associated with them. When multiple navigational properties are involved, EF may not be able to clearly interpret what the relationship should be. In those cases, you will need to configure it manually, as shown in the following code:

```
                  entity.HasOne(d => d.Manufacturer)
                      .WithMany(p => p.Products)
                      .HasForeignKey(d => d.Manufacturerid)
                      .HasConstraintName("product_manufacturerid_id");
```

The preceding code maps the **Manufacturer** entity to **Product** with a 1:n relationship and sets the foreign key column to **product_manufacturerid_id**. Spotting those cases might be tricky; therefore, you should only add manual configurations when an error informs you about such an ambiguity:

```
Unable to determine the relationship represented by navigation property
Entity.NavProperty' of type 'AnotherEntity'. Either manually configure the
relationship, or ignore this property from the model.
```

> **NOTE**
>
> There is no runnable code here; this is just a scaffold of a database.

Now you know what a generated **DbContext** looks like and how to customize models yourself. Without touching model classes, and instead using **ModelBuidler**, it is time to get familiar with doing the opposite, which is generating a database out of the context.

GENERATING DBCONTEXT FROM AN EXISTING DATABASE

For subsequent examples, you will be using the **GlobalFactory2021** database. Just to be sure that what you have made is the same as what the database contains, you'll perform database scaffolding one more time. **Scaffolding** is an operation that takes a database schema (or **DbContext**, in this case) and generates a physical database out of it.

Open the console and run the following:

```
dotnet ef dbcontext scaffold
"Host=localhost;Username=postgres;Password=****;Database=
globalfactory2021" Npgsql.EntityFrameworkCore.PostgreSQL -o Examples/
GlobalFactory2021.
```

For security, do not forget to replace the hardcoded connection string in **DbContext** with the one from the environment variable. The resulting **DbContext** should look like this:

```
◢ GlobalFactory2021
    ▷ C# globalfactory2021Context.cs
    ▷ C# Manufacturer.cs
    ▷ C# Product.cs
```

Figure 6.3: DbContext generated after applying the scaffold command

One of the main advantages of EF is that you can quickly define entities and then create a database out of them. But first, you'll need to learn the code first approach.

Code First and Migrations

Usually, when you need to create a proof of concept, you will create a **DbContext** schema with the models and then generate a database out of that. Such an approach is called **code first**.

In this example, you will use the context you have generated from the **GlobalFactory2021** database and then generate a new database out of it. This approach requires an extra package named **Design**, so make sure it is installed by running the following command:

```
dotnet add package Microsoft.EntityFrameworkCore.Design
```

EF is able to generate a database and have different versioning for it. In fact, it can move from one database version to another. A single database version at any given time is called **migration**. Migrations are needed to ensure that you do not just always recreate databases (after all, you don't want to lose the existing data), but instead, apply them neatly in a secure and trusted way. To add the first migration, from the VS Code terminal, run the following:

```
dotnet ef migrations add MyFirstMigration -c globalfactory2021Context
```

This will generate a migration file:

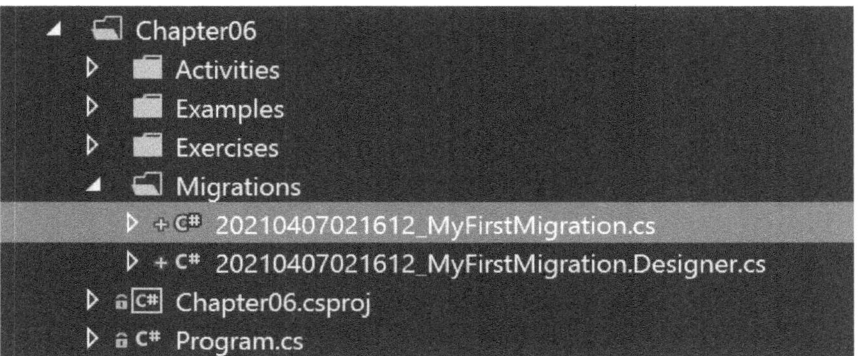

Figure 6.4: New migration with defaults placed under the project Migrations folder in the project root

The migration provides EF with information about the next database schema version and can therefore be used to generate a database from (or apply new changes to an existing database). Note that since you have multiple **DbContext** schemas and EF cannot tell you which context to use, you have to provide one explicitly. It is also worth mentioning that running this command requires selecting a default project, which includes the required context. and placing the migrations in that project's directory.

Why can't you just generate a database right away? When working with data, capturing a change at any given time and being able to go back to a previous version is very important. Even though directly generating a database might sound easy, it is not a viable approach because changes happen all the time. You want to be in control and have a choice to switch between versions at will. The migrations approach also works with code versioning systems, such as Git, because you can see the changes made to your database through a migration file. You will learn more about version control in *Chapter 11*, *Production-Ready C#: from Development to Deployment*.

Before creating a database, make sure you change the database name inside the connection string so that a new database can be created and not overwritten. Creating a new database from a migration you have can be done by running this command:

```
dotnet ef database update -c globalfactory2021context
```

If you open **pgAdmin**, you will see a very familiar view with the **manufacturer** and **product**. However, there is one new table for the migration history:

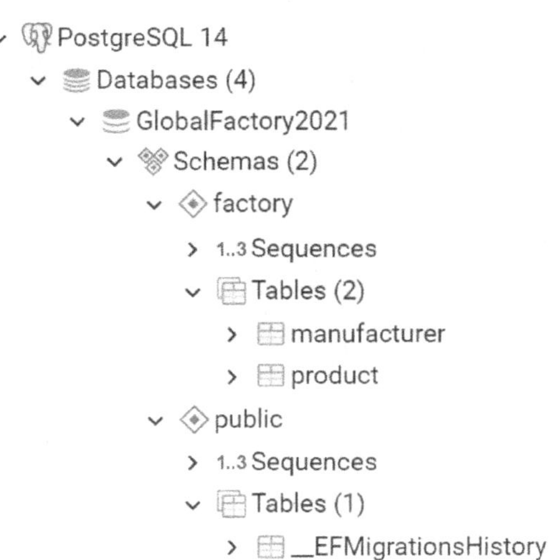

Figure 6.5: Generated database inside pgAdmin browser (simplified view for brevity)

The __**`EFMigrationsHistory`** table lays out all the migrations performed, when they were performed, and the EF version with which they were executed. In the following screenshot, you can see the first migration created as **`MyfirstMigration`**:

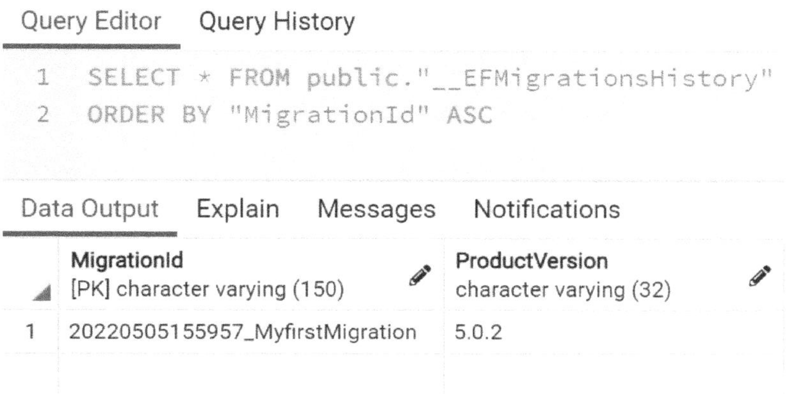

Figure 6.6: EFMigrationsHistory table rows

You might find it strange that a migrations table has only two columns. However, those two columns have all the needed information, such as when, what, and how. Under **`MigrationId`**, the digits before _ refer to the date and time the migration was run. This is followed by the migration name. The **`ProductVersion`** refers to the EF Core version with which the command was executed.

What if you wanted to make changes in your data models? What if you would like the **`manufacturer`** table to also have a date for the foundation? You would need to go through the same flow—add a migration and update the database.

So first, you would add a new property inside a **`Manufacturer`** class:

```
public DateTime FoundedAt { get; set; }
```

Here **`FoundedAt`** is a date. It does not need time associated with it, so you should specify an appropriate SQL Server type that maps to it. You would do this in **`GlobalFactory2021Context`** inside the **`OnModelCreating`** method:

```
entity.Property(e => e.FoundedAt)
    .HasColumnType("date")
```

Now you can add that to a new migration:

```
dotnet ef migrations add AddManufacturerFoundedDate -c
globalfactory2021Context
```

Apply the new migration to the database:

```
dotnet ef database update -c globalfactory2021context
```

This will add a new entry to the migration history:

	MigrationId [PK] character varying (150)	ProductVersion character varying (32)
1	20220505155957_MyfirstMigration	5.0.2
2	20220505162206_AddManufactur...	5.0.2

Figure 6.7: Migration 2 as the new migration created in the migrations table

You should see the new column in the **`manufacturer`** table as follows:

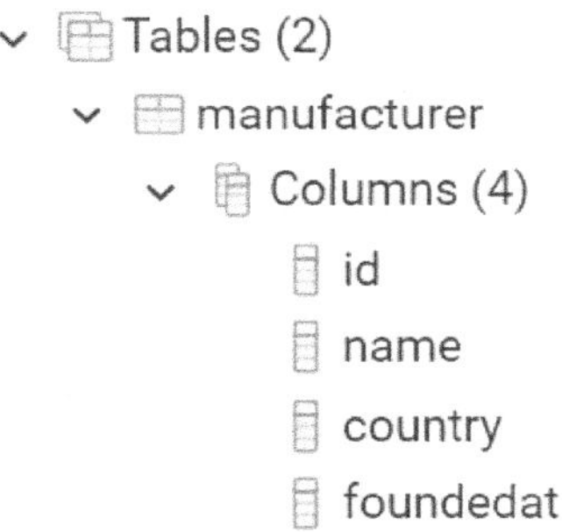

Figure 6.8: The manufacturer table with the new column named as foundedat

Now you know how to apply your models, change them, and generate a database out of the models. So far, you have made the following changes:

- Added the **`FoundedAt`** property and model builder changes.
- Created a migration file.
- Updated a database with that migration file.

Undoing those changes will involve doing the opposite, in this sequence:

- Rolling back database changes (updating the database to the last successful migration).
- Removing the migration file.
- Removing model builder changes.

EF migrations allow you to selectively apply any migration you want. Here, you will be applying the previous migration:

```
dotnet ef database update MyFirstMigration -c globalfactory2021context
```

You will delete the migration file using the following command:

```
dotnet ef migrations remove -c globalfactory2021Context
```

When working with big and complex databases, especially when they are already in production, performing migration using EF tools may become too complex. After all, you do not have full control of the exact script EF will generate for a migration. If you ever need a custom migration script, EF will no longer fit your bill. However, you can always convert whatever EF would do into SQL. You can do this by running the following command:

```
dotnet ef migrations script -c globalfactory2021context
```

This command produces, instead of a C# migration class, a SQL script. Executing a SQL script (often modified) is the preferred way of performing migrations in a production environment.

Those are just some basic yet common scenarios that you will be dealing with when working with databases. Change almost always happens; therefore, you should expect it and be prepared, as you will see in the following exercise.

EXERCISE 6.03: MANAGING PRODUCT PRICE CHANGES

Once again, your manager is impressed with your results. This time, they have asked you to keep track of product price changes. They would like a new table, **`ProductPriceHistory`**, that holds a record of the changes in the price of a product.

The following steps will help you complete this exercise:

1. To track price changes, add a new model, **`ProductPriceHistory`** with the following fields:
    - **`Id`**
    - **`Price`**
    - **`DateOfPrrice`**
    - **`ProductId`**
    - **`Product`**

The code for the new model will be as follows:

```
public class ProductPriceHistory
{
    public int Id { get; set; }
    public decimal Price { get; set; }
    public DateTime DateOfPrice { get; set; }
    public int ProductId { get; set; }

    public Product Product { get; set; }
}
```

2. Next, update the **Product** model so that it includes the historical price changes. So, add a new collection property, **ProductPriceHistory**:

```
public ICollection<ProductPriceHistory> PriceHistory { get; set; }
```

3. Change the **Price** column. **Price** should now be a method that gets the latest price of a product and the full model now looks like this:

```
public partial class Product
{
    public int Id { get; set; }
    public string Name { get; set; }
    public int ManufacturerId { get; set; }

    public decimal GetPrice() => PriceHistory
        .Where(p => p.ProductId == Id)
        .OrderByDescending(p => p.DateOfPrice)
        .First().Price;

    public Manufacturer Manufacturer { get; set; }
    public ICollection<ProductPriceHistory> PriceHistory { get; set;
}
}
```

4. Update **DbContext** to include a new **DbSet** and add the **ProductPriceHistory** configuration to the **OnModelCreating** method, as follows:

```
modelBuilder.Entity<ProductPriceHistory>(entity =>
{
    entity.ToTable("ProductPriceHistory", "Factory");

    entity.Property(e => e.Price)
        .HasColumnType("money");

    entity.Property(e => e.DateOfPrice)
        .HasColumnType("date");
```

   The preceding code provides mappings to a table and column property types. A **Product** has many historical price changes, therefore it forms a 1:n relation with a **PriceHistory**.

5. Just after the preceding code, create a 1:n relation between **Product** and **PriceHistory**:

```
 RelationalForeignKeyBuilderExtensions.
HasConstraintName((ReferenceCollectionBuilder)
        entity.HasOne(d => d.Product)
            .WithMany(p => p.PriceHistory)
            .HasForeignKey(d => d.ProductId), "FK_
ProductPriceHistory_Product");
});
```

6. For the database change to be captured (so that you can apply the change from code to database or roll back), add the **migration** as follows:

```
dotnet ef migrations add AddProductPriceHistory -c
globalfactory2021Contextv3 -o Exercises/Exercise03/Migrations
```

The following will be generated:

Migrations
20210407034203_MyFirstMigration.cs
20210407034203_MyFirstMigration.Designer.cs
20210407034336_AddProductPriceHistory.cs
20210407034336_AddProductPriceHistory.Designer.cs
globalfactory2021Contextv3ModelSnapshot.cs

Figure 6.9: The generated database migrations and extra files

7. In order to apply the migration, run the following command:

```
dotnet ef database update -c globalfactory2021contextv3
```

8. Create a **Demo** by adding some dummy data:

Demo.cs

```
public static class Demo
{
    public static void Run()
    {
        var db = new globalfactory2021Contextv3();
        var manufacturer = new Manufacturer
        {
            Country = "Canada",
            FoundedAt = DateTime.UtcNow,
            Name = "Fake Toys"
        };

        var product = new Product
        {
            Name = "Rubber Sweater",
```

The complete code can be found here: https://packt.link/4FMz4.

Here, you first created a **manufacturer** and its **product** and then added a few price changes. Once the changes were saved, you disconnected from the database (so that you don't work with cached entities). In order to test whether it works, you queried all **"Fake Toys"** manufacturer with their products and their price history.

> **NOTE**
>
> When working with dates, especially in the context of databases or environments that may be shared beyond your local, prefer to use dates without your locale by calling **DateTime.UtcNow**.

9. In order to run this exercise, comment all lines within **`static void Main(string[] args)`** body except **`Exercises.Exercise03.Demo.Run();`** in **`Program.cs`**. You will see the following output:

```
Fake Toys Rubber Sweater 15.5000
```

In the **`Demo`**, you created a **`manufacturer`** with one product which is a toy (**`Rubber Sweater`**). The toy has two prices: **`15.11`** and **`15.50`** (the latest). You then saved that toy in the database, disconnected, and reconnected from that database (making sure that the toy is not cached, but rather fetched), and executed an eager loading-based join.

> **NOTE**
>
> You can find the code used for his exercise at https://packt.link/viVZW.

EF is effective for rapid database development, but for that same reason, it is also very dangerous. Inexperienced developers often rely on the magic that happens behind the scenes and therefore forget that EF cannot magically optimize data models to fit your specific scenario or guess that the intended query should perform better. The following sections will review the main mistakes that people make while working with EF.

PITFALLS OF EF

EF abstracts a lot of details from you, significantly simplifying your work. However, it also introduces the risk of not being aware of what is actually happening. Sometimes, you might achieve what you want, but there may be a chance that you are not optimally achieving your goal. The following are some of the most common mistakes made in EF.

EXAMPLES SETUP

For all the following examples, assume that you will have this line initialized at the start:

```
var db = new GlobalFactory2021Context();
```

Assume, too, that every example will finish with this:

```
db.Dispose();
```

Also, the data itself will be seeded (pre-generated) using the following code:

DataSeeding.cs

```
public static class DataSeeding
{
    public const string ManufacturerName = "Test Factory";
    public const string TestProduct1Name = "Product1     ";
    /// <summary>
    /// Padding should be 13 spaces to the right as per our test data, db and 
filtering requirements
    /// </summary>
    public const string TestProduct2NameNotPadded = "Product2";
    public const decimal MaxPrice = 1000;

    public static void SeedDataNotSeededBefore()
    {
        var db = new globalfactory2021Context();
        var isDataAlreadySeeded = db.Manufacturers.Any(m => m.Name == 
ManufacturerName);
        if (isDataAlreadySeeded) return;
```

The complete code can be found here: https://packt.link/58JTd.

The preceding code creates a **manufacturer** with **10,000** products, but only if that **manufacturer** does not already exist. The **ManufacturerName** will be exactly 13 characters long, and their prices will be random, but no bigger than the maximum price. All of this information is saved to a database before you disconnect from it.

> **NOTE**
>
> This is no runnable code and will be used in all the performance comparison examples.

All the examples will compare two functions achieving the same output. A summary of all the comparisons is done by executing this demo code:

Demo.cs

```
public static class Demo
{
    public static void Run()
    {
        // For benchmarks to be more accurate, make sure you run the seeding before
anything
        // And then restart the application
        // Lazy loading is a prime example of being impacted by this inverting the
intended results.
        DataSeeding.SeedDataNotSeededBefore();
        // Slow-Faster example pairs
        // The title does not illustrate which you should pick
        // It rather illustrates when it becomes a problem.

    CompareExecTimes(EnumerableVsQueryable.Slow, EnumerableVsQueryable.Fast,
"IEnumerable over IQueryable");
    CompareExecTimes(MethodChoice.Slow, MethodChoice.Fast, "equals over ==");
    CompareExecTimes(Loading.Lazy, Loading.Eager, "Lazy over Eager loading");
```

The complete code can be found here: https://packt.link/xE0Df.

Here, you compare in-memory and SQL filtering, lazy and eager loading, tracked and untracked entities, and adding entities one by one as opposed to adding them in bulk. In the paragraphs that follow, you will find the functions being compared, but every comparison will show the following:

- Names of a scenario
- Slow and fast versions for doing the same thing

You will be using a stopwatch to measure execution time and print a formatted comparison after each run. In order to run this example, comment all lines within **static void Main(string[] args)** body except **Examples.PerformanceTraps.Demo.Run();** in **Program.cs**. You can refer to the *Summary of Results* section for the output.

The idea behind these examples is to compare an efficient way of working with EF with a direct equivalent inefficient way. The slow scenario is the inefficient way and the fast (which is the efficient one) is the way it should be done. The next section will detail the efficient way of using EF.

MULTIPLE ADDS

Sometimes, without realizing it at the time, you'll find that you tend to use the most straightforward route while writing programs. For example, to add 100 items, you may use 100 individual addition operations. However, this isn't always the optimal approach and is especially inefficient when you're using EF. Instead of one query for a bulk of 100, you might run a single insert 100 times. As an example, see the following code:

```
for (int i = 0; i < 1000; i++)
{
    var product = new Product
    {
        Name = productName,
        Price = 11,
        ManufacturerId = 2
    };
    db.Products.Add(product);
}
```

This code creates **`1,000`** products and attaches them to **`DbContext`**. What happens is that those **`1,000`** entities inside a **`DbContext`** schema are tracked. Instead of tracking them all as a single batch, you track each individually.

What you want to do, though, is to work with range operations:

- **`AddRange`** or
- **`UpdateRange`**, or
- **`RemoveRange`**

A better version of the preceding code, designed to work in an optimal way with batches, looks like this:

```
var toAdd = new List<Product>();
for (int i = 0; i < 1000; i++)
{
    var product = new Product
    {
        Name = productName,
        Price = 11,
        Manufacturerid = 2
    };
```

```
    toAdd.Add(product);
}
db.Products.AddRange(toAdd);
```

When creating multiple items with the intention to add them to the database, you should first add them to a list. After your list is complete, you can add the items as a batch to **DbSet<Product>**. You still have the problem of multiple adds, but the benefit of it over directly calling a **DbSet<Product>** add is that you no longer hit the change tracker with every add. In order to run this example, comment all lines within **static void Main(string[] args)** body except **Examples.PerformanceTraps.Demo.Run();** in **Program.cs**.

> **NOTE**
>
> You can find the code used for this example at https://packt.link/wPLyB.

The next section will take a look at another pitfall—how to query properly based on equality of properties.

EQUALS OVER ==

The devil lies in the details. C# developers usually do not make this mistake, but if you are moving between languages (especially from Java), you might be doing this when filtering:

```
var filtered = db.Products
    .Where(p => p.Name.Equals(DataSeeding.TestProduct1Name))
    .ToList();
```

For LINQ, it is harmless. However, while using EF, this approach is not recommended. The problem is that EF can convert only some expressions to SQL. Usually, a complex method, such as equals, cannot be converted because it comes from a base object class, which can have multiple implementations. Instead, use a simple equality operator:

```
var filtered = db.Products
    .Where(p => p.Name == DataSeeding.TestProduct1Name)
    .ToList();
```

The problem with the first attempt was that it would first get all products (that is, execute a **get** statement in SQL) and only then the filter would be applied (in memory, in C#). Once again, this is problematic because getting with a filter applied in a database-native language is optimal but getting products in SQL and then filtering in C# is suboptimal. The problem is solved in the second attempt by replacing **Equals** with the equality operator, **==**. In order to run this example, comment all lines within **static void Main(string[] args)** body except **Examples.PerformanceTraps.Demo.Run();** in **Program.cs**.

> **NOTE**
>
> You can find the code used for this example at https://packt.link/js2es.

USING IENUMERABLE OVER IQUERYABLE

Another example involves misunderstanding the concept of **IEnumerable<Product>**:

```
IEnumerable<Product> products = db.Products;
var filtered = products
    .Where(p => p.Name == DataSeeding.TestProduct1Name)
    .ToList();
```

Here, you are getting products by a specific product name. But what happens when you assign a **DbSet<Product>** object to **Ienumerable<Product>** is that the **SELECT *** statement is executed. Therefore, instead of getting only the filtered products that you need, you will first get everything and then manually filter it.

You might wonder why you couldn't filter right away. In some cases, it makes sense to build queries and pass them across methods. But when doing so, you should not execute them until they are completely built. Therefore, instead of **Ienumerable<Product>**, you should use **Iqueryable<Product>**, which is an abstraction of queried entities—an expression that will be converted to SQL after calling **ToList<Product>** or similar. An efficient version of the preceding code would look like this:

```
IQueryable<Product> products = db.Products;
var filtered = products
    .Where(p => p.Name == DataSeeding.TestProduct1Name)
    .ToList();
```

The latter works faster because you apply a filter in SQL and not in memory. In order to run this example, comment all lines within **static void Main(string[] args)** body except **Examples.PerformanceTraps.Demo.Run();** in **Program.cs**.

> **NOTE**
>
> You can find the code used for this example at https://packt.link/ehq6C.

Eager and lazy loading has already been mentioned, but there is still another complexity that is significant enough and should be covered. The next section details them.

LAZY OVER EAGER LOADING

In EF, you have an interesting **n+1 queries** problem. For example, if you get a list of items, then getting the list of their respective manufacturers afterward would result in a SQL query being executed; this would be lazy-loaded. Fortunately, from EF 2.1, this no longer happens by default, and it needs to be enabled explicitly. Assume that in the following examples, you have already enabled it.

Here is a query to get any first item and its manufacturer:

```
var product = db.Products.First();
// Lazy loaded
var manufacturer = product.Manufacturer;
```

Initially, upon looking at this code, you might think that there is no issue, but this small chunk of code executes two SQL queries:

- First, it selects the top product.
- Then it selects the associated manufacturer, along with the manufacturer ID.

To make the code more efficient, you need to explicitly specify that you do want the **Manufacturer** to be included with a product. A better, more efficient version of the code is as follows:

```
var manufacturer = db.Products
    // Eager loaded
    .Include(p => p.Manufacturer)
    .First()
    .Manufacturer;
```

The latter translates to a single query where a join between two tables is made and the first item from one of them is returned. In order to run this example, comment all lines within **`static void Main(string[] args)`** body except **`Examples.PerformanceTraps.Demo.Run();`** in **`Program.cs`**.

> **NOTE**
>
> You can find the code used for this example at https://packt.link/osrEM.

READ-ONLY QUERIES

EF assumes many things when running your queries. In most cases, it gets it right, but there are many cases when you should be explicit and order it not to assume. For example, you could get all the products like this:

```
var products = db.Products
    .ToList();
```

By default, EF will track all retrieved and changed entities. This is useful in some cases, but not always. When you have **read-only queries**, to just get and not modify entities, you would explicitly tell EF to not track any of them. An optimal way of getting products is as follows:

```
var products = db.Products
    .AsNoTracking()
    .ToList();
```

All this code does is run a query against the database and map the results. EF keeps the context clean. In order to run this example, comment all lines within **`static void Main(string[] args)`** body except **`Examples.PerformanceTraps.Demo.Run();`** in **`Program.cs`**.

> **NOTE**
>
> You can find the code used for this example at https://packt.link/rSW1k.

SUMMARY OF RESULTS

The following snippet shows all results from the previous sections, in a tabulated form:

```
IENUMERABLE OVER IQUERYABLE          Scenario1: 75ms,    Scenario2: 31ms
EQUALS OVER ==                       Scenario1: 33ms,    Scenario2: 24ms
LAZY OVER EAGER LOADING              Scenario1: 3ms,     Scenario2: 29ms
READ-ONLY QUERIES                    Scenario1: 40ms,    Scenario2: 10ms
MULTIPLE ADDS                        Scenario1: 8ms,     Scenario2: 8ms
```

Note that the output depends on the machine you are running the database, the data, and more. The point of this comparison is not to give you hard rules of what should be chosen, but rather to show how different approaches might save a lot of computing time.

EF is a powerful tool that allows rapid work with databases; however, you should be careful with how you use it. Do not worry, even if you think you are not sure how the queries work internally, there is still a way to see what happens underneath.

TOOLS TO HELP YOU SPOT PROBLEMS EARLY ON

EF is a toolbox in itself; it allows you to easily hook into it and track what is happening without any external tools. You can enable logging all the EF actions by adding this to the **`OnConfiguring`** method:

```
optionsBuilder.LogTo((s) => Debug.WriteLine(s));
```

If you run any of the example's code, this will log the trace inside an **`output`** window, as follows:

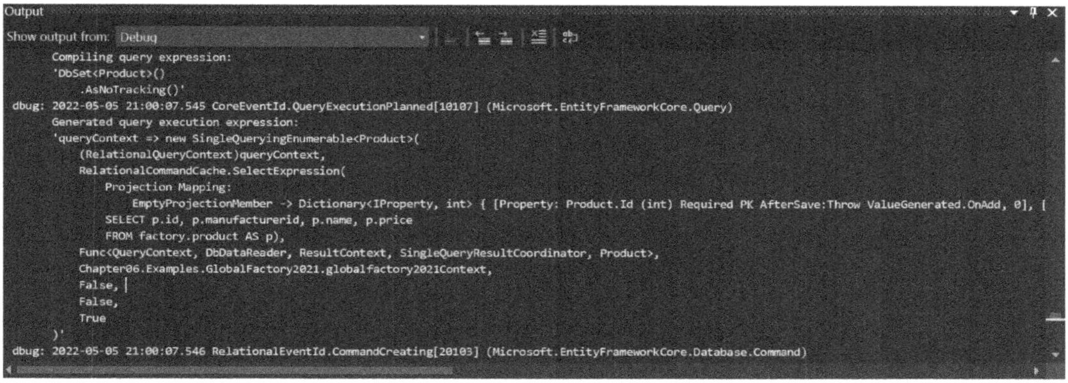

```
Output
Show output from: Debug
      Compiling query expression:
      'DbSet<Product>()
          .AsNoTracking()'
dbug: 2022-05-05 21:00:07.545 CoreEventId.QueryExecutionPlanned[10107] (Microsoft.EntityFrameworkCore.Query)
      Generated query execution expression:
      'queryContext => new SingleQueryingEnumerable<Product>(
          (RelationalQueryContext)queryContext,
          RelationalCommandCache.SelectExpression(
              Projection Mapping:
                  EmptyProjectionMember -> Dictionary<IProperty, int> { [Property: Product.Id (int) Required PK AfterSave:Throw ValueGenerated.OnAdd, 0], [
              SELECT p.id, p.manufacturerid, p.name, p.price
              FROM factory.product AS p),
          Func<QueryContext, DbDataReader, ResultContext, SingleQueryResultCoordinator, Product>,
          Chapter06.Examples.GlobalFactory2021.globalfactory2021Context,
          False,
          False,
          True
      )'
dbug: 2022-05-05 21:00:07.546 RelationalEventId.CommandCreating[20103] (Microsoft.EntityFrameworkCore.Database.Command)
```

Figure 6.10: Debugging output after running the performance pitfalls demo

The image shows what SQL is generated when EF executes the code—specifically selecting all products.

This approach is useful when you want to both fully debug your application and know every step EF makes. It is efficient for spotting queries that you expect to execute as SQL but execute in memory.

In the next section, you will learn about patterns that will help you organize database communication code.

WORKING WITH A DATABASE IN ENTERPRISE

When talking about databases, you usually imagine SQL or another language to talk with them. On top of that, another language (C#, in this case) is most often used to connect to a database to execute SQL queries. If not controlled, C# gets mixed with SQL, and it causes a mess of your code. Over the years, there have been a few patterns refined to implement the communication with a database in a clean way. Two such patterns, namely, Repository and CQRS, are commonly used to this day.

REPOSITORY PATTERN

The **Repository** is a pattern that targets a model and defines all (if needed) possible CRUD operations. For example, if you take a **Product** model, you could have a repository with this interface:

```
public interface IProductRepository
{
    int Create(Product product);
    void Delete(int id);
    void Update(Product product);
    Product Get(int id);
    IEnumerable<Product> Get();
}
```

This is a classical repository pattern where every database operation is abstracted away. This allows you to do pretty much anything you want in a database without worrying about the underlying database or even the technology you use to communicate with the database.

Note that a **Create** method in this case returns an integer. Usually, when writing code, you would segregate methods that change a state from those that query something. In other words, do not try to both get something and change something. However, in this case, it is difficult to achieve because the ID of an entity will be generated by the database. Therefore, if you want to do something with the entity, you will need to get that ID. You could instead return the whole entity, but that is like getting a house when all you need is an address.

Given you want to do the same four operations (create, delete, update, and get), the pattern would look like this:

```
public interface IManufacturerRepository
{
    int Create(Manufacturer product);
    void Delete(int id);
    void Update(Manufacturer product);
    Manufacturer Get(int id);
    IEnumerable<Manufacturer> Get();
}
```

It looks almost the same; the only difference is the targeted entity. Given that you had a very simple application that just does data processing in a very simple way, it would make sense to make these repositories generic:

```
public interface IRepository<TEntity>: IDisposable where TEntity : class
{
    int Create(TEntity productentity);
    void Delete(long id)(int id);
    void Update(TEntity entityproduct);
    TEntity Get(long id)(int id);
    IEnumerable<TEntity> Get();
    void Dispose();
}
```

Here, instead of **Product** or **Manufacturer**, the interface takes a generic **TEntity** that must be a class. You have also inherited an **IDisposable** interface to clean up all the resources that a repository used. This repository is still flawed. So, should you be able to persist any class? In that case, it would be nice to mark the classes that you could persist in.

Yes, you can do that. When talking about a repository, you should realize that even if something is supposed to be saved in a database, that does not mean that it will be saved separately. For example, contact information will always be saved with a person. A person can exist without contact information but contact information cannot exist without a person. Both person and contact information are entities. However, a person is also an aggregate (that is the entity that you will be targeting when adding data to a database), and it can exist by itself. This means that it makes no sense to have a repository for contact information if storing it would violate data integrity. Therefore, you should create a repository not per entity, but per aggregate.

What should every row in a database have? It should have an ID. An entity is a model that you can persist (that is, have an ID); therefore, you can define an interface for it:

```
public interface IEntity
{
    int Id { get; }
}
```

Please note that here you are using a **get**-only property because it does not make sense to set an ID in all cases. However, being able to identify an object (by getting the ID) is critical. Also note that the ID, in this case, is an integer because it is just a simple example and there will not be much data; but in real applications, it is usually either an integer or a GUID. Sometimes, an ID could even be both. In those cases, a consideration to make an entity interface generic (that is, taking generic **TId**) could be made.

What about an aggregate? An **aggregate** is an entity; you would therefore write the following:

```
public interface IAggregate : IEntity
{
}
```

In this scenario, you would then just write **Person: IAggregate, ContactInfo: IEntity**. If you apply the same principles to the two tables you had, you will get **Product: IAggregate, Manufacturer: IAggregate** because the two can be saved separately.

> **NOTE**
>
> There is no runnable code here; however, you will be using it in the upcoming exercise. You can find the code used for this example at https://packt.link/JDLAo.

Writing a repository for every aggregate might become a tedious job, especially if there is no special logic to the way persistence is done. In the upcoming exercise, you will learn how to generalize and reuse repositories.

EXERCISE 6.04: CREATING A GENERIC REPOSITORY

Being coupled to an ORM may make your business logic harder to test. Also, due to persistence being so rooted at the core of most applications, it might be a hassle to change an ORM. For those reasons, you may want to put an abstraction layer in between business logic and a database. If you use **DbContext** as is, you couple yourself to **EntityFramework**.

In this exercise, you will learn how to create a database operations abstraction—a generic repository—that will work on any entity and support create, delete, update, and get operations. Implement those methods one by one:

1. First, create a generic repository class that takes **DbContext** in the constructor:

   ```
   public class Repository<TAggregate>: IRepository<TAggregate> where
   TAggregate: class
   {
       private readonly DbSet<TAggregate> _dbSet;
       private readonly DbContext _context;

       public Repository(DbContext context)
       {
           _dbSet = context.Set<TAggregate>();
           _context = context;
       }
   ```

   The **context.Set<TEntity>()** allows getting a table-model binding and then using it throughout the repository. Another interesting point is that you didn't have to supply a concrete **DbContext** as it uses generic entities, and a generic repository is applicable to every kind of context.

2. To implement a **Create** operation, add a method to insert a single aggregate:

   ```
   public int Create(TAggregate aggregate)
   {
       var added = _dbSet.Add(aggregate);
       _context.SaveChanges();

       return added.Entity.Id;
   }
   ```

3. To implement a **Delete** operation, add a method to delete an aggregate by ID:

```
public void Delete(int id)
{
    var toRemove = _dbSet.Find(id);
    if (toRemove != null)
    {
        _dbSet.Remove(toRemove);
    }

    _context.SaveChanges();
}
```

4. To implement an **Update** operation, add a method to update an entity by overriding the old values with the values of a new entity:

```
public void Update(TAggregate aggregate)
{
    _dbSet.Update(aggregate);
    _context.SaveChanges();
}
```

5. To implement a **Read** operation, add a method to get a single entity by ID:

```
public TAggregate Get(int id)
{
    return _dbSet.Find(id);
}
```

6. A **Read** operation should also support getting all the entities. So, add a method to get all entities:

```
public IEnumerable<TAggregate> Get()
{
    return _dbSet.ToList();
}
```

7. Passing a **DbContext** to a constructor will open a database connection. As soon as you are done using a database, you should disconnect. In order to support a conventional disconnect, implement an **IDisposable** pattern:

```
        public void Dispose()
        {
            _context?.Dispose();
        }
    }
```

8. To test whether the generic repository works, create a new **Run()** method:

```
public static void Run()
{
```

9. Inside the **Run()** method, initialize a new repository for the **Manufacturer** entity:

```
        var db = new FactoryDbContext();
        var manufacturersRepository = new Repository<Manufacturer>(db);
```

10. Test whether the **Create** operation works, by inserting a new **manufacturer** as shown in the following code:

```
        var manufacturer = new Manufacturer { Country = "Lithuania", Name
= "Tomo Baldai" };
         var id = manufacturersRepository.Create(manufacturer);
```

11. Test whether the **Update** operation works, by updating the manufacturer's name as follows:

```
        manufacturer.Name = "New Name";
        manufacturersRepository.Update(manufacturer);
```

12. Test whether the **Read** operation works on a single entity, by retrieving the new manufacturer from a database and print it:

```
        var manufacturerAfterChanges = manufacturersRepository.Get(id);
        Console.WriteLine($"Id: {manufacturerAfterChanges.Id}, " +
                     $"Name: {manufacturerAfterChanges.Name}");
```

You should see the following output:

```
Id: 25, Name: New Name
```

13. Test whether the **Read** operation works on all entities by getting the count of all manufacturers with the following code:

```
    var countBeforeDelete = manufacturersRepository.Get().Count();
```

14. You can test whether the **Delete** operation works by deleting the new manufacturer as follows:

```
    manufacturersRepository.Delete(id);
```

15. In order to see the impact of delete (one less manufacturer is expected), compare the counts as follows:

```
    var countAfter = manufacturersRepository.Get().Count();
    Console.WriteLine($"Before: {countBeforeDelete}, after: 
{countAfter}");
}
```

16. In order to run this exercise, comment all lines within **static void Main(string[] args)** body except **Exercises.Exercise04.Demo.Run();** in **Program.cs**. You should see the following output upon running the **dotnet run** command:

```
Before: 3, after: 2
```

Repositories used to be the way to go (maybe 10-20 years ago) for implementing interactions with a database because these were a well-abstracted way to make calls against a database. An **abstraction** from a database would enable people to change the underlying database provider if needed. If a database changes, only the class that implements the interface will change but whatever consumes the interface will remain unaffected.

Looking back at **DbContext** and **DbSet**, you might ask why those can't be used directly. The answer is that you can, and it serves a similar purpose as repositories do. That is why the repository pattern should only be used if your queries are sufficiently complex (meaning it's several lines long).

> **NOTE**
>
> You can find the code used for this exercise at https://packt.link/jDR0C.

The next section will explore another benefit of EF that is, local database testing.

TESTING DATA PERSISTENCE LOGIC LOCALLY

When developing software, you should always have quality and testability in mind. The problem with database testability is that it often requires a physical machine to host a database somewhere. However, you do not always have access to such a setup, especially at the start of a project.

Thankfully, EF is very flexible and offers a few packages to help out here. There are three main ways of testing with EF—**`InMemory`**, using SQLite, and calling an actual database. You have already seen plenty of demos calling a physical database. Next, you'll explore the other two: In-Memory and SQLite.

IN-MEMORY DATABASE PROVIDER

An **in-memory database provider** is just a bunch of in-memory lists available internally that make no queries whatsoever to a database. Usually, even garbage collection eliminates its state. Before you can continue, just like all other database providers, you will need to add one to your project.

Run the following command:

```
dotnet add package Microsoft.EntityFrameworkCore.InMemory
```

This command enables you to use an in-memory database when supplying **`DbContextOptionsBuilder`** with the **`UseInMemoryDatabase`** option, as done in the following snippet:

```
var builder = new DbContextOptionsBuilder<FactoryDbContext>();
builder.UseInMemoryDatabase(Guid.NewGuid().ToString());
var options = builder.Options;
_db = new FactoryDbContext(options);
```

In this snippet, you've used an options builder and created a new, isolated, in-memory database. The most important part here is the **`builder.UseInMemoryDatabase();`** method, which specifies that an in-memory database should be created. Also, note the **`Guid.NewGuid().ToString()`** argument. This argument is for a database name. In this case, it means that every time you call that line you will generate a unique database name, thus ensuring isolation between the new test databases. If you don't use this argument, you risk affecting a context under the test state. You want to avoid that for testing scenarios. When it comes to testing, starting with a fresh state is the right way to go.

In order to run this example, comment all lines within **`static void Main(string[] args)`** body except **`Examples.TestingDb.Demo.Run();`** in **`Program.cs`**.

> **NOTE**
>
> You can find the code used for this example at https://packt.link/mOodJ.

To test whether a generic repository for manufacturers works (assume that the preceding code will be reused), first create a new repository:

```
var productsRepository = new Repository<Product>(db);
```

The power of this pattern is that a new entity repository is simply specified as a different generic argument. If you wanted to test a manufacturer, you would not need to design a repository class for it. All you would have to do is to initialize a repository with **`Manufacturer`** passed as a generic argument, for example **`new Repository<Manfacturer>(db)`**.

Now, create a test **`product`** and save it:

```
var product = new Product {Name = "Test PP", ManufacturerId = 1, Price =
9.99m};
var id = productsRepository.Create(product);
```

To test the price update method, update **`product.Price`** and call the **`Update`** method:

```
product.Price = 19m;
productsRepository.Update(product);
```

In order to check whether a product was created successfully, call a **`Get`** method and pass the new product **`id`**:

```
var productAfterChanges = productsRepository.Get(id);
```

Type the following to print the product to the console:

```
Console.WriteLine($"Id: {productAfterChanges.Id}, " +
                  $"Name: {productAfterChanges.Name}, " +
                  $"Price: {productAfterChanges.Price}");
```

The output will get displayed as follows:

```
Id: 1, Name: Test PP, Price: 19
```

Now you need to check whether delete works. So, create a new product:

```
var productToDelete = new Product { Name = "Test PP 2", ManufacturerId = 
1, Price = 9.99m };
var idToDelete = productsRepository.Create(productToDelete);
```

Check the current count of products in a repository:

```
var countBeforeDelete = productsRepository.Get().Count();
```

Now delete the product:

```
productsRepository.Delete(idToDelete);
```

Check the count once again, comparing it with the previous one:

```
var countAfter = productsRepository.Get().Count();
Console.WriteLine($"Before: {countBeforeDelete}, after: {countAfter}");
```

In order to run this example, comment all lines within **`static void Main(string[] args)`** body except **`Examples.TestingDb.Demo.Run();`** in **`Program.cs`**. The following output will get displayed:

```
Before: 2, after: 1
```

> **NOTE**
>
> You can find the code used for this example at https://packt.link/DGjf2.

Using an In-Memory provider has its limitations. Up next, you will learn another alternative to testing code depending on the **`DbContext`** with fewer limitations.

SQLITE DATABASE PROVIDER

The problem with in-memory providers is that you cannot run any SQL statements on them. If you do, the code fails. Also, an in-memory provider is all about in-memory data structures and has nothing to do with SQL. SQLite database provider is free from those problems. The only issue it has is that SQLite is a dialect of SQL, so some raw SQL queries of other providers might not work.

To try out SQLite, run the following command in the VS Code terminal:

```
dotnet add package Microsoft.EntityFrameworkCore.Sqlite
```

The installed NuGet allows you to use SQLite provider when creating a **DbContext** schema, like this:

```
var connection = new SqliteConnection("Filename=:memory:");
connection.Open();

var builder = new DbContextOptionsBuilder<FactoryDbContext>();
builder.UseSqlite(connection);
var options = builder.Options;
var db = new FactoryDbContext(options);
db.Database.EnsureCreated();
```

In the preceding snippet, you have created a SQL connection, specifying that an in-memory SQLite database will be used. The **Db.Database.EnsureCreated()** was needed because the database would not always be created using that connection string. In order to run this example, comment all lines within **static void Main(string[] args)** body except **Examples.TestingDb.Demo.Run();** in **Program.cs**.

> **NOTE**
>
> You can find the code used for this example at https://packt.link/rW3JS.

If you were to create **ProductsRepository** and run the same code from the **InMemory** database example, you would get an error: **SQLite Error 19: 'FOREIGN KEY constraint failed'**. This is due to a missing manufacturer with an ID of 1 to which you are trying to link the new test products. This is a prime example of why the EF in-memory provider is not that reliable.

In order to fix this, add the following just before creating a test product:

```
var manufacturer = new Manufacturer() { Id = 1 };
db.Manufacturers.Add(manufacturer);
db.SaveChanges();
```

The only thing to remember is to clean up. After you are done using a database context that was created using a SQL connection, do not forget to dispose of that connection this way:

```
connection.Dispose();
```

At this point, you already know how to use `DbContext` in many different ways in order to communicate with a database. However, a dependency on a third-party library (EF Core) and unit testing maybe be tricky if all depends on a specific ORM. In the next paragraph, you will learn how to escape such a dependency.

A FEW WORDS ON REPOSITORY

The **Repository pattern** works for simple CRUD applications because it can further simplify database interactions. However, given you are using EF, it is already simple enough to interact with a database and another layer of abstraction is not always justified. After all, one of the key reasons why the Repository pattern caught so much attention is that it allows you to escape database interactions. However, the EF in-memory provider allows that too, so there is even less of a reason to use a repository.

The generic repository pattern is a useful abstraction. It abstracts away database interaction under a simple interface. However, for non-trivial scenarios, you are likely to need your custom CRUD operations and then you would create a non-generic repository. In fact, non-generic repositories are the recommended approach (given you want to implement the pattern) because you rarely want all the CRUD methods for all the entities. It is not rare to end up with as little as a single method on a repository. If you use a generic repository, you could still make all methods virtual and override them, but then you will end up overriding all the time or having methods that you don't use. It is less than ideal.

The following section will explore a different pattern that strives to make simple, optimal interactions per database operation—CQRS.

QUERY AND COMMAND HANDLERS PATTERNS

Command Query Responsibility Segregation (**CQRS**) is a pattern that aims to separate reads from writes. Instead of one class for all CRUD operations, you will have one class per CRUD method. On top of that, instead of one entity that fits all, you will have request and query object models dedicated to those specific scenarios. In CQRS, all database operations can be classified into two:

- Command: An operation that changes state (create, update, delete).
- Query: An operation that gets something, without affecting the state.

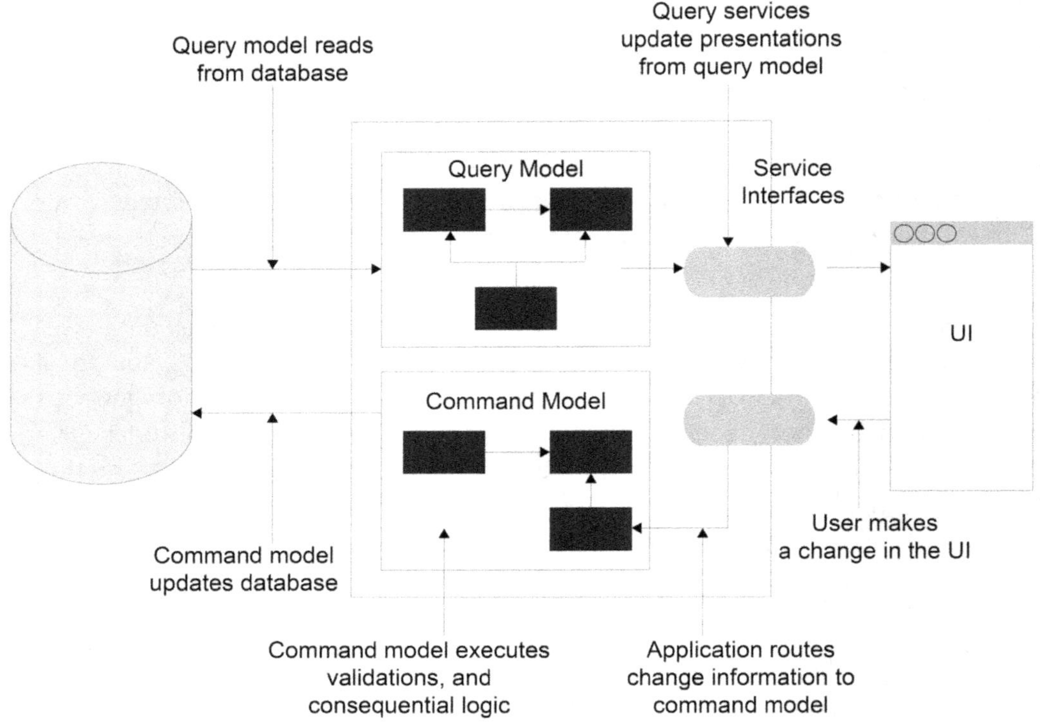

Figure 6.11: CQRS pattern as used by Martin Fowler

> **NOTE**
>
> The original source for this diagram can be found at https://www.martinfowler.com/bliki/CQRS.html.

In order to implement a command handler for creating a product, you would start by defining the command. What does the product need? It needs a name and a price, as well as a manufacturer. The ID for the create command is not needed (because the database generates it) and the manufacturer property can be removed as well because you will not make use of navigational properties. The name of a CQRS operation is made up of three parts—operation name, entity name, and **command** or **query** suffix. You are creating a product; therefore, the model will be called **CreateProductCommand**:

```
public class CreateProductCommand
{
    public string Name { get; set; }
    public decimal Price { get; set; }
    public int ManufacturerId { get; set; }
}
```

Next, you will create a handler of this command. In the constructor, pass the database context. In the **Handle** method, pass **CreateProductCommand**:

CreateProductQueryHandler.cs

```
public class CreateProductCommandHandler
{
    private readonly FactoryDbContext _context;

    public CreateProductCommandHandler(FactoryDbContext context)
    {
        _context = context;
    }

    public int Handle(CreateProductCommand command)
    {
        var product = new Product
        {
            ManufacturerId = command.ManufacturerId,
            Name = command.Name,
```

The complete code can be found here: https://packt.link/xhAVS.

Handlers are simple, single-method objects that implement all that is needed to process a command or a query. In order to test things, you'll also create a **GetProductQueryHandler** class:

```
public class GetProductQueryHandler
{
    private readonly FactoryDbContext _context;

    public GetProductQueryHandler(FactoryDbContext context)
    {
```

```
        _context = context;
    }

    public Product Handle(int id)
    {
        return _context.Products.Find(id);
    }
}
```

The idea is almost the same, except that, in this case, querying is so simple that the optimal model for it is a simple integer. In some scenarios, if you can predict the complexity growing and the query becoming more complex, then even such an integer could go to a model (in order to avoid a breaking change of query format changing completely—from a primitive integer to an object).

In order to see whether the command and query work, you will be using an in-memory database context once again. So, create a command to create a new product, a handler to handle it, execute it, and print the results as follows:

```
var command = new CreateProductCommand { Name = "Test PP", Manufacturerid
= 1, Price = 9.99m };
var commandHandler = new CreateProductCommandHandler(db);
var newProductId = commandHandler.Handle(command);
```

Create a query to get the created product and a handler to execute the query:

```
var query = newProductId;
var queryHandler = new GetProductQueryHandler(db);
var product = queryHandler.Handle(query);
Console.WriteLine($"Id: {product.Id}, " +
                  $"Name: {product.Name}, " +
                  $"Price: {product.Price}");
```

In order to run this example, comment all lines within **`static void Main(string[] args)`** body except **`Examples.Cqrs.Demo.Test();`** in **`Program.cs`**. The output will be displayed as follows:

```
Id: 1, Name: Test PP, Price: 9,99
```

> **NOTE**
>
> You can find the code used for this example at https://packt.link/Ij6J8.

You might have wondered why, after so many demos, the `ProductId` is still `1`. That's because it is an in-memory database—one that you create fresh for a new test every time. Since you are starting with an empty database every time, the first addition of a new entity to a database results in a new item with an ID of 1.

You might wonder if you made some changes to a database or added a column to it, and how it would impact the rest of the codebase and the business logic. The next section will detail these scenarios.

Separating the Database Model from the Business Logic (Domain) Model

Databases often change. However, should that impact the rest of the codebase? Should the fact that a column type changed, or another column was added affect the business logic? There is no straight answer to that. It all depends on the project scope, the resources, and the team's maturity. However, if you are working on a medium or a big project, you should consider segregating the database and domain completely. This does not only mean that different logic should be placed in different projects, but it also means that those projects should be decoupled from one another.

It is okay for a database layer to consume a domain layer, but it is not okay for the domain layer to do the same. If you want a complete separation between the two, you will have to introduce an anti-corruption layer. It is a concept that says not to consume foreign models and instead map them as soon as they hit the public component of that layer. The idea is that all interfaces should be domain-specific (that is, work with domain models). However, for a database communication implementation, internally, you will be working with database entities instead of domain models. This requires mapping one to another (when taking input or returning output).

In cases where database entities change completely, the domain-specific interface will remain the same. Only the mapping will change, which will prevent the database from impacting anything else. It is not an easy thing to grasp and implement for a beginner. It is recommended that you ignore that for now; your personal project scope is not worth the effort and you might not see any benefit.

This concludes the theoretical portion of this section. In the following section, you will put this into practice with an activity.

ACTIVITY 6.01: TRACKING SYSTEM FOR TRUCKS DISPATCHED

A logistics company has hired you to keep track of dispatched trucks. A single dispatch includes the current location of a truck, the truck's ID, and the driver's ID. In this activity, you will create a database for dispatched trucks, seed it with a few dispatches, and prove it works by getting all possible data from it.

You will create two classes (**Truck** and **Person**), which consist of the following objects:

- **Truck**: **Id**, **Brand**, **Model**, **YearOfMaking**
- **Person**: **Id**, **Name**, **DoB**

All tables are stored in the **TruckLogistics** database, in the **TruckLogistics** schema.

Perform the following steps to complete this activity:

1. Create a **Person** class.
2. Create a **Truck** class.
3. Create a **TruckDispatch** class.
4. Create a **TruckDispatchDbContext** schema with three tables.
5. Create a connection string (ideally from environment variables).
6. Add a database migration.
7. Generate a database from the migration.
8. Connect to a database.
9. Seed the database with the initial data.
10. Get all data from the database.
11. Print the results.
12. Dispose of the **TruckDispatchesDbContext** schema (that is, disconnect).

    After completing these steps correctly, you should see the following output:

```
Dispatch: 1 1,1,1 2021-11-02 21:45:42
Driver: Stephen King 2021-07-25 21:45:42
Truck: Scania R 500 LA6x2HHA 2009
```

NOTE

In order to run this activity, comment all lines within `static void Main(string[] args)` body except `Activities.Activity01.Demo.Run();` in `Program.cs`.

The database should look like this:

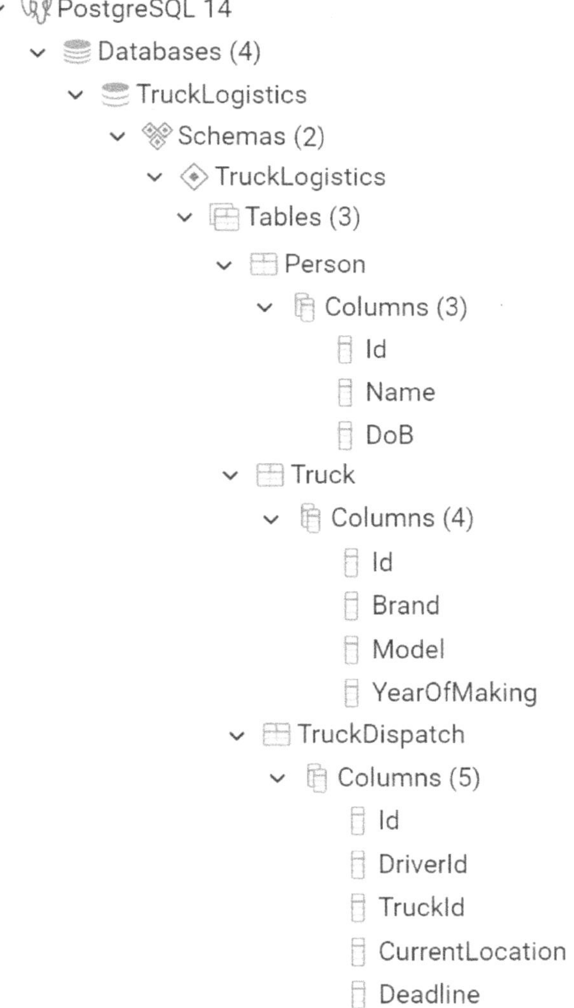

Figure 6.12: Generated TruckLogistics database (simplified for brevity)

And the following migration files (similar, not exact) will be created:

Migrations
20210725183611_InitialMigration.cs
20210725183611_InitialMigration.Designer.cs
TruckDispatchDbContextModelSnapshot.cs

Figure 6.13: Migration files created for the solution

> **NOTE**
>
> The solution to this activity can be found at https://packt.link/qclbF.

With the successful execution of this activity, you should now have solid know-how of how EF is used for rapidly developing solutions integrated with a database.

SUMMARY

In this chapter, you covered the benefits of an ORM and how to talk with a database from C# using the EF Core 6. EF allowed you to abstract a database using `DbContext` and include abstractions to tables, `DbSet`.

You experienced the simplicity of consuming a database using EF, which felt almost the same as writing LINQ queries. The only difference was the initial setup of a connection using a database context. You learned the client input should not be trusted, but ORMs allow you to consume queries with confidence because they take security into consideration and protect you from SQL injection. However, the way you connect to a database (that is, the connection string) has to be secured, and for that reason, you must store it just like any other secret and not hardcode it. You also studied the most common pitfalls when working with EF and tools that could help avoid those pitfalls. This chapter has given you enough skills to create and consume databases using EF.

In the next chapter, you will be focusing more on web applications—what they are, and how to build them.

7

CREATING MODERN WEB APPLICATIONS WITH ASP.NET

OVERVIEW

There are many types of applications in use nowadays and web apps top the list of the most used ones. In this chapter, you will be introduced to ASP.NET, a web framework built with C# and the .NET runtime, made to create web apps with ease. You will also learn the anatomy of a basic ASP.NET application, web application development approaches such as server-side rendering and single-page applications, and how C# helps implement these approaches to build safe, performant, and scalable applications.

INTRODUCTION

In *Chapter 1*, *Hello C#*, you learned that .NET is what brings C# to life, as it contains both a **Software Development Kit** (**SDK**) used to build your code and a runtime that executes the code. In this chapter, you will learn about ASP.NET, which is an open-source and cross-platform framework embedded within the .NET runtime. It is used for building applications for both frontend and backend applications for web, mobile, and IoT devices.

It is a complete toolbox for these kinds of development, as it provides several built-in features, such as lightweight and customizable HTTP pipelines, dependency injection, and support for modern hosting technologies, such as containers, web UI pages, routing, and APIs. A well-known example is Stack Overflow; its architecture is built entirely on top of **ASP.NET**.

The focus of this chapter is to acquaint you with the fundamentals of ASP.NET and to give you both an introduction and an end-to-end overview of web application development with Razor Pages, a built-in toolbox included in ASP.NET to build web apps.

ANATOMY OF AN ASP.NET WEB APP

You'll begin this chapter by creating a new Razor Pages application with ASP.NET. It is just one of the various types of apps that can be created with ASP.NET but will be an effective starting point as it shares and showcases a lot of commonalities with other web application types that can be built with the framework.

1. To create a new Razor Pages app, enter the following commands in the CLI:

```
dotnet new razor -n ToDoListApp
dotnet new sln -n ToDoList
dotnet sln add ./ToDoListApp
```

   Here you are creating a to-do list application with Razor Pages. Once the preceding command is executed, you will see a folder with the following structure:

```
/ToDoListApp
|-- /bin
|-- /obj
|-- /Pages
```

```
    |-- /Properties
    |-- /wwwroot
    |-- appsettings.json
    |-- appsettings.Development.json
    |-- Program.cs
    |-- ToDoListApp.csproj
    |ToDoList.sln
```

2. Open the root folder in Visual Studio Code.

There are some files inside these folders that will be covered in the upcoming sections. For now, consider this structure:

- **bin** is the folder where the final binaries go after the application is built.
- **obj** is the folder where the compiler places intermediate outputs during the build process.
- **Pages** is the folder where the application Razor Pages will be placed.
- **Properties** is a folder containing the **launchSettings.json** file, a file where the run configurations are placed. In this file, you can define some configuration for local run i.e., environment variables and application ports.
- **wwwroot** is the folder where all the static files of the application go.
- **appsettings.json** is a configuration file.
- **appsettings.Development.json** is a configuration file for the Development environment.
- **Program.cs** is the program class that you have seen since *Chapter 1*, *Hello C#*. It is the entry point of an application.

Now that you know that in .NET 6.0, it is the **Program.cs** file, created at the root of the folder, that brings a **WebApplication** to life, you can begin to explore **Program.cs** in greater depth in the next section.

PROGRAM.CS AND THE WEBAPPLICATION

As mentioned earlier, **Program.cs** is the entry point of any C# application. In this section, you will see how a typical **Program** class is structured for an ASP.NET app. Consider the following example of **Program.cs**, which describes a very simple ASP.NET application:

Program.cs

```
var builder = WebApplication.CreateBuilder(args);

// Add services to the container.
builder.Services.AddRazorPages();

var app = builder.Build();

// Configure the HTTP request pipeline.
if (!app.Environment.IsDevelopment())
{
app.UseExceptionHandler("/Error");

// The default HSTS value is 30 days. You may want to change this for production
scenarios, see https://aka.ms/aspnetcore-hsts.
```

The complete code can be found here: https://packt.link/tX9iK.

The first thing done here is the creation of a **WebApplicationBuilder** object. This object contains everything that's needed to bootstrap a basic Web Application in ASP.NET—Configuration, Logging, DI, and Service Registration, Middlewares, and other Host configurations. This Host is the one responsible for the lifetime management of a web application; they set up a web server and a basic HTTP pipeline to process HTTP requests.

As you can see, it is quite impressive how, in a few lines of code, so many things can be done that enable you to run a well-structured web application. ASP.NET does all of that so that you can focus on providing value through the functionalities you will build.

> **NOTE**
>
> Bootstrap is a CSS library for the beautification of web content. You can know more about it at the official website.

MIDDLEWARES

Think of middleware as small pieces of applications that connect to each other to form a pipeline for handling HTTP requests and responses. Each piece is a component that can do some work either before or after another component is executed on the pipeline. They are also linked to each other through a **`next()`** call, as shown in *Figure 7.1*:

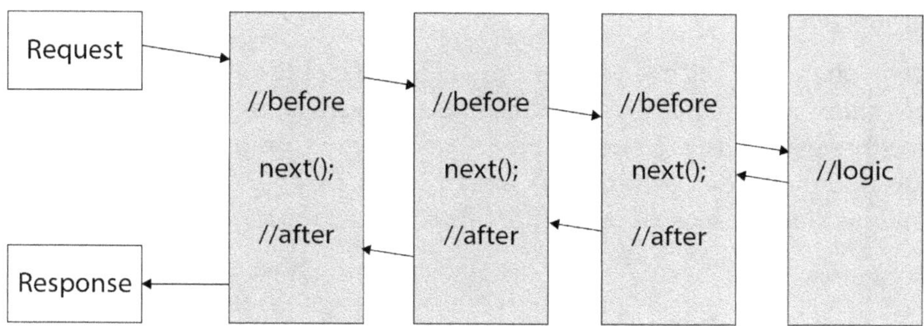

Figure 7.1: The Middleware for an HTTP pipeline

Middleware is a whole universe unto itself. The following list defines the salient features for building a web application:

- The order in which the middleware is placed matters. As they are chained one after another, the placement of each component impacts the way the pipeline is processed.
- The **`before`** logic, as shown in *Figure 7.1*, is executed until the endpoint is finally reached. Once the endpoint is reached, the pipeline continues to process the response using the **`after`** logic.
- **`next()`** is a method call that will execute the next middleware in the pipeline, before executing the **`after`** logic of the current middleware.

In ASP.NET applications, middleware can be defined in the **`Program.cs`** file after the **`WebApplicationBuilder`** calls the **`Build`** method with a **`WebApplication?`** object as a result of this operation.

The application you created in the *Program.cs and the WebApplication* section, already contains a set of middlewares placed for new boilerplate Razor Pages applications that will be called sequentially when an **HTTP request** arrives.

This is easily configurable because the **WebApplication** object contains a generic **UseMiddleware<T>** method. This method allows you to create middleware to embed into the HTTP pipeline for requests and responses. When used within the **Configure** method, each time the application receives an incoming request, this request will go through all the middleware in the order the requests are placed within the **Configure** method. By default, ASP.NET provides basic error handling, **autoredirection** to HTTPS, and serves static files, along with some basic routing and authorization.

However, you might notice in your **Program.cs** file, of the *Program.cs and the WebApplication* section, there are no **UseMiddleware<>** calls. That's because you can write extension methods to give a more concise name and readability to the code, and the ASP.NET framework already does it by default for some built-in middlewares. For instance, consider the following example:

```
using Microsoft.AspNetCore.HttpsPolicy;

public static class HttpsPolicyBuilderExtensions
{
public static IApplicationBuilder UseHttpsRedirection(this WebApplication
app)
      {
            app.UseMiddleware<HttpsRedirectionMiddleware>();
            return app;
}
}
```

Here, a sample of the built-in **UseHttpsRedirection** extension method is used for enabling a redirect middleware.

LOGGING

Logging might be understood as the simple process of writing everything that is done by an application to an output. This output might be the console application, a file, or even a third-party logging monitor application, such as the ELK Stack or Grafana. Logging has an important place in assimilating the behavior of an application, especially with regard to error tracing. This makes it an important concept to learn.

One thing that enables ASP.NET to be an effective platform for enterprise applications is its modularity. Since it is built on top of abstractions, any new implementation can be easily done without loading too much into the framework. The **logging** abstractions are some of these.

By default, the **WebApplication** object created in **Program.cs** adds some logging providers on top of these logging abstractions, which are **Console**, **Debug**, **EventSource**, and **EventLog**. The latter—**EventLog**—is an advanced feature specific to the Windows OS only. The focus here will be the **Console** logging provider. As the name suggests, this provider will output all the logged information to your application console. You'll learn more about it later in this section.

As logs basically write everything your application does, you might wonder whether these logs will end up being huge, especially for large-scale apps. They might be, but an important thing while writing application logs is to grasp the **severity** of the log. There might be some information that is crucial to log, such as an unexpected exception. There might also be information that you would only like to log to a development environment, to know some behaviors better. That said, a log in .NET has seven possible log levels, which are:

- **Trace** = 0
- **Debug** = 1
- **Information** = 2
- **Warning** = 3
- **Error** = 4
- **Critical** = 5
- **None** = 6

Which level is output to the provider is defined via variables set either as environment variables or via the **appSettings.json** file in the **Logging:LogLevel** section, as in the following example:

```
{
  "Logging": {
    "LogLevel": {
      "Default": "Information",
      "ToDoListApp": "Warning",
      "ToDoListApp.Pages": "Information"
    }
  }
}
```

In this file, there are log categories, which are either the **`Default`** category or part of the namespace of the type that wants to set the log. That is exactly why these namespaces exist. For instance, you could set two different levels of logging for files inside the namespace.

In the preceding example configuration, the entire **`ToDoListApp`** is a set namespace to write logs only with **`LogLevel`** equal to or higher than **`Warning`**. You are also specifying that, for the **`ToDoListApp.Pages`** category/ namespace, the application will write all logs with a level equal to or higher than **`Information`**. This means that the changes on a more specific namespace override the settings that were set at a higher level.

This section showed you how to configure log levels for an application. With this knowledge, you can now grasp the concept of DI, as discussed in the following section.

DEPENDENCY INJECTION

Dependency Injection (**DI**) is a technique supported natively by the ASP. NET framework. It is a form of achieving a famous concept in object-oriented programming called **Inversion of Control** (**IoC**).

Any component that an object requires to function can be termed a **dependency**. In the case of a class, this might refer to parameters that need to be constructed. In the case of a method, it might be the method that parameters need for the execution. Using IoC with dependencies means delegating the responsibility of creating a class to the framework, instead of doing everything manually.

In *Chapter 2, Building Quality Object-Oriented Code*, you learned about **interfaces**. Interfaces are basically a common form of establishing a contract. They allow you to focus on what the outcome is of a call, rather than how it is executed. When you use IoC, your dependencies can now be interfaces instead of concrete classes. This allows your classes or methods to focus on the contracts established by these interfaces, instead of implementation details. This brings the following advantages:

- You can easily replace implementations without affecting any class that depends on the contracts.
- It decouples the application boundaries and modules, as the contracts usually do not need any hardened dependencies.
- It makes testing easier, allowing you to create these explicit dependencies as mocks, or fakes, and focus on behavior instead of real implementation details.

Imagine now that to create the middleware of your application, you need to construct each of their dependencies, and you have a lot of middleware chained to each other on the constructor. Clearly, this would be a cumbersome process. Also, testing any of this middleware would be a tedious process, as you would need to rely on every single concrete implementation to create an object.

By injecting dependencies, you tell the compiler how to construct a class that has its dependencies declared on the constructor. The DI mechanism does this at runtime. This is equivalent to telling the compiler that whenever it finds a dependency of a certain type, it should resolve it using the appropriate class instance.

ASP.NET provides a native DI container, which stores the information pertaining to how a type should be resolved. You'll next learn how to store this information in the container.

In the **Program.cs** file, you'll see the call **builder.Services.AddRazorPages()**. The **Services** property is of type **IServiceCollection** and it holds the entire set of dependencies—also known as services—that is injected into the container. A lot of the required dependencies for an ASP.NET application to run are already injected in the **WebApplication.CreateBuilder(args)** method called at the top of the **Program.cs** file. This is true, for instance, for some native logging dependencies as you will see in the next exercise.

EXERCISE 7.01: CREATING CUSTOM LOGGING MIDDLEWARE

In this exercise, you will create custom logging middleware that will output the details and the duration of an HTTP request to the console. After creating it, you will place it in the HTTP pipeline so that it is called by every request your application receives. The purpose is to give you a first practical introduction to the concepts of middleware, logging, and DI.

The following steps will help you complete this exercise:

1. Create a new folder called **Middlewares**.
2. Inside this folder, create a new class named **RequestLoggingMiddleware**.
3. Create a new private readonly field named **RequestDelegate** and initialize this field inside the constructor:

```
private readonly RequestDelegate _next;
public RequestLoggingMiddleware(RequestDelegate next)
{
    _next = next;
}
```

This is the reference that ASP.NET gathers as the next middleware to be executed on the HTTP pipeline. By initializing this field, you can call the next registered middleware.

4. Add a **using** statement to the **System.Diagnostics** namespace so that a special class named **Stopwatch** can be added It will be used to measure the request time length:

```
using System.Diagnostics;
```

5. Create a private **readonly ILogger** field. The **ILogger** interface is the default interface provided by .NET to manually log information.
6. After that, place a second parameter inside the constructor for the **ILoggerFactory** type. This interface is another one provided by .NET that allows you to create **ILogger** objects.
7. Use the **CreateLogger<T>** method from this factory to create a logger object:

```
private readonly ILogger _logger;
private readonly RequestDelegate _next;

public RequestLoggingMiddleware(RequestDelegate next, ILoggerFactory
loggerFactory)
{
    _next = next;
    _logger = loggerFactory.CreateLogger<RequestLoggingMiddleware>();
}
```

Here, **T** is a generic parameter that refers to a type, which is the log category, as seen in the *Logging* section. In this case, the category will be the type of the class where the logging will be done that is, the **RequestLoggingMiddleware** class.

8. Once the fields have been initialized, create a new method with the following signature:

```
public async Task InvokeAsync(HttpContext context) { }
```

9. Inside this method, declare a variable called **Stopwatch** and assign the **Stopwatch.StartNew()** value to it:

```
var stopwatch = Stopwatch.StartNew();
```

   The **Stopwatch** class is a helper that measures the execution time from the moment the **.StartNew()** method is called.

10. After this variable, write a **try-catch** block with code to call the next request as well as a call to the **.Stop()** method from the **stopwatch** to measure the elapsed time that the **_next()** call took:

```
using System.Diagnostics;

namespace ToDoListApp.Middlewares;

public class RequestLoggingMiddleware
{
    private readonly ILogger _logger;
    private readonly RequestDelegate _next;

    public RequestLoggingMiddleware(RequestDelegate next, 
ILoggerFactory loggerFactory)
    {
        _next = next;
        _logger = loggerFactory.
CreateLogger<RequestLoggingMiddleware>();
    }
```

   You can also deal with a possible exception here. So, it is better to wrap these two calls inside a **try-catch** method.

11. In the **Program.cs** file, call the custom middleware by placing the declaration as follows:

```
var app = builder.Build();

// Configure the HTTP request pipeline.
app.UseMiddleware<RequestLoggingMiddleware>();
```

   Write it in the line right below where the **app** variable is assigned.

12. Finally, in the **Program.cs** file, place a **using** statement to **ToDoListApp.Middlewares**:

Program.cs

```
using ToDoListApp.Middlewares;

var builder = WebApplication.CreateBuilder(args);

// Add services to the container.
builder.Services.AddRazorPages();

var app = builder.Build();

// Configure the HTTP request pipeline.
app.UseMiddleware<RequestLoggingMiddleware>();

if (!app.Environment.IsDevelopment())
{
    app.UseExceptionHandler("/Error");
```

The complete code can be found here: https://packt.link/tX9iK.

13. To see the application running on your web browser and its output in the Visual Studio Code, type the following command at the address bar:

```
localhost:####
```

Here **####** represents the port number. This would be different for different systems.

14. After pressing enter, the following screen gets displayed:

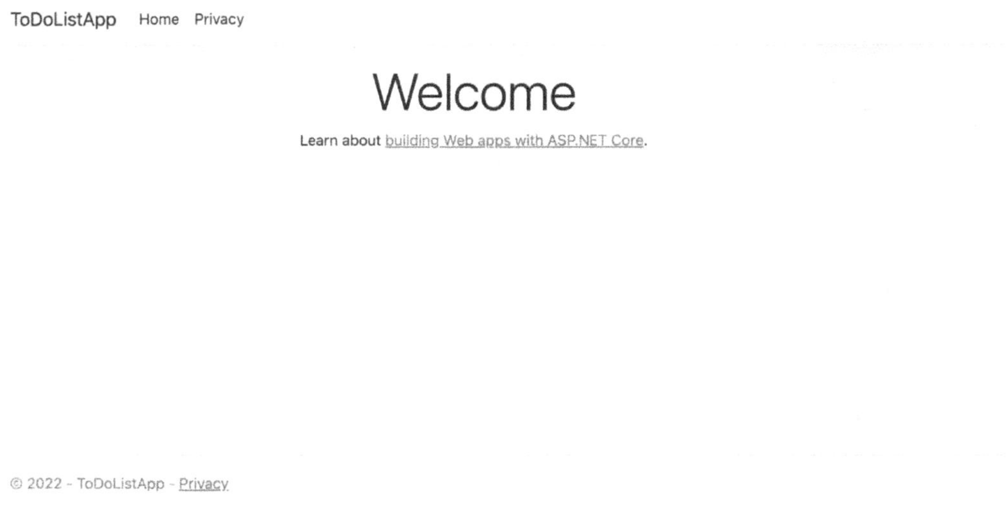

ToDoListApp Home Privacy

Welcome

Learn about building Web apps with ASP.NET Core.

© 2022 - ToDoListApp - Privacy

Figure 7.2: Application running on the browser

15. Perform *Step 13* each time after executing the exercise/ activity in VS Code.
16. Press **`Control+C`** inside the VS code terminal to break the task before executing another exercise/ activity.
17. After executing the application in your browser, you'll see a similar output in the Visual Studio Code terminal:

```
info: ToDoListApp.Middlewares.RequestLoggingMiddleware[0]
      HTTP GET request for path / with status 200 executed in 301 ms
info: ToDoListApp.Middlewares.RequestLoggingMiddleware[0]
      HTTP GET request for path /lib/bootstrap/dist/css/bootstrap.
min.css with status 200 executed in 18 ms
info: ToDoListApp.Middlewares.RequestLoggingMiddleware[0]
      HTTP GET request for path /css/site.css with status 200
executed in 1 ms
info: ToDoListApp.Middlewares.RequestLoggingMiddleware[0]
      HTTP GET request for path /favicon.ico with status 200 executed
in 1 ms
```

You will observe that the output on the console logs messages with an elapsed time of HTTP requests coming in the middleware pipelines. Since you've declared it with your methods, it should take the execution time considering all the pipeline chains.

In this exercise, you created your first middleware—the **`RequestLoggingMiddleware`**. This middleware measures the execution time of an HTTP request, in your HTTP pipeline. By placing it right before all other middlewares, you will be able to measure the entire execution time of a request that goes through the entire middleware pipeline.

> **NOTE**
>
> You can find the code used for this exercise at https://packt.link/i04Iq.

Now imagine you have 10 to 20 middleware for the HTTP pipeline, each has its own dependencies, and you must manually instantiate each middleware. IoC comes in handy in such cases by delegating to ASP.NET the instantiation of these classes, as well as injecting their dependencies. You have already seen how to create custom middleware that uses the native ASP.NET logging mechanism with DI.

In ASP.NET, logging and DI are powerful mechanisms that allow you to create very detailed logs for an application. This is possible, as you've seen, through **logger** injection via constructors. For these loggers, you can create an object of a log category in two ways:

- As shown in the exercise, one way is the injection of **ILoggerFactory**. You could call the **CreateLogger(categoryName)** method, which receives a string as an argument. You could also call the **CreateLogger<CategoryType>()** method, which receives a generic type. This approach is preferable as it sets the category for the **logger** as the full name of the type (including the namespace).
- Another way would be through the injection of **ILogger<CategoryType>**. In this case, the category type is usually the type of the class where you are injecting the logger, as seen in the previous exercise. In the previous exercise, you could replace the injection of **ILoggerFactory** with **ILogger<RequestLoggingMiddleware>** and assign this new injected dependency directly to the **ILogger** private field as follows:

```
private readonly ILogger _logger;
private readonly RequestDelegate _next;

public RequestLoggingMiddleware(RequestDelegate next, ILogger<
RequestLoggingMiddleware> logger)
{
    _next = next;
    _logger = logger;
}
```

You now know that logging and DI are powerful mechanisms that allow you to create very detailed logs for an application. Before moving to Razor pages, it is important to learn about the life cycle of an object within an application. This is called the dependency lifetimes.

DEPENDENCY LIFETIMES

Before moving on to the next and main topic of this chapter, it is important to talk about dependency lifetimes. All the dependencies used in the previous exercise were injected via the constructor. But the resolution of these dependencies was only possible because ASP.NET registers these dependencies beforehand, as mentioned in the **Program.cs** section. In the following code, you can see an example of code built into ASP.NET that deals with the logging dependency registration, by adding the **ILoggerFactory** dependency to the services container:

LoggingServiceCollectionExtensions.cs

```
public static IServiceCollection AddLogging(this IServiceCollection services,
Action<ILoggingBuilder> configure)
{T
if (services == null)
     {
     throw new ArgumentNullException(nameof(services));
     }

     services.AddOptions();

     services.TryAdd(ServiceDescriptor.Singleton<ILoggerFactory, LoggerFactory>());

services.TryAdd(ServiceDescriptor.Singleton(typeof(ILogger<>), typeof(Logger<>)));

services.TryAddEnumerable(ServiceDescriptor.
Singleton<IConfigureOptions<LoggerFilterOptions>>(new
DefaultLoggerLevelConfigureOptions(LogLevel.Information)));

configure(new LoggingBuilder(services));

return services;
}
```

The complete code can be found here: https://packt.link/g4JPp.

> **NOTE**
>
> The preceding code is an example from a standard library and built into ASP.NET that deals with the logging dependency registration.

A lot is going on here, but the two important things to consider are as follows:

- The method here is **TryAdd**, which registers a dependency on the DI container.
- The **ServiceDescriptor.Singleton** method is what defines a dependency lifetime. This is the final important concept of the *Dependency Injection* section of this chapter.

A dependency lifetime describes the life cycle of an object within an application. ASP.NET has three default lifetimes that can be used to register a dependency:

- **Transient**: Objects with this lifetime are created every time they are requested and disposed of after use. This is effective for **stateless dependencies**, which are dependencies that do not need to keep the state when they are called. For instance, if you need to connect to an HTTP API to request some information, you can register a dependency with this lifetime, since HTTP requests are stateless.

- **Scoped**: Objects with a scoped lifetime are created once for each client connection. For instance, in an HTTP request, a **scoped dependency** will have the same instance for the entire request, no matter how many times it is called. This dependency carries some state around for a certain amount of time. At the end of the connection, the dependency is disposed of.
- **Singleton**: Objects with a singleton lifetime are created once for an entire application's lifetime. Once they are requested, their instance will be carried on while the application is running. This kind of lifetime should be considered carefully as it might consume a lot of memory.

As mentioned before, the manual registration of these dependencies can be done in the **`ConfigureServices`** method located in the **`Startup`** class. Every new dependency that is not provided and automatically registered by ASP.NET should be manually registered there and knowing about these lifetimes is important as they allow the application to manage the dependencies in different ways.

You have learned that the resolution of these dependencies was only possible because ASP.NET registers three default lifetimes that can be used to register a dependency. You will now move on to Razor pages that enable the construction of page-based applications with all the capabilities provided and powered by ASP.NET.

RAZOR PAGES

Now that you have covered the main aspects pertaining to an ASP.NET application, you'll continue to build the application that you started at the beginning of the chapter. The goal here is to build a to-do list application, where you can easily create and manage a list of tasks on a Kanban-style board.

Earlier sections have referenced Razor Pages, but what exactly is it? **Razor Pages** is a framework that enables the construction of page-based applications with all the capabilities provided and powered by ASP.NET. It was created to enable the building of dynamic data-driven applications with a clear separation of concerns that is, having each method and class with separate but complementary responsibilities.

BASIC RAZOR SYNTAX

Razor Pages uses Razor syntax, a syntax powered by Microsoft that enables a page to have static HTML/ CSS/ JS, C# code, and custom **tag helpers**, which are reusable components that enable the rendering of HTML pieces in pages.

If you look at the `.cshtml` files generated by the `dotnet new` command that you ran in the first exercise, you will notice a lot of HTML code and, inside this code, some methods, and variables with a `@` prefix. In Razor, as soon as you write this symbol, the compiler detects that some C# code will be written. You're already aware that HTML is a markup language used to build web pages. Razor uses it along with C# to create powerful markup combined with server-rendered code.

If you want to place a block of code, it can be done within brackets like:

```
@{ … }
```

Inside this block, you are allowed to do basically everything you can do with C# syntax, from local variable declarations to loops and more. If you want to put a **`static @`**, you have to escape it by placing two `@` symbols for it to be rendered in HTML. That happens, for instance, in email IDs, such as **`james@@bond.com`**.

FILE STRUCTURE

Razor Pages end with the **`.cshtml`** extension and might have another file, popularly called the **code-behind** file, which has the same name but with the **`.cshtml.cs`** extension. If you go to the root folder of your application and navigate to the **`Pages`** folder, you will see the following structure generated upon the creation of a page:

```
|-- /Pages
|---- /Shared
|------ _Layout.cshtml
|------ _ValidationScriptsPartial.cshtml
|---- _ViewImports.cshtml
|---- _ViewStart.cshtml
|---- Error.cshtml
|---- Error.cshtml.cs
|---- Index.cshtml
|---- Index.cshtml.cs
|---- Privacy.cshtml
|---- Privacy.cshtml.cs
```

The **`Index`**, **`Privacy`**, and **`Error`** pages are automatically generated after project creation. Briefly look at the other files here.

The **/Shared** folder contains a shared **Layout** page that is used by default in the application. This page contains some shared sections, such as navbars, headers, footers, and metadata, that repeat in almost every application page:

_Layout.cshtml

```
<!DOCTYPE html>
<html lang="en">
<head>
    <meta charset="utf-8" />
    <meta name="viewport" content="width=device-width, initial-scale=1.0" />
    <title>@ViewData["Title"] - ToDoListApp</title>
    <link rel="stylesheet" href="~/lib/bootstrap/dist/css/bootstrap.min.css" />
    <link rel="stylesheet" href="~/css/site.css" asp-append-version="true" />
    <link rel="stylesheet" href="~/ToDoListApp.styles.css" asp-append-
version="true" />
</head>
<body>
    <header>
        <nav class="navbar navbar-expand-sm navbar-toggleable-sm navbar-light bg-
white border-bottom box-shadow mb-3">
            <div class="container">
                <a class="navbar-brand" asp-area="" asp-page="/Index">ToDoListApp</
a>
```

The complete code can be found here: https://packt.link/2Hb8r.

Keeping these shared sections in a single file makes reusability and maintainability easier. If you look at this **Layout** page generated in your boilerplate, there are some things worth highlighting:

- By default, a Razor Pages app is generated using Twitter Bootstrap for design—a library used for writing beautiful, simple, and responsive websites—and jQuery for basic scripting. This can be customized for each application, as those are just static files.
- There is a special **RenderBody()** method that indicates where the generated HTML for the application pages will be placed.
- Another method, named **RenderSection()**, is useful for rendering predefined sections per page. It is useful, for instance, when some static file, such as an image, script, or stylesheet, is needed only for some pages. In this way, you can place these files inside specific sections only in the pages where they are needed and call the **RenderSection** method at the level of the HTML you want them to be rendered. This is done on the **_Layout.cshtml** page.

The **_ViewImports.cshtml** file is another important file; it enables the application pages to share common directives and reduces effort by placing these directives on every page. It is where all the global using namespaces, tag helpers, and global **Pages** namespaces are defined. Some of the directives this file supports are as follows:

- **@namespace**: Used to set the base namespace for **Pages**.
- **@inject**: Used to place DI within the page.
- **@model**: Includes **PageModel**, a class that will determine what information the page will handle.
- **@using**: Similar to the **.cs** files, this directive allows you to fully qualify namespaces at the top level of a Razor page to avoid repeating these namespaces throughout the code.

The **_ViewStart.cshtml** file is used to place code that will be executed at the start of each page call. On this page, you define the **Layout** property while setting the **Layout** page.

Now that you are familiar with the basics of Razor Pages, it is time to start working on your application and dive into some more interesting topics. You will start by creating the basic structure of the to-do list application.

EXERCISE 7.02: CREATING A KANBAN BOARD WITH RAZOR

The goal of this exercise will be to start the to-do application creation with its first component—a Kanban board. This board is used for controlling workflows, where people can divide their work into cards and move these cards between different statuses, such as To Do, Doing, and Done. A popular application that uses this is Trello. The same **ToDoListApp** project created in the *Exercise 7.01* will be used throughout this chapter to learn new concepts and incrementally evolve the application, including in this exercise. Perform the following steps:

1. Navigate to the root folder of your application and create a folder named **Models**.
2. Inside the **Models** folder, create a new **enum** called **ETaskStatus** with the **ToDo**, **Doing**, and **Done** options:

```
public enum ETaskStatus
{
ToDo,
Doing,
```

```
Done
}
```

3. Again, in the **Models** folder, create a new class called **ToDoTask** that will be used to create a new task for your to-do list with the following properties:

```
namespace ToDoListApp.Models;

public class ToDoTask
{
    public Guid Id { get; set; }
    public DateTime CreatedAt { get; set; }
    public DateTime? DueTo { get; set; }
    public string Title { get; set; }
    public string? Description { get; set; }
    public ETaskStatus Status { get; set; }
}
```

4. Create two constructors as shown here for the **ToDoTask** class:

ToDoTask.cs

```
namespace ToDoListApp.Models;

public class ToDoTask
{
    public ToDoTask()
    {
        CreatedAt = DateTime.UtcNow;
        Id = Guid.NewGuid();
    }

    public ToDoTask(string title, ETaskStatus status) : this()
    {
        Title = title;
        Status = status;
    }
```

The complete code can be found here: https://packt.link/nFk00.

Create one with no parameters to set the default values for the **Id** and **CreatedAt** properties, and the other with lowercase-named parameters for the preceding class to initialize the **Title** and **Status** properties.

The **Pages/ Index.cshtml** is automatically generated in your application boilerplate. It is this page that will be the entry point of your application.

5. Now, customize it by editing the file **Pages/ Index.cshtml.cs** and replacing the boilerplate code with the code shown as follows:

Index.cshtml.cs

```
using System.Collections.Generic;
using Microsoft.AspNetCore.Mvc.RazorPages;
using ToDoListApp.Models;

namespace ToDoListApp.Pages;

public class IndexModel : PageModel
{
    public IList<ToDoTask> Tasks { get; set; } = new List<ToDoTask>();

    public IndexModel()
    {
    }
```

The complete code can be found here: https://packt.link/h8mni.

Basically, this code fills your model. Here, the **OnGet** method of **PageModel** is used to tell the application that when the page is loaded, it should fill the model with the properties assigned to **Task**

6. Replace the code within **Pages/ Index.cshtml** with the code shown as follows in order to create your Kanban board with the task cards:

Index.cshtml

```
@page
@using ToDoListApp.Models
@model IndexModel
@{
    ViewData["Title"] = "My To Do List";
}

<div class="text-center">
    <h1 class="display-4">@ViewData["Title"]</h1>
    <div class="row">
        <div class="col-4">
            <div class="card bg-light">
                <div class="card-body">
                    <h6 class="card-title text-uppercase text-truncate py-
2">To Do</h6>
                    <div class="border border-light">
```

The complete code can be found here: https://packt.link/IhELU.

This page is your view. It shares the properties from the **Pages/ Index.cshtml.cs** class (also called the **code-behind class**). When you assign a value to the **Tasks** property in the code-behind class, it becomes visible to the view. With this property, you can populate the HTML from the page.

7. Now, run your application with the **`dotnet run`** command. You will see the following on the **`Index`** page when the application is loaded on the browser:

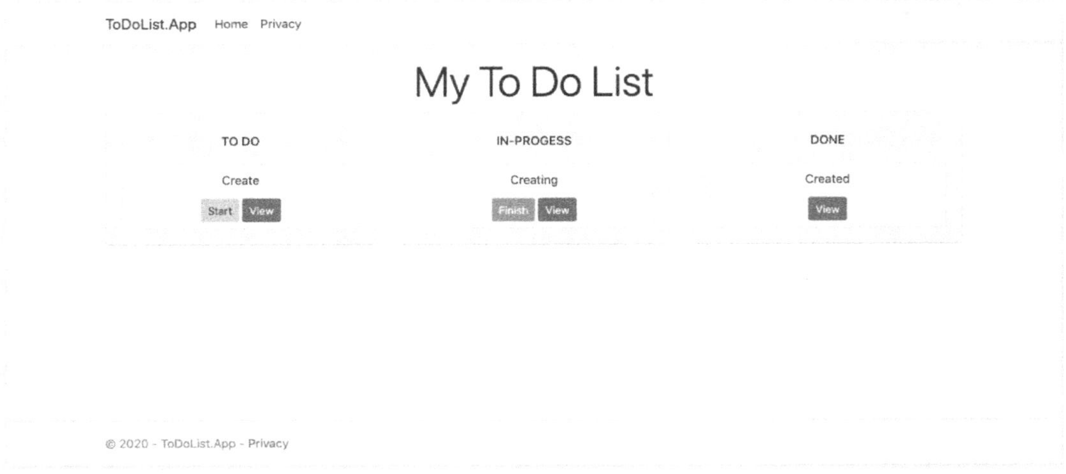

Figure 7.3: Displaying your first application, the Kanban board

Notice that, for now, the application does not contain any logic. What you built here is simply a UI powered by the **`PageModel`** data.

> **NOTE**
>
> You can find the code used for this exercise at https://packt.link/1PRdq.

As you saw in *Exercise 7.02*, for every page created there are two main types of files which are a **`.cshtml`** and a **`.cshtml.cs`** file. These files form the foundations of each Razor page. The next section will detail about this difference in the filename suffix and how these two files complement each other.

PAGEMODEL

In the **`Index.cshtml.cs`** file that you created in *Exercise 7.02*, you might have noticed that the class inside it inherits from the **`PageModel`** class. Having this code-behind class provides some advantages—such as a clear separation of concerns between the client and the server—and this makes maintenance and development easier. It also enables you to create both unit and integration tests for the logic placed on the server. You will learn more about testing in *Chapter 10, Automated Testing*.

A **PageModel** may contain some properties that are bound to the view. In *Exercise 7.02*, the **IndexModel** page has a property that is a **List<ToDoTask>**. This property is then populated when the page loads on the **OnGet()** method. So how does populating happen? The next section will discuss the life cycle of populating properties and using them within **PageModel**.

THE LIFE CYCLE WITH PAGE HANDLERS

Handler methods are a core feature of Razor Pages. These methods are automatically executed when the server receives a request from the page. In *Exercise 7.02*, for instance, the **OnGet** method will be executed each time the page receives a **GET** request.

By convention, the handler methods will answer according to the HTTP verb of the request. So, for instance, if you wanted something to be executed after a **POST** request, you should have an **OnPost** method. Also, after a **PUT** request, you should have an **OnPut** method. Each of these methods has an asynchronous equivalent, which changes the method's signature; an **Async** suffix is added to the method name, and it returns a **Task** property instead of **void**. This also makes the **await** functionality available for the method.

There is, however, one tricky scenario in which you may want a form to perform multiple actions with the same HTTP verb. In that case, you could perform some confusing logic on the backend to handle different inputs. Razor Pages, however, provides you with a functionality right out of the box called **tag helpers**, which allows you to create and render HTML elements on the server before placing them on the client. The anchor tag helper has an attribute called **asp-page-handler** that allows you to specify the name of the handler being called on the server. Tag helpers will be discussed in the next section, but for now, consider the following code as an example. The code contains an HTML form containing two submit buttons, to perform two different actions—one for creating an order, and the other for canceling an order:

```
<form method="post">
    <button asp-page-handler="PlaceOrder">Place Order</button>
    <button asp-page-handler="CancelOrder">Cancel Order</button>
</form>
```

On the server side, you only need to have two handlers, one for each action, as shown in the following code:

```
public async Task<IActionResult> OnPostPlaceOrderAsync()
{
    // ...
}

public async Task<IActionResult> OnPostCancelOrderAsync()
{
    // ...
}
```

Here, the code behind the page matches the value of the **form** method and the **asp-page-handler** tag on the **.cshtml** file to the method name on the code-behind file. That way, you can have multiple actions for the same HTTP verb in the same form.

A final note on this subject is that in this case, the method name on the server should be written as:

```
On + {VERB} + {HANDLER}
```

This is written with or without the **Async** suffix. In the previous example, the **OnPostPlaceOrderAsync** method is the **PlaceOrder** handler for the **PlaceOrder** button, and **OnPostCancelOrderAsync** is the handler for the **CancelOrder** button.

RENDERING REUSABLE STATIC CODE WITH TAG HELPERS

One thing you might have noticed is that the HTML written previously is lengthy. You created Kanban cards, lists, and a board to wrap it all. If you take a closer look at the code, it has the same pattern repeated all the way through. That raises one major problem, maintenance. It is hard to imagine having to handle, maintain, and evolve all this plain text.

Fortunately, tag helpers can be immensely useful in this regard. They are basically components that render static HTML code. ASP.NET has a set of built-in tag helpers with custom server-side attributes, such as anchors, forms, and images. Tag helpers are a core feature that helps make advanced concepts easy to handle, such as model binding, which will be discussed a little further ahead.

Besides the fact that they add rendering capabilities to built-in HTML tags, they are also an impressive way to achieve reusability on static and repetitive code. In the next exercise, you will learn how to create a customized tag helper.

EXERCISE 7.03: CREATING REUSABLE COMPONENTS WITH TAG HELPERS

In this exercise, you are going to improve upon your work in the previous one. The improvement here will be to simplify the HTML code by moving part of this code that could be reused to custom tag helpers.

To do so, perform the following steps:

1. Open the **_ViewImports.cshtml** file, which was created with your application.
2. Add the following lines to the end with the content to define custom tag helpers **@addTagHelper** directive:

```
@addTagHelper *, Microsoft.AspNetCore.Mvc.TagHelpers
@addTagHelper *, ToDoListApp
```

   In the preceding code, you added all the custom tag helpers that exist within this namespace using the asterisk (*****).

3. Now, create a new folder under the project's root (**ToDoApp**) called **TagHelpers**.
4. Create a new class inside this folder called **KanbanListTagHelper.cs**.
5. Make this class inherit from the **TagHelper** class:

```
namespace ToDoListApp.TagHelpers;
```

6. This inheritance is what allows ASP.NET to identify both built-in and custom tag helpers.
7. Now place a **using** statement for the **Microsoft.AspNetCore.Razor.TagHelpers** namespace:

```
using Microsoft.AspNetCore.Razor.TagHelpers;

namespace ToDoListApp.TagHelpers;

public class KanbanListTagHelper : TagHelper
{
}
```

8. For the **KanbanListTagHelper** class, create two string properties, called **Name** and **Size**, with getters and setters:

```
using Microsoft.AspNetCore.Razor.TagHelpers;

namespace ToDoListApp.TagHelpers;

public class KanbanListTagHelper : TagHelper
{
    public string? Name { get; set; }
    public string? Size { get; set; }
}
```

9. Override the base asynchronous **ProcessAsync (TagHelperContext context**, **TagHelperOutput)** output method with the following code:

KanbanListTagHelper.cs

```
public override async Task ProcessAsync(TagHelperContext context, 
TagHelperOutput output)
{
      output.TagName = "div";
      output.Attributes.SetAttribute("class", $"col-{Size}");

      output.PreContent.SetHtmlContent(
      $"<div class=\"card bg-light\">"
           + "<div class=\"card-body\">"
           + $"<h6 class=\"card-title text-uppercase text-truncate py-
2\">{Name}</h6>"
           + "<div class \"border border-light\">");

      var childContent = await output.GetChildContentAsync();
      output.Content.SetHtmlContent(childContent.GetContent());
```

The complete code can be found here: https://packt.link/bjFlk.

Every tag helper has a standard HTML tag as an output. That is why, at the beginning of your methods, the **TagName** property was called from the **TagHelperOutput** object to specify the HTML tag that will be used as output. Additionally, you can set the attributes for this HTML tag by calling the **Attributes** property and its **SetAttribute** method from the **TagHelperOutput** object. That is what you did right after specifying the HTML output tag.

10. Now, create another class named **KanbanCardTagHelper.cs** with the same inheritance and namespace using a statement such as the previous one:

```
namespace ToDoListApp.TagHelpers;

using Microsoft.AspNetCore.Razor.TagHelpers;

public class KanbanCardTagHelper: TagHelper
{
    public string? Task { get; set; }
}
```

For this class, create a **string** property with public getters and setters named **Task.**

11. In this new class, override the base synchronous **Process(TagHelperContext context, TagHelperOutput output)** method. Within this method, write the following code:

```
public override void Process(TagHelperContext context,
TagHelperOutput output)
{
     output.TagName = "div";
     output.Attributes.SetAttribute("class", "card");

     output.PreContent.SetHtmlContent(
     "<div class=\"card-body p-2\">"
          + "<div class=\"card-title\">");

     output.Content.SetContent(Task);

     output.PostContent.SetHtmlContent(
     "</div>"
          + "<button class=\"btn btn-primary btn-sm\">View</button>"
          + "</div>");

output.TagMode = TagMode.StartTagAndEndTag;
}
```

An important concept to know about is how the HTML content is placed within the tag helper. As you can see, the code uses three different properties from the **TagHelperOutput** object to place the content:

- **PreContent**
- **Content**
- **PostContent**

The pre-and post-properties are useful to set the content right before and after that you want to generate. A use case for them is when you want to set up fixed content as **div** containers, headers, and footers.

Another thing you did here was set how the tag helper will be rendered through the **Mode** property. You used **TagMode.StartTagAndEndTag** as a value because you used a **div** container as a tag output for the tag helper, and **div** elements have both start and end tags in HTML. If the output tag were some other HTML element, such as email, which is self-closing, you would use **TagMode.SelfClosing** instead.

12. Finally, go to the **Index.cshtml** file under the Pages folder and replace the HTML created in *Exercise 7.02* with the tag helpers to make your code concise:

Index.cshtml

```
@page
@using ToDoListApp.Models
@model IndexModel
@{
    ViewData["Title"] = "My To Do List";
}

<div class="text-center">
    <h1 class="display-4">@ViewData["Title"]</h1>
    <div class="row">
        <kanban-list name="To Do" size="4">
            @foreach (var task in Model.Tasks.Where(t => t.Status ==
ETaskStatus.ToDo))
            {
                <kanban-card task="@task.Description">
                </kanban-card>
```

The complete code can be found here: https://packt.link/YIgdp.

13. Now run the application with the following command:

```
dotnet run
```

14. In your browser, navigate to the localhost:#### address provided by the Visual Studio console output just like you did in the last exercise:

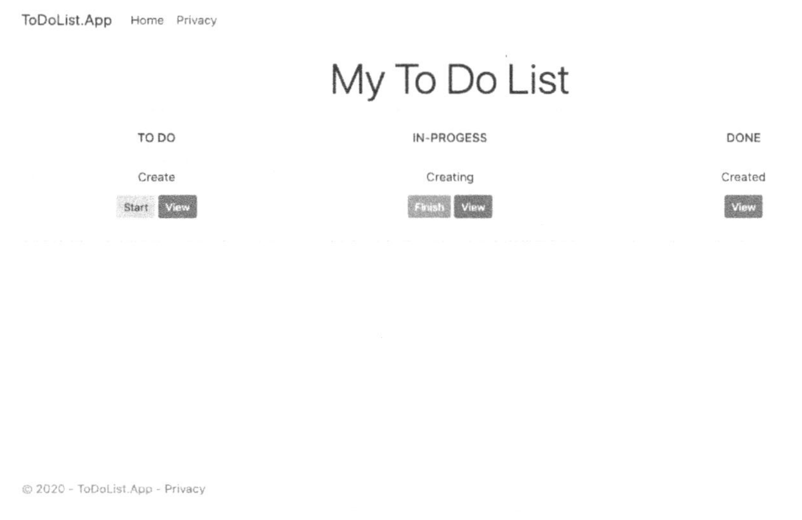

Figure 7.4: The frontend displayed in the browser

You will see the same result at the frontend that you had before, as shown in *Figure 7.3*. The improvement is in the fact that even though the output is the same, you have now a much more modular and concise code to maintain and evolve.

> **NOTE**
>
> You can find the code used for this exercise at https://packt.link/YEdiU.

In this exercise, you used tag helpers to create reusable components that generate static HTML code. You can see now that the HTML code is much cleaner and more concise. The next section will detail about creating interactive pages by linking what's on the Code Behind with your HTML view using the concept of model binding.

MODEL BINDING

So far, you have covered concepts that helped create a foundation for the to-do app. As a quick recap, the main points are as follows:

- **`PageModel`** is used to add data to a page.
- Tag helpers add custom static rendering to the HTML generated by the server.
- Handler methods define the way a page interacts with an HTTP request.

One final overarching concept that is central to building Razor Pages applications is **model binding**. The data used as arguments in handler methods and passed through the page model is rendered through this mechanism. It consists of extracting data in key/ value pairs from HTTP requests and placing them in either the client-side HTML or the server-side code, depending on the direction of the binding that is, whether the data is moving from client to server or from server to client.

This data might be placed in routes, forms, or query strings and is binding to .NET types, either primitive or complex. *Exercise 7.04* will help clarify how the model binding works when coming from the client to the server.

EXERCISE 7.04: CREATING A NEW PAGE TO SUBMIT TASKS

The goal of this exercise is to create a new page. It will be used to create new tasks that will be displayed on the Kanban board. Perform the following steps to complete this exercise:

1. Inside the project root folder, run the following commands:

```
dotnet add package Microsoft.EntityFrameworkCore
dotnet add package Microsoft.EntityFrameworkCore.Sqlite
dotnet add package Microsoft.EntityFrameworkCore.Design
```

2. At the root of the project, create a new folder named **`Data`** with a **`ToDoDbContext`** class inside it. This class will inherit from Entity Framework's **`DbContext`** and will be used to access the database.

3. Now add the following code in it:

```
using Microsoft.EntityFrameworkCore;
using ToDoListApp.Models;


namespace ToDoListApp.Data;

public class ToDoDbContext : DbContext
{
    public ToDoDbContext(DbContextOptions<ToDoDbContext> options) :
base(options)
    {

    }

    public DbSet<ToDoTask> Tasks { get; set; }
}
```

4. Update your **Program.cs** file to match the following:

Program.cs

```
using Microsoft.EntityFrameworkCore;
using ToDoListApp.Data;
using ToDoListApp.Middlewares;

var builder = WebApplication.CreateBuilder(args);

// Add services to the container.
builder.Services.AddRazorPages();

builder.Services.AddDbContext<ToDoDbContext>(opt => opt.UseSqlite("Data
Source=Data/ToDoList.db"));

var app = builder.Build();

// Configure the HTTP request pipeline.
app.UseMiddleware<RequestLoggingMiddleware>();
```

The complete code can be found here: https://packt.link/D4M8o.

This change will register the **DbContext** dependencies within the DI container, as well as sets up the database access.

5. Run the following commands on the terminal to install the **dotnet ef** tool. This is a CLI tool that will help you to iterate with database helpers, such as schema creation and update:

```
dotnet tool install --global dotnet-ef
```

6. Now, build the application and run the following commands on the terminal:

```
dotnet ef migrations add 'FirstMigration'
dotnet ef database update
```

These commands will create a new migration that will create the schema from your database and apply this migration to your database.

7. After the migration has run and the database is updated, create a new folder called **Tasks** inside the **Pages** folder.

8. Move the Index page files—**index.cshtml** and **index.cshtml.cs**—to the **Tasks** folder.

9. Next, replace the **AddRazorPages** call in the **Program.cs** with the following call:

```
builder.Services.AddRazorPages(opt =>
{
    opt.Conventions.AddPageRoute("/Tasks/Index", "");
});
```

This will add a convention for the page routes to be called.

10. Replace the header tag inside the **_Layout.cshtml** file (under **Pages/Shared/**) to create a shared **navbar** for the application:

```
<header>
        <nav class="navbar navbar-expand-sm navbar-toggleable-sm
navbar-light bg-white border-bottom box-shadow mb-3">
            <div class="container">
                <a class="navbar-brand" asp-area="" asp-page="/
Index">MyToDos</a>
                <button class="navbar-toggler" type="button"
data-toggle="collapse" data-target=".navbar-collapse" aria-
controls="navbarSupportedContent"
                        aria-expanded="false" aria-label="Toggle
navigation">
                    <span class="navbar-toggler-icon"></span>
                </button>
                <div class="navbar-collapse collapse d-sm-inline-flex
flex-sm-row-reverse">
                    <ul class="navbar-nav flex-grow-1">
```

```
                    <li class="nav-item">
                        <a class="nav-link text-dark" asp-area=""
asp-page="/tasks/create">Create Task</a>
                    </li>
                </ul>
            </div>
        </div>
    </nav>
</header>
```

This **navbar** will allow you to access the newly created page.

11. Create the **Create.cshtml** page (under **Pages/Tasks/**) and add the following code:

Create.cshtml

```
@page "/tasks/create"
@model CreateModel
@{
    ViewData["Title"] = "Task";
}

<h2>Create</h2>
<div>
    <h4>@ViewData["Title"]</h4>
    <hr />
    <dl class="row">
        <form method="post" class="col-6">
            <div class="form-group">
                <label asp-for="Task.Title"></label>
                <input asp-for="Task.Title" class="form-control" />
```

The complete code can be found here: https://packt.link/2NjdN.

This should contain a form that will use a **PageModel** class to create the new tasks. For each form input field, an **asp-for** attribute is used inside the **input** tag helper. This attribute is responsible for filling the HTML input with a proper value in the **name** attribute.

Since you are binding to a complex property inside the page model named **Task**, the name value is generated with the following syntax:

```
{PREFIX}_{PROPERTYNAME} pattern
```

Here **PREFIX** is the complex object name on the **PageModel**. So, for an ID of a task, an input with **name="Task_Id"** is generated on the client-side and the input is populated with the **value** attribute having the **Task.Id** property value that comes from the server. In the case of the page, as you are creating a new task, the field does not come previously populated. That is because with the **OnGet** method you assigned a new object to the **Task** property of the **PageModel** class.

12. Now, create the code-behind page, named **CreateModel.cshtml.cs** (placed in **Pages/Tasks/**):

Create.cshtml.cs

```
using Microsoft.AspNetCore.Mvc;
using Microsoft.AspNetCore.Mvc.RazorPages;
using ToDoListApp.Data;
using ToDoListApp.Models;

namespace ToDoListApp.Pages.Tasks;

public class CreateModel : PageModel
{
    private readonly ToDoDbContext _context;

    public CreateModel(ToDoDbContext context)
    {
        _context = context;
    }
```

The complete code can be found here: https://packt.link/06ciR.

When posting a form, all the values inside the form are placed in the incoming **HttpRequest**. The call to **TryUpdateModelAsync** tries to populate an object with these values that the request brought from the client-side. Since the form is created with the **name** attribute in the input element with the format that has been explained previously, this method knows how to extract these values and bind them to the object. Put simply, that is the magic behind model binding.

13. Now, replace the code of **Index.cshtml** (under **Pages/Tasks/**) with the following:

Index.cshtml

```
@page
@using ToDoListApp.Models
@model IndexModel
@{
    ViewData["Title"] = "My To Do List";
}

<div class="text-center">
    @if (TempData["SuccessMessage"] != null)
    {
        <div class="alert alert-success" role="alert">
            @TempData["SuccessMessage"]
        </div>
    }
    <h1 class="display-4">@ViewData["Title"]</h1>
```

The complete code can be found here: https://packt.link/hNOTx.

This code adds a section that introduces an alert to be displayed if there is an entry with the **SuccessMessage** key in the **TempData** dictionary.

14. Finally, add some display and validation rules via data annotations to the **Models/ToDoTask.cs** class properties:

ToDoTask.cs

```
using System.ComponentModel;
using System.ComponentModel.DataAnnotations;

namespace ToDoListApp.Models;

public class ToDoTask
{
    public ToDoTask()
    {
        CreatedAt = DateTime.UtcNow;
        Id = Guid.NewGuid();
    }

    public ToDoTask(string title, ETaskStatus status) : this()
    {
```

The complete code can be found here: https://packt.link/yau4p.

Here the **`Required`** data annotation over the property is to ensure that this property is set with a valid value. In this exercise, you added persistence with Entity Framework Core and SQLite and created a new page that creates a task for the to-do application, finally saving it into the database.

15. Now run the code in VS Code.
16. To see the output on your web browser, type the following command on the address bar:

```
Localhost:####
```

Here **####** represents the port number. This would be different for different systems.

After pressing enter, the following screen is displayed:

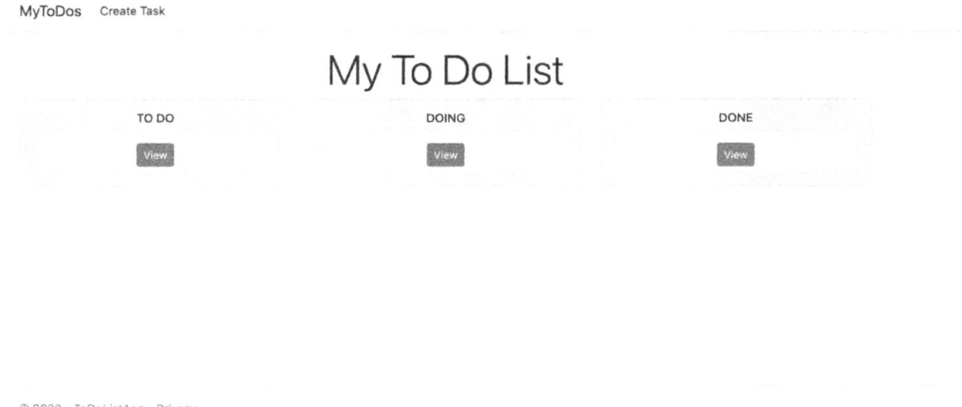

Figure 7.5: Home page with Create Task button in the navigation bar

17. Click on the **`Create Task`** button, and you'll see the page you just created to insert new cards into your Kanban Board:

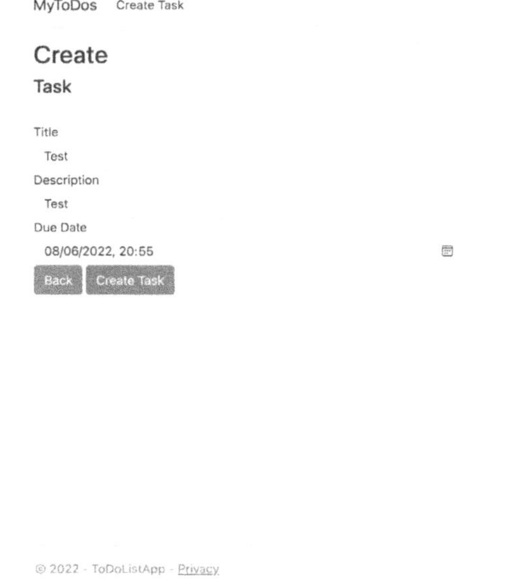

Figure 7.6: The Create Task page

> **NOTE**
>
> You can find the code used for this exercise at https://packt.link/3FPaG.

Now, you'll take a deep dive into how model binding brings it all together, enabling you to transport data back and forth between the client and the server. You will also know more about validations in the next section.

VALIDATION

Validating data is something you will often need to do while developing an application. Validating a field may either mean that it is a required field or that it should follow a specific format. An important thing you may have noticed in the final part of the previous exercise is that you placed some **`[Required]`** attributes on top of some model properties in the final step of the last exercise. Those attributes are called **data annotations** and are used to create server-side validations. Moreover, you can add some client-side validation combined with this technique.

Note that in *Step 10* of *Exercise 7.04*, the frontend has some span tag helpers with an **asp-validation-for** attribute pointing to the model properties. There is one thing that binds this all together—the inclusion of the _ **ValidationScriptsPartial.cshtml** partial page. Partial pages are a subject discussed in the next section, but for now, it is enough to know that they are pages that can be reused inside other ones. The one just mentioned includes default validation for the pages.

With those three placed together (that is, the required annotation, the **asp-validation-for** tag helper, and the **ValidationScriptsPartial** page), validation logic is created on the client-side that prevents the form from being submitted with invalid values. If you want to perform the validation on the server, you could use the built-in **TryValidateModel** method, passing the model to be validated.

DYNAMIC BEHAVIOR WITH PARTIAL PAGES

So far, you have built a board to display tasks and a way to create and edit them. Still, there is one major feature for a to-do application that needs adding—a way to move tasks across the board. You can start as simple as moving one way only—from to-do to doing, and from doing to done.

Until now, your task cards were built using tag helpers. However, tag helpers are rendered as **static** components and do not allow any dynamic behavior to be added during rendering. You could add tag helpers directly to your page, but you would have to repeat it for every board list. That is exactly where a major Razor Pages feature comes into play and that is **Partial Pages**. They allow you to create reusable page code snippets in smaller pieces. That way, you can share the base page dynamic utilities and still avoid duplicate code in your application.

This concludes the theoretical portion of this section. In the following section, you will put this into practice with an exercise.

EXERCISE 7.05: REFACTORING A TAG HELPER TO A PARTIAL PAGE WITH CUSTOM LOGIC

In this exercise, you will create a partial page to replace **KanbanCardTagHelper** and add some dynamic behavior to your task's cards, such as changing content based on custom logic. You will see how partial pages help in reducing duplicate code and make it more easily reusable. Perform the following steps to complete this exercise:

1. Inside the **Pages/Tasks** folder, create a new file called **_TaskItem.cshtml** with the following content:

   _TaskItem.cshtml

```
@model ToDoListApp.Models.ToDoTask

<form method="post">
    <div class="card">
        <div class="card-body p-2">
            <div class="card-title">
                @Model.Title
            </div>
            <a class="btn btn-primary btn-sm" href="/tasks/@Model.Id">View</a>
            @if (Model.Status == Models.ETaskStatus.ToDo)
            {
                <button type="submit" class="btn btn-warning btn-sm" href="@
Model.Id" asp-page-handler="StartTask" asp-route-id="@Model.Id">
                    Start
                </button>
```

   The complete code can be found here: https://packt.link/aUOcj.

   The **_TaskItem.cshtml** is basically a partial page that contains the **.cshtml** code of a card from the Kanban board.

2. Now, replace the code within the **Index.cshtml.cs** file with the following code that can read the saved tasks from the database and place the actions you created on the partial page:

   Index.cshtml.cs

```
using System;
using System.Collections.Generic;
using System.Linq;
using Microsoft.AspNetCore.Mvc;
using Microsoft.AspNetCore.Mvc.RazorPages;
using ToDoListApp.Data;
using ToDoListApp.Models;

namespace ToDoListApp.Pages
{
    public class IndexModel : PageModel
    {
        private readonly ToDoDbContext _context;

        public IndexModel(ToDoDbContext context)
```

   The complete code can be found here: https://packt.link/Tqgup.

This code creates handler methods for the three HTTP requests—a **GET request** and two **POST requests**. It also places the logic to be executed on these handlers. You will read values from the database with GET and save them back with POST.

3. Finally, update the **`Index.cshtml`** page with the following code to replace the use of tag helpers by the partial Razor page with your Kanban cards:

`Index.cshtml`

```
@page
@using ToDoListApp.Models

@model IndexModel
@{
    ViewData["Title"] = "MyToDos";
}

<div class="text-center">

    @if(TempData["SuccessMessage"] != null)
    {
        <div class="alert alert-success" role="alert">
            @TempData["SuccessMessage"]
        </div>
```

`The complete code can be found here:` https://packt.link/9SRsY.

Doing so, you'll notice how much duplicate code gets eliminated.

4. Now run the application with the following command:

```
dotnet run
```

5. Next click at the Create Task button and fill the form. After a Task is created, you'll see a confirmation message, as shown in *Figure 7.7*.

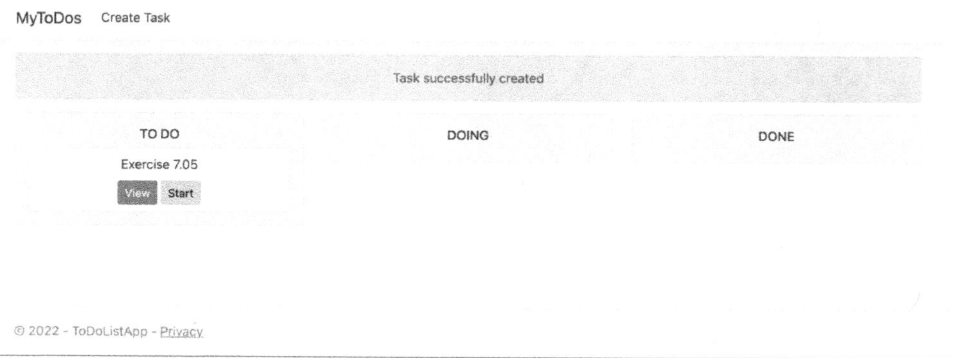

Figure 7.7: The Home screen after a Task creation

> **NOTE**
>
> If you have created some tasks in the previous screen, the screen display might be different on your system.

In this exercise, you created an almost fully functional to-do application in which you can create tasks and save them to the database, and even log your requests to see how long they take.

> **NOTE**
>
> You can find the code used for this exercise at https://packt.link/VVT4M.

Now, it is time to work on an enhanced feature through an activity.

ACTIVITY 7.01: CREATING A PAGE TO EDIT AN EXISTING TASK

Now it's time to enhance the previous exercise with a new and fundamental feature that is, to move tasks across the Kanban board. You must build this application using the concepts covered in this chapter such as model binding, tag helpers, partial pages, and DI.

To complete this activity, you need to add a page to edit the tasks. The following steps will help you complete this activity:

1. Create a new file called **`Edit.cshtml`** with the same form as **`Create.cshtml`**.
2. Change the route at the page directive to receive **`"/tasks/{id}"`**.
3. Create the code-behind file that loads a task by the **`OnGet ID`** from the **`DbContext`** schema. If the ID does not return a task, redirect it to the **`Create`** page.
4. On the Post form, recover the task from the database, update its values, send a success message, and redirect to the Index view afterward.

The output of a page is displayed as follows:

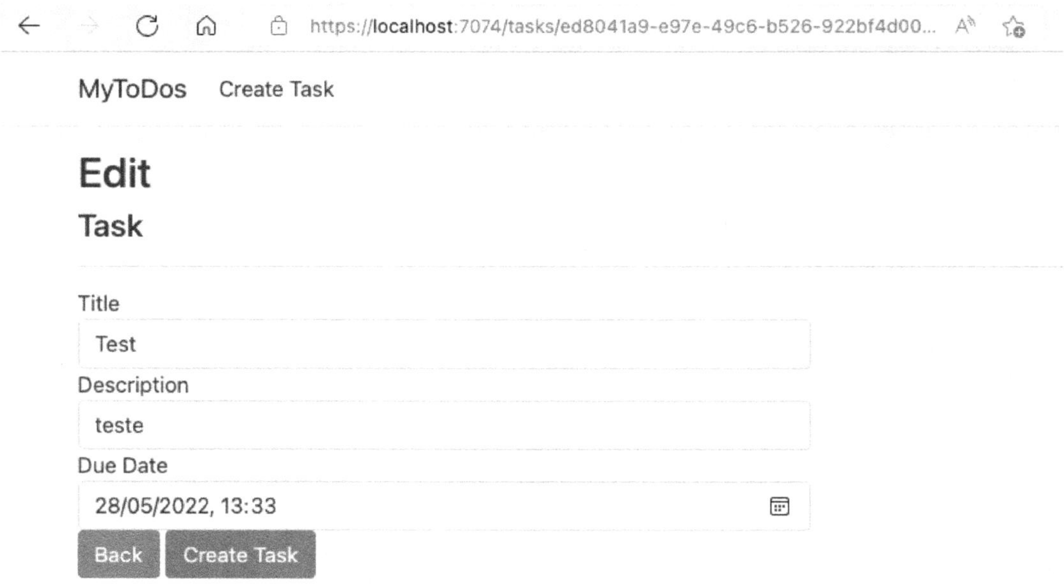

https://localhost:7074/tasks/ed8041a9-e97e-49c6-b526-922bf4d00...

MyToDos Create Task

Edit

Task

Title

Test

Description

teste

Due Date

28/05/2022, 13:33

Back Create Task

Figure 7.8: The Edit Task Page as output to the activity

> **NOTE**
>
> The solution to this activity can be found at https://packt.link/qclbF.

With the examples and activity so far, you now know how to develop pages with Razor. In the next section, you will learn how to work with a tool that has an even smaller scope of isolated and reusable logic called view components.

VIEW COMPONENTS

So far, you have seen two ways of creating reusable components to provide better maintenance and reduce the amount of code and that is tag helpers and partial pages. While a tag helper produces mainly static HTML code (as it translates a custom tag into an existing HTML tag with some content inside it), a **`partial page`** is a small Razor page inside another Razor page that shares the page data-binding mechanism and can perform some actions such as form submission. The only downside to **`partial pages`** is that the dynamic behavior relies on the page that contains it.

This section is about another tool that allows you to create reusable components that is, view components. **View components** are somewhat similar to partial pages, as they also allow you to provide dynamic functionality and have logic on the backend. However, they are even more powerful as they are self-contained. This self-containment allows them to be developed independently of the page and be fully testable on their own.

There are several requirements for creating view components, as follows:

- The custom component class must inherit from **`Microsoft.AspNetCore.Mvc.ViewComponent`**.
- It must either have the **`ViewComponent`** suffix in the class name or be decorated with the **`[ViewComponent]`** attribute.
- This class must implement either a **`IViewComponentResult Invoke()`** synchronous method or a **`Task<IViewComponentResult> InvokeAsync()`** asynchronous method (when you need to call async methods from within).
- The result of both previous methods is typically the **`View(model)`** method with the view component model as an argument. On the frontend, the default view filename should, by convention, be called **`Default.cshtml`**.
- For the view to be rendered, it must be located in either **`Pages/Components/{MY_COMPONENT_NAME}/Default.cshtml`** or **`/Views/Shared/Components/{MY_COMPONENT_NAME}/Default.cshtml`**.
- If not located in any of the preceding paths, the location of the view must be explicitly passed as an argument on the **`View`** method returned in the **`Invoke`** or **`InvokeAsync`** methods.

This concludes the theoretical portion of this section. In the following section, you will put this into practice with an exercise.

EXERCISE 7.06: CREATING A VIEW COMPONENT TO DISPLAY TASK STATISTICS

In this exercise, you will create a view component that allows you to display some statistics regarding delayed tasks on the navbar of the application. Working through this exercise, you will learn the basic syntax of view components and how to place them in Razor Pages. Perform the following steps to do so:

1. Under the root of the **ToDoListApp** project, create a new folder called **ViewComponents**.
2. Inside this folder, create a new class called **StatsViewComponent**:

```
namespace ToDoListApp.ViewComponents;

public class StatsViewComponent
{
}
```

3. Again, inside the **ViewComponents** folder, create a new class named **StatsViewModel** with two public **int** properties, named **Delayed** and **DueToday**:

```
namespace ToDoListApp.ViewComponents;

public class StatsViewModel
{
    public int Delayed { get; set; }
    public int DueToday { get; set; }
}
```

4. Edit the **StatsViewComponent** class to inherit from the **ViewComponent** class that is contained in the **Microsoft.AspNetCore.Mvc** namespace:

```
using Microsoft.AspNetCore.Mvc;

public class StatsViewComponent : ViewComponent
{
}
```

5. Inject **ToDoDbContext** via a constructor initializing a **private readonly** field:

```
public class StatsViewComponent : ViewComponent
{
    private readonly ToDoDbContext _context;
    public StatsViewComponent(ToDoDbContext context) => _context =
context;
}
```

   Place the proper **using** namespaces.

6. Create a method named **InvokeAsync** with the following signature and content:

   StatsViewComponent.cs

```
using ToDoListApp.Data;
using Microsoft.AspNetCore.Mvc;
using Microsoft.EntityFrameworkCore;
using System.Linq;

namespace ToDoListApp.ViewComponents;

public class StatsViewComponent : ViewComponent
{
    private readonly ToDoDbContext _context;
    public StatsViewComponent(ToDoDbContext context) => _context = context;

    public async Task<IViewComponentResult> InvokeAsync()
    {
        var delayedTasks = await _context.Tasks.Where(t =>
```

   The complete code can be found here: https://packt.link/jl2Ue.

   This method will use **ToDoDbContext** to query the database and retrieve the delayed tasks, as well as the ones that are due on the current day.

7. Now under the **Pages** folder, create a new folder called **Components**.

8. Under it make another folder called **Stats**.

9. Then, inside the **Stats** folder, create a new file called **default.cshtml** with the following content:

```
@model ToDoListApp.ViewComponents.StatsViewModel

<form class="form-inline my-2 my-lg-0">
    @{
         var delayedEmoji = Model.Delayed > 0 ? "😨" : "😀";
         var delayedClass = Model.Delayed > 0 ? "btn-warning" :
"btn-success";
```

```
        var dueClass = Model.DueToday > 0 ? "btn-warning" :
"btn-success";
    }
    <button type="button" class="btn @delayedClass my-2 my-sm-0">
        <span class="badge badge-light">@Model.Delayed</span> Delayed
Tasks @delayedEmoji
    </button>
     
    <button type="button" class="btn @dueClass my-2 my-sm-0">
        <span class="badge badge-light">@Model.DueToday</span> Tasks
Due Today ⏰
    </button>
</form>
```

The **default.cshtml** will contain the view part of the view component class. Here, you are basically creating a **.cshtml** file based on a model specified.

10. Finally, in **_Layout.cshtml** (under **Pages/Shared/**), add a call to the **ViewComponent** by adding the **<vc:stats></vc:stats>** tag inside your navbar. Replace the page code with the following:

_Layout.cshtml

```
<!DOCTYPE html>
<html lang="en">

<head>
    <meta charset="utf-8" />
    <meta name="viewport" content="width=device-width, initial-scale=1.0" />
    <title>@ViewData["Title"] - ToDoListApp</title>
    <link rel="stylesheet" href="~/lib/bootstrap/dist/css/bootstrap.min.css"
/>
    <link rel="stylesheet" href="~/css/site.css" asp-append-version="true" />
    <link rel="stylesheet" href="~/ToDoListApp.styles.css" asp-append-
version="true" />
</head>

<body>
    <header>
        <nav class="navbar navbar-expand-sm navbar-toggleable-sm navbar-light
bg-white border-bottom box-shadow mb-3">
```

The complete code can be found here: https://packt.link/DNUBC.

11. Run the application to see your navbar as shown in *Figure 7.8*:

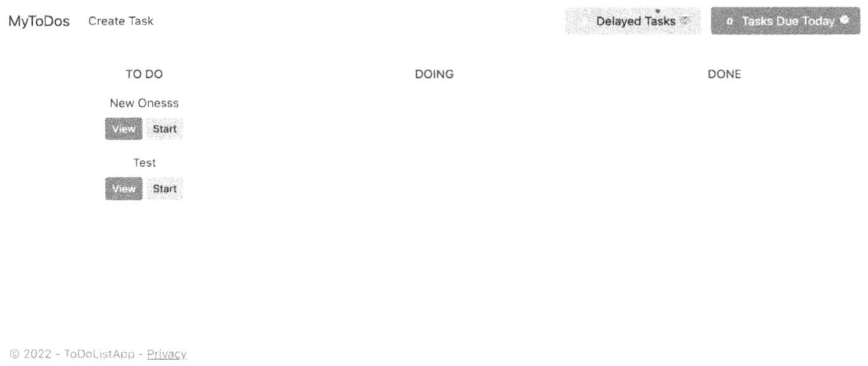

Figure 7.9: The Task stats view component

In this exercise, you created your first view component which is a task stat displayed right on your navbar. As you may have noticed, one efficient thing about view components that distinguishes them from partial pages is that they are independent of the page they are displayed on. You build both your frontend and backend all self-contained inside the component, with no external dependencies on the page.

> **NOTE**
>
> You can find the code used for this exercise at https://packt.link/j9eLW.

This exercise covered view components, which allow you to display some statistics regarding delayed tasks on the navbar of the application. With this knowledge, you will now complete an activity wherein you will work in a view component to show a log history.

ACTIVITY 7.02: WRITING A VIEW COMPONENT TO DISPLAY TASK LOG

As the final step of this chapter, this activity will be based on a common task in real-world applications—to have a log of user activities. In this case, you will write every change the user does to a field to the database and display it in a view. To do so, you would need to use a view component.

The following steps will help you complete this activity:

1. Create a new class under the **Models** folder named **ActivityLog**. This class should have the following properties: **Guid Id**, **String EntityId**, **DateTime Timestamp**, **String Property**, **String OldValue**, and **String NewValue**.
2. Create a new **DbSet<ActivityLog>** property for this model under **ToDoDbContext**.
3. Under your **DbContext**, create a method to generate activity logs for the modified properties of **Entries** under the Entity Framework's **ChangeTracker** with **EntityState.Modified**.
4. Override **SaveChangesAsync()** in **DbContext**, by adding the generated logs to **DbSet** right before calling the **base** method.
5. Create a new Entity Framework Core migration and update the database to support this migration.
6. Create the **ViewComponent** class, which should load all logs for a given **taskId** passed on the invocation and return them to the **ViewComponent**.
7. Create the **ViewComponent** view, which should take a collection of **ActivityLog** as a model and display them in a Bootstrap table, if any exists. If no logs are recorded, show an alert saying that no logs are available.
8. Add the view component to the **Edit** page, passing the **taskId** property.

9. Run the application and check the final output by opening a task's details. You will see a box on the right with your activity logs or a message with no logs, if there are no activity logs recorded, for that task yet.

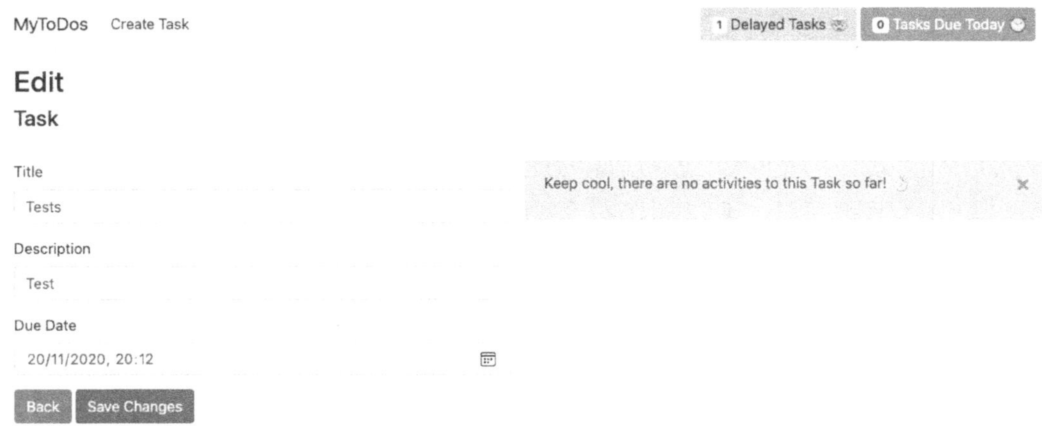

Figure 7.10: The Activity log being displayed with no logs

In this activity, you were able to create an isolated view component with completely new functionality that's decoupled from a page, allowing it to work on a single feature at a time.

> **NOTE**
>
> The solution to this activity can be found at https://packt.link/qclbF.

SUMMARY

In this chapter, you learned the foundations of building a modern web application with C# and Razor Pages. You focused on important concepts at the beginning of the chapter, such as middleware, logging, DI, and configuration. Next, you used Razor Pages to create CRUD models along with Entity Framework and used some more advanced features, such as custom tag helpers, partial pages, and view components, which enable you to create more easily maintainable application features.

Finally, you saw how ASP.NET model binding works so that there can be a two-way data binding between the client and the server. By now, you should have an effective foundation for building modern web applications with ASP.NET and Razor Pages on your own.

Over the next two chapters, you will learn about building and communicating with APIs.

8

CREATING AND USING WEB API CLIENTS

OVERVIEW

In this chapter, you will step into the world of HTTP practice by making calls to Web APIs. You will interact with Web APIs in a variety of ways using a web browser, your own HTTP client, and NuGet packages. You will learn the basics of security involved in Web APIs, use PayPal to make sandbox payments, and explore cloud services such as Azure Text Analytics and Azure Blob storage.

By the end of this chapter, you will be able to read HTTP requests and response messages, make calls to any Web API, and create your own HTTP client to simplify your work with complex APIs. You will also be able to dissect and learn both incoming HTTP requests and outgoing HTTP responses in any form and use development tools in the Chrome browser to inspect traffic moving back and forth when browsing your favorite websites.

INTRODUCTION

The **World Wide Web** (**WWW**) (or just the **web**) is a big store of all sorts of documents (XML, JSON, HTML, MP3, JPG, etc.) accessible through **Uniform Resource Locators** (**URLs**). A document in the context of the web is often called a **resource**. Some resources do not change. They are stored somewhere, and with every request, the same resource will be returned. Such resources are called **static**. Other resources are **dynamic**, which means they will be generated on demand.

Communication on the web happens through protocols. In the context of retrieving documents, you use **Hypertext Transfer Protocol** (**HTTP**). **Hypertext** is a special text that holds a link to a resource on the web. Clicking on it opens the resource it points to. HTTP is based on a **client-server architecture**. In simple terms, a **client** sends requests, and the **server** responds. An example of this in practice is the communication between a browser (client) and a website (hosted on a server). Usually, a single server serves many clients:

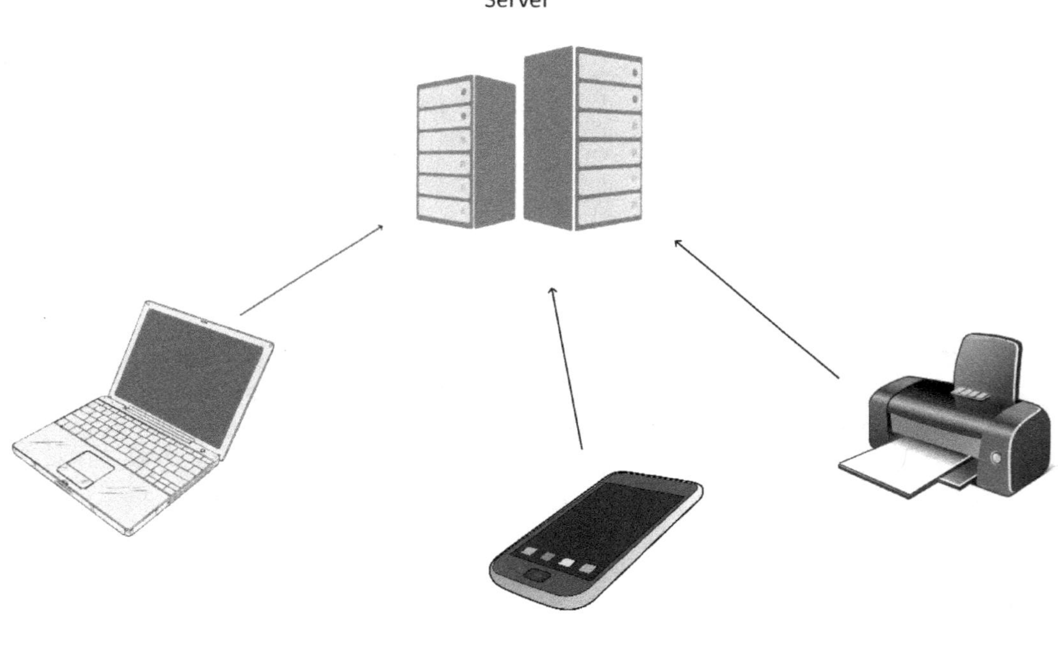

Figure 8.1: Client-server architecture

When you navigate to a website, you send an **`HTTP GET`** request, and the server responds by displaying the relevant site content in the browser. **`GET`** is an **HTTP verb**—a method identifying how a request should be treated. Common HTTP verbs are the following:

- **`GET`**: Get a resource.
- **`POST`**: Create a resource or send a complex query.
- **`PUT`**: Update all resource fields.
- **`PATCH`**: Update a single field.
- **`DELETE`**: Remove a resource.

BROWSER

A modern browser is more than just a tool to access content on the internet. It includes tools to dissect elements of a website, inspect traffic, and even execute code. This side of a browser is called **developer tools**. The exact key binds may vary but pressing **`F12`** or **`Control + Shift + I`** should call up the Developer Tools tab. Perform the following steps to get to know it better:

1. Open Google Chrome or any other browser.
2. Navigate to google.com. Press the keys **`Control + Shift + I`**.

3. Go to **Network** (**1**). The following window should be displayed:

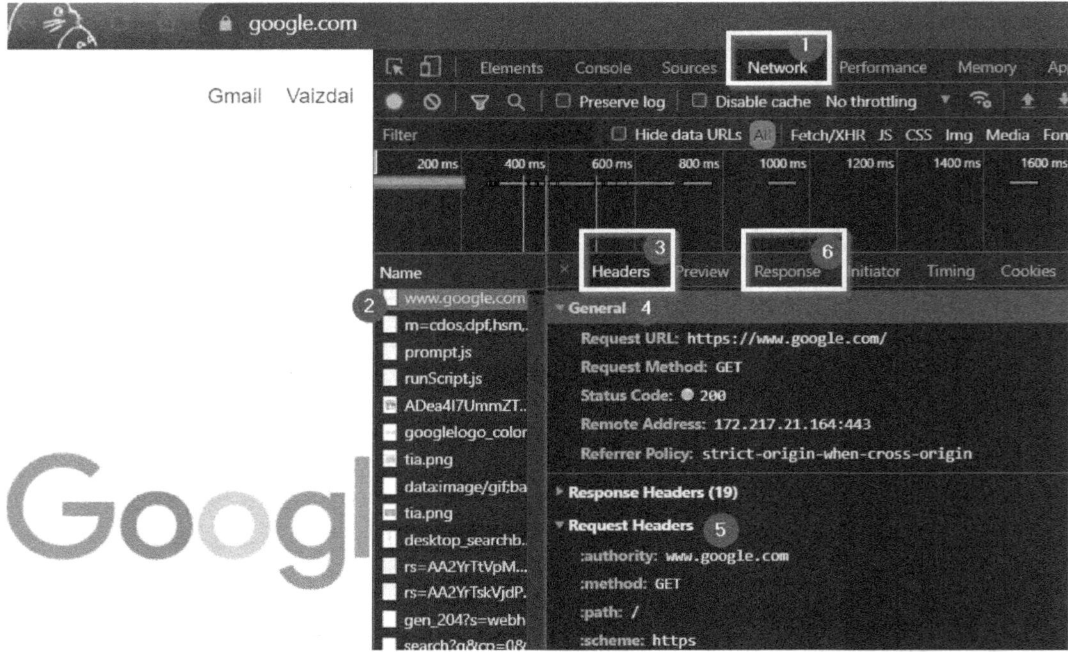

Figure 8.2: Chrome with developer tools open with google.com loaded

4. Select the first entry, www.google.com (**2**).
5. Click **Headers** (**3**).
6. In the **General** (**4**) section, you can observe the effects when you navigated to google.com. The first thing that happened was **HTTP GET** request was sent to https://www.google.com/.
7. In the **Request Headers** section (**5**), you can see the metadata sent with the request.
8. To see how Google responded, click the **Response** section (**6**).

This flow is called the client-server architecture, and the following applies:

- The client is the Chrome browser that sends a request to google.com.
- The server is a machine(s) hosting google.com that responds with google.com website contents.

Web API

An **Application Programming Interface** (**API**) is an interface through which you can call some functionality using code. It could be a class or an interface in C#, or a browser (you can interact with it through code provided by its own interface), but in the context of HTTP, it is a web service. A **web service** is an API hosted on a remote machine that is accessible through HTTP. An access point used to invoke a single piece of functionality on a **Web API** is called an **endpoint**. The most commonly used Web API type is RESTful.

RESTful API

A **Representational State Transfer** (**REST**) API is an API built on the following six principles. Four principles are a given whatever framework you use implementing a RESTful API, and, as a client, they should be expected:

- **Client-server**: A connection is made between a client and server. The client sends a request in order to get a response from a server.
- **Stateless**: The server will be able to process requests regardless of prior requests. This means that each request should contain all the information, rather than relying on a server to remember what happened before.
- **Cacheable**: The ability to specify which requests can be cached using HTTP methods or headers.
- **Code on demand** (optional): REST allows scripts to be downloaded and executed on the client side. Back when the internet was made mostly of static pages, this was useful, but nowadays it is either not needed or is seen as a security risk.

However, the other two principles (Client-server and Stateless) depend on you, and thus you will want to pay more attention to them. A **layered system** is a system made of layers, and each layer communicates only with the layer directly below it. A typical example of this is a three-tier architecture, where you separate presentation, business logic, and the data store. From a practical point of view, this means that a RESTful API (business logic layer) should not send HTML as a response because the responsibility for rendering output lies with the client (the presentation layer).

The last principle is called a **uniform interface**. It defines a set of rules for an API:

- Identification of resources:

  Some examples of these are get all instances of a resource (**/resource**), create a resource (**/resource**), get a single resource (**/resource/id**), and get all instances of a subresource in a resource (**/resource/subresource/**).

- Manipulation of resources through these representations:

  Resources are manipulated using HTTP verbs representing **Create, Read, Update, and Delete** (**CRUD**)—**GET**, **UPDATE**, **PUT**, **PATCH**, **DELETE**.

- Self-descriptive messages:

  A response that includes all the required information, without any extra documentation, and indicates how the message should be processed (headers, mime type, etc.).

- **Hypermedia as the engine of application state** (**HATEOAS**):

  Hyperlinks are included in response to all the related resources so that you can navigate to them. This guideline is usually ignored.

REST is not the same as HTTP. REST is a set of guidelines, while HTTP is a protocol. The two might be confused because HTTP constraints heavily overlap with REST constraints (methods, headers, etc.). However, a RESTful API does not have to use HTTP to be RESTful, and at the same time HTTP can violate REST constraints by using a session or query parameters to provide actions to perform. A RESTful API can work with both XML and JSON data formats. However, almost all scenarios involve JSON.

POSTMAN

Postman is one of the most popular tools used for testing different kinds of Web APIs. It is easy to set up and use. Postman, just like a browser, acts as an HTTP client. In order to download Postman, go to https://www.postman.com/. You will need to sign up and then download the installer. Once you have installed Postman, perform the following steps:

1. Open Postman.
2. Create your workspace by clicking **Workspaces** and then click on **Create Workspace**.
3. In the new window, go to the **Collections** tab (**2**) and click the **Create new Collection** (**+**) button (**3**).

4. Create a **`New Collection`** (**4**).
5. Click on **`Add a request`** (**5**):

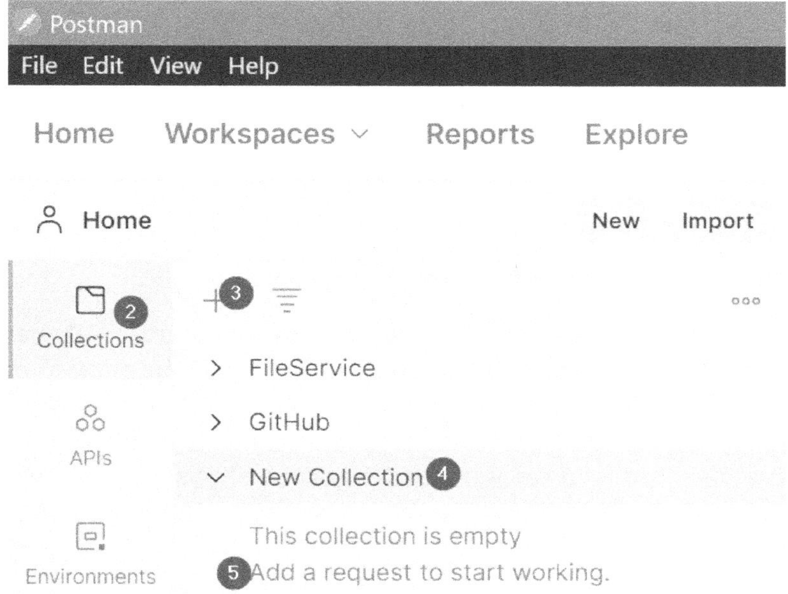

Figure 8.3: New Postman collection without requests

A new request window will open.

6. Click the edit symbol beside **`New Request`** and name the new request **`Users`** (**6**).
7. Select the **`GET`** HTTP verb and copy-paste the URL https://api.github.com/users/github-user (**7**).

> **NOTE**
>
> Here, and in all places that follow, replace **`github-user`** with your own GitHub username.

8. Click the **`Send`** button (**8**).

9. Now scroll down to see the response result returned (**9**):

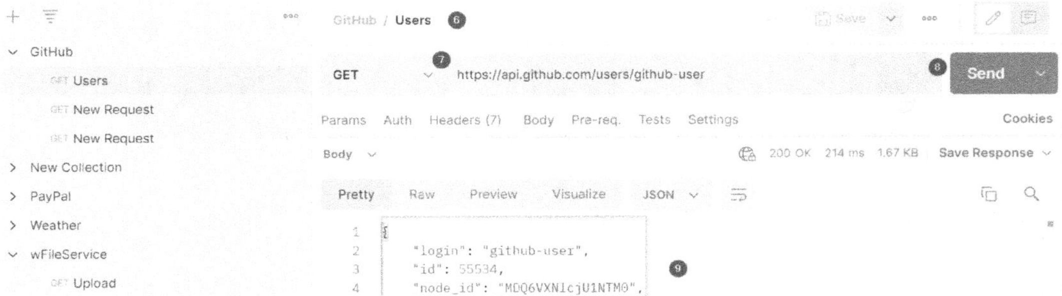

Figure 8.4: GET GitHub user request in Postman

Postman is superior to a browser when it comes to acting as an **HTTP client**. It is focused on forming **HTTP requests** and displays response information compactly, offering multiple output formats. In Postman, you can use multiple environments, set up pre-and post-conditions for requests, automated calls, and a lot more, but going through those advanced features is beyond the scope of this chapter. For now, it is enough to learn that Postman is a go-to tool for testing Web APIs by hand.

CLIENT

REST requires communication between a client and a server. In the previous examples, the client role was taken by either a browser or Postman. However, neither a browser nor Postman can replace a client in your code. Instead, you'll need to create an HTTP request using C#.

Popular Web APIs often have a client created for you (in most common languages as well). The purpose of a Web API client is to simplify interactions with the underlying API. For example, instead of sending a **`DELETE`** request on an endpoint that does not support it and getting the response **`Method Not Allowed`**, you won't even have such an option on a custom client.

OCTOKIT

Octokit is a GitHub API client. It exposes a C# class through which you can pass objects to make calls to GitHub. The benefit of such a client is that you don't need to worry about which headers to pass or how to name things so that they are properly serialized. An API client handles all that for you.

You can install the Octokit client in your project by running the following command in the VS Code terminal or command prompt:

```
dotnet add package Octokit
```

Once you have the Octokit client installed, you can use it to create a GitHub client, as follows:

```
var github = new GitHubClient(new ProductHeaderValue("Packt"));
```

In the preceding snippet, you needed a new **ProductHeaderValue** because GitHub expects a **UserAgent** header. As mentioned earlier, custom HTTP clients prevent a mistake from happening before you can even make a request. In this case, not providing a **UserAgent** header (through **ProductHeaderValue**) is not an option.

To see whether the client works, try to get information on the username **github-user**:

```
const string username = "github-user";
var user = await github.User.Get(username);
```

> **NOTE**
>
> In GitHub, **github-user** is displayed as **Almantask**. It is better to change it to your individual GitHub username for the code to work.

To print the date when the user was created, type the following code:

```
Console.WriteLine($"{username} created profile at {user.CreatedAt}");
```

You will see the following output:

```
github-user created profile at 2018-06-22 07:51:56 +00:00
```

Every method available on the **GitHub API** is also available on **GitHub client Octokit**. You don't need to worry about the endpoint, mandatory headers, a response, or the request format; it is all defined by the strongly typed client.

> **NOTE**
>
> You can find the code used for this example at https://packt.link/DK2n2.

API Key

With many public free APIs, you may be faced with concerns such as the following:

- How can you control an overwhelming number of requests?
- At what point should which client be charged?

If all these public APIs offered only anonymous access, you would not be able to identify the clients or determine how many calls each of them has made. An **API key** serves as the most basic means of authentication (identifying the client) and authorization (granting them access to do something with an API). Simply put, an API key allows you to call an API. Without it, you would have little to no access to an API.

To help you grasp the use of API keys better, the next section will look at a Web API that requires one, that is, Azure Text Analytics.

Azure Text Analytics

Azure Text Analytics is an Azure API used to analyze text in the following ways:

- Identify named entities (people, events, organizations)
- Interpret the mood of the text (positive, negative, neutral)
- Produce a summary of a document or highlight key phrases
- Process unstructured medical data, such as recognizing people, classifying diagnoses, and so on

In order to demonstrate the Azure Text Analytics API, you will focus on **sentimental analysis**. This is the process of evaluating text according to a positive, negative, or neutral confidence score:

- The score of 1, which means 100%, is the probability that the prediction (negative, positive, neutral) is correct.

- The score of 0, which means 0%, is an impossible prediction.

> **NOTE**
>
> Using Azure Text Analytics is free until you analyze more than 5,000 words per 30 days.

Before you begin coding, you'll need to set up Azure Text Analytics on the Azure cloud. After all, you need both an endpoint and an API key to make a call to this API.

> **NOTE**
>
> Make sure you have set up an Azure subscription. If you don't have one, go to https://azure.microsoft.com/en-gb/free/search and follow the instructions there to create a **free** subscription. An Azure free trial offers many services for free. Some of those services will remain free even after a year. A student subscription is an option for getting Azure credits and free services for a longer period. A credit or debit card is required to create an Azure subscription; however, you won't be charged unless you exceed the given Azure credits of the free service limitations.

One way in which Azure Text Analytics could be used to sort positive and negative feedback is by determining whether what you wrote sounds passive-aggressive or friendly. To see this in action, follow the steps to create a small application that analyzes any text you input into a console:

1. First, go to https://portal.azure.com/#create/Microsoft.CognitiveServicesTextAnalytics.

2. Click **Continue to create your resource** without using any additional features:

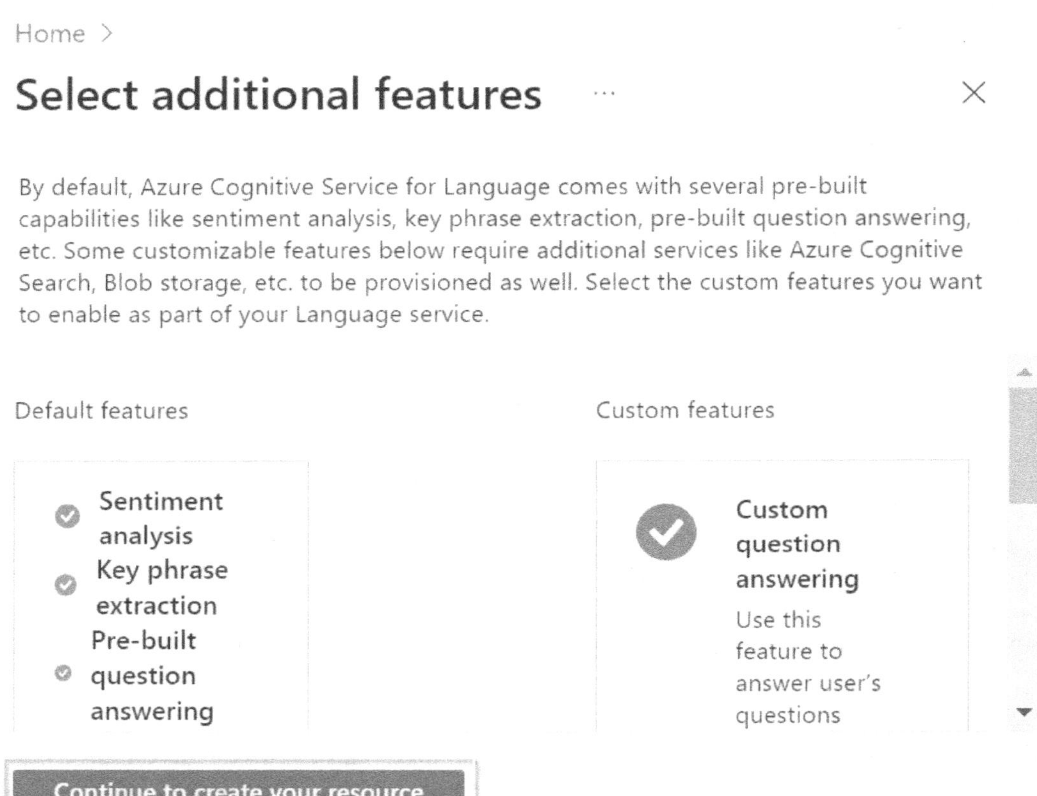

Figure 8.5: Azure Text Analytics resource creation

3. In the Create text analytics window, click the **Basics** tab. This is the first tab opened at the start of the creation of a new resource.

4. Select an option in the **Subscription** and **Resource group** fields:

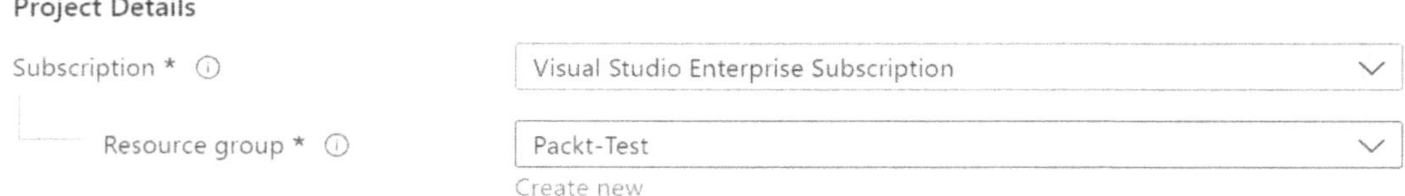

Figure 8.6: Entering the project details for new resource creation

5. Then, select the region, for example, **North Europe**.
6. Enter the name, for example, **Packt-Test**.
7. After that, select the **Free F0** pricing tier and click the **Review + create** button:

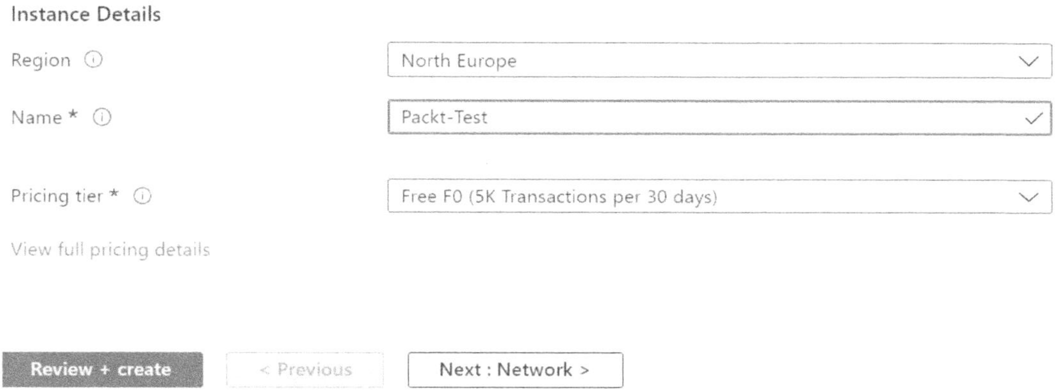

Instance Details

Region ⓘ North Europe

Name * ⓘ Packt-Test

Pricing tier * ⓘ Free F0 (5K Transactions per 30 days)

View full pricing details

Review + create | < Previous | Next : Network >

Figure 8.7: Azure Text Analytics pricing tier

A new window gets displayed confirming your input.

8. Click the **Create** option. The Text Analytics API will start deploying. After the deployment of the service is done, a new window will open saying **Your deployment is complete**.
9. Click on the **Go to resource** button:

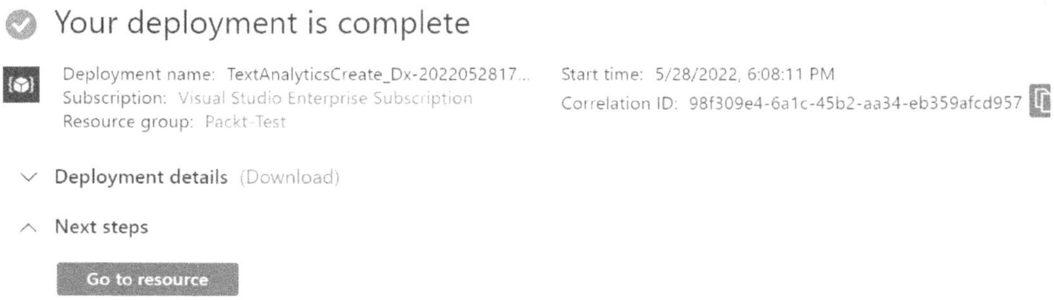

Your deployment is complete

Deployment name: TextAnalyticsCreate_Dx-2022052817...
Subscription: Visual Studio Enterprise Subscription
Resource group: Packt-Test

Start time: 5/28/2022, 6:08:11 PM
Correlation ID: 98f309e4-6a1c-45b2-aa34-eb359afcd957

Deployment details (Download)

Next steps

Go to resource

Figure 8.8: The Text Analytics API showing the deployment as complete

The Text Analytics resource window gets displayed.

10. Click the **Keys and Endpoint** option. You will see the **Endpoint** option along with **KEY 1** and **KEY 2** to make calls to this API. You can choose from either of the keys:

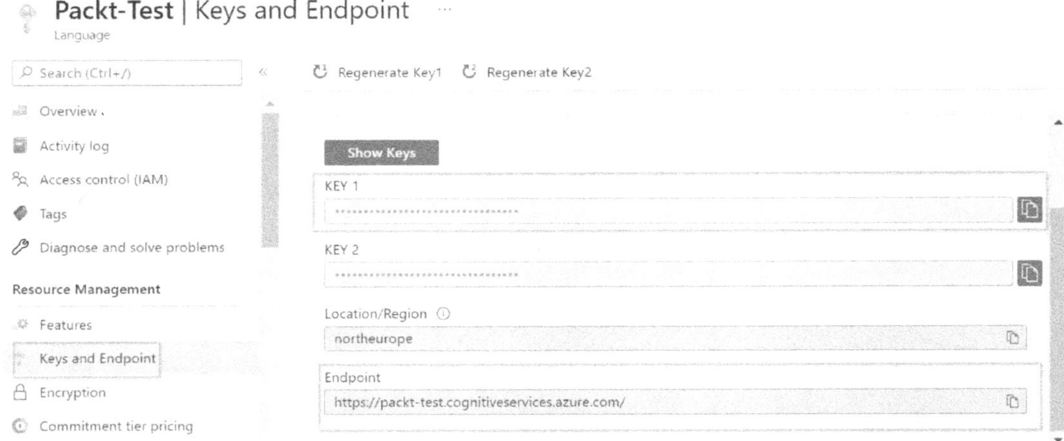

Figure 8.9: Azure Text Analytics quick start window with API key hyperlink

11. Keep track of **KEY 1** (an API key). An API key is a secret and should not be exposed in plain text. You will once again be using the environment variables to store it.

    Create an environment variable with key and value pair. The value will be the endpoint API key required to connect to Azure Text Analytics. To help identify the missing environment variable, use a helper class. The **GetOrThrow** method will get a user environment variable, and if it doesn't exist, will throw an exception:

```
public static class EnvironmentVariable
{
    public static string GetOrThrow(string environmentVariable)
    {
        var variable = Environment.
GetEnvironmentVariable(environmentVariable, EnvironmentVariableTarget.
User);
        if (string.IsNullOrWhiteSpace(variable))
        {
            throw new ArgumentException($"Environment variable
{environmentVariable} not found.");
        }
```

```
                return variable;
            }
        }
```

12. Keep track of the **Endpoint** option. You will use it in the upcoming exercise to call the API you have just deployed.

This section helped you to set up Azure Text Analytics on the Azure cloud, in addition to setting both an endpoint and an API key to make a call to the API. In the following exercise, you will be using the Azure Text Analytics client to make calls to the API.

EXERCISE 8.01: PERFORMING SENTIMENTAL TEXT ANALYSIS ON ANY TEXT

Azure Text Analytics is just another REST API. Once again, you make HTTP calls to it and get a response. This time, you will send over a text to get its sentimental analysis. Do another practice run of using a strongly typed client and make calls to a RESTful API from C#.

Using a recently deployed Azure Text Analytics service (**Pack-Test**, in this case), perform sentimental analysis on any text you want. Perform the following steps to complete this exercise:

1. Install the **Azure.AI.TextAnalytics** NuGet package to get an Azure Text Analytics API client as follows:

```
dotnet add package Azure.AI.TextAnalytics
```

2. Add the **TextAnalysisApiKey** environmental variable.
3. Then add the **TextAnalysisEndpoint** environmental variable.
4. Create a **Demo** class and add references to the two environmental variables that you have recently added:

```
public class Demo
{
    private static string TextAnalysisApiKey { get; } =
EnvironmentVariable.GetOrThrow("TextAnalysisApiKey");
    private static string TextAnalysisEndpoint { get; } =
EnvironmentVariable.GetOrThrow("TextAnalysisEndpoint");
```

These properties are used to hide sensitive values of the API key and endpoint.

5. Create a new **BuildClient** method to build an API client:

```
static TextAnalyticsClient BuildClient()
{
    var credentials = new AzureKeyCredential(TextAnalysisApiKey);
    var endpoint = new Uri(TextAnalysisEndpoint);
    var client = new TextAnalyticsClient(endpoint, credentials);

    return client;
}
```

The API client requires both a base URL—a kind of **Unified Resource Identifier** (**URI**)—and an API key to operate, both of which are passed to it during initialization.

6. Using the client, create the **PerformSentimentalAnalysis** method to analyze the text:

```
private static async Task<DocumentSentiment>
PerformSentimentalAnalysis(TextAnalyticsClient client, string text)
{
    var options = new AnalyzeSentimentOptions { IncludeOpinionMining
= true };
    DocumentSentiment documentSentiment = await client.
AnalyzeSentimentAsync(text, options: options);

    return documentSentiment;
}
```

Here, you are using the configuration object **AnalyzeSentimentOptions** to extract targets and opinions on them. The client has both the **AnalyzeSentimentAsync** and **AnalyzeSentiment** methods. For public client libraries, exposing both async and non-async versions of the same method is a very common scenario. After all, not everyone will be comfortable with an async API. However, when making calls to another machine (DB, API, and similar) it's best to use an async API. This is because an async call will not block the thread on which the call is made while it is waiting for a response from an API.

7. Now create a **DisplaySentenceSymmary** function to display the sentence's overall evaluation:

```
private static void DisplaySentenceSummary(SentenceSentiment
sentence)
{
    Console.WriteLine($"Text: \"{sentence.Text}\"");
    Console.WriteLine($"Sentence sentiment: {sentence.Sentiment}");
    Console.WriteLine($"Positive score: {sentence.ConfidenceScores.
Positive:0.00}");
    Console.WriteLine($"Negative score: {sentence.ConfidenceScores.
Negative:0.00}");
    Console.WriteLine($"Neutral score: {sentence.ConfidenceScores.
Neutral:0.00}{Environment.NewLine}");
}
```

8. Create a **DisplaySentenceOpinions** function to display the message **Opinions** for every target in a sentence:

```
private static void DisplaySentenceOpinions(SentenceSentiment
sentence)
{
    if (sentence.Opinions.Any())
    {
        Console.WriteLine("Opinions: ");
        foreach (var sentenceOpinion in sentence.Opinions)
        {
            Console.Write($"{sentenceOpinion.Target.Text}");
            var assessments = sentenceOpinion
                .Assessments
                .Select(a => a.Text);
            Console.WriteLine($" is {string.Join(',',
assessments)}");
            Console.WriteLine();
        }
    }
}
```

The target of a sentence is a subject that has an opinion (grammatical modifier) applied to it. For example, with the sentence, **a beautiful day**, **day** would be a target and **beautiful** an opinion.

9. To perform a sentimental analysis on text typed in a console, create a **SentimentAnalysisExample** method:

```
static async Task SentimentAnalysisExample(TextAnalyticsClient
client, string text)
{
    DocumentSentiment documentSentiment = await
PerformSentimentalAnalysis(client, text);
    Console.WriteLine($"Document sentiment: {documentSentiment.
Sentiment}\n");

    foreach (var sentence in documentSentiment.Sentences)
    {
        DisplaySentenceSummary(sentence);
        DisplaySentenceOpinions(sentence);
    }
}
```

The analysis text, in the preceding code snippet, evaluates the overall text's sentiment and then breaks it down into sentences, evaluating each.

10. To demonstrate how your code works, create a static **Demo.Run** method:

```
public static Task Run()
{
    var client = BuildClient();
    string text = "Today is a great day. " +
                       "I had a wonderful dinner with my family!";
    return SentimentAnalysisExample(client, text);
}
```

With the environment variable set correctly, the following output should be displayed:

```
Document sentiment: Positive

Text: "Today is a great day."
Sentence sentiment: Positive
Positive score: 1,00
Negative score: 0,00
Neutral score: 0,00
```

```
Text: "I had a wonderful dinner with my family!"
Sentence sentiment: Positive
Positive score: 1,00
Negative score: 0,00
Neutral score: 0,00

Opinions:
dinner is wonderful
```

You did not hardcode the value of an API key here because an API key, exposed publicly, poses a risk of being used not the way it was intended to. If stolen, it could have disastrous consequences (for example, being overused, creating a false resource, leaking data, deleting data, etc.). That is why when dealing with secrets, use the minimal possible countermeasures, that is, environmental variables.

Another benefit of environment variables is the ability to have a different value in different environments (local, integration, system test, production, etc.). Different environments often use different resources. So, pointing to those resources through environment variables will not require any changes to the code.

In order to run this exercise, go to https://packt.link/GR27A and comment all lines within the **`static void Main(string[] args)`** body, except **`await Exercises.Exercise01.Demo.Run();`**. Similarly, uncomment the respective exercises'/examples'/activities' code lines in **`Program.cs`** before executing each of them.

> **NOTE**
>
> You can find the code used for this exercise at https://packt.link/y1Bqy.

This exercise is just one of the many in which you consumed a public Web API. Azure is full of services like this. Calling an API using a strongly typed client is simple; however, not all APIs have one. In the next section, you will learn how to create your own Web API client.

YOUR OWN CLIENT

So far, you've only used a premade client to consume a Web API. However, for less popular APIs, there will not be any client for you to use. In those cases, you will have to make HTTP calls yourself. In .NET, the way of making calls has evolved quite a lot. If you don't want any third-party libraries, you can use the **HttpClient** class.

HTTPCLIENT

In this section, you'll repeat the GitHub **Users** example (from the *Postman* section), but this time using **HttpClient**. The flow for this is quite simple and is described for you in detail in the following example:

1. Within the **GitHttp** static class, create the **GetUser** method:

```
public static async Task GetUser()
```

2. Within the **GitExamples** method, first, create a client:

```
client = new HttpClient { BaseAddress = new Uri("https://api.github.
com") };
client.DefaultRequestHeaders.Add("User-Agent", "Packt");
```

Creating a client almost always involves specifying a specific base URL. Often, Web APIs require mandatory headers to be passed, or else they will invalidate the request (**400 Bad Request**). For GitHub, you need to send the **User-Agent** header identifying the client that calls the API. Adding the **Packt** user agent header to default headers will send that header with every request to the client.

3. You then create a request as follows:

```
const string username = "github-user"; //replace with your own
var request = new HttpRequestMessage(HttpMethod.Get, new Uri($"users/
{username}", UriKind.Relative));
```

Remember to replace **github-user** with your own GitHub username. Here, you've specified that you want to create a **GET** request. You did not specify a full path, but rather only the endpoint you want to hit; therefore, you had to flag **UriKind** as **Relative**.

4. Next, send a request using the client:

```
var response = await client.SendAsync(request);
```

There is only an async version of sending an HTTP request message, so you need to wait for it. The result of sending **HttpRequestMessage** is **HttpResponseMessage**.

5. Then, deserialize the content to a usable object as follows:

```
var content = await response.Content.ReadAsStringAsync();
var user = JsonConvert.DeserializeObject<User>(content);
```

Deserializing is the act of converting a structured text such as JSON into in-memory objects. For this, you need to convert the content to a string and then deserialize it. You could use a user model from Octokit NuGet. Since you are already making custom calls, you might as well use a custom model. For the bare minimum (only the fields you use), your model could look like this:

```
public class User
{
    public string Name { get; set; }
    [JsonProperty("created_at")]
    public DateTime CreatedAt { get; set; }
}
```

The line **[JsonProperty("created_at")]**, above **public DateTime CreatedAt { get; set; }**, binds the JSON field to the C# property. This binding is needed because the names don't match.

If you want to create your own client (for making GitHub calls), it's your responsibility to expose all data that the API returns and not just the data you may need for a particular scenario by letting the consumer choose.

6. Use the message from a previous call from Postman to get the GitHub user response body to generate models to deserialize to. In this case, the response message is as follows (message truncated for clarity):

```
{
   "login":"github-user",
   "id":40486932,
   "node_id":"MDQ6VXNlcjQwNDg2OTMy",
   "name":"Kaisinel",
   "created_at":"2018-06-22T07:51:56Z",
   "updated_at":"2021-08-12T14:55:29Z"
}
```

There are many tools available that can convert JSON to the C# model.

7. In this case, use https://json2csharp.com/ to convert JSON to the C# model code.
8. Copy the response (**`GET github/user`**) and go to https://json2csharp.com/.
9. Paste the response into the textbox on the left and click the **`Convert`** button:

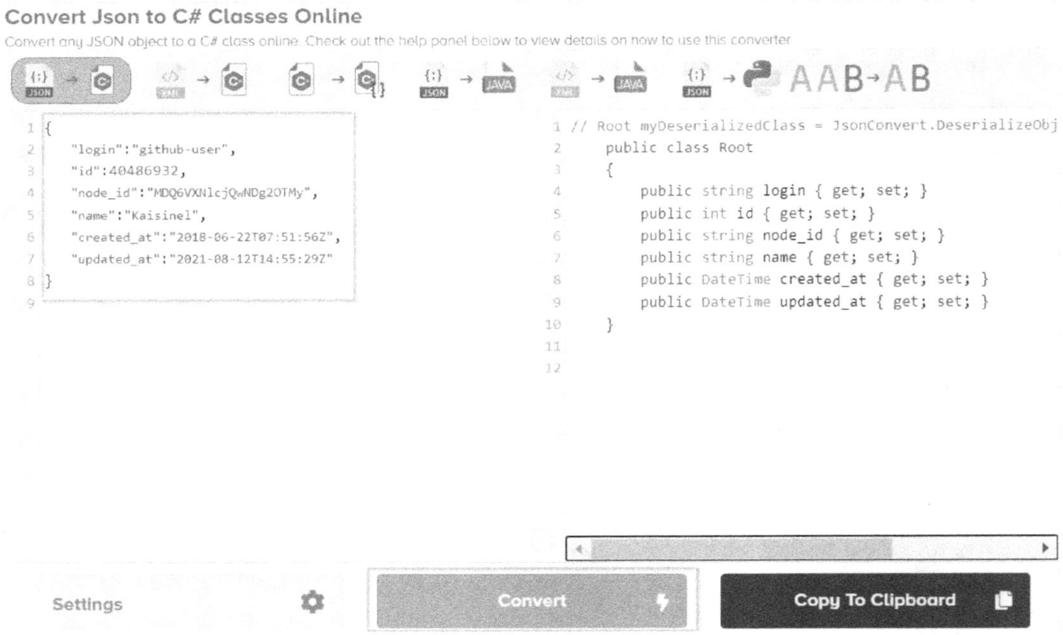

Figure 8.10: Converting JSON to the C# model code

The left side displays a model for the JSON, while the right side displays the code (C# class) that is generated from JSON.

10. Copy the content on the right and paste it into your code:

```
public class Root
{
    public string login { get; set; }
    public int id { get; set; }
    public string node_id { get; set; }
    public string name { get; set; }
    public DateTime created_at { get; set; }
    public DateTime updated_at { get; set; }
}
```

This is your model. Observe in the preceding code that **Root** is an unreadable class name. This is because the converter didn't have a way to know what class JSON represents. The **Root** class represents a user; therefore, rename it **User**.

Lastly, the converter was probably created prior to .NET 5, which is why it didn't have a feature for records. A record is a great class for serialization purposes and a great candidate for a **data transfer object** (**DTO**). A DTO is a class that has no logic but simply data, and sometimes attributes for binding serialization. The benefits you get are the following:

- Value equality
- **ToString** will return properties and their values
- The ability to define them with a less verbose syntax

So, use a record for defining DTOs in your applications whenever possible.

11. Rename the (**Root** to **User**) and change the type from **class** to **record**. The code line looks like this with no changes needed to the properties:

```
public record User
```

12. Finally, run the following line of code:

```
Console.WriteLine($"{user.Name} created profile at {user.CreatedAt}");
```

The output gets displayed as follows:

```
Kaisinel created profile at 2018-06-22 07:51:56
```

In order to run this exercise, go to https://packt.link/GR27A and comment all lines within the **static voidMain(string[] args)** body, except **await Examples.GitHttp.Demo.Run();**. Similarly, uncomment the respective exercises'/examples'/activities' code lines in **Program.cs** before execution.

> **NOTE**
>
> You can find the code used for this example at https://packt.link/UPxmW.

Now that you have seen the benefits of using the **HttpClient** class in lieu of third-party libraries, you can now explore the **IDisposable** pattern in the following section.

HTTPCLIENT AND IDISPOSABLE

HttpClient implements the **IDisposable** pattern. In general, right after you are done using an object that implements **IDisposable**, you should clean up and call the **Dispose** method or wrap the calls within a **using** block. However, **HttpClient** is special in that you should not frequently create and dispose of it all over again. The problem with disposing and re-initializing **HttpClient** is that **HttpClient** manages connections it makes to other APIs and disposing of **HttpClient** does not properly close those connections (or sockets).

The most dangerous part about that is that you will not notice any difference in developing your application locally, due to the massive number of connections available. However, when deploying an application to a live environment, you risk running out of free socket connections. Once again, avoid calling a **Dispose** method and reinitializing **HttpClient**. If you must, use **HttpClientFactory**. Not only does **HttpClientFactory** manage the lifetime of socket connections by managing **HttpClientMessageHandler** (the component responsible for sending the HTTP request and receiving the response) but it also provides logging capability, allows centralized management of clients' configuration, supports injecting middleware to clients, etc. The mentioned benefits are important if you use **HttpClient** in an enterprise setting. You can learn more about **HttpClientFactory** in *Chapter 9, Creating API Services*.

Ideally, you should have one static **HttpClient**, which you can reuse for calls to Web APIs throughout your application. However, you should not have a single **HttpClient** for everything. The point about not disposing of **HttpClient** and having a static one is not a hard rule. If you call many different APIs, they will have their own base addresses, mandatory headers, and so on. Having a single object for all is not a viable scenario.

The requests you've handled so far were publicly accessible and did not have security. However, expensive or private operations in Web APIs are usually protected. Typically, protection is set up using an Authorization header. In many cases, an Authorization header involves some sort of an ID and secret. In the case of the GitHub API, it involves a client ID and client secret. But to get them, you will need to create an OAuth app.

Before you can do this though, you need to get familiar with OAuth.

OAuth

OAuth is an open-standard authorization protocol that allows delegating access on behalf of a user. This section will explore two examples:

- Real-life analogy
- API analogy

Real-Life Analogy

Imagine a child at school. The teacher of that child is organizing a trip to another city. A permission slip from the parents is needed. The parents give a note: *It's okay for my child to go to place X.* The child gives the note to the teacher and gets permission to travel to a field trip to destination X.

API Analogy

Many applications are interconnected, with integrations to each other. For example, the famous social platform Discord allows you to display whatever accounts you have on other social media. But to do that, you need to connect to the platform of social media you want to display. For example, when you are on Discord and try to link a Twitter account, you will be required to log in on Twitter. A login will require a certain scope of access (your profile name, in this case). A successful login is proof that access is given, and Discord will be able to display your profile information on Twitter on your behalf.

OAuth App for GitHub

Returning to the subject of GitHub, what is an **OAuth app**? It is a registration for a single point of security. It acts as your application identity. A GitHub user might have zero or more applications. As mentioned before, an OAuth app includes a client ID and secret. Through them, you can use the GitHub API. In other words, you can set it up to request access to secure features of GitHub, such as changing your personal data on GitHub.

GitHub has an interesting API limitation. If more than 60 unauthenticated requests come from the same IP, it will block subsequent requests for up to an hour. However, the rate limitation can be removed by authorizing requests. That is the prime reason why you will be using authorization for an otherwise public endpoint.

OAuth usually involves two client applications:

- One that requests permission on behalf of someone
- Another that grants that permission

Therefore, when setting up OAuth, you will most likely be required to create a URL to return to after the permission is granted from the client that can grant access. Setting up an OAuth app on GitHub involves these steps:

1. In the top-right corner, click on your profile picture and click **Settings**:

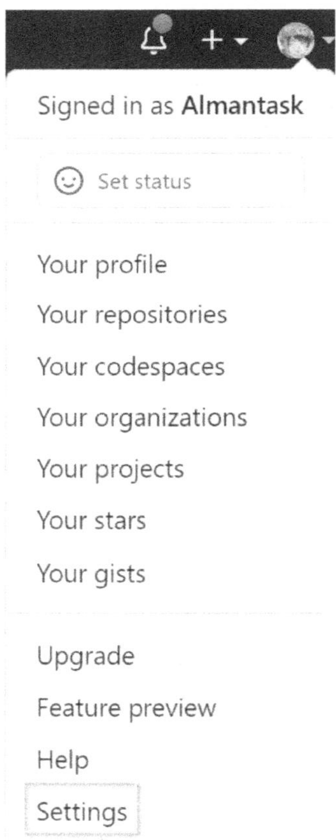

Figure 8.11: Account settings in GitHub

2. On the left side, scroll down almost to the bottom of the menu and click the **`Developer settings`** option:

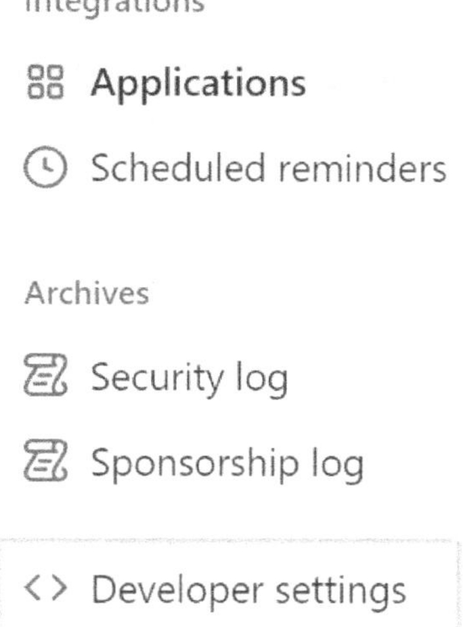

Figure 8.12: Developer settings in GitHub

3. Now select the **`Oauth Apps`** option:

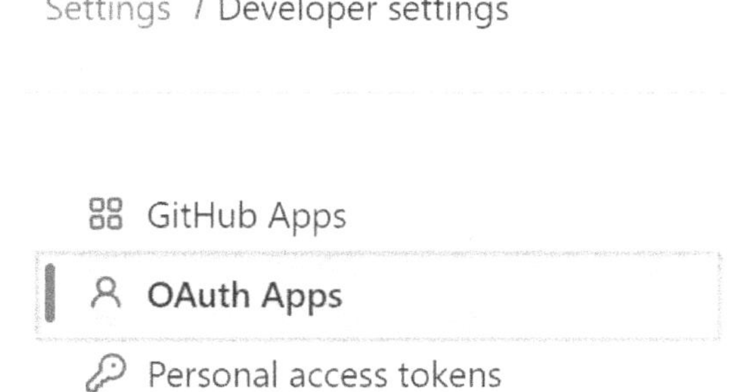

Figure 8.13: Selecting OAuth apps in Developer settings in GitHub

4. Then click the **`Register a new application`** button:

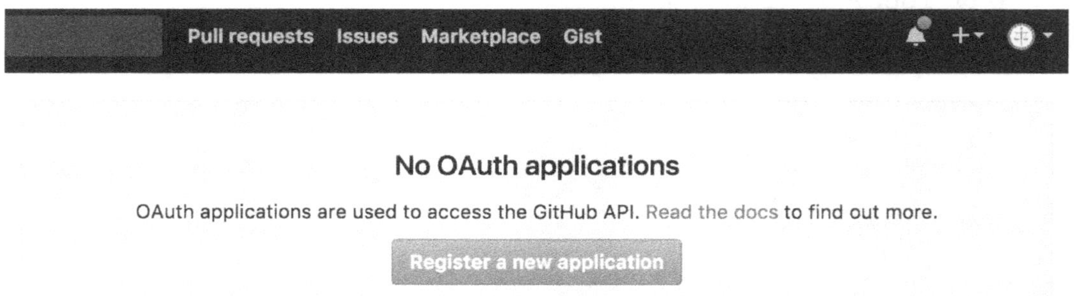

Figure 8.14: Creating a new OAuth app in GitHub

> **NOTE**
>
> If you have previously created an OAuth app, then this window will display all those listed. In order to create a new one, you will have to click **`New OAuth App`**.

5. In the next window, you will complete the form. Start by filling in **`Application name`** (**5**). Avoid using special characters.
6. Next, fill in **`Homepage URL`** (**6**).

   This URL usually points to a website that describes the use of OAuth for a particular case and why it is required. Even if you don't have a website that describes such a case, you can type a placeholder URL (in this case, **`myapp.com`**). The field accepts anything as long as it is a valid URL.
7. Fill in the **`Authorization callback URL`** (**7**) field. This can be whatever you want. Here, **`myapp.com/home`** is used. Use a valid callback URL.

8. Click **`Register application`** (8):

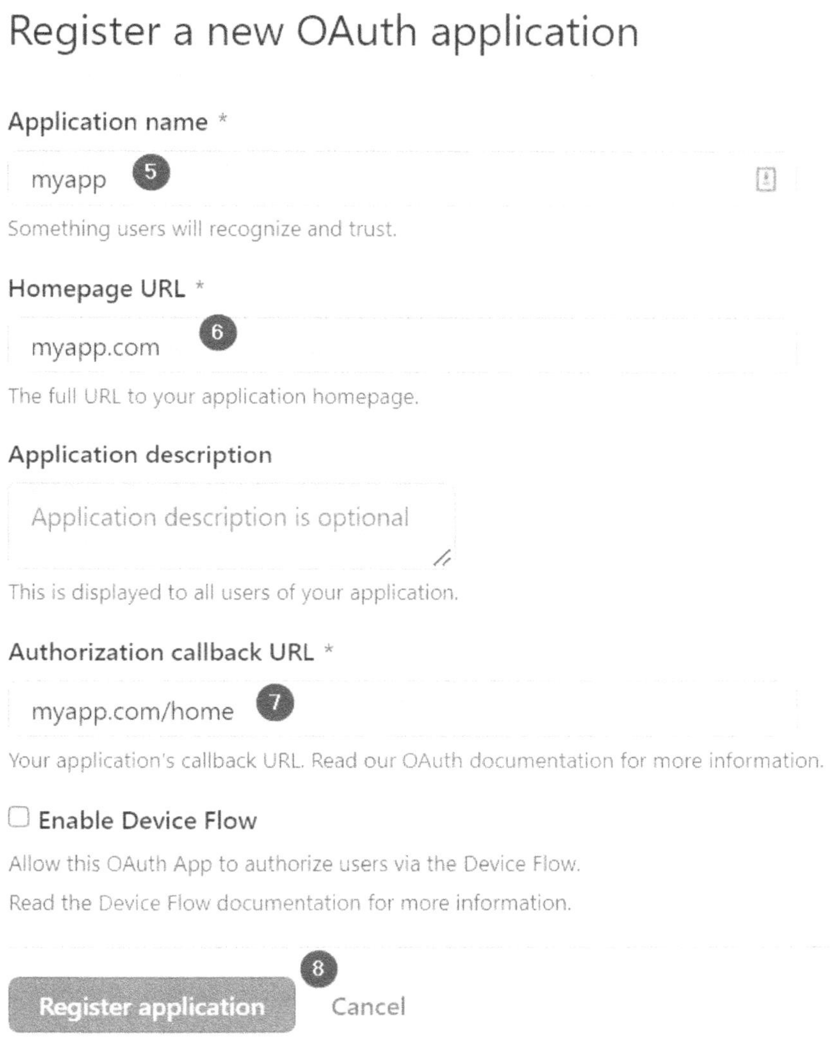

Register a new OAuth application

Application name *

myapp 5

Something users will recognize and trust.

Homepage URL *

myapp.com 6

The full URL to your application homepage.

Application description

Application description is optional

This is displayed to all users of your application.

Authorization callback URL *

myapp.com/home 7

Your application's callback URL. Read our OAuth documentation for more information.

Enable Device Flow

Allow this OAuth App to authorize users via the Device Flow.

Read the Device Flow documentation for more information.

8

Register application Cancel

Figure 8.15: New OAuth app window in GitHub

9. In the new window, you will see **`Client ID`** and **`Client secrets`**:

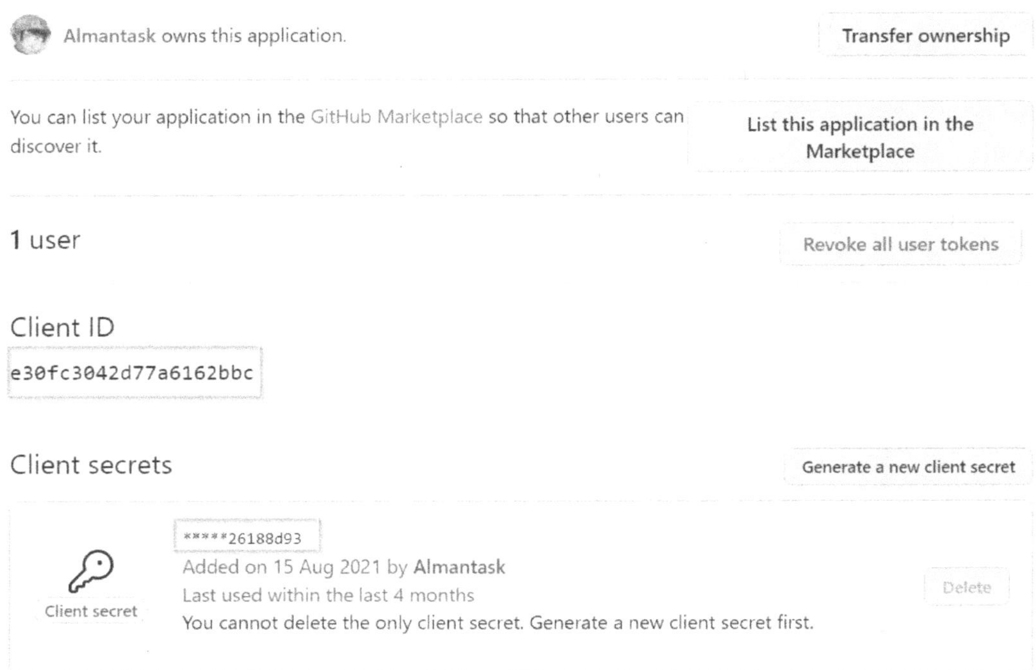
Packt-Test
Almantask owns this application.
Transfer ownership
You can list your application in the GitHub Marketplace so that other users can discover it.
List this application in the Marketplace
1 user
Revoke all user tokens
Client ID
e30fc3042d77a6162bbc
Client secrets
Generate a new client secret
*****26188d93
Added on 15 Aug 2021 by Almantask
Last used within the last 4 months
You cannot delete the only client secret. Generate a new client secret first.
Client secret
Delete

Figure 8.16: Details of a new OAuth app on GitHub with app credentials—Client ID and Client secrets

It is best to store a client secret in a safe place for future reference because you will see it only once on GitHub. If you forget it, you will have to create a new secret and delete the old one.

Now you have successfully created an OAuth app on GitHub. The client secret is partly hidden in this screenshot for a reason. You should never expose it publicly. In order to use it in a demo, you will use environmental variables first to hide them.

10. So, store the values in environmental variables **`GithubClientId`** and **`GithubSecret`**.

11. Then expose the two through static properties in **Demo.cs** (explained earlier) as follows:

```
private static string GitHubClientId { get; } = Environment.
GetEnvironmentVariable("GithubClientId", EnvironmentVariableTarget.
User);
private static string GitHubSecret { get; } = Environment.
GetEnvironmentVariable("GithubSecret", EnvironmentVariableTarget.
User);
```

This section covered the steps to set up an OAuth app in GitHub that can be used to request access to secure features of GitHub, such as changing your personal data. With this knowledge, you can now use a client ID and client secret to create authorized calls on the GitHub API, as demonstrated in the following section.

AUTHORIZATION HEADER

Authorization headers come in three forms—basic, API key (or personal access token), and third-party authentication. The GitHub API does not allow an unlimited number of calls from the same source. Like the Azure Text Analytics client, it uses an API key as well. However, in this case, the API key is used for rate limiting (how many calls you can make in an hour). For anonymous calls, it only allows 60 calls an hour. However, by using a valid Authorization header, the amount is increased to 5,000.

In the following example, you'll make one more call than the rate limit allows (60 + 1 = 61). That way, you will get user information 61 times. For that to happen, you will also make sure that the **CacheControl** header is set to **NoCache** because you don't want a request to be ignored after 60 consecutive calls:

```
public static async Task GetUser61Times()
{
    const int rateLimit = 60;
    for (int i = 0; i < rateLimit + 1; i++)
    {
        const string username = "github-user";
        var request = new HttpRequestMessage(HttpMethod.Get, new
Uri($"users/{username}", UriKind.Relative));
        request.Headers.CacheControl = new CacheControlHeaderValue()
{NoCache = true};
```

```
        var response = await client.SendAsync(request);
        if (!response.IsSuccessStatusCode)
        {
            throw new Exception(response.ReasonPhrase);
        }
```

This block of code is an adaptation of the **GetUser** method from the *HttpClient* section. There are three main adjustments here:

- The first is that everything in a loop runs 61 times.
- You have also added an error handler, which means if a response is not a success, you will print an error message returned by the API.
- Lastly, you add a **CacheControl** header to ignore caching (because you do want 61 calls to the server).

Running this code results in an error message on the sixty-first call, which proves the API rate limitation (the error message has been truncated for clarity):

```
60) Kaisinel created profile at 2018-06-22 07:51:56
Unhandled exception. System.Exception: rate limit exceeded
```

To fix this, you will need to add an **Authorization** header (you will add it just under the **CacheControl** header):

GitHttp.cs

```
public static async Task GetUser61Times(string authHeader)
{
    const int rateLimit = 60;
            for (int i = 0; i < rateLimit + 1; i++)
            {
                const string username = "github-user"; // replace with your own
                var request = new HttpRequestMessage(HttpMethod.Get, new
Uri($"users/{username}", UriKind.Relative));
               request.Headers.CacheControl = new CacheControlHeaderValue(){NoCache
= true};
               request.Headers.Add("Authorization", authHeader);

               var response = await client.SendAsync(request);
                if (!response.IsSuccessStatusCode)
                {
                    throw new Exception(response.ReasonPhrase);
                }
```

The complete code can be found here: https://packt.link/1C5wb.

Due to GitHub's limitations on anonymous calls (for example, the fact that you can make only 60 requests per hour to get user profile information), you will find it more efficient to provide an Authorization header so that you are identified and therefore released from such strict constraints. In the examples that follow, you will get an authorization token that you will feed to this method, thus showing how authorization will help you overcome the rate limit.

When running the demo code placed at https://packt.link/Uz2BL, it is recommended that you run one example at a time (i.e., uncomment one line and comment the rest within the **Run** method). This is because the **Demo.cs** file is a mix of authorized and anonymous calls, and you might get unexpected results. However, keep the line where you get a token as it may be required by individual examples.

At the end of this section, you should have grasped the logic behind the Authorization header and its three forms—basic, API key (or personal access token), and third-party authentication—and learned that, like the Azure Text Analytics client, the GitHub API uses an API key. Now you can move on to basic authentication.

BASIC AUTHENTICATION

Basic authentication involves a username and password. The two are usually combined in a single string and encoded using the following format:

```
Basic username:password
```

Here is the code used to generate an authorization taken for basic authentication:

```
public static string GetBasicToken()
{
    var id = GitHubClientId;
    var secret = GitHubSecret;
    var tokenRaw = $"{id}:{secret}";
    var tokenBytes = Encoding.UTF8.GetBytes(tokenRaw);
    var token = Convert.ToBase64String(tokenBytes);

    return "Basic " + token;
}
```

Use a username and password to get a basic token. Then pass it to the **GetUser61Times** method:

```
var basicToken = GitExamples.GetBasicToken();
await GitExamples.GetUser61Times(basicToken);
```

Calling **GetUser61Times** no longer displays an error because the rate limitation is avoided by supplying an Authorization header.

> **NOTE**
>
> You can find the code used for this example at https://packt.link/Uz2BL and https://packt.link/UPxmW.

The next section will cover the more specialized API key and personal access token, which are similar as they both grant access to otherwise protected data.

API KEY AND PERSONAL ACCESS TOKEN

A personal access token is limited to personal data. However, an API key can be used for the whole API. Other than the scope of what can be accessed, the two have no difference in how they are used. You can add an API key or a personal access token to an Authorization header as is.

But, of course, to use an access token of a certain API, you first need to create it. You can do this through the following steps:

1. Go to GitHub's **Developer settings** option under **Settings** window.
2. Navigate to **Personal access tokens** (**1**).
3. Select **Generate new token** button (**2**):

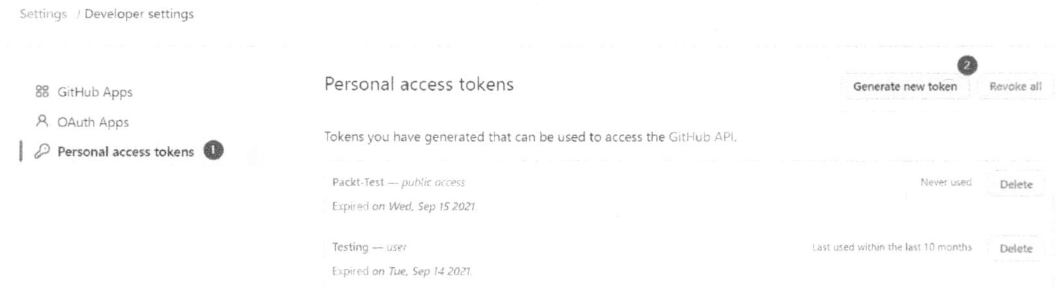

Figure 8.17: Creating a new personal access token

4. Next, enter your GitHub password.
5. Add a note (this can be anything) and scroll down. This screen will help you to modify user data, so check the **user** checkbox (**4**) to get access to it.

6. Click the **`Generate token`** button (**5**):

☑ user (4)	Update ALL user data
read:user	Read ALL user profile data
user:email	Access user email addresses (read-only)
user:follow	Follow and unfollow users
☐ delete_repo	Delete repositories
☐ write:discussion	Read and write team discussions
☐ read:discussion	Read team discussions
☐ admin:enterprise	Full control of enterprises
☐ manage_runners:enterprise	Manage enterprise runners and runner-groups
☐ manage_billing:enterprise	Read and write enterprise billing data
☐ read:enterprise	Read enterprise profile data
☐ admin:gpg_key	Full control of public user GPG keys
☐ write:gpg_key	Write public user GPG keys
☐ read:gpg_key	Read public user GPG keys

(5) Generate token   Cancel

Figure 8.18: Scope of access configured for a personal access token

In the new window, you will see all the personal access tokens, along with the newly added ones:

Personal access tokens   Generate new token   Revoke all

Tokens you have generated that can be used to access the GitHub API.

Make sure to copy your personal access token now. You won't be able to see it again!

✓ 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   Delete

Figure 8.19: A new personal access token created on GitHub

> **NOTE**
>
> Remember that you will see the value of a token only once. So, make sure you copy and store it securely. Also, be aware that the personal access token expires after a month, at which point you need to regenerate it.

7. Create an environmental variable called **GitHubPersonalAccess**.
8. Add the personal access token to **Demo.cs**:

```
private static string GitHubPersonAccessToken { get; } =
Environment.GetEnvironmentVariable("GitHubPersonalAccess",
EnvironmentVariableTarget.User);
```

9. Run the following code:

```
await GetUser61Times(GitHubPersonAccessToken);
```

   You will observe that calling the **GetUser61Times** method does not fail.

Access tokens, authorization tokens, API keys, and JWTs (which will be further covered in the following sections) are different means to prove to an API that you have been granted access to it and have rights to a resource you want. But regardless of which specific kind of authorization you use, they will usually all go to the same place—that is, the Authorization header.

The next section will detail an authorization protocol called OAuth2.

THIRD-PARTY AUTHENTICATION—OAUTH2

GitHub is an example of an **authorization server**. It allows access to a resource or functionality in the name of the owner. For example, updating the user's employment status is only available to a logged-in user. However, this can be done directly given the user has been granted the access to do so. A program getting access on behalf of someone is what **OAuth2** is all about.

Perform the following steps to modify the user's employment status:

1. Navigate to this URL or send an HTTP **GET** request:

```
https://github.com/login/oauth/authorize?client_
id={{ClientId}}&redirect_uri={{RedirectUrl}}
```

Here, **`{{ClientId}}`** and **`{{RedirectUrl}}`** are the values that you have set in the OAuth2 GitHub app.

> **NOTE**
>
> Replace the placeholders **`{{ClientId}}`** and **`{{RedirectUrl}}`** with the ones from your GitHub OAuth app.

The following screen prompts you to log in to your GitHub app:

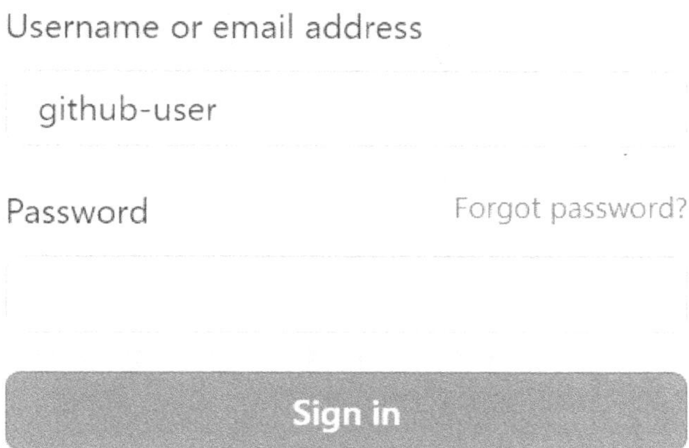

Figure 8.20: Signing in to OAuth2 GitHub app

2. Complete **`Username`** and **`Password`**.
3. Next, click the **`Sign in`** button to log in.

   After a successful login, you will be redirected to a URL specified in your OAuth2 app.

4. Create a request for the token by sending an HTTP **POST** request to a URI in the following format:

```
{tokenUrl}?client_id={clientId}&redirect_uri={redirectUri}&client_
secret={secret}&code={code}:
```

The code for it is as follows:

```
private static HttpRequestMessage CreateGetAccessTokenRequest()
{
    const string tokenUrl = "https://github.com/login/oauth/access_
token";
    const string code = "2ecab6ecf412f28f7d4d";
    const string redirectUri = "https://www.google.com/";
    var uri = new Uri($"{tokenUrl}?client_
id={GitHubClientId}&redirect_uri={redirectUri}&client_
secret={GitHubSecret}&code={code}");
    var request = new HttpRequestMessage(HttpMethod.Post, uri);
    return request;
}
```

In this case, the redirect URL was https://www.google.com. The URI you ended up with was https://www.google.com/?code=a681b5126b4d0ba160ba. The **code=** part is the code needed to get the **OAuth** access token. The token is returned in the following format:

```
access_token=gho_
bN0J89xHZqhKOUhI5zd5xgsEZmCKMb3WXEQL&scope=user&token_type=bearer
```

5. Before this token can be used, you need to parse it from the response. So, create a function to parse the token response:

```
private static Dictionary<string, string> ConvertToDictionary(string
content)
{
    return content
        .Split('&')
        .Select(kvp => kvp.Split('='))
        .Where(kvp => kvp.Length > 1)
        .ToDictionary(kvp => kvp[0], kvp => kvp[1]);
}
```

This takes every = property and puts it into a dictionary. The string before = is a key and the string after = is a value.

6. Use the **GetToken** function to create and send a request and parse a response, then format the token and return it:

```
private static async Task<string> GetToken()
{
    HttpRequestMessage request = CreateGetAccessTokenRequest();

    var response = await client.SendAsync(request);
    var content = await response.Content.ReadAsStringAsync();

    Dictionary<string, string> tokenResponse =
ConvertToDictionary(content);

    // ValidateNoError(tokenResponse);

    var token = $"{tokenResponse["token_type"]}
{tokenResponse["access_token"]}";
    return token;
}
```

Here, you created a request, sent it to a client, parsed the response as a token, and then returned it. **ValidateNoError** is commented out for now. You will come back to it later. The returned token should look something like this:

```
bearer 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
```

This token is a **bearer token**, which is a token generated by an authorization server (in this case, GitHub) that grants access to GitHub on behalf of you (or any other username used for logging in to GitHub). You can use it to send requests that require special access. For example, update the employment status of a user.

7. To update the employment status of a user, use the **UpdateEmploymentStatus** function:

```
public static async Task UpdateEmploymentStatus(bool isHireable,
string authToken)
{
    var user = new UserFromWeb
    {
        hireable = isHireable
    };
    var request = new HttpRequestMessage(HttpMethod.Patch, new Uri("/
user", UriKind.Relative));
    request.Headers.Add("Authorization", authToken);
```

```
    var requestContent = JsonConvert.SerializeObject(user, new
JsonSerializerSettings { NullValueHandling = NullValueHandling.Ignore
});
    request.Content = new StringContent(requestContent, Encoding.
UTF8, "application/json");
    var response = await client.SendAsync(request);
    var responseContent = await response.Content.ReadAsStringAsync();
    Console.WriteLine(responseContent);
}
```

This block of code sets the user's property **isHireable** to **true** and prints the updated user information. The important part here is content; when sending **PUT**, **PATCH**, or a **POST** request, you often need a **body** with a request (or **content** in other words).

The act of converting an in-memory object into structured text (for example, JSON) is called **serialization**. In this case, a body is a user update. You send a **PATCH** request because you only want to change the updated values. If a value is not provided in the content, it should not change. That's the key difference between a **PATCH** and **POST** request—a successful request overrides all values (even if you don't provide them).

You used **new JsonSerializerSettings { NullValueHandling = NullValueHandling.Ignore }** in order to avoid providing **null** values. This is because you do not want to update all the fields; just the ones you have supplied.

When creating HTTP content, you also need to supply a MIME type (a type of media sent over with the request). It is needed so that the server has a hint for how it is expected to process the request. A MIME type follows this format:

```
type/subtype
```

In this case, **application/json** means that the client should expect JSON from a server. **application** is the most common MIME type, which means binary data.

There is also **StringContent**, which is a type of serialized content, usually as JSON or XML. Alternatively, you could use **StreamContent** or **ByteContent**, but those are slightly rarer and are used when performance or the volume of data is of concern.

The following code shows the full demo:

```
public static async Task Run()
{
    var oathAccessToken = await GitExamples.GetToken();
    await GitExamples.UpdateEmploymentStatus(true, oathAccessToken);
}
```

In the **GetToken** method (used in *Step 6* of the *Third-Party Authentication (OAuth2)* section), there was one commented line of code, **ValidateNoError**. Uncomment it and implement the **GetToken** method, because you won't always get a successful response, and parsing a token in that case will fail (i.e., it won't exist). Therefore, it is always a good idea to validate the server response and throw an exception when the unexpected happens. Look at the following GitHub error format:

```
error=bad_verification_code&error_
description=The+code+passed+is+incorrect+or+expired.&error_
uri=https%3A%2F%2Fdocs.github.com%2Fapps%2Fmanaging-oauth-
apps%2Ftroubleshooting-oauth-app-access-token-request-
errors%2F%23bad-verification-code
```

It is not very readable. **ValidateNoError** will format the response and throw that as an exception, instead of letting it fail silently:

```
private static void ValidateNoError(Dictionary<string, string>
tokenResponse)
{
    if (tokenResponse.ContainsKey("error"))
    {
        throw new Exception(
            $"{tokenResponse["error"].Replace("_", " ")}. " +
            $"{tokenResponse["error_description"].Replace("+", "
")}");
    }
}
```

If you run the code again and it fails for the same reasons, the error message will now read as follows:

```
bad verification code. The code passed is incorrect or expired.
```

This section covered the basics of how to send HTTP requests with some sort of security in place. In the sections that follow (*Restsharp* and *Refit*), you will create clients using third-party libraries to remove some of the boilerplate code required by **HttpClient**.

> **NOTE**
>
> You can find the code used for this example at https://packt.link/UPxmW.

REQUEST IDEMPOTENCY

An **idempotent** HTTP request is a request that always results in the same outcome. Only **GET**, **PUT**, and **PATCH** requests are idempotent because they either make no change or make the same change all over again, but that change does not ever cause an error and results in the same data. **DELETE** is not idempotent because deleting an already deleted item will produce an error. **POST** may or may not be idempotent, but that solely depends on the implementation.

PUT, PATCH, OR POST

The difference between **PUT**, **PATCH**, and **POST** can be summed up as follows:

- **PUT** is used for overriding fields in a model. Even if a single value is explicitly provided, the whole model will have the unprovided values (or at least that's the expectation). For example, if you wanted to update user details by first getting the old details and then sending a modified version, you would use **PUT**.
- **PATCH** is used for updating only a single value that was provided explicitly. For example, if you wanted to update a username, it would make sense to send **PATCH** over a **PUT** request.
- **POST** is used for creating items or sending a complex query. Either way, the default expectation of this verb is to have side effects. For example, if you wanted to create a user, you would use a **POST** request.

EXERCISE 8.02: HTTPCLIENT CALLING A STAR WARS WEB API

You might be familiar with Star Wars. There are movies, games, and TV series. However, did you know that it also has multiple APIs to retrieve data? The upcoming exercise will introduce you to a different format of an API and will make you familiar with deserializing slightly more complex responses.

In this exercise, you will create a strongly typed API client that will, under the hood, use **`HttpClient`**. The client will be used to return Star Wars movies. You will be using **Star Wars API** (**SWAPI**) (https://swapi.dev/). The required endpoint is https://swapi.dev/api/films/. Perform the following steps to complete this exercise:

1. Create a new class to hold **`HttpClient`** with a base URL:

```
public class StarWarsClient
    {
        private readonly HttpClient _client;

        public StarWarsClient()
        {
            _client = new HttpClient {BaseAddress = new Uri("https://
swapi.dev/api/")};
        }
```

   This will act as a strongly typed API client.

   > **NOTE**
   >
   > The / at the end of the URI indicates that more text will be appended to the URI (after **`api`** rather than after **`dev`**).

2. Create a type for representing a movie:

Film.cs

```
public record Film
{
    public string Title { get; set; }
    public int EpisodeId { get; set; }
    public string OpeningCrawl { get; set; }
    public string Director { get; set; }
    public string Producer { get; set; }
    [JsonProperty("release_date")]
    public string ReleaseDate { get; set; }
    public string[] Characters { get; set; }
    public string[] Planets { get; set; }
    public string[] Starships { get; set; }
    public string[] Vehicles { get; set; }
    public string[] Species { get; set; }
    public DateTime Created { get; set; }
```

The complete code can be found here: https://packt.link/tjHLa.

This is a class you will use for deserializing movies within a response. The **ReleaseDate** property has **[JsonProperty("release_date")]** above it to specify that the **"release_date"** JSON field will map to the **ReleaseDate** C# property.

3. Create a type for storing results:

```
public record ApiResult<T>
{
    public int Count { get; set; }

    public string Next { get; set; }

    public string Previous { get; set; }

    [JsonProperty("results")]
    public T Data { get; set; }
}
```

This is also a type for deserializing a movie response; however, the Star Wars API returns results in paginated format. It contains **Previous** and **Next** properties pointing to previous and next pages. For example, if you don't provide the page you want, it will return a value of **null**. However, the next property will point to the next page only if there are any elements left (otherwise it will also be **null**). Querying the API using next or previous as a URI will return the resources of that page. You used the **JsonProperty** attribute above **T Data** to provide JSON-to-property mapping because the property and JSON names do not match (the JSON field name is **results** while **Data** is the property name).

> **NOTE**
>
> You could have changed **ApiResult** to have the **Results** property instead of **Data**. However, **ApiResult.Results** is a bit confusing. When writing code, instead of ease of automation (in this case, serialization), choose ease of maintainability and readability. For this reason, the name chosen in *Step 3* is different but clearer.

4. Now, create a method to get multiple films:

```
public async Task<ApiResult<IEnumerable<Film>>> GetFilms()
{
```

   You've returned a task so that others can await this method. Almost all HTTP calls will be **async Task**.

5. Create an HTTP request to get all movies:

```
var request = new HttpRequestMessage(HttpMethod.Get, new Uri("films",
UriKind.Relative));
```

   The URI is relative because you're calling it from **HttpClient** that already has a base URI set.

6. To query the Star Wars API for movies, send this request:

```
var response = await _client.SendAsync(request);
```

7. It returns **HttpResponseMessage**. There are two important parts to this: status code and response body. C# has a method to determine whether there were any errors based on the status code. To handle errors, use the following code:

```
if (!response.IsSuccessStatusCode)
{
      throw new HttpRequestException(response.ReasonPhrase);
}
```

   Error handling is important because a failed HTTP request will often result in an error status code rather than an exception. It's recommended you do something similar before trying to deserialize the response body as, if it fails, you might get an unexpected body.

8. Now, call the **ReadAsStringAsync** method:

```
var content = await response.Content.ReadAsStringAsync();
var films = JsonConvert.DeserializeObject<ApiResult<Film>>(content);

    return films;
}
```

The response has content that is more likely to be a kind of stream. To convert **HttpContent** to a string, call the **ReadAsStringAsync** method. This returns a string (JSON), which allows you to convert JSON to a C# object and deserialize the results. Lastly, you get the results by deserializing the response content body and converting it all to **ApiResult<Film>**.

9. For a demo, create the client and use it to get all the Star Wars films, then print them:

```
public static class Demo
{
    public static async Task Run()
    {
        var client = new StarWarsClient();
        var filmsResponse = await client.GetFilms();
        var films = filmsResponse.Data;
        foreach (var film in films)
        {
            Console.WriteLine($"{film.ReleaseDate} {film.Title}");
        }
    }
}
```

If everything is fine, you should see the following result:

```
1977-05-25 A New Hope
1980-05-17 The Empire Strikes Back
1983-05-25 Return of the Jedi
1999-05-19 The Phantom Menace
2002-05-16 Attack of the Clones
2005-05-19 Revenge of the Sith
```

This exercise illustrates how to create strongly typed HTTP clients for simplicity.

> **NOTE**
>
> You can find the code used for this exercise at https://packt.link/2CHpb.

You might have noticed that sending an HTTP request and using an HTTP client is very similar to the way a simple text file is sent to the GitHub API. Even if it was different, endpoints throughout the same API usually share the same requirements. However, if you manually craft an HTTP request every time you need to call an API, you are not being very efficient. A better way is to create something reusable. A common approach is to create **`BaseHttpClient`**. You will put this into practice in the following activity.

ACTIVITY 8.01: REUSING HTTPCLIENT FOR THE RAPID CREATION OF API CLIENTS

The problem with **`HttpClient`** is that you still have to manage many things by yourself:

- Error handling
- Serializing and deserializing
- Mandatory headers
- Authorization

When working in a team or on a bigger project, you are likely to be making more than just one HTTP call. The consistency and same requirements between different calls need to be managed.

The aim of this activity is to show one of many ways you can simplify working with repetitive HTTP calls. You will be using the **`BaseHttpClient`** class, which you will create first. The class will generalize error handling and deserializing responses and requests, which will significantly simplify different HTTP calls that you make. Here, you will learn how to implement a base client by rewriting **`StarWarsClient`** using **`BaseHttpClient`**.

Perform the following steps to complete this activity:

1. Create a base **`HttpClient`** class. A base client wraps **`HttpClient`**. Therefore, you will hold a private reference to it and allow it to be created from a URL. The inner **`HttpClient`** often also includes base headers, but they are not required in this case.

2. Define a way to create requests for every method. For brevity, stick to a **`GET`** request. Within a **`GET`** request, it is a common practice to define the default headers, but once again, it is not mandatory in this example.

3. Create a method to send requests and include error handling and deserialization.
4. In SWAPI, if you are querying multiple results, you get back **ApiResult<IEnumerable<T>>** for pagination. Create a **SendGetManyRequest** method.
5. Use the base client you have created and simplify the client from *Exercise 8.02*.
6. Run the code through the same demo code but using the new version of **StarWarsClient**.
7. If you run the demo once again with the new **StarWarsClient**, you should see the same films returned:

```
1977-05-25 A New Hope
1980-05-17 The Empire Strikes Back
1983-05-25 Return of the Jedi
1999-05-19 The Phantom Menace
2002-05-16 Attack of the Clones
2005-05-19 Revenge of the Sith
```

In order to run this activity, go to https://packt.link/GR27A and comment all lines within the **static void Main(string[] args)** body, except **await Activities.Activity01.Demo.Run();**.

> **NOTE**
>
> The solution to this activity can be found at https://packt.link/qclbF.

Reusing **HttpClient** like that is very useful because it removes code duplication. However, calling a Web API and removing duplicate code is a common problem and is likely to be solved in some way by some libraries. The following section will explore how to simplify calls to a Web API using two popular NuGet packages:

- RestSharp
- Refit

RESTSHARP

The idea behind **RestSharp** is very similar to the base **HttpClient**—reducing code duplicity. It simplifies the creation of a request and provides a lot of the utility for making HTTP calls. Redo **StarWarsClient** using **RestSharp**, but first, you'll install the **RestSharp** NuGet:

```
dotnet add package RestSharp
```

Now create a client that is very similar to the one you created in *Activity 8.01*:

```
    public class StarWarsClient
    {
        private readonly RestClient _client;

        public StarWarsClient()
        {
            _client = new RestClient("https://swapi.dev/api/");
        }
```

Having **RestSharp** created gives you a response serialization out of the box. It is also able to guess which HTTP method you will use:

```
        public async Task<ApiResult<IEnumerable<Film>>> GetFilms()
        {
            var request = new RestRequest("films");
            var films = await _client.
GetAsync<ApiResult<IEnumerable<Film>>>(request);

            return films;
        }
    }
```

You passed the minimum required information to make an HTTP request (calling films, returning **ApiResult<IEnumerable<Film>>**) and the rest is done. This is very much like the base client you wrote previously.

> **NOTE**
>
> **ApiResult** is the same type used in *Exercise 8.02*.

However, if you run this code against your demo, you will notice that the **Data** property (on JSON) comes back as **null**. This is because you had a **JsonProperty** attribute on the **response** and **film** classes. RestSharp uses a different serializer, which does not know about those attributes. To make it work, you could either change all the attributes to what RestSharp comprehends or use the same serializer as before. You are using **Newtonsoft.Json** and, in order to use that in RestSharp, you need to call the **UseSerializer** method, selecting **JsonNetSerializer**:

```
        public StarWarsClient()
        {
            _client = new RestClient("https://swapi.dev/api/");
            _client.UseSerializer(() => new JsonNetSerializer());
        }
```

On running the demo, the following output gets displayed:

```
1977-05-25 A New Hope
1980-05-17 The Empire Strikes Back
1983-05-25 Return of the Jedi
1999-05-19 The Phantom Menace
2002-05-16 Attack of the Clones
2005-05-19 Revenge of the Sith
```

The results are the same as those in *Exercise 8.02*; however, the difference is using the **Newtonsoft** serializer in the preceding example. **RestSharp** is probably the best abstraction for **HttpClient** as it minimizes the amount of code you need to write to make HTTP calls even while keeping its similarities with **HttpClient**.

> **NOTE**
>
> You can find the code used for this example at https://packt.link/f5vVG.

The example aims to communicate with Web APIs using HTTP requests. Even though the demo files look the same, they are using either a different library or design pattern. In the following activity, you will practice consuming more APIs using RestSharp.

ACTIVITY 8.02: THE COUNTRIES API USING RESTSHARP TO LIST ALL COUNTRIES

The address https://restcountries.com/v3/ is a public web API that provides a list of all existing countries. Suppose that using that API, you need to get a list of all countries, find a country by its capital city (for example, Vilnius), and find all the countries that speak in a given language (for example, Lithuanian). You need to print only the first two country names, their regions, and their capitals, and implement a strongly typed client to access this API using **RestSharp**.

The aim of this activity is to make you feel more comfortable using third-party libraries (**RestSharp**) when making HTTP calls. Using third-party libraries often saves a lot of time. It allows you to reuse something that is already available.

Perform the following steps to complete this activity:

1. Create a base client class using the URL https://restcountries.com/v3/.

> **NOTE**
>
> Navigating to https://restcountries.com/v3/ will return the HTTP status code **404** with a **Page Not Found** message. This is because the base API URI doesn't contain any information on a resource; it is yet to be completed and is just the beginning of a full URI for a resource.

2. Create models for serialization.
3. Use the example https://restcountries.com/v3/name/peru to get a response.
4. Copy the response and then use a class generator, such as https://json2csharp.com/, to make models out of JSON (response).
5. Within the client, create the following methods: **Get**, **GetByCapital**, and **GetByLanguage**.
6. Create a demo calling all three methods.
7. Print the countries within each response.

   The result should be as follows:

```
All:
Aruba Americas Oranjestad
Afghanistan Asia Kabul

Lithuanian:
```

```
Lithuania Europe Vilnius

Vilnius:
Lithuania Europe Vilnius
```

> **NOTE**
>
> The solution to this activity can be found at https://packt.link/qclbF.

You now know that RestSharp simplifies the creation of a request and provides a lot of the utilities for making HTTP calls. The next section will help you practice using Refit, which is another way to consume an API.

REFIT

Refit is the smartest client abstraction because it generates a client from an interface. All you have to do is provide an abstraction:

1. To use the **`Refit`** library, first install the **`Refit`** NuGet:

```
dotnet add package Refit
```

2. To create a client in Refit, first create an interface with HTTP methods:

```
public interface IStarWarsClient
{
    [Get("/films")]
    public Task<ApiResult<IEnumerable<Film>>> GetFilms();
}
```

   Please note that the endpoint here is **`/films`** rather than **`films`**. If you run the code with **`films`**, you will get an exception suggesting that you change the endpoint with a preceding **/**.

3. To resolve the client, simply run the following code:

```
var client = RestService.For<IStarWarsClient>("https://swapi.dev/
api/");
```

   On running the demo, the following output gets displayed:

```
1977-05-25 A New Hope
1980-05-17 The Empire Strikes Back
1983-05-25 Return of the Jedi
```

```
1999-05-19 The Phantom Menace
2002-05-16 Attack of the Clones
2005-05-19 Revenge of the Sith
```

The results are the same as the ones you saw in *Exercise 8.02*; however, the difference is in the implementation.

> **NOTE**
>
> You can find the code used for this example at https://packt.link/cqkH5.

Use Refit only when your scenarios are trivial. Though Refit might seem like the easiest solution, it comes with its own complications when you need custom authorization for more complex scenarios. You will simplify the solution further in the following activity.

ACTIVITY 8.03: THE COUNTRIES API USING REFIT TO LIST ALL COUNTRIES

The more different ways you know of doing the same thing, the easier you can make a choice and pick the best tool for the job. Different teams may use different tools and Refit is quite a unique, minimalistic approach that you may encounter. Others may say it complicates work because there is too much hidden in the client interface (less code often does not mean that you can grasp the code easily). It doesn't matter whether you are for Refit or against it; it's good to have practiced things first-hand and formed your own opinion. This activity will help you do exactly that. Here, you will access the Countries API to display all countries, countries by their language, and by their capital city.

The aim of this activity is to show how practical Refit can be for rapid prototyping when it comes to consuming simple APIs. The steps for this are as follows:

1. Create models for serialization. For that, use the example https://restcountries.com/v3/name/peru to get a response.
2. Now copy the response.
3. Then use a class generator, such as https://json2csharp.com/, to make models out of JSON (response).
4. Define an interface with methods: **`Get`**, **`GetByCapital`**, and **`GetByLanguage`**.

5. Create a demo printing a country name, region, and country status.

   The result will be displayed as follows:

```
All:
Aruba Americas Oranjestad
Afghanistan Asia Kabul

Lithuanian:
Lithuania Europe Vilnius

Vilnius:
Lithuania Europe Vilnius
```

> **NOTE**
>
> The solution to this activity can be found at https://packt.link/qclbF.

.NET has a few other native ways of creating HTTP requests, and for that, you can use **HttpWebRequest** or **WebClient**. The two are not deprecated and it is fine to use them, but they are older alternatives compared to the newer **HttpClient**. The next section covers all these.

In the following section, you'll find out about other libraries that solve the problem of code duplication when using **HttpClient**.

OTHER WAYS OF MAKING HTTP REQUESTS

Refit and RestSharp are just two of many libraries solving the problem of code duplication when using **HttpClient**. Flurl and TinyRest are another two popular alternatives. New libraries are created every year and they are ever evolving. There is no one best way that suits all scenarios. To be sure you make the right choice, you'll want to do a little research first as there are some pitfalls to these alternatives to consider.

HttpClient was designed for the lowest-level HTTP calls in .NET. It is the safest option because it is well-documented, tested, and allows the most freedom. Though there are many libraries that are much simpler to use than **HttpClient**, they often target basic scenarios (no authorization, no dynamically set headers). When it comes to creating advanced HTTP calls, they often turn out to be quite complicated.

When it comes to choosing which client to use, first go for the one provided natively by the API. If there is no client for the API, think about the complexity and scope of your project. For simple, small-scope projects, use whatever NuGet **HttpClient** alternative you find the most convenient. But if the scope of a project is big and the calls are complex, use the native **HttpClient** offered by the framework.

In the next exercise, you will implement an example where using Refit will turn it into a complication. To fix that complication, you will use both **HttpClient** and RestSharp.

EXERCISE 8.03: A STRONGLY TYPED HTTP CLIENT FOR TESTING PAYMENTS IN A PAYPAL SANDBOX

A common scenario in programming is making payments. However, during the development stage, you don't want to use a real bank account and thus look for ways to process payments in a test environment—that is, a **sandbox**. In this exercise, you will learn how to call a payments sandbox API. You will use **PayPal's sandbox** API (https://developer.paypal.com/docs/api/orders/v2/) to create an order and get the order that you have created.

This exercise will use **Refit** for the client interface and the implementation resolution. It will also use **HttpClient** to provide a way of getting **auth** headers for Refit. Lastly, you will use RestSharp to get an access token from within **HttpClient**. Perform the following steps to complete this exercise:

1. Go to https://www.paypal.com/tt/webapps/mpp/account-selection.
2. Create a PayPal account (either personal or business).
3. Choose your location and click the **Get Started** button.
4. Provide your mobile number.

5. Click the **`Next`** button and enter the code.
6. Set up your profile by entering an email address and password.
7. Provide your address details.
8. Now link your credit or debit card. You can also do this for free by following the instructions given at https://www.paypal.com/tt/webapps/mpp/account-selection.

> **NOTE**
>
> Creating an account on PayPal is free. The linking of credit (or debit) card requirement is just a part of account creation, and it doesn't charge you. The payment gets refunded as soon as the authentication is confirmed.

9. Now log out of the account and go to https://developer.paypal.com/developer/accounts/.
10. Click the **`Log in to Dashboard`** button and proceed ahead:

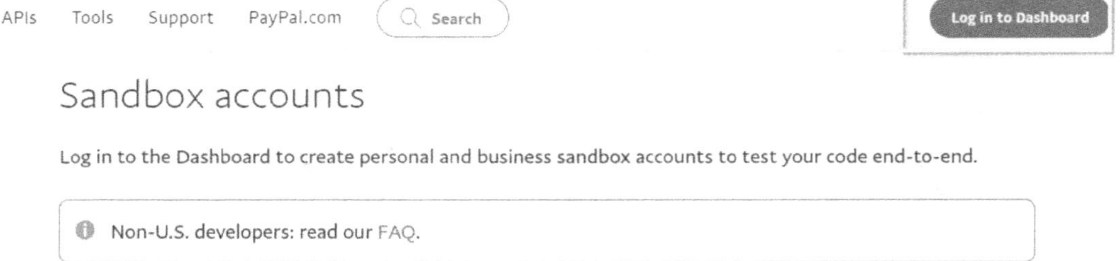

Figure 8.21: Log in to the PayPal dashboard to manage both sandbox and live environments

11. Then enter the requested credentials and proceed to the next screen.

12. Click the **`Accounts`** option under the **`Sandbox`** option. You will see two test accounts created for you:

Figure 8.22: Sandbox PayPal accounts for testing

You will use these accounts to do testing in the next steps.

> **NOTE**
>
> The PayPal sandbox is free.

13. Go to https://developer.paypal.com/developer/applications to get your client ID and secret. Just like the GitHub example, PayPal uses an OAuth app to provide you with a client ID and a secret.

14. For one of the default accounts, PayPal also generates a default OAuth app. So, click the **`Sandbox`** tab and select **`Default Application`**:

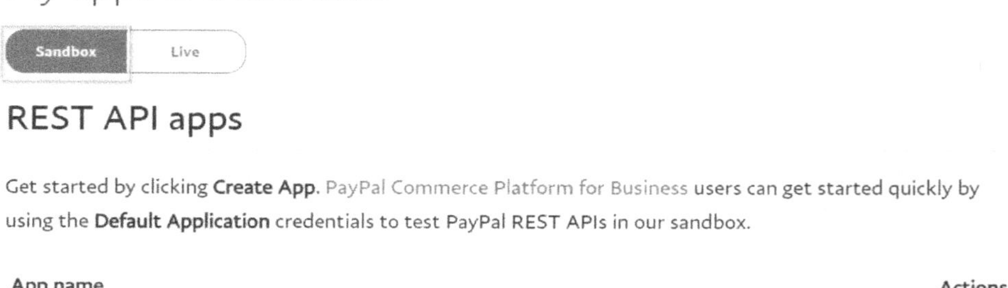

Figure 8.23: OAuth app creation for PayPal

15. In the new window, inspect both **`Client ID`** and **`Secret`**.
16. Take note of both and store them in environmental variables:

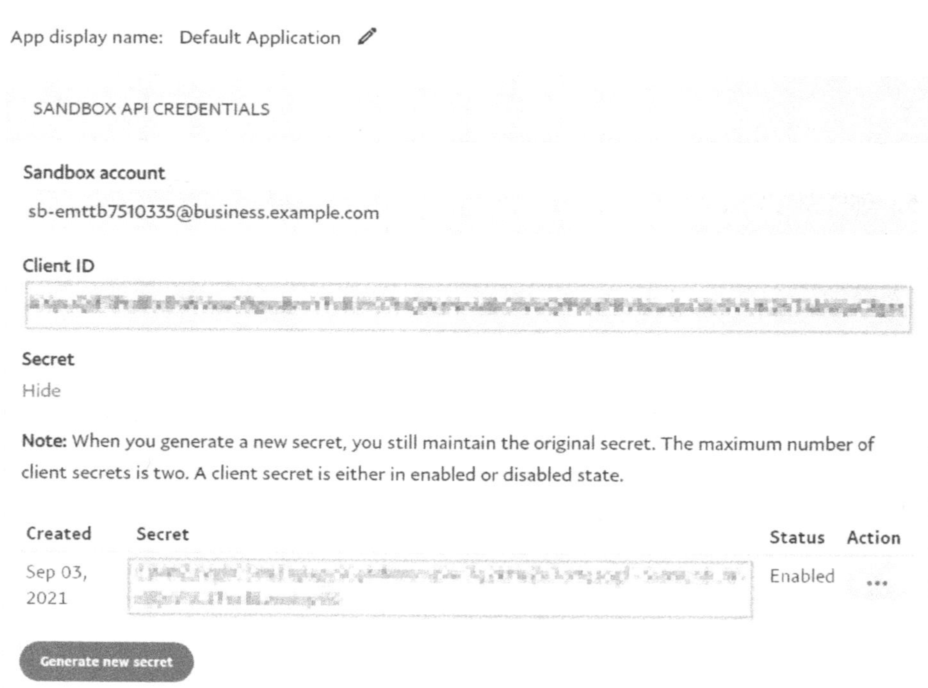

Figure 8.24: Default application details displaying Client ID and Secret

17. Create properties for accessing the PayPal client ID and secret in a new empty class, **Exercise03.AuthHeaderHandler.cs**:

```
public static string PayPalClientId { get; } = EnvironmentVariable.
GetOrThrow("PayPalClientId");
public static string PayPalSecret { get; } = EnvironmentVariable.
GetOrThrow("PayPalSecret");
```

Here, the **EnvironmentVariable.GetOrThrow** helper methods are used to get the user's environment variable or throw it if it doesn't exist. You will use these properties to make a connection to the sandbox PayPal API.

> **NOTE**
>
> You can find the code used for environment variables at https://packt.link/y2MCy.

18. In the **Demo.cs** class, add a **const** variable for the **BaseAddress** of a PayPal sandbox:

```
public const string BaseAddress = "https://api.sandbox.paypal.com/";
```

BaseAddress will be used for initializing different clients (RestSharp and Refit) with the PayPal URL.

19. Use **Refit** to create a client with **CreateOrder** and **GetOrder** methods:

```
public interface IPayPalClient
{
    [Post("/v2/checkout/orders")]
    public Task<CreatedOrderResponse> CreateOrder(Order order);

    [Get("/v2/checkout/orders/{id}")]
    public Task<Order> GetOrder(string id);
}
```

To get a sample request, refer to the documentation of the API that you want to call. Usually, they have an example request. In this case, the PayPal **CreateOrder** request can be found at https://developer.paypal.com/docs/api/orders/v2/:

```
{
   "intent":"CAPTURE",
   "purchase_units":[
```

```
      {
        "amount":{
          "currency_code":"USD",
          "value":"100.00"
        }
      }
    ]
  }
```

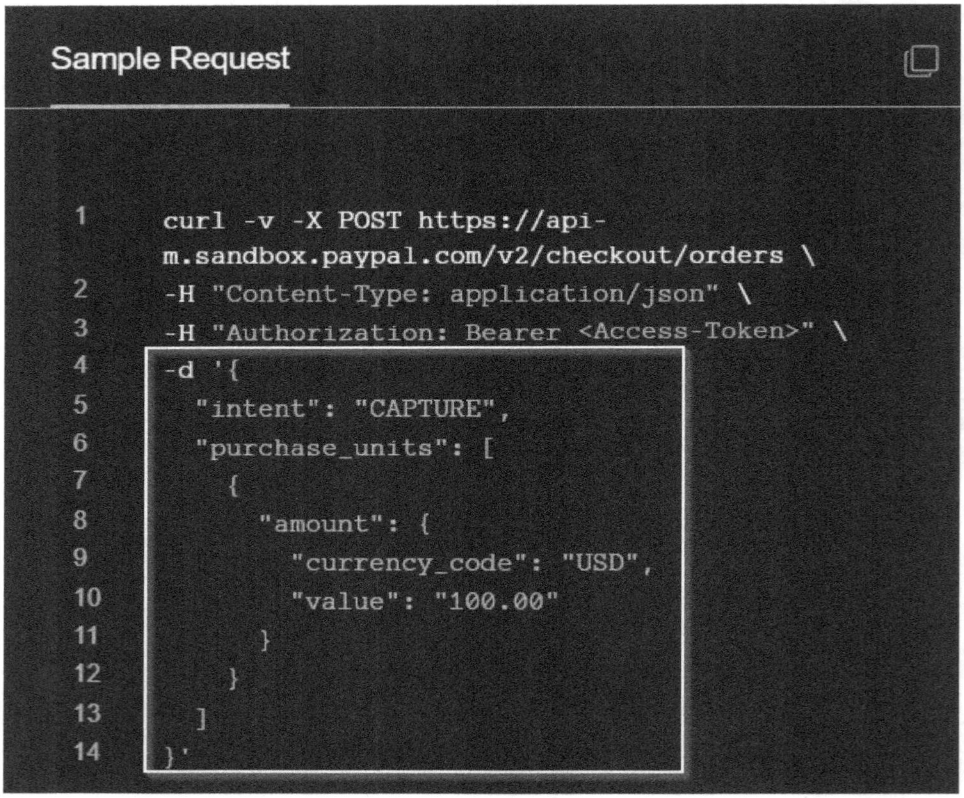
Sample Request
1 curl -v -X POST https://api-m.sandbox.paypal.com/v2/checkout/orders \
2 -H "Content-Type: application/json" \
3 -H "Authorization: Bearer <Access-Token>" \
4 -d '{
5 "intent": "CAPTURE",
6 "purchase_units": [
7 {
8 "amount": {
9 "currency_code": "USD",
10 "value": "100.00"
11 }
12 }
13]
14 }'

Figure 8.25: PayPal CreateOrder example request with highlighted body

In *Figure 8.25*, `-d` is an argument and does not belong to the request body.

20. Use https://json2csharp.com/ and generate C# classes out of the JSON. The corresponding C# classes will be generated for you.

21. Rename **RootObject** to **Order** and change all classes to the **record** type because it's a more suitable type for DTO:

IPayPalClient.cs

```
public record Order
{
    public string intent { get; set; }
    public Purchase_Units[] purchase_units { get; set; }
}

public record Name
{
    public string name { get; set; }
}

public record Purchase_Units
{
    public Amount amount { get; set; }
    public Payee payee { get; set; }
```

The complete code can be found here: https://packt.link/GvEZ8.

22. Using the same PayPal docs (https://developer.paypal.com/docs/api/orders/v2/), copy the example response:


```
{
    "id": "7XS70547FW3652617",
    "intent": "CAPTURE",
    "status": "CREATED",
    "purchase_units": [
        {
            "reference_id": "default",
            "amount": {
                "currency_code": "USD",
                "value": "100.00"
            },
            "payee": {
                "email_address": "sb-emttb7510335@business.example.
com",
                "merchant_id": "7LSF4RYZLRB96"
            }
        }
    ],
    "create_time": "2021-09-04T13:01:34Z",
    "links": [
```

```
        {
            "href": "https://api.sandbox.paypal.com/v2/checkout/
orders/7XS70547FW3652617",
            "rel": "self",
            "method": "GET"
        }
    ]
}
```

23. Use https://json2csharp.com/ and generate C# classes out of the JSON. Here, you will get classes very similar to the ones from request JSON. The only difference is the response (simplified for brevity):

```
public class CreateOrderResponse
{
    public string id { get; set; }
}
```

24. Use **AuthHeaderHandler** to fetch an access token when you make a request and make sure it inherits **DelegatingHandler**:

```
public class AuthHeaderHandler : DelegatingHandler
{
```

To make calls to PayPal, you will need an **auth** header with every request. The **auth** header value is retrieved from yet another endpoint. Refit cannot just add a header on a whim. You can, however, set up Refit using a custom **HttpClient** with a custom **HttpMessageHandler** that fetches an access token whenever you make a request. The **AuthHeaderHandler** is used for that reason.

DelegatingHandler is a class that allows intercepting **HttpRequest** when it's being sent and doing something before or after it. In this case, before you send an HTTP request, you will fetch an **auth** header and add it to the request sent.

25. Now, override **SendRequest** by adding a bearer token to **AuthenticationHeader**:

```
protected override async Task<HttpResponseMessage>
SendAsync(HttpRequestMessage request, CancellationToken
cancellationToken)
{
                var accessToken = await
GetAccessToken(CreateBasicAuthToken());
                request.Headers.Authorization = new
```

```
AuthenticationHeaderValue("Bearer", accessToken);

                return await base.SendAsync(request,
cancellationToken).ConfigureAwait(false);
}
```

26. To get an access token, you first need to get an OAuth token using basic **auth** (the client ID and secret):

```
 private static string CreateBasicAuthToken()
      {
                var credentials = Encoding.GetEncoding("ISO-8859-1").
GetBytes(PayPalClientId + ":" + PayPalSecret);
                var authHeader = Convert.ToBase64String(credentials);

                return "Basic " + authHeader;
      }
```

27. Getting an access token will require an **auth** token. Use the **RestSharp** client and add an **Authorization** header to the request.
28. Next, set **content-type** to **application/x-www-form-urlencoded** as per the PayPal API spec.
29. Add the body content **grant_type=client_credentials** as follows:

```
            private static async Task<string> GetAccessToken(string
authToken)
            {
                var request = new RestRequest("v1/oauth2/token");
                request.AddHeader("Authorization", authToken);
                request.AddHeader("content-type", "application/x-www-
form-urlencoded");
                request.AddParameter("application/x-www-form-
urlencoded", "grant_type=client_credentials", ParameterType.
RequestBody);
```

30. Execute the preceding request and return the response using the private nested class **Response** to simplify your work:

```
                var response = await RestClient.
ExecuteAsync<Response>(request, Method.POST);

                return response.Data.access_token;
            }
        private class Response
        {
```

```
            public string access_token { get; set; }
        }
    }
```

Why is the nested class needed? Here, the access token is nested within the response. It's not just a string that it returns, but rather an object. To parse it yourself from JSON would be a little complicated. However, you already know how to deserialize objects. So, even if it's just one property, deserializing still helps.

31. Now, create **RestClient** for the **GetAccessToken** method. Do so in the **AuthHandler** class:

```
private static readonly RestClient RestClient = new
RestClient(baseAddress);
```

32. In the **Demo** class, create the method **Run**:

```
public static async Task Run()
        {
```

33. Resolve a **Refit** client with a custom **AuthHeaderHandler** provider:

```
            var authHandler = new AuthHeaderHandler {InnerHandler =
new HttpClientHandler() };
            var payPalClient = RestService.For<IPayPalClient>(new
HttpClient(authHandler)
                {
                    BaseAddress = new Uri(baseAddress)
                });
```

34. Assuming that a payment was made by creating an **Order** object, run the following code:

```
var order = new Order
            {
                intent = "CAPTURE",
                purchase_units = new[]
                {
                    new Purchase_Units
                    {
                        amount = new Amount
                        {
                            currency_code = "EUR", value = "100.00"
                        }
```

```
                }
            }
        };
```

35. Now, call PayPal API and create an order endpoint with the order you've just created.

36. Get the created order to see if it works and print the retrieved order payment information:

```
var createOrderResponse = await payPalClient.CreateOrder(order);
var payment = await payPalClient.GetOrder(createOrderResponse.id);
var pay = payment.purchase_units.First();
Console.WriteLine($"{pay.payee.email_address} - " +
                  $"{pay.amount.value}" +
                  $"{pay.amount.currency_code}");
```

With the environment variables set correctly, you should see the following output:

```
sb-emttb7510335@business.example.com - 100.00EUR
```

As mentioned earlier, this is a sandbox API. However, a switch to a live environment with real money would just be a matter of setting up new PayPal accounts in that environment and calling a different endpoint: https://api-m.paypal.com.

> **NOTE**
>
> You won't be able to access https://api-m.paypal.com because it is for production PayPal use and is paid. However, that should be the only change in code (a different base URI) when you are ready to move on to real integration with PayPal.

Please make sure you have the environment variables set and are using your own client and secret. Otherwise, some unhandled exception errors may be displayed.

> **NOTE**
>
> You can find the code used for this exercise at https://packt.link/cFRq6.

You now know how to do simple CRUD operations with Web APIs. However, you have only worked with text so far. So, will calling an API with an image be any different? Find that out in the next activity.

ACTIVITY 8.04: USING AN AZURE BLOB STORAGE CLIENT TO UPLOAD AND DOWNLOAD FILES

Azure Blob Storage is a cloud service on Azure for storing different files (logs, images, music, and whole drives). Before you can use any Azure Storage services, you will need a storage account. **Blobs** are just files, but they cannot be directly stored within an account; instead, they need a container.

An Azure Storage Container is like a directory where other files are stored. However, unlike a directory, a container cannot contain other containers. Use an Azure Storage Account to create two containers, upload an image and a text file, and then download the uploaded files locally. All this will be done in your own client, which wraps around the Azure Blob storage client.

The aim of this activity is to familiarize yourself with working on files through cloud storage while putting all that you have learned so far to the test. Perform the following steps to complete this activity:

1. Navigate to **`Azure Storage Accounts`**.
2. Create a new Azure Storage Account.
3. Store a blob storage access key in environmental variables with the name **`BlobStorageKey`**.
4. Install the **`Azure Blob Storage`** client.
5. Create the **`FilesClient`** class for storing fields for blobs client and default container client (where blobs will be stored by default).
6. Create a constructor to initialize the two clients (to support access to different containers).
7. Add a method to create a container or get an existing one if it already exists.
8. Create a method to upload a file to a specific container.
9. Create a method to download a file from a specific container.
10. Create a **`Demo`** class with paths to download and upload directories.
11. Add test data, namely the two files—that is, an image and a text file (*Figure 8.26*, *Figure 8.27*, and *Figure 8.28*):

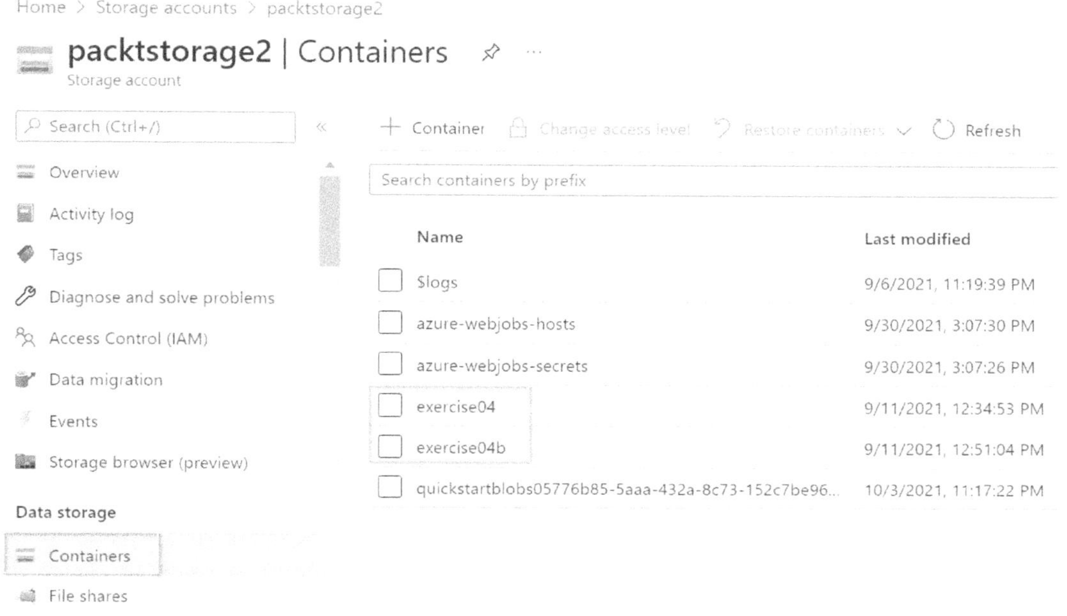

Figure 8.26: Two Azure Storage containers, exercise04 and exercise04b, in your storage account

Text file:

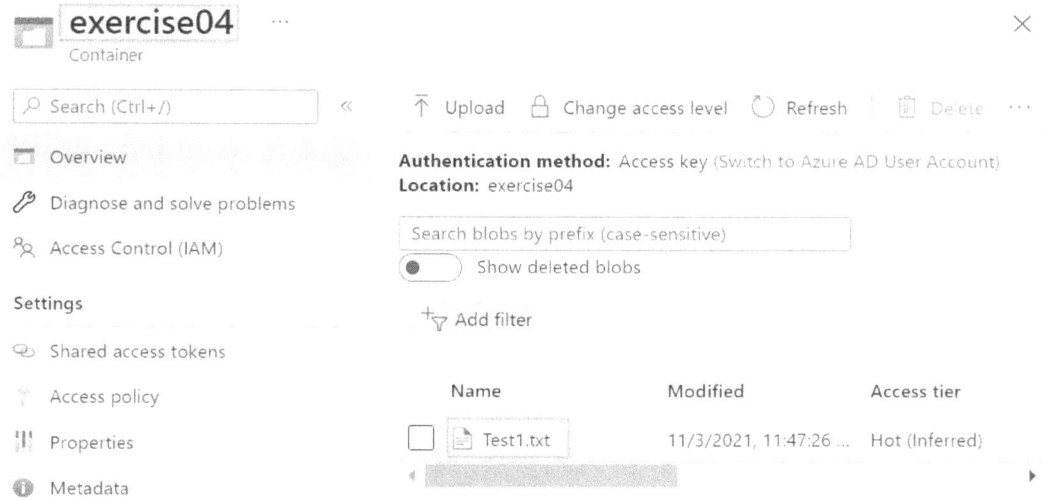

Figure 8.27: Test1.txt file uploaded in exercise04 container

Image file:

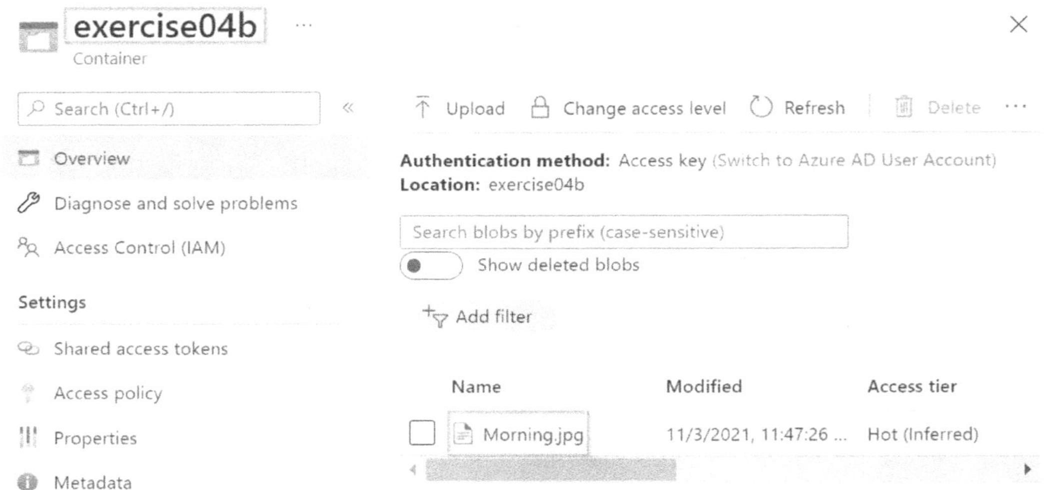

Figure 8.28: Morning.jpg file uploaded in exercise04b container

12. Create the method **Run** to upload a text file and then download it locally.
13. Run the code. If you did everything correctly, you should see the following output with both files downloaded locally:

Figure 8.29: Morning.jpg and Test1.txt files downloaded from the two containers after the demo code execution

> **NOTE**
>
> The solution to this activity can be found at https://packt.link/qclbF.

It is near impossible to create a perfect client that is suitable for everyone. Therefore, even when there is a solution for some problem given to you, you will often still need to further abstract it away, adapting it to solve exactly the problem you have. The problem you had was in uploading and downloading a file from and to a specific folder. To solve the problem, you abstracted away layers of clients exposing just two functions—one for uploading a file and another for downloading a file.

SUMMARY

No matter the kind of programmer you are, there will be many scenarios in which you will have to consume a web service. There are different kinds of services online, but the most common type is RESTful. REST is just a set of guidelines and should therefore not be mixed up with HTTP. REST APIs are simple, self-documented, well-structured, and are currently a golden standard of Web APIs. However, in most cases in the context of RESTful APIs, a request is sent over HTTP and your message contains JSON.

The main tool for making HTTP calls using C# is **`HttpClient`**, however, before you try to implement HTTP calls yourself, you should look for a NuGet package of the Web API you're trying to consume. Azure Blob storage, Azure Text Analytics, PayPal, and GitHub are just a few examples of Web APIs.

In this chapter, you learned about a lot of functionality on the web that is done for you. It's not hard to consume; all you need to know now is how to communicate with the third-party RESTful Web APIs. In the next chapter, you will learn how to create your own RESTful web services using the ASP.NET Core Web API template as well as being introduced to Azure Functions and the special tools Swagger and NuGet.

9

CREATING API SERVICES

OVERVIEW

In modern software development, most logic is served through distinct web services. This is essential to be able to both call and make new web services as a developer.

In this chapter, you will be creating your own RESTful web service using the ASP.NET Core Web API template. You will learn not only how to do it but also some of the best practices for designing and building a Web API. You will also learn how to protect an API using Azure Active Directory (AAD), centralize error handling, troubleshoot errors, generate documentation, and more.

By the end of this chapter, you will be able to create professional Web APIs that are secured with AAD, hosted on the cloud, scalable, and able to serve thousands of users.

INTRODUCTION

ASP.NET Core is a part of the .NET Core framework that is targeted at creating web apps. Using it, you can create both frontend (such as Razor or Blazor) and backend (such as Web API or gRPC) applications. However, in this chapter, you will be focusing on creating RESTful Web APIs. Creating a new web service for the first time might sound like a daunting task, but don't worry too much; for most scenarios, there is a template to get you started. In this chapter, you will create a few Web APIs using ASP.NET Core 6.0.

ASP.NET CORE WEB API

In *Chapter 8, Creating and Using Web API Clients*, you learned how to call RESTful APIs. In this chapter, you will be making one. Web API is a template for creating RESTful Web APIs in .NET. It contains routing, **Dependency Injection** (**DI**), an example controller, logging, and other useful components to get you started.

CREATING A NEW PROJECT

In order to create a new Web API, follow these steps:

1. Create a new directory.
2. Name it after a project you want to create.
3. Navigate to that directory using the `cd` command.
4. Execute the following at the command line:

   ```
   dotnet new webapi
   ```

   That is all it takes to get started.

5. To see whether this is executing as expected, run the following and see your application come to life (*Figure 9.1*):

```
dotnet run --urls=https://localhost:7021/
```

```
PS C:\Users\ITWORK\source\repos\The-C-Sharp-Workshop\MyProject> dotnet run
Building...
info: Microsoft.Hosting.Lifetime[14]
      Now listening on: https://localhost:7021
```

Figure 9.1: Terminal window showing the port the application is hosted on

In *Figure 9.1*, you will see port 7021 for the **`https`** version of the application. There may be multiple ports, especially if you are hosting both **`HTTP`** and **`HTTPs`** versions of an application. However, the key thing to remember is that you can the port where an application runs (for example, through the command line).

A **port** is a channel through which you allow a certain application to be called by all other applications. It is a number that appears after a base URL and it allows a single application through. Those applications don't have to be outsiders; the same rules also apply to internal communication.

Localhost refers to an application hosted locally. Later in this chapter, you will configure the service to bind to whatever port you want.

> **NOTE**
>
> There are 65,535 ports available on a single machine. Ports zero through 1023 are called well-known ports because usually, the same parts of the system listen on them. Typically, if a single application is hosted on one machine, the port will be 80 for **`http`** and 443 for **`https`**. If you are hosting multiple applications, the ports will vary drastically (usually starting from port 1024).

WEB API PROJECT STRUCTURE

Every Web API is made of at least two classes—**Program** and one or more controllers (**WeatherForecastController** in this case):

- **Program**: This is the **starting point** of an application. It serves as a low-level runner of an application and manages dependencies.
- **Controller**: This is a **REST API endpoint** in .NET. It usually follows a pattern of **[Model]Controller**. In this example case, **WeatherForecastController** will be called using a **/weatherforecast** endpoint.

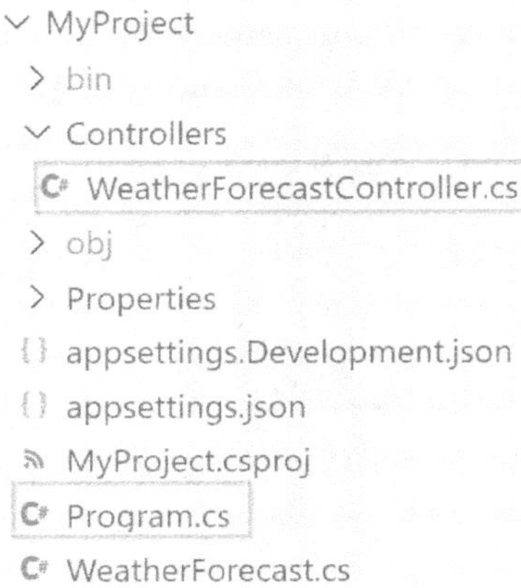

Figure 9.2: The newly created MyProject structure in VS Code with key parts highlighted

AN IN-DEPTH LOOK AT WEATHERFORECASTCONTROLLER

The controller from the default template is preceded by two attributes:

- **[ApiController]**: This attribute adds common, convenient (yet opinionated) Web API functionality.
- **[Route("[controller]")]**: This attribute is used to provide a routing pattern of a given controller.

For example, in cases where these attributes are absent or the request is complex, you would need to validate an incoming HTTP request yourself without routing out of the box:

```
[ApiController]
[Route("[controller]")]
public class WeatherForecastController : ControllerBase
{
```

This controller has **/WeatherForecast** as the route. The route is usually made of the word that precedes the word **Controller** unless specified otherwise. When developing APIs professionally, or when you have a client- and server-side application, it is recommended to preappend **/api** to the route, making it **[Route("api/[controller]")]**.

Next, you'll learn about the controller class declaration. Common controller functions come from a derived **ControllerBase** class and a few components (usually a logger) and services. The only interesting bit here is that, instead of **Ilogger**, you use **ILogger<WeatherForecastController>**:

```
    private readonly ILogger<WeatherForecastController> _logger;

    public WeatherForecastController(ILogger<WeatherForecastController> 
logger)
    {
        _logger = logger;
    }
```

The reason behind using the generic part is solely for getting the context from the place where the log was called. Using a generic version of a logger, you use a fully qualified name of a class that is supplied as a generic argument. Calling **logger.Log** will prefix it with a context; in this case, it will be **Chapter09.Service.Controllers.WeatherForecastController[0]**.

Lastly, look at the following controller method:

```
    [HttpGet]
    public IEnumerable<WeatherForecast> Get()
    {
        return new List<WeatherForecast>(){new WeatherForecast()};
    }
}
```

The **[HttpGet]** attribute binds the **Get** method with the root controller endpoint's (**/WeatherForecast**) HTTP GET method. There is a version of that attribute for every HTTP method, and they are **HttpGet**, **HttpPost**, **HttpPatch**, **HttpPut**, and **HttpDelete**. To check whether the service works, run the application using the following command:

```
dotnet run --urls=https://localhost:7021/
```

Here, the **-urls=https://localhost:7021/** argument is not a requirement. This argument simply makes sure that the port picked by .NET is the same as is indicated in this example during execution.

To see the output, navigate to **https://localhost:7021/weatherforecast/** in the browser. This will return a single default **WeatherForecast** upon calling HTTP GET:

[{"date":"0001-01-01T00:00:00","temperatureC":0,"temperatureF":32,"summary":null}].

> **NOTE**
>
> When **https://localhost:7021/weatherforecast/** displays an error message (**localhost refused to connect**), it means that the application is likely running, but on a different port. So, always remember to specify a port as described in the *Creating a New Project* section (*Step 5*).

RESPONDING WITH DIFFERENT STATUS CODES

Find out what status codes can **public IEnumerable<WeatherForecast> Get()** respond with. Using the following steps, you can play around with it and inspect what happens in the browser:

1. Navigate to **https://localhost:7021/weatherforecast/** in the browser.
2. Click on **More tools**.

3. Select the **Developer tools** option. Alternatively, you can use the **F12** key to launch the developer tools.

4. Next, click on the **Network** tab.

5. Click on the **Headers** tab. You will see that **https://localhost:7021/weatherforecast/** responds with **200 Status Code**:

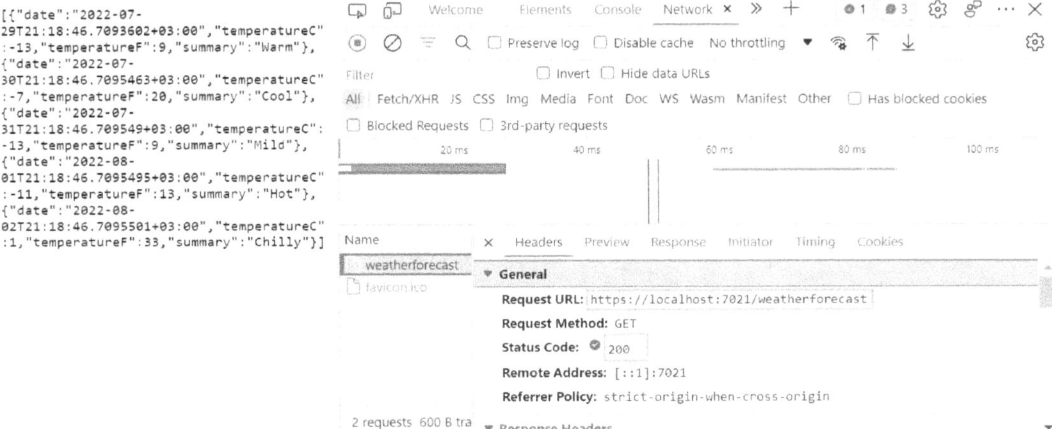

Figure 9.3: Dev tools Network tab—inspecting response headers of a successful response

6. Create a new endpoint called **GetError** that throws an exception if a rare circumstance arises while a program is running:

```
[HttpGet("error")]
public IEnumerable<WeatherForecast> GetError()
{
    throw new Exception("Something went wrong");
}
```

7. Now, call **https://localhost:7021/weatherforecast/error**. It responds with a status code of **500**:

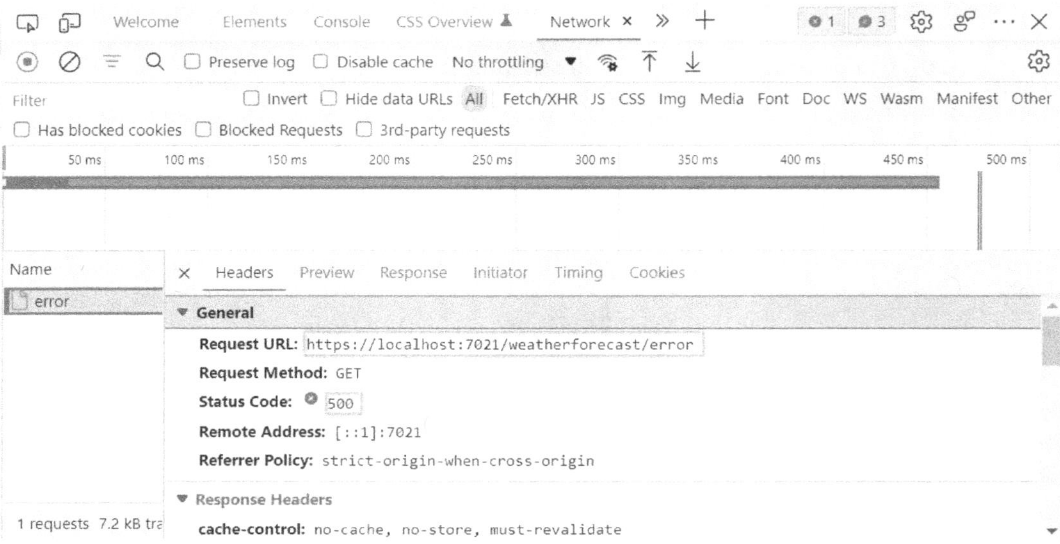

Figure 9.4: Dev tools Network tab—inspecting a response with an exception

What should you do if you want a different status code to be returned? For that, the **BaseController** class contains utility methods for returning any kind of status code you require. For example, if you wanted to explicitly return an OK response, instead of returning a value right away, you could return **Ok(value)**. However, if you try changing the code, you will get the following error:

```
Cannot implicitly convert type 'Microsoft.AspNetCore.Mvc.OkObjectResult'
to 'Chapter09.Service.Models.WeatherForecast'
```

This does not work because you do not return an HTTP status code from a controller; you either return some value or throw some error. To return any status code of your choice, you need to change the return type. For that reason, a controller should never have a return type of some value. It should always return the **IActionResult** type—a type that supports all status codes.

Create one more method for getting the weather for any day of the week. If the day is not found (a value less than **1** or more than **7**), you will explicitly return **404 – not found**:

```
[HttpGet("weekday/{day}")]
public IActionResult GetWeekday(int day)
{
```

```
    if (day < 1 || day > 7)
    {
        return NotFound($"'{day}' is not a valid day of a week.");
    }

    return Ok(new WeatherForecast());
}
```

Here, you added one new **`{day}`** at the end of the endpoint. This is a placeholder value, which comes from a matching function argument (in this case, **`day`**). Rerunning the service and navigating to **`https://localhost:7021/weatherforecast/weekday/8`** will result in a **`404 - not found`** status code because it is more than the max allowed day value, which is **`7`**:

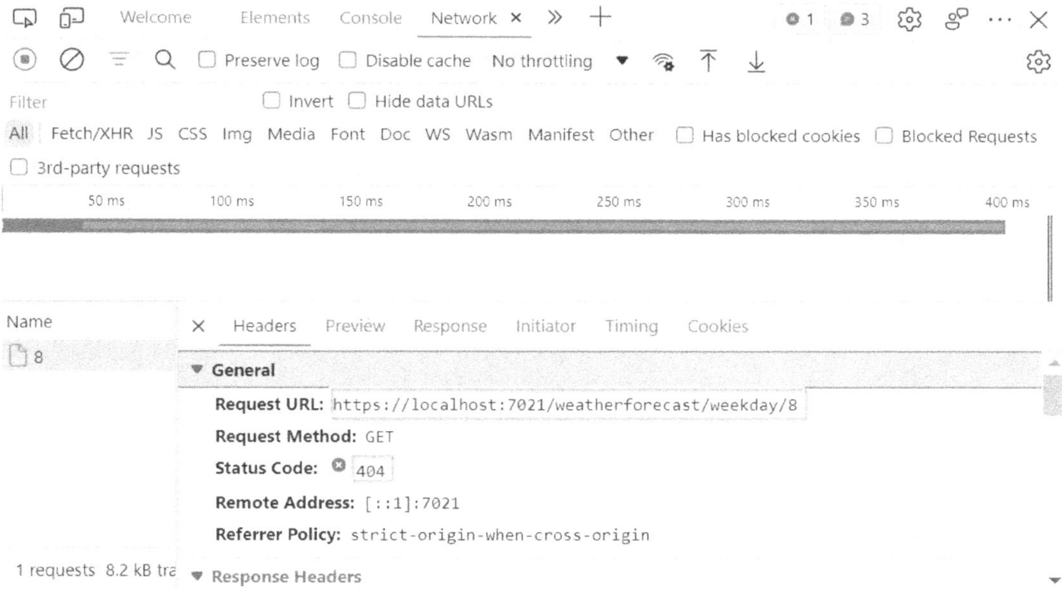

Figure 9.5: The response to finding a weather forecast for a non-existent day of the week

> **NOTE**
>
> You can find the code used for this example at https://packt.link/SCudR.

This concludes the theoretical portion of this topic. In the following section, you will put this into practice with an exercise.

EXERCISE 9.01: .NET CORE CURRENT TIME SERVICE

Once you have managed to run a Web API once, adding new controllers should be trivial. Often, whether a service is running or not, it is checked using the most basic logic; whether it is returning OK or getting the current **DateTime** value. In this exercise, you will create a simple current time service returning the current time in ISO standard. Perform the following steps to do so:

1. Create a new controller called **TimeController** to get the local time and further add functions for testing purposes:

```
[ApiController]
[Route("[controller]")]
public class TimeController : ControllerBase
{
```

   The controller shown here isn't just for testing; it acts as business logic too.

2. Add an endpoint for HTTP GET called **GetCurrentTime** that points to the **time/current** route. You will use it to get the current time:

```
    [HttpGet("current")]
    public IActionResult GetCurrentTime()
    {
```

3. Return the current **DateTime** converted to a string in ISO format:

```
        return Ok(DateTime.Now.ToString("o"));
    }
}
```

4. Navigate to **https://localhost:7021/time/current** and you should see the following response:

```
2022-07-30T15:06:28.4924356+03:00
```

As mentioned in the *Web API Project Structure* section, you can use the endpoint to determine whether a service is running or not. If it is running, then you will get the **DateTime** value, which you saw in the preceding output. If it is not running, then you would get a response with a status code of **404 - not found**. If it is running but with problems, then you would get the **500** status code.

> **NOTE**
>
> You can find the code used for this exercise at https://packt.link/OzaTd.

So far, all your focus was on a controller. It's time you shift your attention to another crucial part of a Web API—the **`Program`** class.

BOOTSTRAPPING A WEB API

The **`Program`** class wires up the whole API together. In layman's terms, you register the implementations for all the abstractions used by controllers and add all the necessary middleware.

DEPENDENCY INJECTION

In *Chapter 2, Building Quality Object-Oriented Code*, you explored the concept of DI. In *Chapter 7, Creating Modern Web Applications with ASP.NET*, you had a look at an example of DI for logging services. In this chapter, you will get hands-on experience in DI and the **Inversion of Control** (**IoC**) container—a component used to wire up and resolve all the dependencies in a central place. In .NET Core and later, the default container is **`Microsoft.Extensions.DependencyInjection`**. You will learn more about that a bit later.

PROGRAM.CS AND MINIMAL API

The simplest Web API in .NET 6 looks like this:

```
// Inject dependencies (DI)
var builder = WebApplication.CreateBuilder(args);
builder.Services.AddControllers();

// Add middleware
var app = builder.Build();

if (builder.Environment.IsDevelopment())
{
    app.UseDeveloperExceptionPage();
}

app.MapControllers();

app.Run();
```

This is a minimal API because it makes use of the top-level statements feature. Prior to .NET 6, you would have two methods within a **Startup** class (**Configure** and **ConfigureService**) and a **Program** class. Now you have a single file, **Program.cs**, and no classes or methods. You can still use the old way of starting an application. In fact, .NET 6 will generate similar classes under the hood. However, if you are making a new app in .NET 6, then using a minimal API should be preferred.

Break down the preceding code snippet. To start the application, you first need to build it. So, you will create a builder using the following line of code:

```
var builder = WebApplication.CreateBuilder(args);
```

builder.Services specifies which services are to be injected. In this case, you registered the implementation of the controllers. So, here you have just one controller calling—that is, **WeatherForecastController**:

```
builder.Services.AddControllers();
```

When you use **builder.Build()**, you can access the **app** object and further configure the application by adding middleware. For example, to add controller routing, call the following:

```
app.MapControllers();
```

Lastly, **builder.Environment.IsDevelopment()** checks whether the environment is developed. If it is developed, it calls **app.UseDeveloperExceptionPage();**, which adds detailed errors when something fails.

Logging is not mentioned anywhere; yet you still use it. A common pattern is to group all the related injections under the same extension method for **IServiceCollection**. An example of an extension method for all the controller-related functionality, including logging, is the **AddControllers** method.

You already saw the logging messages sent through the console logger right after you ran the API. Under the hood, the **builder.Services.AddLogging** method is called. This method clears all the logging providers:

```
builder.Services.AddLogging(builder =>
{
    builder.ClearProviders();
});
```

If you run the application now, you will not see anything appear in the console (*Figure 9.6*):

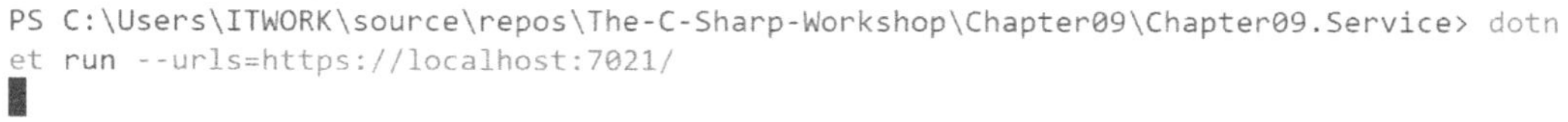

```
PS C:\Users\ITWORK\source\repos\The-C-Sharp-Workshop\Chapter09\Chapter09.Service> dotn
et run --urls=https://localhost:7021/
```

Figure 9.6: Running an application with no logs displayed

However, if you modify **AddLogging** to include **Console** and **Debug** logging in the following way, you will see the logs as in *Figure 9.7*:

```
builder.Services.AddLogging(builder =>
{
    builder.ClearProviders();
    builder.AddConsole();
    builder.AddDebug();
});
```

Now, add an error logging functionality to the error endpoint of **WeatherForecastController**. This will throw an exception when a rare circumstance arises while a program is running:

```
[HttpGet("error")]
public IEnumerable<WeatherForecast> GetError()
{
    _logger.LogError("Whoops");
    throw new Exception("Something went wrong");
}
```

Restart the API with the following command:

```
dotnet run --urls=https://localhost:7021/
```

Now, call **https://localhost:7021/weatherforecast/error** and this will show the logged message (compare *Figure 9.6* and *Figure 9.7*):

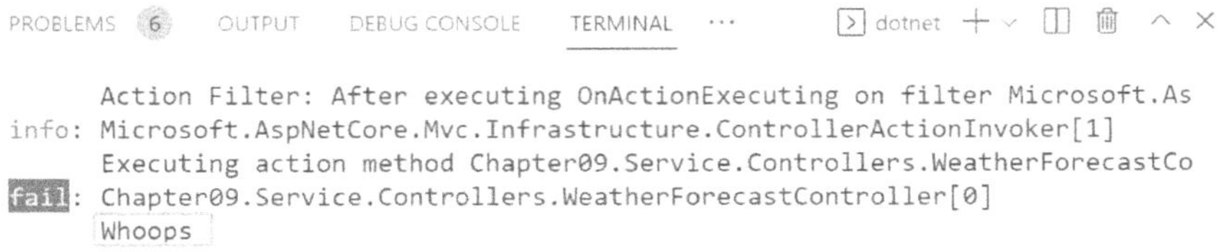

```
PROBLEMS 6   OUTPUT   DEBUG CONSOLE   TERMINAL   ...   dotnet

      Action Filter: After executing OnActionExecuting on filter Microsoft.As
info: Microsoft.AspNetCore.Mvc.Infrastructure.ControllerActionInvoker[1]
      Executing action method Chapter09.Service.Controllers.WeatherForecastCo
fail: Chapter09.Service.Controllers.WeatherForecastController[0]
      Whoops
```

Figure 9.7: The error message, Whoops, displayed on the terminal

THE INNER WORKINGS OF THE ADDLOGGING METHOD

How does the **AddLogging** method work? The decompiled code of the **AddLogging** method looks like this:

```
services.AddSingleton<ILoggerFactory, LoggerFactory>();
```

It is best practice not to initialize loggers by yourself. **ILoggerFactory** provides that functionality as a single place from which you may create loggers. While **ILoggerFactory** is an interface, **LoggerFactory** is an implementation of that interface. **AddSingleton** is a method that specifies that a single instance of **LoggerFactory** will be created and used whenever **ILoggerFactory** is referenced.

Now the question arises: why wasn't **ILoggerFactory** used in a controller? **ILoggerFactory** is used under the hood when resolving an implementation of a controller. When exposing a controller dependency such as a **logger**, you no longer need to care about how it gets initialized. This is a great benefit because it makes the class holding a dependency both more simple and more flexible.

If you do want to use **ILoggerFactory** instead of **Ilogger**, you could have a constructor accepting the factory, as follows:

```
public WeatherForecastController(ILoggerFactory logger)
```

You can then use it to create a **logger**, as follows:

```
_logger = logger.CreateLogger(typeof(WeatherForecastController).
FullName);
```

This latter **logger** functions the same as the former.

This section dealt with the **AddSingleton** method for managing service dependencies in a central place. Proceed to the next section to solve dependency complexities with DI.

THE LIFETIME OF AN INJECTED COMPONENT

The **AddSingleton** method is useful because complex applications have hundreds, if not thousands, of dependencies often shared across different components. It would be quite a challenge to manage the initialization of each. DI solves that problem by providing a central place for managing dependencies and their lifetimes. Before proceeding further, you'll need to learn more about DI lifetimes.

There are three injected object lifetimes in .NET:

- Singleton: Object initialized once per application lifetime
- Scoped: Object initialized once per request
- Transient: Object initialized every time it is referenced

To better illustrate DI and different service lifetimes, the next section will refactor the existing **WeatherForecastController** code.

DI EXAMPLES WITHIN A SERVICE

A **service** is a holder for logic at the highest level. By itself, a controller should not do any business logic and just delegate a request to some other object that is able to handle it. Apply this principle and refactor the **GetWeekday** method using DI.

First, create an interface for the service to which you will move all the logic. This is done to create an abstraction for which you will later provide an implementation. An abstraction is needed because you want to move out as much logic as possible from the controller into other components:

```
public interface IWeatherForecastService
{
    WeatherForecast GetWeekday(int day);
}
```

As you move a portion away from a controller, you would like to handle error scenarios as well. In this case, if a provided day is not between **1** and **7**, you will return a **404 - not found** error. However, at the service level, there is no concept of HTTP status codes. Therefore, instead of returning an HTTP message, you will be throwing an exception. For the exception to be handled properly, you will create a custom exception called **NoSuchWeekdayException**:

```
public class NoSuchWeekdayException : Exception
{
    public NoSuchWeekdayException(int day)
        : base($"'{day}' is not a valid day of a week.") { }
}
```

Next, create a class that implements the service. You will move your code here:

```
public class WeatherForecastService : IWeatherForecastService
{
    public WeatherForecast GetWeekday(int day)
    {
        if (day < 1 || day > 7)
        {
            throw new NoSuchWeekdayException(day);
        }

        return new WeatherForecast();
    }
}
```

The only difference here as compared to the previous code is that, instead of returning **NotFound**, you have used **throw new NoSuchWeekdayException**.

Now, inject the service into a controller:

```
private readonly IWeatherForecastService _weatherForecastService;
private readonly Ilogger _logger;

public WeatherForecastController(IloggerFactory logger,
IWeatherForecastService weatherForecastService)
{
    _weatherForecastService = weatherForecastService;
    _logger = logger.CreateLogger(typeof(WeatherForecastController).
FullName);
}
```

The cleaned-up controller method, in the *Responding with Different Status Codes* section, with minimum business logic, now looks like this:

```
[HttpGet("weekday/{day}")]
public IActionResult GetWeekday(int day)
{
    try
    {
        var result = _weatherForecastService.GetWeekday(day);
        return Ok(result);
    }
    catch(NoSuchWeekdayException exception)
    {
```

```
            return NotFound(exception.Message);
        }
    }
```

It might still seem like the same code; however, the key point here is that the controller no longer does any business logic. It simply maps results from the service back to an HTTP response.

> **NOTE**
>
> In the *Error Handling* section, you will return to this and further remove code from the controller, making it as light as possible.

If you run this code, you would get the following exception when calling any of the controller's endpoints:

```
Unable to resolve service for type 'Chapter09.Service.Examples.
TemplateApi.Services.IweatherForecastService' while attempting
to activate 'Chapter09.Service.Examples.TemplateApi.Controllers.
WeatherForecastController'
```

This exception shows that there is no way that **WeatherForecastController** can figure out the implementation for **IWeatherForecastService**. So, you need to specify which implementation fits the needed abstraction. For example, this is done inside the **Program** class as follows:

```
builder.Services.AddSingleton<IWeatherForecastService,
WeatherForecastService>();
```

The **AddSingleton** method reads this as for the **abstraction** of **IWeatherForecastService**, **register** the **WeatherForecastService implementation**. In the following paragraphs, you will learn how exactly it works.

Now that you have a service to be injected, you can explore what effect each injection has on service calls when calling the following controller method. For that point, you will slightly modify **WeatherForecastService** and **WeatherForecastController**.

Within **WeatherForecastService**, do the following:

1. Inject a **logger**:

```
        private readonly ILogger<WeatherForecastService> _logger;

        public WeatherForecastService(ILogger<WeatherForecastService>
logger)
        {
            _logger = logger;
        }
```

2. When the service is initialized, log a random **Guid** that changes the constructor to look like this:

```
        public WeatherForecastService(ILogger<WeatherForecastService>
logger)
        {
            _logger = logger;
            _logger.LogInformation(Guid.NewGuid().ToString());
        }
```

Within **WeatherForecastController**, do the following:

1. Inject the second instance of **WeatherForecastService**:

```
    public class WeatherForecastController : ControllerBase
    {
        private readonly IWeatherForecastService _
weatherForecastService1;
        private readonly IWeatherForecastService _
weatherForecastService2;
        private readonly ILogger _logger;

        public WeatherForecastController(ILoggerFactory
logger, IWeatherForecastService weatherForecastService1,
IWeatherForecastService weatherForecastService2)
        {
            _weatherForecastService1 = weatherForecastService1;
            _weatherForecastService2 = weatherForecastService2;
            _logger = logger.
CreateLogger(typeof(WeatherForecastController).FullName);
        }
```

2. Call both instances when getting a weekday:

```
        [HttpGet("weekday/{day}")]
        public IActionResult GetWeekday(int day)
        {
            try
            {
                var result = _weatherForecastService1.
GetWeekday(day);
                result = _weatherForecastService1.GetWeekday(day);
                return Ok(result);
            }
            catch (NoSuchWeekdayException exception)
            {
                return NotFound(exception.Message);
            }
        }
```

The **GetWeekday** method is called twice because it will help illustrate DI lifetimes better. Now it is time to explore different DI lifetimes.

SINGLETON

Register the service as a singleton in **Program.cs** in the following way:

```
builder.Services.AddSingleton<IWeatherForecastService,
WeatherForecastService>();
```

After calling the application, you will see the following logs generated while running the code:

```
info: Chapter09.Service.Services.WeatherForecastService[0]
      2b0c4e0c-97ff-4472-862a-b6326992d9a6
info: Chapter09.Service.Services.WeatherForecastService[0]
      2b0c4e0c-97ff-4472-862a-b6326992d9a6
```

If you call the application again, you will see the same GUID logged:

```
info: Chapter09.Service.Services.WeatherForecastService[0]
      2b0c4e0c-97ff-4472-862a-b6326992d9a6
info: Chapter09.Service.Services.WeatherForecastService[0]
      2b0c4e0c-97ff-4472-862a-b6326992d9a6
```

This proves that the service was initialized only once.

SCOPED

Register the service as scoped in **Program.cs** in the following way:

```
builder.Services.AddScoped<IWeatherForecastService,
WeatherForecastService>();
```

After calling the application, you will see the following logs generated while running the code:

```
info: Chapter09.Service.Services.WeatherForecastService[0]
      921a29e8-8f39-4651-9ffa-2e83d2289f29
info: Chapter09.Service.Services.WeatherForecastService[0]
      921a29e8-8f39-4651-9ffa-2e83d2289f29
```

On calling **WeatherForecastService** again, you will see the following:

```
info: Chapter09.Service.Services.WeatherForecastService[0]
      974e082d-1ff5-4727-93dc-fde9f61d3762
info: Chapter09.Service.Services.WeatherForecastService[0]
      974e082d-1ff5-4727-93dc-fde9f61d3762
```

This is a different GUID that has been logged. This proves that the service was initialized once per request, but a new instance was initialized on a new request.

TRANSIENT

Register the service as transient in **Program.cs** in the following way:

```
builder.Services.AddTransient<IWeatherForecastService,
WeatherForecastService>();
```

After calling the application, you should see the following in the logs generated while running the code:

```
info: Chapter09.Service.Services.WeatherForecastService[0]
      6335a0aa-f565-4673-a5c4-0590a5d0aead
info: Chapter09.Service.Services.WeatherForecastService[0]
      4074f4d3-5e50-4748-9d6f-15fb6a782000
```

That there are two different GUIDs logged proves that both services were initialized using different instances. It is possible to use DI and IoC outside of the Web API. DI through IoC is just another library with a few extras given by the Web API template.

> **NOTE**
>
> If you want to use IoC outside of ASP.NET Core, install the following NuGet (or other IoC container): **Microsoft.Extensions.DependencyInjection**.

TRYADD

So far, you have wired implementations to their abstractions using an **Add[Lifetime]** function. However, that is not the best practice in most cases. Usually, you'll want a single implementation to be wired for a single abstraction. However, if you repeatedly call **Add[Lifetime]**, for example, the **AddSingleton** function, you will create a collection of implementing instances (duplicates) underneath. This is rarely the intention and therefore you should protect yourself against that.

The cleanest way to wire dependencies is through the **TryAdd[Lifetime]** method. In the case of a duplicate dependency, it will simply not add a duplicate. To illustrate the difference between the two versions of DIs, compare the injected service counts using different methods. Here, you will inject two identical services as a singleton.

Here you are using the **Add[Lifetime]** service as a singleton:

```
builder.Services.AddSingleton<IWeatherForecastService,
WeatherForecastService>();
Debug.WriteLine("Services count: " + services.Count);
builder.services.AddSingleton<IWeatherForecastService,
WeatherForecastService>();
Debug.WriteLine("Services count: " + services.Count);
```

The command will display the following output:

```
Services count: 156
Services count: 157
```

Here you are using the **TryAdd[Lifetime]** service as a singleton:

```
builder.Services.TryAddSingleton<IWeatherForecastService,
WeatherForecastService>();
Debug.WriteLine("Services count: " + services.Count);
builder.Services.TryAddSingleton<IWeatherForecastService,
WeatherForecastService>();
Debug.WriteLine("Services count: " + services.Count);
```

The command will display the following output:

```
Services count: 156
Services count: 156
```

Observe that **Add[Lifetime]** added a duplicate in the output, while **TryAdd[Lifetime]** did not. Since you don't want duplicate dependencies, it's recommended that you use the **TryAdd[Lifetime]** version.

You can do an injection for a concrete class as well. Calling **builder.Services.AddSingleton<WeatherForecastService, WeatherForecastService>();** is a valid C# code; however, it does not make much sense. DI is used to inject an implementation into an abstraction. This will not work when bootstrapping the service because the following error will be displayed:

```
Unable to resolve a controller
```

The error occurs because there is still an abstraction-implementation binding to be provided. It would only work if a concrete implementation, rather than an abstraction, were exposed in the constructor of the controller. In practice, this scenario is rarely used.

You have learned that the cleanest way of wiring dependencies is through the **TryAdd[Lifetime]** method. You will now create a service that accepts primitive arguments (**int** and **string**) and see how it manages its non-primitive dependencies in an IoC container.

MANUAL INJECTION USING AN IOC CONTAINER

There are scenarios in which you will need to create an instance of a service before injecting it. An example use case could be a service with primitive arguments in a constructor, in other words, a weather forecast service for a specific city with a configured interval for forecast refreshes. So, here, you cannot inject a string or an integer, but you can create a service with an integer and a string and inject that instead.

Modify **WeatherForecastService** with the said features:

```
public class WeatherForecastServiceV2 : IWeatherForecastService
{
    private readonly string _city;
    private readonly int _refreshInterval;

    public WeatherForecastService(string city, int refreshInterval)
    {
        _city = city;
        _refreshInterval = refreshInterval;
    }
```

Return to the **Program** class and try to inject a service for **New York** with a refresh interval of **5** (hours):

```
builder.Services.AddSingleton<IWeatherForecastService,
WeatherForecastService>(BuildWeatherForecastService);
static WeatherForecastServiceV2
BuildWeatherForecastService(IServiceProvider _)
{
    return new WeatherForecastServiceV2("New York", 5);
}
```

In order to inject the service, as always, you use a version of the **builder.Services.Add[Lifetime]** method. However, on top of that, you provided an argument—a delegate specifying how a service should be created. The service provider can be accessed by calling the **BuildServices** method on **IServiceCollection**. This delegate takes **IServiceProvider** as input and uses it to build a new service.

In this case, you did not use it and thus named the argument after the discard operator (_). The remaining contents of the function are just a simple return with the values from the previous paragraph (for brevity, you will not add any extra logic to use the new values). If you had a more complex service, for example, a service that requires another service, you could call the **.GetService<ServiceType>** method from **IServiceProvider**.

Build and **Create** are two common method names. However, they should not be used interchangeably. Use **Build** when building a single dedicated object, while **Create** is used when the intention is to produce many objects of diverse types.

> **NOTE**
>
> You can find the code used for this example at https://packt.link/fBFRQ.

EXERCISE 9.02: DISPLAYING CURRENT TIME IN A COUNTRY API TIME ZONE

In this exercise, you are tasked with creating a Web API that provides the date and time at different time zones of UTC. Through a URL, you will pass a number between **-12** and **+12** and return the time in that time zone.

Perform the following steps:

1. Create an interface called **ICurrentTimeProvider** with a method called **DateTime GetTime(string timezone)**:

```
public interface ICurrentTimeProvider
{
    DateTime GetTime(string timezoneId);
}
```

2. Create a class called **CurrentTimeUtcProvider** implementing **ICurrentTimeProvider** to implement the logic required for the application:

```
public class CurrentTimeUtcProvider : ICurrentTimeProvider
{
```

3. Implement the method of converting the current **DateTime** to **Utc** and then offsetting that based on the time zone passed:

```
    public DateTime GetTime(string timezoneId)
    {
        var timezoneInfo = TimeZoneInfo.
FindSystemTimeZoneById(timezoneId);
        var time = TimeZoneInfo.ConvertTimeFromUtc(DateTime.UtcNow,
timezoneInfo);

        return time;
    }
}
```

4. Create a **CurrentTimeProviderController** controller to make sure it accepts **ICurrentTimeProvider** in the constructor:

```
[ApiController]
[Route("[controller]")]
public class CurrentTimeController : ControllerBase
{
    private readonly ICurrentTimeProvider _currentTimeProvider;

    public CurrentTimeController(ICurrentTimeProvider 
currentTimeProvider)
    {
        _currentTimeProvider = currentTimeProvider;
    }
```

5. Create an **HttpGet** endpoint called **IActionResult Get(string timezoneId)**, which calls the current time provider and returns the current time:

```
    [HttpGet]
    public IActionResult Get(string timezoneId)
    {
        var time = _currentTimeProvider.GetTime(timezoneId);
        return Ok(time);
    }
}
```

Please note that **{timezoneId}** is not specified in the **HttpGet** attribute. This is because the pattern is used for REST parts on an endpoint; however, in this scenario, it is passed as an argument of a query string. If a string contains whitespaces or other special characters, it should be encoded before being passed. You can URL-encode a string using this tool: https://meyerweb.com/eric/tools/dencoder/.

6. In the **Program** class, inject the service:

```
builder.Services.AddSingleton<ICurrentTimeProvider, 
CurrentTimeUtcProvider>();
```

Here, you injected the service as a singleton because it is stateless.

7. Call the **`https://localhost:7021/CurrentTime?timezone=[yourtimezone]`** endpoint with a **`timezoneid`** value of your choice. For example, you can call the following endpoint: **`https://localhost:7021/CurrentTime?timezoneid=Central%20Europe%20Standard%20Time`**.

   You will get the response showing the date and time at that time zone:

```
"2021-09-18T20:32:29.1619999"
```

> **NOTE**
>
> You can find the code used for this exercise at https://packt.link/iqGJL.

OPENAPI AND SWAGGER

OpenAPI is a **REST API** description format. It is a specification of an API with the endpoints it has, the authentication methods it supports, the arguments it accepts, and the example requests and responses it informs. The REST API works with both JSON and XML formats; however, JSON is chosen frequently. **Swagger** is a collection of tools and libraries implementing the OpenAPI standard. Swagger generates two things:

- A web page to make calls to your API
- Generate client code

In .NET, there are two libraries for working with Swagger:

- **`NSwag`**
- **`Swashbuckle`**

USING SWAGGER SWASHBUCKLE

In this section, you will use **`Swashbuckle`** to demonstrate one of many ways to test APIs and generate API documentation. So, install the **`Swashbuckle.AspNetCore`** package by running the following command:

```
dotnet add package Swashbuckle.AspNetCore
```

Just before the **builder.Build()** call, add the following line of code in **Program.cs**:

```
builder.Services.AddSwaggerGen();
```

This injects the Swagger services needed to generate the Swagger schema and the documentation test page.

After **builder.Build()** in **Program.cs**, add the following:

```
app.UseSwagger();
app.UseSwaggerUI(c => { c.SwaggerEndpoint("/swagger/v1/swagger.json", "My API V1"); });
```

The first line supports reaching the OpenAPI Swagger specification and the second one allows accessing the specification on a user-friendly web page.

Now, run the program as follows:

```
dotnet run --urls=https://localhost:7021/
```

When you navigate to **https://localhost:7021/swagger/**, you will see the following screen:

Figure 9.8: A user-friendly Swagger endpoint

Clicking on any of the endpoints will allow you to send an HTTP request to them. This page can be configured to include common information about the project, such as the contact information, licenses it is under, description, terms of services, and more.

The benefits of Swagger do not end here. If you had comments, you could include them on this page as well. You could also include all the possible response types that the endpoint produces. You can even include example requests and set them as defaults when calling an API.

Create a new endpoint to save a weather forecast and then another one to retrieve it. Document both the methods one by one. So, first, update the **IWeatherForecastService** interface to include the two new methods, **GetWeekday** and **GetWeatherForecast**, as follows:

```
    public interface IWeatherForecastService
    {
        WeatherForecast GetWeekday(int day);
        void SaveWeatherForecast(WeatherForecast forecast);
        WeatherForecast GetWeatherForecast(DateTime date);
    }
```

Next, add implementations of those methods to **WeatherForecastService**. To save the weather forecast, you will need storage, and the simplest storage would be **IMemoryCache**. Here, you will need a new field for **IMemoryCache**:

```
private readonly IMemoryCache _cache;
```

Now, update the constructor to inject **IMemoryCache**:

```
public WeatherForecastService(ILogger<WeatherForecastService> logger,
string city, int refreshInterval, IMemoryCache cache)
        {
            _logger = logger;
            _city = city;
            _refreshInterval = refreshInterval;
            _serviceIdentifier = Guid.NewGuid();
            _cache = cache;
        }
```

Then, create the **SaveWeatherForecast** method to save a weather forecast:

```
        public void SaveWeatherForecast(WeatherForecast forecast)
        {
            _cache.Set(forecast.Date.ToShortDateString(), forecast);
        }
```

Create a **GetWeatherForecast** method to get a weather forecast:

```
        public WeatherForecast GetWeatherForecast(DateTime date)
        {
            var shortDateString = date.ToShortDateString();
            var contains = _cache.TryGetValue(shortDateString, out var
entry);
            return !contains ? null : (WeatherForecast) entry;
        }
```

Now, go back to **WeatherForecastController** and create an endpoint for each method so that you can test it using the HTTP requests:

```
        [HttpGet("{date}")]
        public IActionResult GetWeatherForecast(DateTime date)
        {
            var weatherForecast = _weatherForecastService1.
GetWeatherForecast(date);
            if (weatherForecast == null) return NotFound();
            return Ok(weatherForecast);
        }

        [HttpPost]
        public IActionResult SaveWeatherForecast(WeatherForecast
weatherForecast)
        {
            _weatherForecastService1.
SaveWeatherForecast(weatherForecast);
            return CreatedAtAction("GetWeatherForecast", new { date =
weatherForecast.Date.ToShortDateString()}, weatherForecast);
        }
```

Please note that when creating a new weather forecast, you return a **CreatedAtAction** result. This returns an HTTP status code of **201** with a URI used to get the created resource. It was specified that, in order to get the created forecast later, you can use **GetWeatherForecast**. The anonymous **new { date = weatherForecast.Date.ToShortDateString()}** object specifies the arguments needed to call that action. You passed **Date.ToShortDateString()** and not just a date because a full **DateTime** contains more than what you need. Here, you need only a date; therefore, you explicitly cut what you don't need.

Document each method by describing what it does and what status codes it can return. You will then add this information above each endpoint:

```
        /// <summary>
        /// Gets weather forecast at a specified date.
        /// </summary>
        /// <param name="date">Date of a forecast.</param>
        /// <returns>
        /// A forecast at a specified date.
        /// If not found - 404.
        /// </returns>
        [HttpGet("{date}")]
        [ProducesResponseType(StatusCodes.Status404NotFound)]
        [ProducesResponseType(StatusCodes.Status200OK)]
        public IActionResult GetWeatherForecast(DateTime date)

        /// <summary>
        /// Saves a forecast at forecast date.
        /// </summary>
        /// <param name="weatherForecast">Date which identifies a
forecast. Using short date time string for identity.</param>
        /// <returns>201 with a link to an action to fetch a created
forecast.</returns>
        [HttpPost]
        [ProducesResponseType(StatusCodes.Status201Created)]
        public IActionResult SaveWeatherForecast(WeatherForecast
weatherForecast)
```

You have now added XML docs to the two endpoints. Using **`ProducesResponseType`**, you specified what status codes the endpoints could return. If you refresh the Swagger page, you will see the **`SaveWeatherForecast`** endpoint in Swagger:

Figure 9.9: SaveWeatherForecast endpoint in Swagger

If you refresh the Swagger page, you will see the **GetWeatherForecast** endpoint in Swagger:

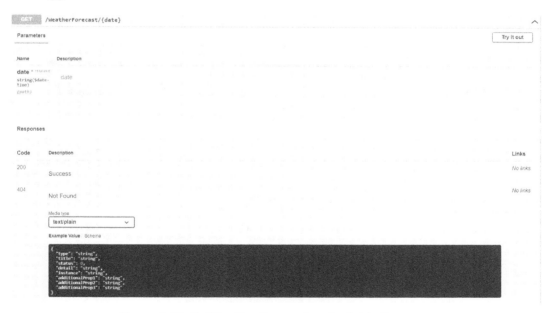

Figure 9.10: GetWeatherForecast endpoint in Swagger

You can see the status code addition, but where did the comments go? By default, Swagger does not pick XML docs. You need to specify what it has to do by configuring your project file. To do so, add the following piece of code inside **<Project>** below the property group of a target framework:

```
<PropertyGroup>
  <GenerateDocumentationFile>true</GenerateDocumentationFile>
  <NoWarn>$(NoWarn);1591</NoWarn>
</PropertyGroup>
```

Chapter09 > Chapter09.Service > Chapter09.Service.csproj

```
<Project Sdk="Microsoft.NET.Sdk.Web">

  <PropertyGroup>
    <TargetFramework>net6</TargetFramework>
  </PropertyGroup>

  <PropertyGroup>
    <GenerateDocumentationFile>true</GenerateDocumentationFile>
    <NoWarn>$(NoWarn);1591</NoWarn>
  </PropertyGroup>
```

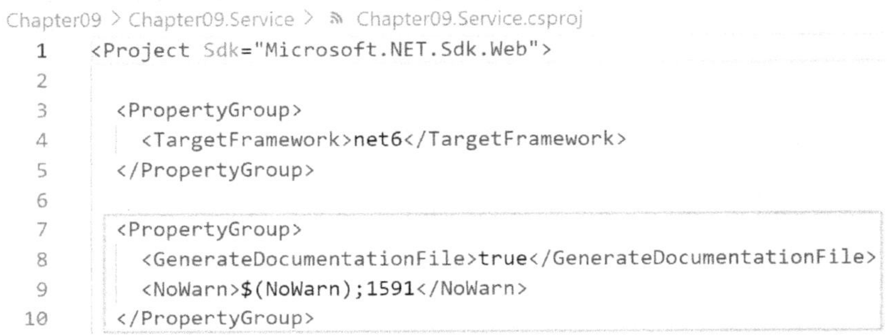

Figure 9.11: Swagger configuration to include XML docs

Lastly, go to the **Program.cs** file and replace **service.AddSwaggerGen()** with this:

```
            builder.Services.AddSwaggerGen(cfg =>
            {
                var xmlFile = $"{Assembly.GetExecutingAssembly().
GetName().Name}.xml";
                var xmlPath = Path.Combine(AppContext.BaseDirectory,
xmlFile);
                cfg.IncludeXmlComments(xmlPath);
            });
```

This is the last piece of code needed to include XML comments in the Swagger docs. Now, refresh the page and you should see the comments included:

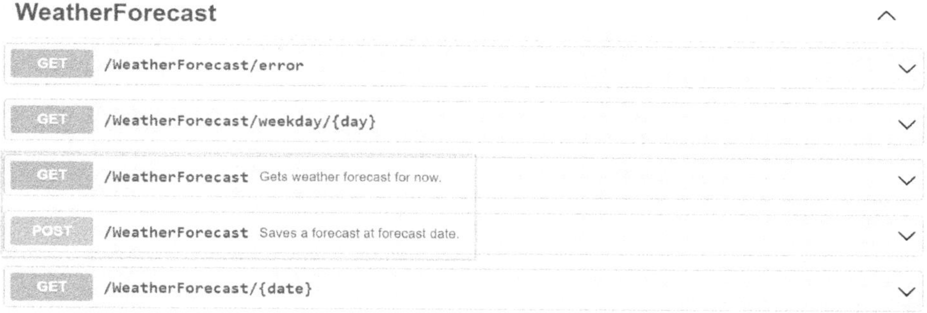

Figure 9.12: WeatherForecast Swagger docs with XML docs included

> **NOTE**
>
> You can find the code used for this example at https://packt.link/iQK5X.

There is a lot more that you can do with Swagger; you can include an example request and response and give default values to parameters. You can even create your own API specification standards and decorate a project namespace to apply the same conventions to every controller and their endpoints, but that is beyond the scope of this book.

The last thing to mention is the ability to generate a client out of the Swagger docs. To do so, follow these steps:

1. In order to download the **swagger.json** OpenAPI documentation artifact, navigate to **https://localhost:7021/swagger/v1/swagger.json**.
2. Right-click anywhere on the page and select the **Save as** option.
3. Then, press the **Enter** key.
4. Next, you will use this JSON to generate client code. So, register and log in to https://app.swaggerhub.com/home (you can use your GitHub account).

5. In the new window, click the **`Create New`** button (**1**):

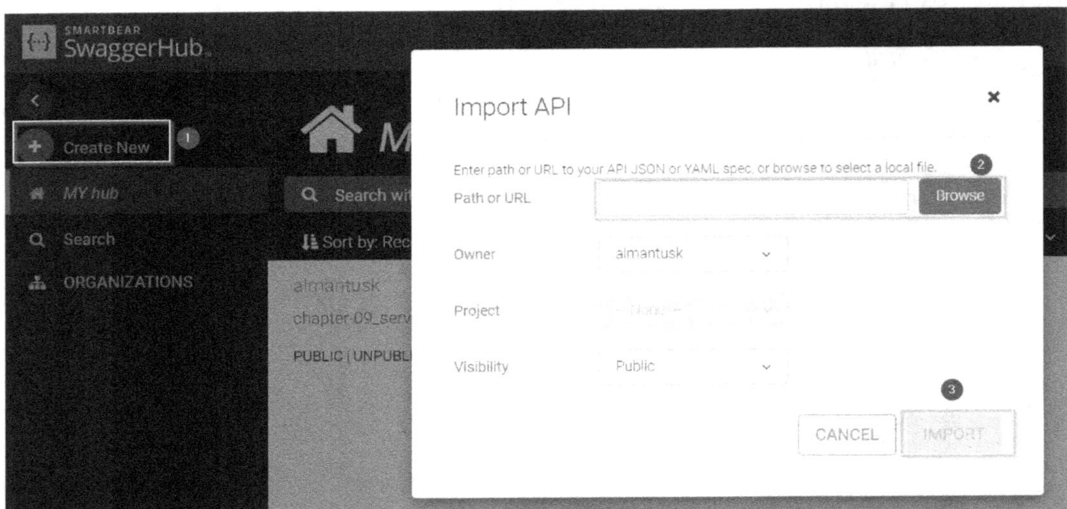

Figure 9.13: SwaggerHub and the Import API window

6. Select the **`Import and document API`** option.
7. Select the Swagger file you have just downloaded by clicking the **`Browse`** button (**2**).
8. Then, hit the **`UPLOAD FILE`** button:

> **NOTE**
>
> When you select the file, the **`IMPORT`** button (**3** in *Figure 9.13*) changes to the **`UPLOAD FILE`** button (**3** in *Figure 9.14*).

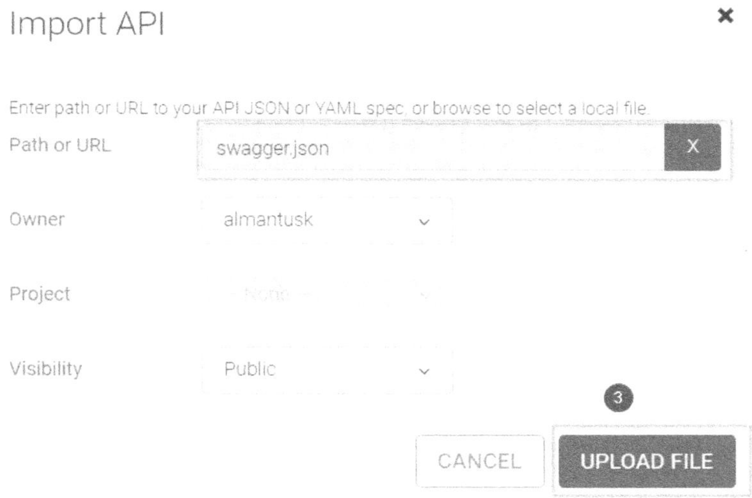

Figure 9.14: SwaggerHub IMPORT button changed to UPLOAD FILE button

9. On the next screen, leave the name of the service and the version with default values.
10. Next, click the **IMPORT DEFINITION** button:

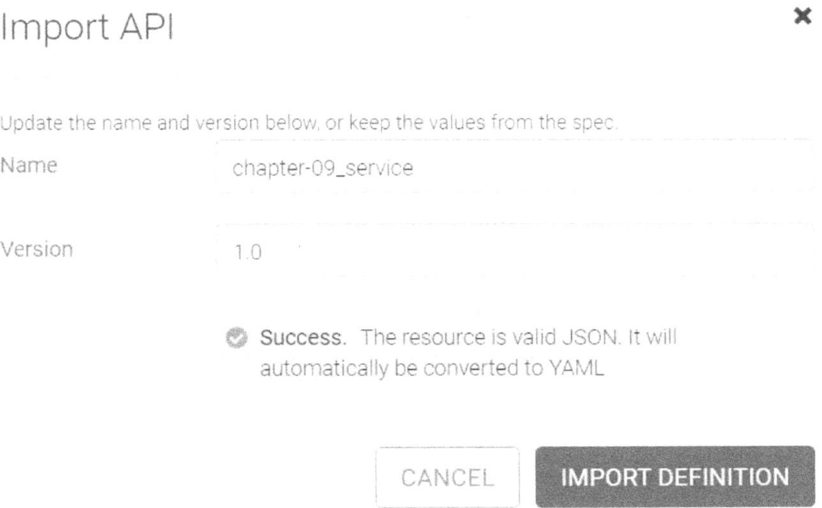

Figure 9.15: SwaggerHub import Swagger service definition

11. Now that the **Swagger.json** API scheme is imported, you can use it to generate a strongly typed C# client code to call the API. So, click the **Export** option (**1**).
12. Then, click the **Client SDK** option (**2**).
13. Select the **csharp** option (**3**):

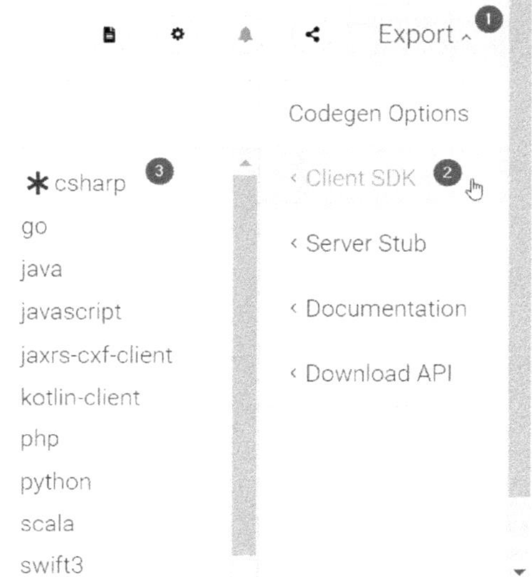

Figure 9.16: Exporting a new client in the C# client from SwaggerHub

A **csharp-client-generated.zip** file will be downloaded.

14. Extract the **csharp-client-generated.zip** file.

15. Navigate the extracted folder and open the **`IO.Swagger.sln`** file. You should see the following:

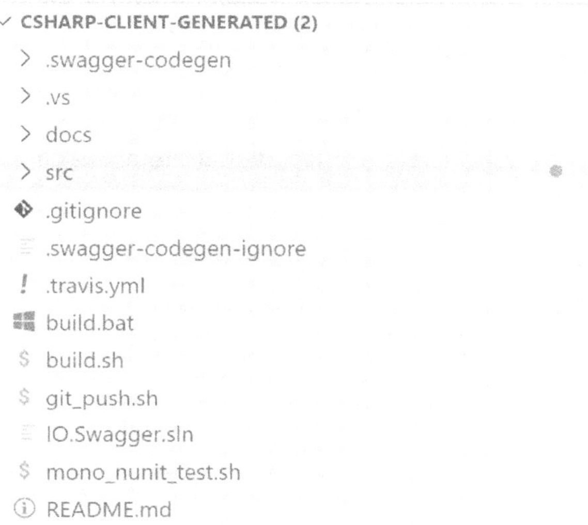

Figure 9.17: Files generated for the client using SwaggerHub

The generated client code not only has a strongly typed HTTP client but also includes tests. It also has a **`README.md`** file on how to call the client and many more common development scenarios.

Now, the question that arises is whether you should use Swagger when you already have Postman. While Postman is one of the most popular tools used for testing different kinds of Web APIs, Swagger is so much more than just a client to test whether the API works. Primarily, Swagger is a tool for documenting the API. From a conventional code, it allows you to generate all that you might need to:

- Test page
- Test the client code
- Test the documentation page

Till now, you have learned that Swagger is a collection of tools and libraries implementing OpenAPI standards that are helpful for testing and documenting your APIs. You can now proceed to grasp error handling.

ERROR HANDLING

You have already learned that the code within a controller should be as minimalistic as possible due to it being the highest level in code (direct call). Specific error handling should not be included in the controller code because it adds complexity to already-complex code. Fortunately, there is a way to map exceptions to HTTP status codes and set up all of them in one place—that is, via the **Hellang.Middleware.ProblemDetails** package. To do so, first install the package by running this command:

```
dotnet add package Hellang.Middleware.ProblemDetails
```

Map **NoSuchWeekdayException** to HTTP status code **404**. In the **Program.cs** file, before **builder.Build()**, add the following code:

```
            builder.Services.AddProblemDetails(opt =>
            {
                opt.MapToStatusCode<NoSuchWeekdayException>(404);
                opt.IncludeExceptionDetails = (context, exception) =>
false;
            });
```

This not only converts an exception to the right status code but also uses **ProblemDetails**—a standard response model based on RFC 7807—to provide faults in an HTTP response. Also, this excludes exception details in the error message.

When developing a service locally, knowing what went wrong is invaluable. However, exposing the stack trace and other information needed to determine the error can expose exploits of your Web API. Thus, it's better to hide it when moving toward the release stage. By default, the **Hellang** library already excludes the exception details in upper environments, so it is better that you don't include that line. For demo purposes and a simplified response message, it was included here.

Before you build a demo, you also need to turn off the default developer exceptions page because it overrides the exceptions in **ProblemDetails**. Simply remove the following block of code from the **Configure** method:

```
        if (builder.Environment.IsDevelopment())
        {
            app.UseDeveloperExceptionPage();
        }
```

Since you already have a central place for handling **NoSuchWeekdayException**, you can simplify the controller method for getting **WeatherForecast** for a given date:

```
[HttpGet("weekday/{day}")]
public IActionResult GetWeekday(int day)
{
    var result = _weatherForecastService.GetWeekday(day);
    return Ok(result);
}
```

When calling the endpoint with an invalid day value (for example, **9**), you get the following response:

```
{
    "type": "/weatherforecast/weekday/9",
    "title": "Not Found",
    "status": 404,
    "traceId": "|41dee286-4c5efb72e344ee2d."
}
```

This centralized error handling approach allows the controllers to be rid of all the **try-catch** blocks.

> **NOTE**
>
> You can find the code used for this example at https://packt.link/CntW6.

You can now map exceptions to HTTP status codes and set them all up in one place. This next section will take a look at another addition to an API, which is request validation.

REQUEST VALIDATION

Another useful addition to an API is request validation. By default, ASP.NET Core has a request validator based on the required attributes. However, there might be complex scenarios where a combination of properties results in an invalid request or a custom error message for which validation is required.

.NET has a great NuGet package for that: **FluentValidation.AspNetCore**. Perform the following steps to learn how to carry out request validation. Before you continue, install the package by running the following command:

```
dotnet add package FluentValidation.AspNetCore
```

This package allows registering custom validators per model. It makes use of existing ASP.NET Core middleware, so all you must do is inject a new validator. Create a validator for **WeatherForecast**.

A validator should inherit the **AbstractValidator** class. This is not obligatory, but it is highly recommended because it implements the common methods for functionality and has a default implementation for generic validation:

```
public class WeatherForecastValidator :
AbstractValidator<WeatherForecast>
```

Through a generic argument, you specified that this is a validator for **WeatherForecast**.

Next is the validation itself. This is done in a constructor of a validator:

```
        public WeatherForecastValidator()
        {
            RuleFor(p => p.Date)
                .LessThan(DateTime.Now.AddMonths(1))
                .WithMessage("Weather forecasts in more than 1 month of
future are not supported");

            RuleFor(p => p.TemperatureC)
                .InclusiveBetween(-100, 100)
                .WithMessage("A temperature must be between -100 and +100
C.");
        }
```

FluentValidation is a .NET library and is all about fluent API, with self-explanatory methods. Here, you require a weather forecast date to be no more than one month in the future. The next validation is to have the temperature between **-100 C** and **100 C**.

If you ping your API through Swagger, the following request gets displayed:

```
{
  "date": "2022-09-19T19:34:34.511Z",
  "temperatureC": -111,
  "summary": "string"
}
```

The response will be displayed as follows:

```
{
  "type": "https://tools.ietf.org/html/rfc7231#section-6.5.1",
  "title": "One or more validation errors occurred.",
  "status": 400,
  "traceId": "|ade14b9-443aaaf79026feec.",
  "errors": {
    "Date": [
      "Weather forecasts in more than 1 month of future are not 
supported"
    ],
    "TemperatureC": [
      "A temperature must be between -100 and +100 C."
    ]
  }
}
```

You don't have to use **`FluentValidation`**, especially if your API is simple and does not have complex rules. But in an enterprise setting, it is highly recommended that you do use it because the level of detail you can add to your validation is unlimited.

You learned about **`FluentValidation`** and the scenarios where it is useful. The next section will touch upon the two options for reading configuration in ASP.NET.

> **NOTE**
>
> You can find the code used for this example at https://packt.link/uOGOe.

CONFIGURATION

In ASP.NET Core Web API, you have two options for reading configuration:

- **IConfiguration**: This is a global configuration container. Even though it allows access to all the configuration properties, injecting it directly into other components is inefficient. This is because it is weakly typed and has a risk of you trying to get a non-existing configuration property.
- **IOptions**: This is strongly typed and convenient because the configuration is fragmented into just the pieces that a component needs.

You can choose either of the two options. It is best practice to use **IOptions** in ASP.NET Core, as the configuration examples will be based on it. Whichever option you choose, you need to store the configuration in the **appsettings.json** file.

Move the hardcoded configuration from a constructor (weather forecast city and refresh interval) and move it into a configuration section in the **appsettings.json** file:

```
"WeatherForecastConfig": {
  "City": "New York",
  "RefreshInterval":  5
}
```

Create a model representing this configuration section:

```
public class WeatherForecastConfig
{
    public string City { get; set; }
    public int RefreshInterval { get; set; }
}
```

You no longer have to inject the two primitive values into the component. Instead, you will inject **IOptions<WeatherForecastConfig>**:

```
public WeatherForecastService(Ilogger<WeatherForecastService> logger,
Ioptions<WeatherForecastConfig> config, ImemoryCache cache)
```

Before the JSON section is useable, you need to bind to it. This can be done by finding the section through **IConfiguration** (via the **builder.Configuration** property):

```
builder.Services.Configure<WeatherForecastConfig>(builder.Configuration.
GetSection(nameof(WeatherForecastConfig)));
```

In this case, **WeatherForecastConfig** has a matching section in the configuration file. Therefore, **nameof** was used. So, **nameof** should be preferred when using the alternative **string** type. That way, if the name of a type changes, the configuration will change consistently (or else the code won't compile).

Remember the **BuildWeatherForecastService** method you used previously? The beauty of it all is that the method can be removed altogether because the service can be created without the need for custom initialization. If you compile and run the code, you will get the same response.

> **NOTE**
>
> You can find the code used for this example at https://packt.link/xoB0K.

ASP.NET Core Web API is just a collection of libraries on top of the .NET Core framework. You can use **appsettings.json** in other types of applications as well. It is better to use individual libraries regardless of the project type you choose. In order to use the configuration through JSON, all you need to do is to install the following NuGet packages:

- **Microsoft.Extensions.Configuration**
- **Microsoft.Extensions.Configuration.EnvironmentVariables**
- **Microsoft.Extensions.Configuration.FileExtensions**
- **Microsoft.Extensions.Configuration.Json**
- **Microsoft.Extensions.Options**

In this section, you learned how to use **IConfiguration** and **IOptions**. Your API is now ready, and it already includes many standard components of a typical Web API. The next section will detail how you can handle this complexity in code.

DEVELOPMENT ENVIRONMENTS AND CONFIGURATION

Applications often need to have two environments—production and development. You want the application development environment to have premade settings, more detailed error messages (if possible), more detailed logging, and lastly, debugging enabled. All of that is not needed for a production environment and you would want to keep it clean.

Other than the build configuration, you manage environments through different configuration files. The **appsettings.json** file is a base configuration file and is used across all environments. This configuration file should contain the configuration you would like for production.

The **Appsettings.development.json** file is a configuration file that will be applied when you build your application in debug mode. Here, **applied** doesn't mean a complete overwrite of settings; **appsettings.json** will still be used with the development settings overriding the matching sections. A common example is described here.

Say **appsettings.json** has the following:

```
{
  "Logging": {
    "LogLevel": {
      "Default": "Information",
      "Microsoft": "Information",
      "Microsoft.Hosting.Lifetime": "Information"
    }
  },
  "AllowedHosts": "*",
  "WeatherForecastConfig": {
    "City": "New York",
    "RefreshInterval": 5
  },
  "WeatherForecastProviderUrl": "https://community-open-weather-
map.p.rapidapi.com/",
  "AzureAd": {
    "Instance": "https://login.microsoftonline.com/",
    "ClientId": "2d8834d3-6a27-47c9-84f1-0c9db3eeb4bb",
    "TenantId": "ddd0fd18-f056-4b33-88cc-088c47b81f3e",
    "Audience": "api://2d8834d3-6a27-47c9-84f1-0c9db3eeb4bb"
  }
}
```

And **appsettings.development.json** has the following:

```
{
  "Logging": {
    "LogLevel": {
      "Default": "Trace",
      "Microsoft": "Trace",
```

```
      "Microsoft.Hosting.Lifetime": "Trace"
    }
  }
}
```

Then, the settings used will be the merged file with override matching sections, as shown here:

```
{
  "Logging": {
    "LogLevel": {
      "Default": "Trace",
      "Microsoft": "Trace",
      "Microsoft.Hosting.Lifetime": "Trace"
    }
  },
  "AllowedHosts": "*",
  "WeatherForecastConfig": {
    "City": "New York",
    "RefreshInterval": 5
  },
  "WeatherForecastProviderUrl": "https://community-open-weather-map.p.rapidapi.com/",
  "AzureAd": {
    "Instance": "https://login.microsoftonline.com/",
    "ClientId": "2d8834d3-6a27-47c9-84f1-0c9db3eeb4bb",
    "TenantId": "ddd0fd18-f056-4b33-88cc-088c47b81f3e",
    "Audience": "api://2d8834d3-6a27-47c9-84f1-0c9db3eeb4bb"
  }
}
```

In the next section, you will learn how to manage DI more cleanly.

BOOTSTRAPPING

Complexity needs to be handled and the complexity referred to here is the `Program` class. You'll need to break it out into smaller pieces and form a Bootstrapping directory specifying the components the service is made of.

When breaking down code within `Program.cs`, it is recommended to use a **fluent API pattern**. This is a pattern where you can chain multiple function calls from a single root object. In this case, you will create several extension methods for the `IServiceCollection` type and chain all the module injections one by one.

To reduce the complexity of the **Program** class, move the DI of different logical sections into different files. Each step that follows will do just that. So, split the controller and API baseline setup to a new file named **ControllersConfigurationSetup.cs**:

```
public static class ControllersConfigurationSetup
{
    public static IserviceCollection AddControllersConfiguration(this
IserviceCollection services)
    {
        services
            .AddControllers()
            .AddFluentValidation();
        return services;
    }
}
```

Now, move the code for logging to a new file named **LoggingSetup.cs**:

```
public static class LoggingSetup
{
    public static IServiceCollection AddLoggingConfiguration(this
IServiceCollection services)
    {
        services.AddLogging(builder =>
        {
            builder.ClearProviders();
            builder.AddConsole();
            builder.AddDebug();
        });
        return services;
    }
}
```

Next, move the request validation logic to a new file named **RequestValidatorsSetup.cs**:

```
public static class RequestValidatorsSetup
{
    public static IServiceCollection AddRequestValidators(this
IServiceCollection services)
    {
        services.AddTransient<Ivalidator<WeatherForecast>,
WeatherForecastValidator>();
```

```
            return services;
        }
    }
```

Move the Swagger setup logic to a new file named **SwaggerSetup.cs**:

```
    public static class SwaggerSetup
    {
        public static IServiceCollection AddSwagger(this
IServiceCollection services)
        {
            services.AddSwaggerGen(cfg =>
            {
                var xmlFile = $"{Assembly.GetExecutingAssembly().
GetName().Name}.xml";
                var xmlPath = Path.Combine(AppContext.BaseDirectory,
xmlFile);
                cfg.IncludeXmlComments(xmlPath);
            });
            return services;
        }
    }
```

Move the injection of the **WeatherForecast**-related classes' code to a new file named **WeatherServiceSetup.cs**:

```
    public static class WeatherServiceSetup
    {
        public static IServiceCollection AddWeatherService(this
IServiceCollection services, IConfiguration configuration)
        {
            services.AddScoped<IWeatherForecastService,
WeatherForecastService>(BuildWeatherForecastService);
            services.AddSingleton<ICurrentTimeProvider,
CurrentTimeUtcProvider>();
            services.AddSingleton<ImemoryCache, MemoryCache>();
            services.Configure<WeatherForecastConfig>(configuration.
GetSection(nameof(WeatherForecastConfig)));
            return services;
        }

        private static WeatherForecastService
BuildWeatherForecastService(IserviceProvider provider)
        {
            var logger = provider
                .GetService<IloggerFactory>()
```

```
                    .CreateLogger<WeatherForecastService>();
                var options = provider.
GetService<Ioptions<WeatherForecastConfig>>();
                return new WeatherForecastService(logger, options, provider.
GetService<ImemoryCache>());
            }
        }
```

Finally, move the exception mapping of HTTP status codes to a new file named **`ExceptionMappingSetup.cs`**:

```
        public static class ExceptionMappingSetup
        {
            public static IServiceCollection AddExceptionMappings(this
IServiceCollection services)
            {
                services.AddProblemDetails(opt =>
                {
                    opt.MapToStatusCode<NoSuchWeekdayException>(404);
                });

                return services;
            }
        }
```

Now move all the new classes under **`/Bootstrap`** folder:

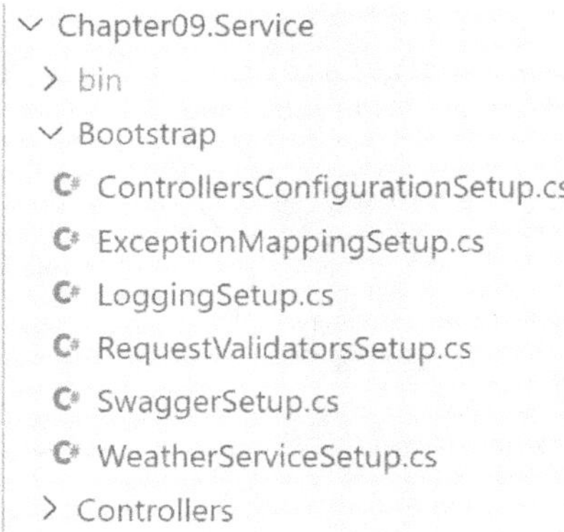

Figure 9.18: Bootstrap folder with the fragmented services injection

Figure 9.18 displays the **Bootstrap** folder. This project structure itself demonstrates what the API is made up of. So, DI becomes as simple as the following:

```
builder.Services
    .AddControllersConfiguration()
    .AddLoggingConfiguration()
    .AddRequestValidators()
    .AddSwagger()
    .AddWeatherService(builder.Configuration)
    .AddExceptionMappings();
```

In some cases, you may want to pass the configuration or environment from a builder to other bootstrap methods or app methods multiple times. If you find yourself repeatedly calling **builder.X**, then consider storing each property in a local variable, as shown here:

```
var services = builder.Services;
var configuration = builder.Configuration;
var environment = builder.Environment;
```

With this, you will no longer repeatedly access the builder and will instead be able to use the needed builder properties directly. This is especially useful if you migrate from .NET Core to .NET 6. **Environment** and **Configuration** used to be properties of a **Program** class, while **Services** would be injected into the **ConfigureServices** method. In .NET 6, **Services** is accessed through a **builder** object. However, with this approach, you can still use those properties or arguments as they were.

From now on, when referring to services, environments, or configurations, you will assume that you are accessing them from **builder.Services**, **builder.Environment**, and **builder.Configuration**, accordingly.

> **NOTE**
>
> You can find the code used for this example at https://packt.link/iQK5X.

CALLING ANOTHER API

A working product is usually made of many APIs communicating with each other. To communicate effectively, one web service often needs to call another service. For example, a hospital may have a website (frontend) that calls a Web API (backend). This Web API orchestrates things by making calls to a booking Web API, a billing Web API, and a staff Web API. A staff Web API may make calls to an inventory API, holidays API, etc.

RAPIDAPI

As discussed in *Chapter 8, Creating and Using Web API Clients*, there are various ways of making HTTP calls to other services (though HTTP is not the only way to call another service). This time, you will try to get weather forecasts from an existing API and format it in your way. For doing so, you will use the RapidAPI Weather API, which can be found at https://rapidapi.com/visual-crossing-corporation-visual-crossing-corporation-default/api/visual-crossing-weather/.

> **NOTE**
>
> RapidAPI is a platform that supports many APIs. The site https://rapidapi.com/visual-crossing-corporation-visual-crossing-corporation-default/api/visual-crossing-weather/ is just one example. Many of the APIs present there are free; however, be aware that an API that is free today might become paid tomorrow. If that happens by the time you read this chapter, go through the examples, and explore the *Weather APIs* section at https://rapidapi.com/category/Weather. You should be able to find similar alternatives there.

This API requires a GitHub account for use. Perform the following steps to use the RapidAPI Weather API:

1. Log in to the website https://rapidapi.com/community/api/open-weather-map/.

> **NOTE**
>
> You can navigate to https://rapidapi.com/community/api/open-weather-map/ only if you are logged in. So, signup at https://rapidapi.com/ and create an account. This is required if you need an API key. Next login and select **`Weather`** category and choose **`Open Weather`** link.

After you log in to the website, you will see the following window:

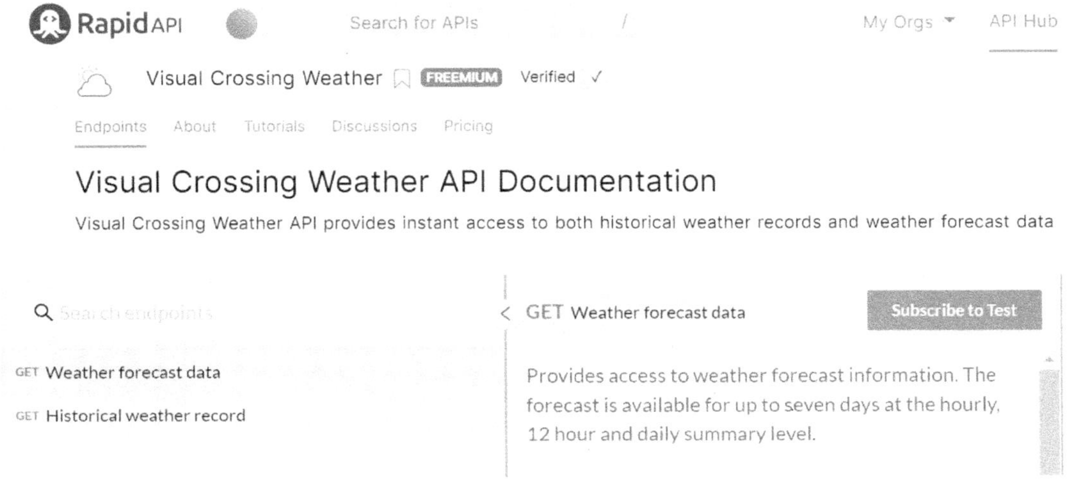

Figure 9.19: Unsubscribed test page of the Visual Crossing Weather API on rapidapi.com

2. Click the **`Subscribe to Test`** button to get access (for free) to making calls to the Web API. A new window will open.

3. Select the **`Basic`** option, which will allow you to make 500 calls a month to that API. For educational purposes, the basic plan should be enough:

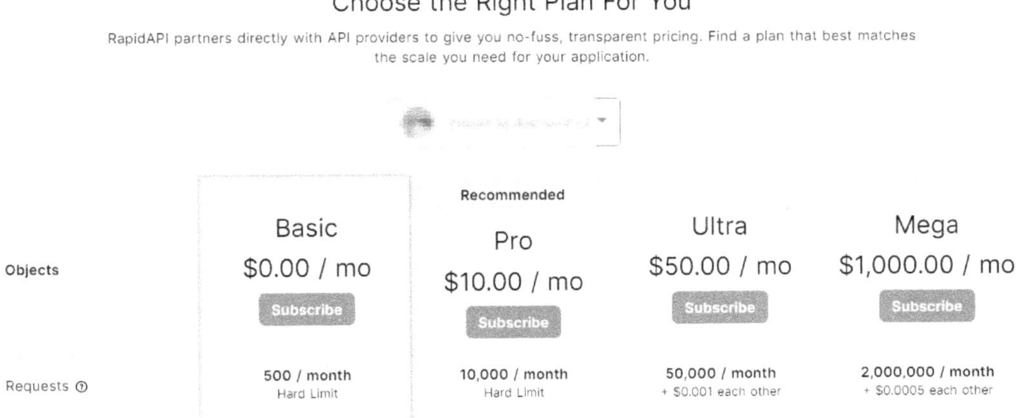

Figure 9.20: RapidAPI subscription fees with a free Basic plan highlighted

You will be redirected to the test page with the **`Test Endpoint`** button available (instead of the **`Subscribe to Test`** button).

4. Now, configure the request. The first configuration asks you to enter the intervals for getting the weather forecast. You want an hourly forecast, so enter **`1`** hour beside **`aggregateHours`** (**1**).

5. Next up is the **`location`** address (**2**).

   In *Figure 9.21*, you can observe that the city, state, and country are specified. These fields ask you to enter your address. However, typing your city name would also work.

6. Choose the default **`contentType`** option as **`csv`** for this API (**3**):

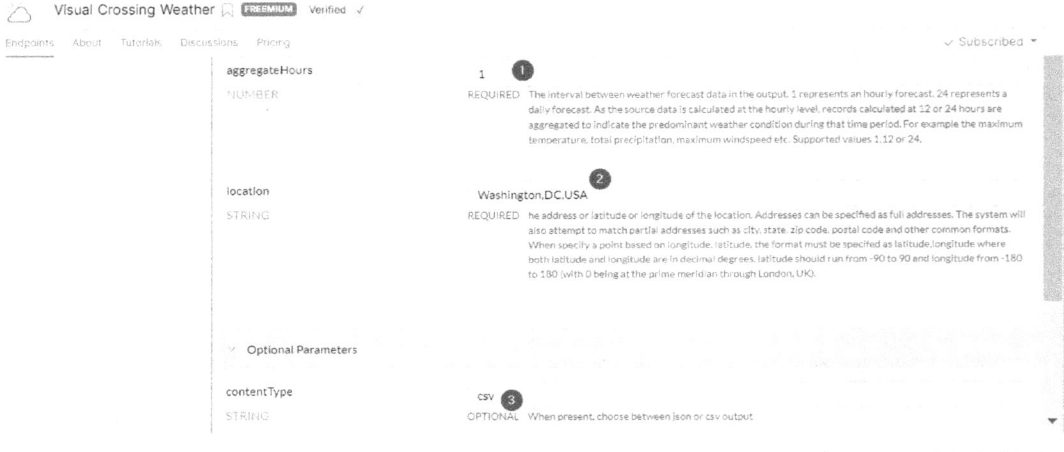

Figure 9.21: GET weather forecast data request configuration

This API is interesting because it allows you to return data in different formats—JSON, XML, and CSV. It is still a Web API and not so RESTful because the data response type is natively CSV. If you choose JSON, it will look unnatural and significantly more difficult to work with.

7. On the next screen, click **`Code Snippets`** (**1**) and then **`(C#) HttpClient`** (**2**) to see the example client code generated for you.

8. Next, click **`Test Endpoint`** (**3**) to send a request.

9. Click the **`Results`** tab (**4**) to view the response (in *Figure 9.22*, other endpoints are collapsed):

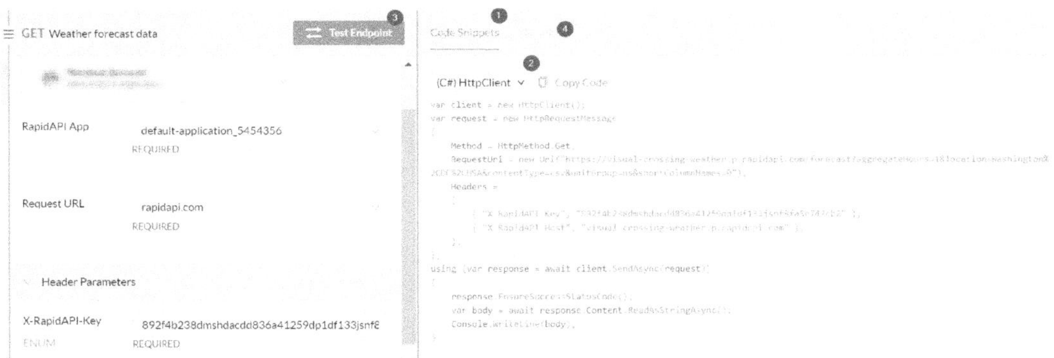

Figure 9.22: rapidapi.com with test request page and example code in C# for making the request

This window provides a nice API. It is also a great way to learn how to make calls to it by giving multiple examples of creating clients using a variety of languages and technologies.

As always, you will not initialize this client directly in a client but inject the client somehow. In *Chapter 8, Creating and Using Web API Clients*, it was mentioned that to have a static **`HttpClient`** over one constantly disposed is an efficient practice. However, for a Web API, there is an even better alternative—**`HttpClientFactory`**.

10. Before you do all that, you need to prepare a few things. First, update the **`appsettings.json`** file with the inclusion of the base URL of an API:

```
"WeatherForecastProviderUrl": "https://visual-crossing-
weather.p.rapidapi.com/"
```

Next, you will need to create another class for fetching the weather details from the said API. For that purpose, you will need an API key. You can find it in the example code snippet on the API website:

Code Snippets

(C#) HttpClient ∨ Copy Code

```
var client = new HttpClient();
var request = new HttpRequestMessage
{
    Method = HttpMethod.Get,
    RequestUri = new Uri("https://visual-crossing-weather.p.rapidapi.com/forecast?aggregateHours=1&location=Washington%
2CDC%2CUSA&contentType=csv&unitGroup=us&shortColumnNames=0"),
    Headers =
    {
        { "X-RapidAPI-Key", "892f4b238dmshdacdd836a41259dp1df133jsnf8fa5e742cb2" },
        { "X-RapidAPI-Host", "visual-crossing-weather.p.rapidapi.com" },
    },
};
using (var response = await client.SendAsync(request))
{
    response.EnsureSuccessStatusCode();
    var body = await response.Content.ReadAsStringAsync();
    Console.WriteLine(body);
}
```

Figure 9.23: RapidAPI API key in the example code snippet

11. Save the API key as an environment variable because it is a secret and storing secrets in code is bad practice. So, name it as **`x-rapidapi-key`**.
12. Lastly, the returned weather forecast might be quite different from yours. You can see the example response by clicking the **`Test Endpoint`** button:

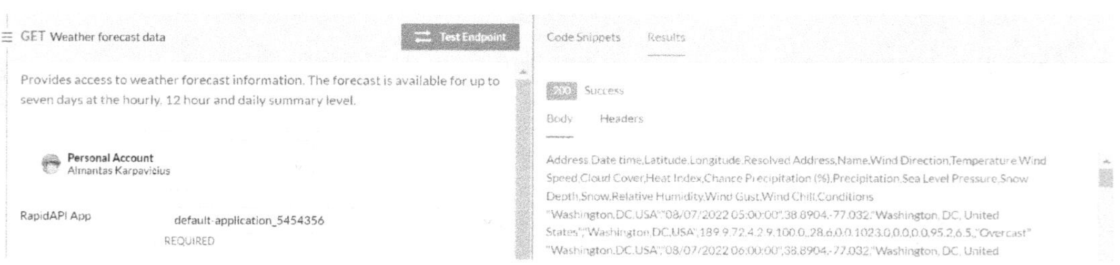

Figure 9.24: RapidAPI example response from GET current weather data endpoint

13. Copy the results received after clicking the **`Test Endpoint`** button.
14. Paste the results in https://toolslick.com/generation/code/class-from-csv.
15. Give the class name as **`WeatherForecast`** and leave the rest of the settings as the defaults.

16. Finally, press the **GENERATE** button:

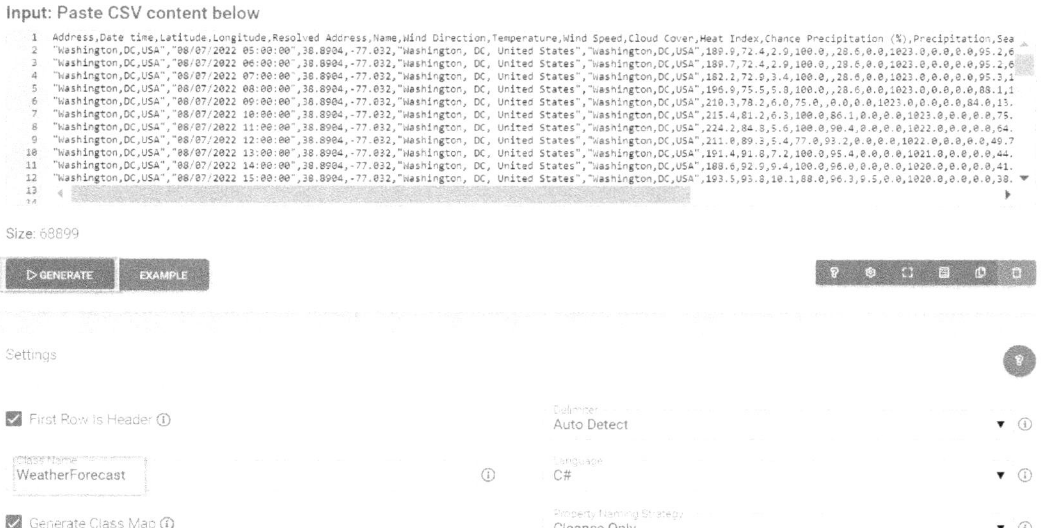

Figure 9.25: Response content pasted to https://toolslick.com/generation/code/class-from-csv

This will create two classes, **WeatherForecast** and **WeatherForecastClassMap**:

Output: Generated Classes

```
public class WeatherForecast
{
    public string Address { get; set; }
    public DateTime Datetime { get; set; }
    public double Latitude { get; set; }
    public double Longitude { get; set; }
    public string WindDirection { get; set; }
    public string Temperature { get; set; }
    public string WindSpeed { get; set; }
    public string Conditions { get; set; }
}

public class WeatherForecastClassMap : ClassMap<WeatherForecast>
```

Figure 9.26: Generated data model and mapping classes (simplified for brevity)

WeatherForecast represents the object to which the data from this API will be loaded.

17. Create a file called **WeatherForecast.cs** under the **Dtos** folder (DTO will be described in detail in the *DTO and Mapping Using AutoMapper* section) and paste the class there.

18. Remove the bits that do not have a connection to an already-existing **WeatherForecast** model. The cleaned-up model will look as follows:

```
public class WeatherForecast
{
    public DateTime Datetime { get; set; }
    public string Temperature { get; set; }
    public string Conditions { get; set; }
}
```

You should know that **WeatherForecastClassMap** is a special class. It is used by the **CsvHelper** library, which is used for parsing CSV files. You could parse CSV files yourself; however, **CsvHelper** makes it a lot easier to parse.

19. To use **CsvHelper**, install its NuGet package:

```
dotnet add package CsvHelper
```

WeatherForecastCsv represents a mapping from a CSV to a C# object.

20. Now, create a file called **WeatherForecastClassMap.cs** under the **ClassMaps** folder and paste the class there.

21. Keep only the mappings that match the **WeatherForecast** class that was edited in *Step 17*:

```
public class WeatherForecastClassMap : ClassMap<WeatherForecast>
{
    public WeatherForecastClassMap()
    {
        Map(m => m.Datetime).Name("Date time");
        Map(m => m.Temperature).Name("Temperature");
        Map(m => m.Conditions).Name("Conditions");
    }
}
```

> **NOTE**
>
> You can find the code used for this example at https://packt.link/dV6wX and https://packt.link/mGJMW.

In the previous section, you learned how to get weather forecasts from an existing API and format them your way using the RapidAPI Weather API. Now it is time to proceed to the service client and use the models created, along with the settings, parse the API response, and return the current time weather.

Service Client

Now you have all the ingredients that are needed to create the provider class. You learned in *Chapter 8, Creating and Using Web API Clients*, that when communicating with another API, it's best to create a separate component for it. So, here you will start from an interface abstraction, **IWeatherForecastProvider**:

```
public interface IWeatherForecastProvider
{
    Task<WeatherForecast> GetCurrent(string location);
}
```

Next, create an implementation of that interface—that is, a class taking **HttpClient** for DI:

```
public class WeatherForecastProvider : IWeatherForecastProvider
    {
        private readonly HttpClient _client;

        public WeatherForecastProvider(HttpClient client)
        {
            _client = client;
        }
```

To implement an interface, start with writing a method definition for getting the current weather:

```
public async Task<WeatherForecast> GetCurrent(string location)
{
```

Next, create a request to call HTTP GET with a relative URI for getting a forecast of the CSV type at a given location:

```
var request = new HttpRequestMessage
{
    Method = HttpMethod.Get,
    RequestUri = new
Uri($"forecast?aggregateHours=1&location={location}&contentType=csv",
UriKind.Relative),
};
```

Now, send a request and verify that it was a success:

```
using var response = await _client.SendAsync(request);
response.EnsureSuccessStatusCode();
```

If the status code is not in the range of **`200-300`**, the **`response.EnsureSuccessStatusCode();`** throws an exception. Set up a CSV reader to prepare for deserializing weather forecasts:

```
var body = await response.Content.ReadAsStringAsync();
using var reader = new StringReader(body);
using var csv = new CsvReader(reader, CultureInfo.InvariantCulture);
csv.Context.RegisterClassMap<WeatherForecastClassMap>();
```

You are adding a **`using`** statement to **`StringReader`** and **`CsvReader`** because both implement the **`IDisposable`** interface for disposing unmanaged resources. This happens when you use the **`using`** statement within a function after it returns.

Lastly, deserialize the forecasts:

```
var forecasts = csv.GetRecords<WeatherForecast>();
```

This way, you request the API to return forecasts starting from today and stopping a few days in the future with 1-hour intervals. The first returned forecast is the forecast of the current hour—that is, the forecast that you need:

```
return forecasts.First();
}
```

Now, you will use **`Newtonsoft.Json`** for deserialization. Install the following package to do so:

```
dotnet add package Microsoft.AspNetCore.Mvc.NewtonsoftJson
```

Update the **`AddControllersConfiguration`** method by appending the following line on the services object:

```
.AddNewtonsoftJson();
```

This line replaces the default serializer with **`Newtonsoft.Json`**. Now, **`Newtonsoft.Json`** doesn't have to be used; however, it is a much more popular and complete library for serialization compared to the default one.

> **NOTE**
>
> You can find the code used for this example at https://packt.link/jmSwi.

Till now, you have learned how to create a service client and make basic HTTP calls using it. It's effective for grasping the basics; however, the classes the API uses should be coupled with the classes of the APIs it consumes. In the next section, you will learn how to decouple the API from third-party API models using a DTO and mapping via **`AutoMapper`**.

DTO AND MAPPING USING AUTOMAPPER

The weather forecast model from RapidAPI is a **Date Transfer Object** (**DTO**)—a model used just for transferring data and convenient serialization. RapidAPI may change its data model and, if that happens, the DTO will change as well. If you are just presenting the data you had received and don't need to perform any logical operations on it, then any change may be alright.

However, you will usually apply business logic to a data model. You already know that references to a data model are scattered across multiple classes. With every change to a DTO, a class may have to change as well. For example, the DTO property that was called **`weather`** has now changed to **`weathers`**. Another example is of a property that was previously called **`description`** will now be called a **`message`**. So, renaming a DTO property like this will require you to make changes everywhere they are referenced. The bigger the project, the worse of an issue this becomes.

The advice of the SOLID principles is to avoid such changes (refer to *Chapter 2, Building Quality Object-Oriented Code*). One of the ways to achieve this is by having two kinds of models—one for domain and the other for outside calls. This will require a mapping between foreign objects (coming from outside APIs) into your own.

Mapping can be done either manually or by using some popular libraries. One of the most popular mapping libraries is **AutoMapper**. It allows you to map from one object to another using property names. You can also make your own mappings. Now, you will use this library to configure a mapping between a weather forecast DTO and a weather forecast model.

So, first install NuGet:

```
dotnet add package AutoMapper.Extensions.Microsoft.DependencyInjection
```

This library allows you to inject **AutoMapper** into **ServiceCollection**. Here, **AutoMapper** uses the **Profile** class to define a mapping.

A new mapping should inherit the **Profile** class. So, inside the constructor of the new profile, use a **CreateMap** method to provide a mapping:

```
    public class WeatherForecastProfile : Profile
    {
        public WeatherForecastProfile()
        {
            CreateMap<Dtos.WeatherForecast, Models.WeatherForecast>()
```

Next, in order to map every property from the **CreateMap** method, call the **ForMember** method and specify how to do a mapping:

```
                .ForMember(to => to.TemperatureC, opt => opt.MapFrom(from
=> from.main.temp));
```

Here, the value of **TemperatureC** comes from **main.temp** inside the DTO.

For the other property, you will concatenate all the weather descriptions into one string and call that a summary (**BuildDescription**):

```
        private static string BuildDescription(Dtos.WeatherForecast
forecast)
        {
            return string.Join(",",
                forecast.weather.Select(w => w.description));
        }
```

Now, use the lambda method, **ForMember**, when building a weather forecast summary mapping:

```
.ForMember(to => to.Summary, opt => opt.MapFrom(from =>
BuildDescription(from)))
```

Create a **MapperSetup** class and inject **AutoMapper** from the **AddModelMappings** method to provide different mapping profiles:

```
public static class MapperSetup
{
    public static IServiceCollection AddModelMappings(this
IServiceCollection services)
    {
        services.AddAutoMapper(cfg =>
        {
            cfg.AddProfile<WeatherForecastProfile>();
        });

        return services;
    }
}
```

Append **.AddModelMappings()** to the **services** object calls. With this, you can call **mapper.Map<Model.WeatherForecast>(dtoForecast);**.

> **NOTE**
>
> You can find the code used for this example at https://packt.link/fEfdw and https://packt.link/wDqK6.

The **AutoMapper** mapping library allows you to map from one object to another by default mapping matching property names. The next section will detail how you can use DI to reuse **HttpClient**.

HTTPCLIENT DI

Continuing with DI, you now want to get into the habit of using the fragmented **ConfigureServices** approach. So, first, create a class called **HttpClientsSetup** and then create a method for adding the configured **HttpClients**:

```
    public static class HttpClientsSetup
    {
        public static IServiceCollection
AddHttpClients(IServiceCollection services)
        {
```

Next, for the injection itself, use the **AddHttpClient** method:

```
services.AddHttpClient<IWeatherForecastProvider,
WeatherForecastProvider>((provider, client) =>
            {
```

In the preceding section, it was mentioned that the keys should be hidden and stored in environment variables. To set a default start URI of every call, set **BaseAddress** (**WeatherForecastProviderUrl** used in *Step 10* of the *RapidAPI* section).

To append the API key on every request, get the API key that you stored in environment variables and assign it to default headers as **x-rapidapi-key**:

```
                client.BaseAddress = new
Uri(config["WeatherForecastProviderUrl"]);
                var apiKey = Environment.GetEnvironmentVariable("x-
rapidapi-key", EnvironmentVariableTarget.User);
                client.DefaultRequestHeaders.Add("x-rapidapi-key",
apiKey);
            });
```

To finish the injection-builder pattern, you need to return the **services** object, as follows:

```
return services;
```

Now, go back to **services** in **Program** and append the following:

```
.AddHttpClients(Configuration)
```

To integrate the client you have just set up, go to **WeatherForecastService**, and inject the **mapper** and **provider** components:

```
public WeatherForecastService(..., IWeatherForecastProvider provider,
IMapper mapper)
```

Change the **GetWeatherForecast** method to either get the cached forecast of this hour or fetch a new one from the API:

```
        public async Task<WeatherForecast> GetWeatherForecast(DateTime
date)
        {
            const string DateFormat = "yyyy-MM-ddthh";
            var contains = _cache.TryGetValue(date.ToString(DateFormat),
out var entry);
            if(contains){return (WeatherForecast)entry;}

            var forecastDto = await _provider.GetCurrent(_city);
            var forecast = _mapper.Map<WeatherForecast>(forecastDto);
```

```
            forecast.Date = DateTime.UtcNow;

            _cache.Set(DateTime.UtcNow.ToString(DateFormat), forecast);

            return forecast;
        }
```

This method, just like the preceding one, first tries to get a value from the cache. If the value exists, then the method returns a value. However, if the value does not exist, the method calls the API for the preconfigured city, maps the DTO forecast to the model forecast, and saves it in the cache.

If you send an HTTP GET request to **`https://localhost:7021/WeatherForecast/`**, you should see the following response:

```
{"date":"2021-09-21T20:17:47.410549Z","temperatureC":25,"temperatureF":76
,"summary":"clear sky"}
```

Calling the same endpoint results in the same response. However, the response times are significantly faster due to the cache being used rather than repeating a call to the forecast API.

> **NOTE**
>
> You can find the code used for this example at https://packt.link/GMFmm.

This concludes the theoretical portion of this topic. In the following section, you will put this into practice with an exercise.

EXERCISE 9.03: PERFORMING FILE OPERATIONS BY CALLING AZURE BLOB STORAGE

A common task with a Web API is to perform a variety of operations on files, such as download, upload, or delete. In this exercise, you will reuse a portion of **`FilesClient`** from *Activity 8.04* of *Chapter 8, Building Quality Object-Oriented Code*, to serve as a baseline client for calling Azure Blob storage and call its methods via REST endpoints to do the following operations on a file:

- Download a file.
- Get a shareable link with expiration time.
- Upload a file.
- Delete a file.

Perform the following steps to do so:

1. Extract an interface for **FilesClient** and call it **IFilesService**:

```
public interface IFilesService
    {
        Task Delete(string name);
        Task Upload(string name, Stream content);
        Task<byte[]> Download(string filename);
        Uri GetDownloadLink(string filename);
    }
```

   The new interface is simplified as you will work on a single container. However, as per the requirements, you have added a few new methods: **Delete**, **Upload**, **Download**, and **GetDownloadLink**. The **Download** method is for downloading a file in its raw form—that is, bytes.

2. Create a new class called **Exercises/Exercise03/FilesService.cs**.
3. Copy the following parts of https://packt.link/XC9qG there.
4. Rename **Client** to **Service**.
5. Also change the **Exercise04** reference (used in *Chapter 8, Building Quality Object-Oriented Code*) to **Exercise03** (a new one to be used for this chapter):

FilesService.cs

```
public class FilesService : IFilesService
    {
        private readonly BlobServiceClient _blobServiceClient;
        private readonly BlobContainerClient _defaultContainerClient;

        public FilesClient()
        {
            var endpoint = "https://packtstorage2.blob.core.windows.net/";
            var account = "packtstorage2";
            var key = Environment.GetEnvironmentVariable("BlobStorageKey", 
EnvironmentVariableTarget.User);
            var storageEndpoint = new Uri(endpoint);
            var storageCredentials = new StorageSharedKeyCredential(account, 
key);
            _blobServiceClient = new BlobServiceClient(storageEndpoint, 
storageCredentials);
            _defaultContainerClient = CreateContainerIfNotExists("Exercise03).
Result;
        }

        private async Task<BlobContainerClient> 
CreateContainerIfNotExists(string container)
```

You can find the complete code here: https://packt.link/fNQAX.

The constructor initializes **`blobServiceClient`** to get **`blobClient`**, which allows you to do operations in the *Exercice03* directory in the Azure Blob Storage Account. If the folder doesn't exist, **`blobServiceClient`** will create it for you:

```
        {
            var lowerCaseContainer = container.ToLower();
            var containerClient = _blobServiceClient.
GetBlobContainerClient(lowerCaseContainer);
            if (!await containerClient.ExistsAsync())
            {
                containerClient = await _blobServiceClient.
CreateBlobContainerAsync(lowerCaseContainer);
            }

            return containerClient;
        }
```

> **NOTE**
>
> For the preceding step to work, you will need an Azure Storage Account. So, refer to *Activity 8.04* of *Chapter 8*, *Building Quality Object-Oriented Code*.

6. Create the **`ValidateFileExists`** method to validate whether a file exists in the storage, else throw an exception (a small helper method that did not exist before):

```
private static void ValidateFileExists(BlobClient blobClient)
{
    if (!blobClient.Exists())
    {
        throw new FileNotFoundException($"File {blobClient.Name} in
default blob storage not found.");
    }
}
```

7. Now, create the **Delete** method to delete a file:

```
public Task Delete(string name)
{
    var blobClient = _defaultContainerClient.GetBlobClient(name);
    ValidateFileExists(blobClient);

    return blobClient.DeleteAsync();
}
```

   Here, you will first get a client for the file and then check whether the file exists. If not, then you will throw a **FileNotFoundException** exception. If the file exists, then you will delete the file.

8. Create the **UploadFile** method to upload a file:

```
public Task UploadFile(string name, Stream content)
{
    var blobClient = _defaultContainerClient.GetBlobClient(name);
    return blobClient.UploadAsync(content, headers);
}
```

   Once again, you first get a client that allows you to perform operations on a file. Then, feed the content and headers to it to upload.

9. Create the **Download** method to download a file in bytes:

```
        public async Task<byte[]> Download(string filename)
        {
            var blobClient = _defaultContainerClient.
GetBlobClient(filename);
            var stream = new MemoryStream();
            await blobClient.DownloadToAsync(stream);

            return stream.ToArray();
        }
```

This method creates a memory stream and downloads the file to it. Please note that this is not going to work on large files.

> **NOTE**
>
> If you would like to learn more on how to process large files, please refer to https://docs.microsoft.com/en-us/aspnet/core/mvc/models/file-uploads?view=aspnetcore-6.0#upload-large-files-with-streaming.

There is a way to present raw downloaded bytes as an image or JSON, rather than as generic downloadable content. With an HTTP request or response, you can send a header specifying the way the content should be interpreted. This header is called **Content-Type**. Each application will process this differently. In the context of Swagger, **image/png** will be displayed as an image, while **application/json** will be shown as JSON.

10. Create a **GetUri** method to get a URI of **blobClient**:

```
private Uri GetUri(BlobClient blobClient)
{
    var sasBuilder = new BlobSasBuilder
    {
        BlobContainerName = _defaultContainerClient.Name,
        BlobName = blobClient.Name,
        Resource = "b",
        ExpiresOn = DateTimeOffset.UtcNow.AddHours(1)
    };
    sasBuilder.SetPermissions(BlobSasPermissions.Read);

    var sasUri = blobClient.GenerateSasUri(sasBuilder);
    return sasUri;
}
```

Getting a URI requires the use of **BlobSasBuilder**, through which you can generate a shareable URL to a blob. Through the builder, specify the kind of resource you are trying to share (**"b"** stands for blob) and the expiry time. You need to set the permissions (to read) and pass the **sasBuilder** builder to the **blobClient** client to generate **sasUri**.

11. Now, use a filename to create a file download link:

```
        public Uri GetDownloadLink(string filename)
        {
            var blobClient = _defaultContainerClient.
GetBlobClient(filename);
            var url = GetUri(blobClient);

            return url;
        }
```

12. Inside the **ExceptionMappingSetup** class and the **AddExceptionMappings** method, add the following mapping:

```
opt.MapToStatusCode<FileNotFoundException>(404);
```

13. Create an extension method to inject a module of **FileUploadService**:

```
public static class FileUploadServiceSetup
{
    public static IServiceCollection AddFileUploadService(this
IServiceCollection services)
    {
        services.AddScoped<IFilesService, FilesService>();
        return services;
    }
}
```

An extension method is a simplified way of showing a new method to an existing interface.

14. Append it to **services** in **Program.cs** to use the **FileUploadService** module:

```
.AddFileUploadService();
```

15. Now, create a controller for files:

```
    [Route("api/[controller]")]
    [ApiController]
    public class FileController : ControllerBase
    {
```

Controller creation is standard on MVC architecture, and this allows users to access **FileService** through HTTP requests.

16. Then, inject **IFilesService** to provide an interface through which file-related functionality could be accessed:

```
        private readonly IFilesService _filesService;

        public FileController(IFilesService filesService)
        {
            _filesService = filesService;
        }
```

17. Next, create an endpoint to delete a file:

```
        [HttpDelete("{file}")]
        public async Task<IActionResult> Delete(string file)
        {
            await _filesService.Delete(file);

            return Ok();
        }
```

18. Create an endpoint to download a file:

```
    [HttpGet("Download/{file}")]
        public async Task<IActionResult> Download(string file)
        {
            var content = await _filesService.Download(file);
            return new FileContentResult(content, "application/octet-
stream ");
        }
```

19. Create an endpoint for getting a shareable file download link:

```
        [HttpGet("Link")]
        public IActionResult GetDownloadLink(string file)
        {
            var link = _filesService.GetDownloadLink(file);
            return Ok(link);
        }
```

20. Create an endpoint for uploading a file:

```
        [HttpPost("upload")]
        public async Task<IActionResult> Upload(IFormFile file)
        {
            await _filesService.UploadFile(file.FileName, file.
```

```
OpenReadStream());

            return Ok();
        }
```

IFormFile is a common way of passing small files to a controller. However, from **IFormFile**, you need file contents as a stream. You can get this using the **OpenReadStream** method. Swagger allows you to use the File Explorer window to choose the file you want to upload.

21. Now you run the API.

    Your Swagger documentation will have a new section with the controller methods. Here are the responses of each:

    - Upload file request:

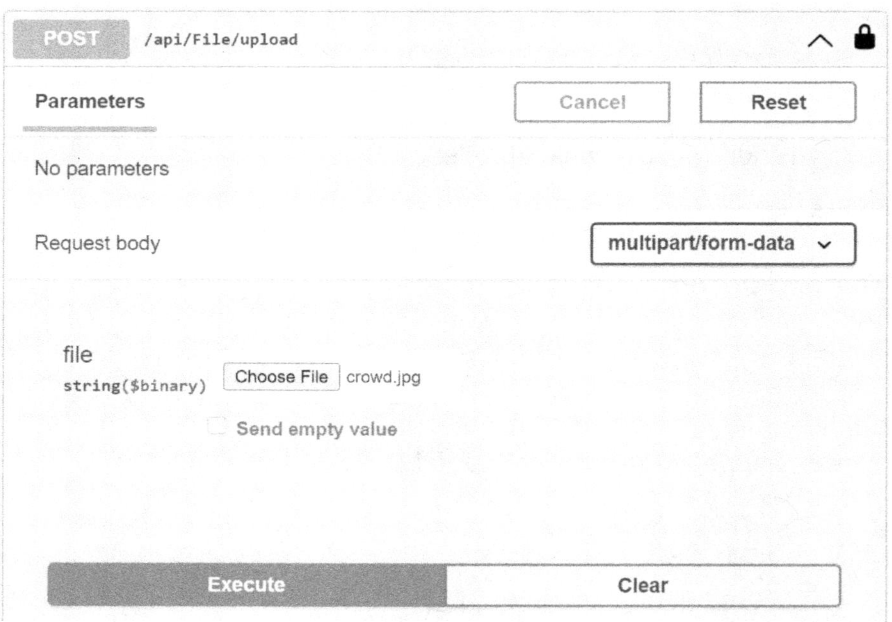

Figure 9.27: Upload file request in Swagger

- Upload file response:

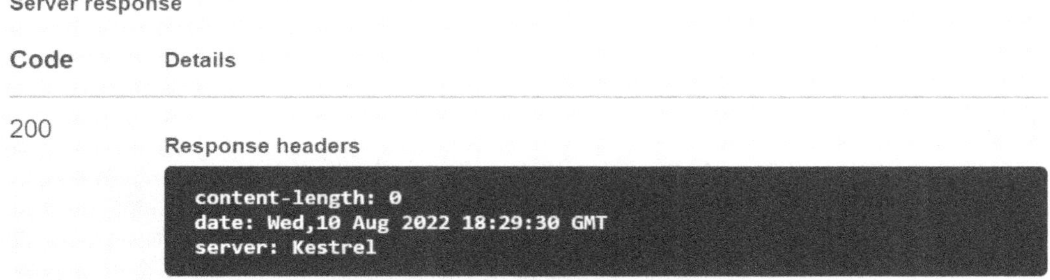

Figure 9.28: Upload file response in Swagger

- Get download link request:

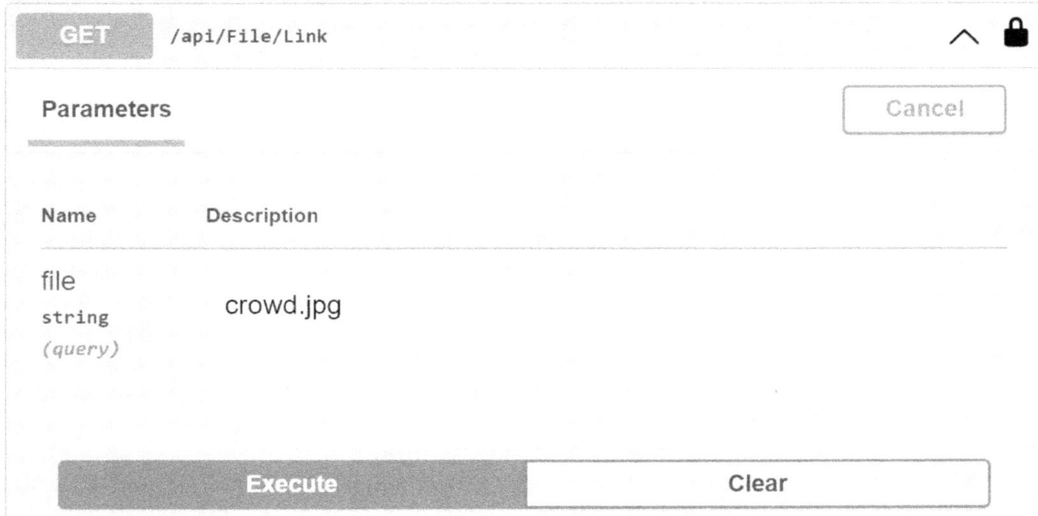

Figure 9.29: Get download link request in Swagger

- Get download link response:

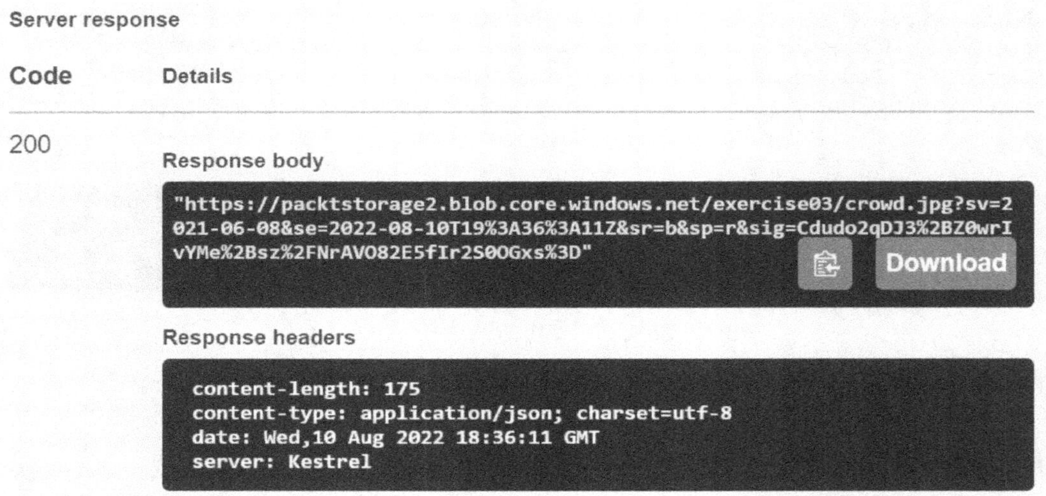

Figure 9.30: Get download link response in Swagger

- Download file request:

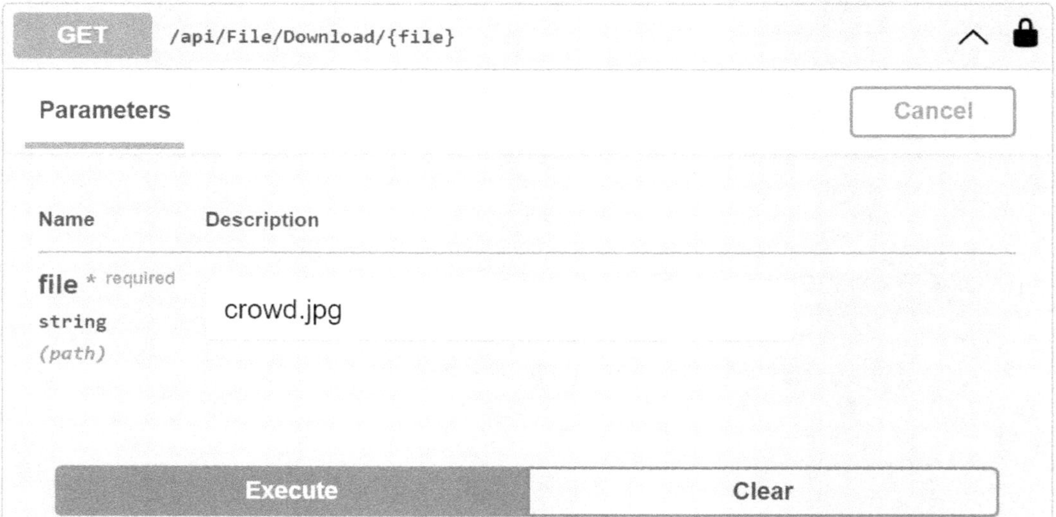

Figure 9.31: Download file request in Swagger

- Download file response:

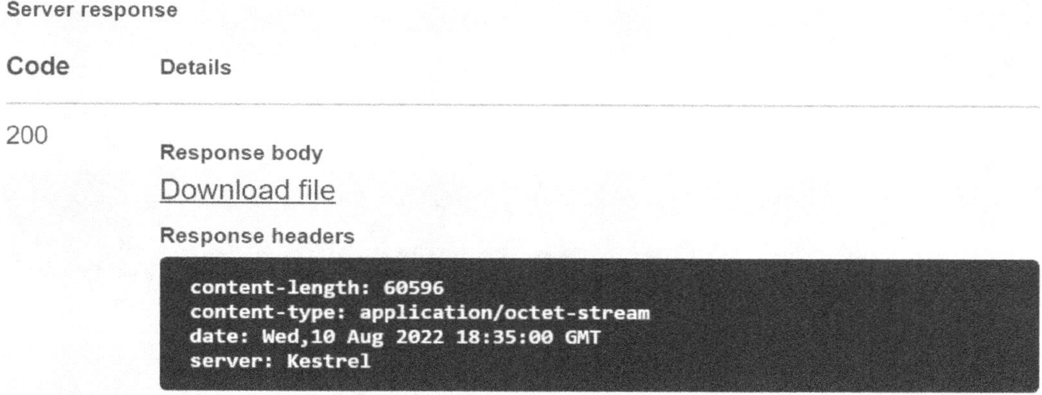

Figure 9.32: Download file response in Swagger

- Delete file request:

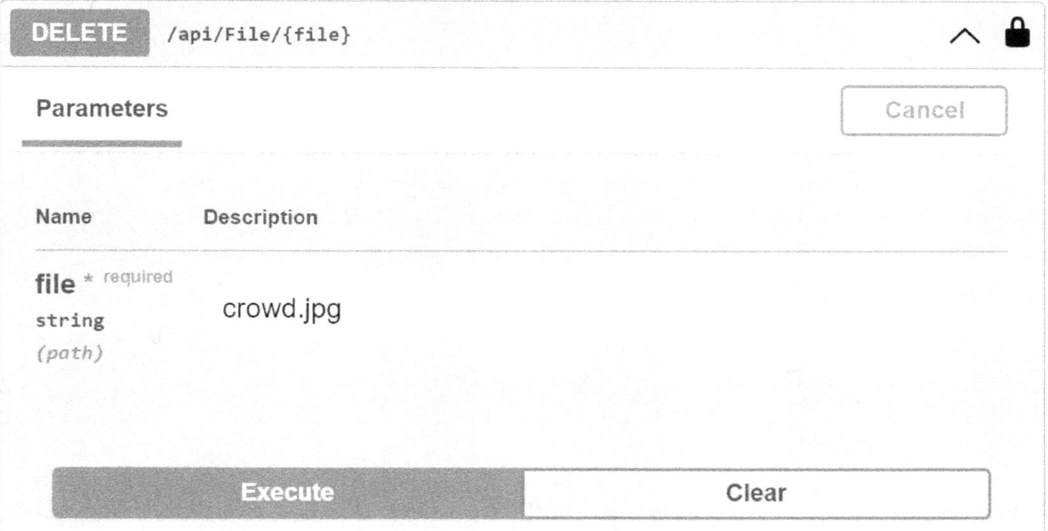

Figure 9.33: Delete file request in Swagger

- Delete file response:

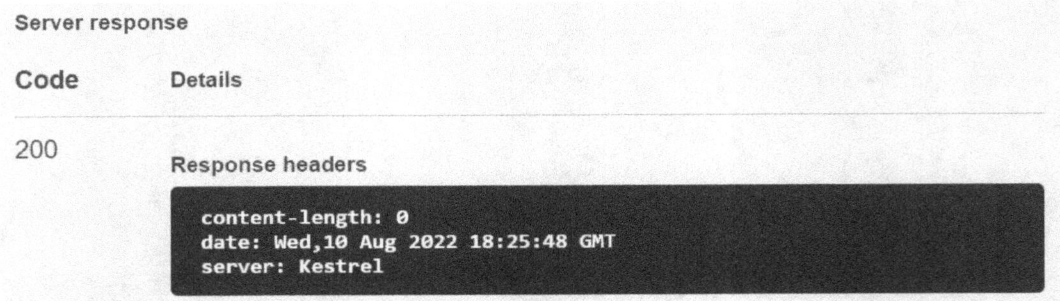

Figure 9.34: Delete file response in Swagger

This exercise illustrated the remaining aspects of what you can do with a Web API.

> **NOTE**
>
> You can find the code used for this exercise at https://packt.link/cTa4a.

The volume of functionality you can serve through the web is immense. However, this comes with its own big problem. How do you ensure that your API is consumed only by the intended identities? In the next section, you will explore how to secure a Web API.

SECURING A WEB API

Every now and then, you'll hear about a major security breach on the news. In this section, you will learn how to protect a public API using AAD.

AZURE ACTIVE DIRECTORY

Azure Active Directory (**AAD**) is Microsoft's cloud identity and access management service that is used to sign in to well-known applications, such as Visual Studio, Office 365, and Azure, and to internal resources. AAD uses OpenID to provide user identity through a JavaScript Web Token.

JWT

A **JavaScript Web Token** (**JWT**) is a collection of personal data encoded and sent over as a mechanism of authentication. A single field encoded in a JWT is called a **claim**.

OpenID Connect

OpenID Connect (**OIDC**) is the protocol used for getting the **ID token**, which provides user identity or an access token. It's a layer on top of OAuth 2 to get an identity.

OAuth serves as a means of getting an access token on behalf of some user. With OIDC, you get an identity; this has a role and access comes from that role. When a user wants to log in to a website, OpenID might require them to input their credentials. This might sound exactly the same as OAuth; however, don't mix the two. OpenID is all about acquiring and verifying the user's identity and granting access coming with a role. OAuth, on the other hand, gives access to a user to do a limited set of functionalities.

A real-life analogy would be as follows:

- OpenID: You come to an airport and present your passport (which is issued by the government) confirming your role (passenger) and identity that way. You are **granted** a **passenger** role and allowed to board an airplane.
- OAuth: You come to an airport and the staff asks you to take part in an emotional state tracking event. With your **consent**, the staff (**others**) at the airport can now track more of your personal data.

The following is a summary:

- OpenID provides authentication and **verifies who you are**.
- OAuth is authorization that allows others to do **things on your behalf**.

APPLICATION REGISTRATION

The first step in securing a Web API using Azure is to create an application registration in AAD. Perform the following steps to do so:

1. Navigate to **`Azure Active Directory`** by typing **`active dir`** in the search bar:

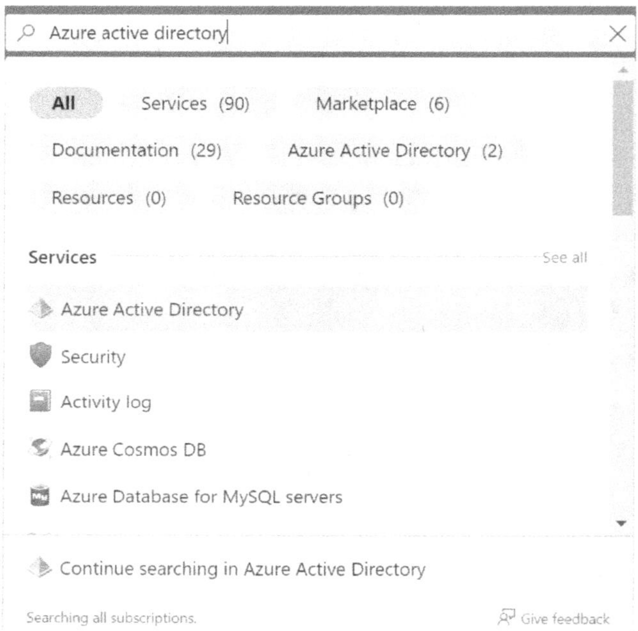

Figure 9.35: Azure Active Directory being searched in portal.azure

2. In the new window, click the **`App registrations`** option (**1**).
3. Then, click the **`New registration`** button (**2**):

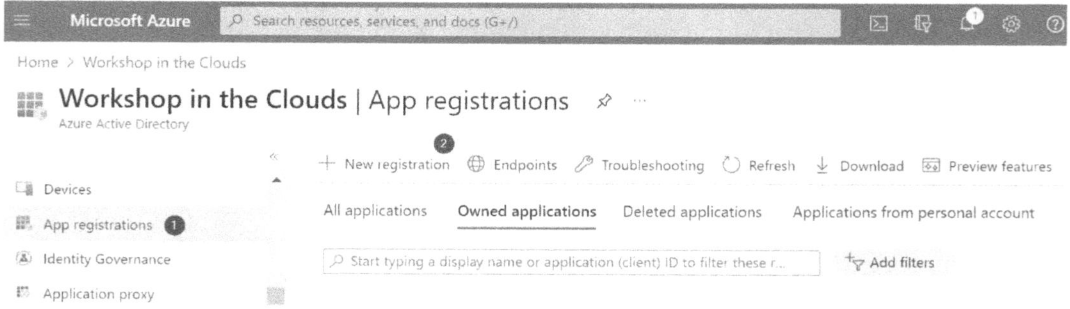

Figure 9.36: Azure app registration

4. In the new window, enter **`Chapter09WebApi`** as the name.
5. Keep the other settings as the default and click the **`Register`** button:

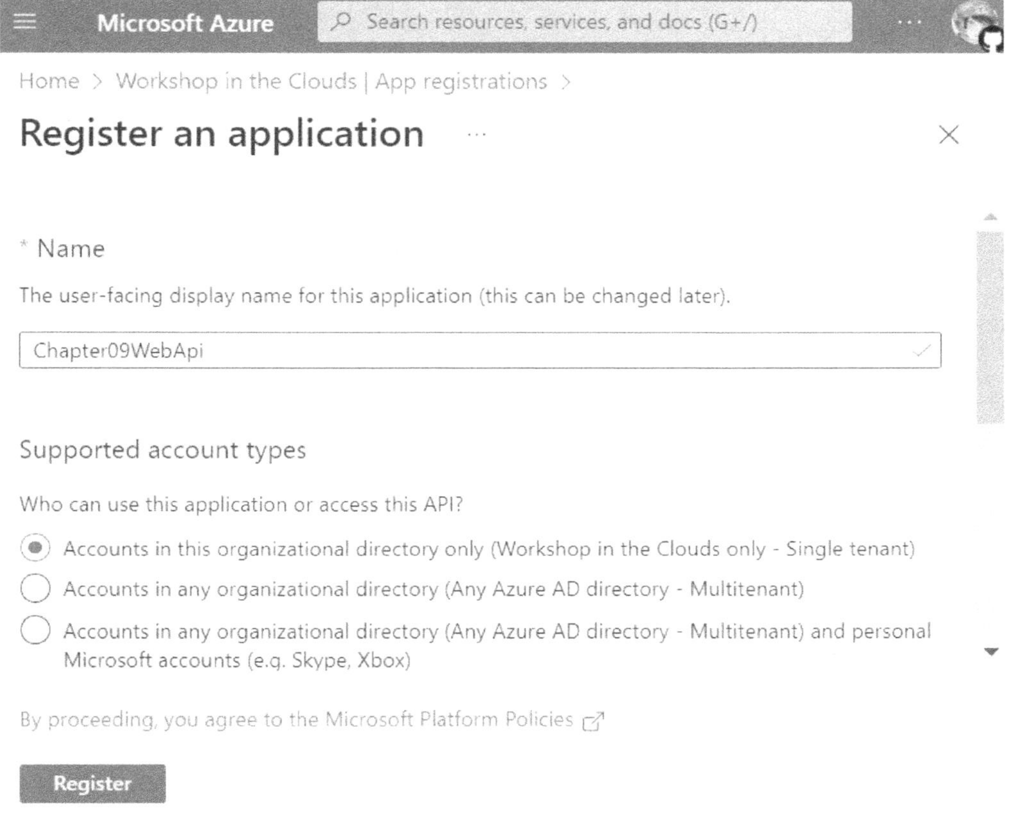

Figure 9.37: The new app registration named Chapter09WebApi

6. To access an API, you need at least one scope or role. In this example, you will create a scope called **`access_as_user`**.
7. Scopes in general can be used to control which part of an API is accessible to you. For the scope to be available for all users, you will need to select **`Admins and users`**.

8. In this trivial example, given the token is valid, you will allow access to everything. So, select the **`Access all as a user`** option. The exact values of the other fields do not matter:

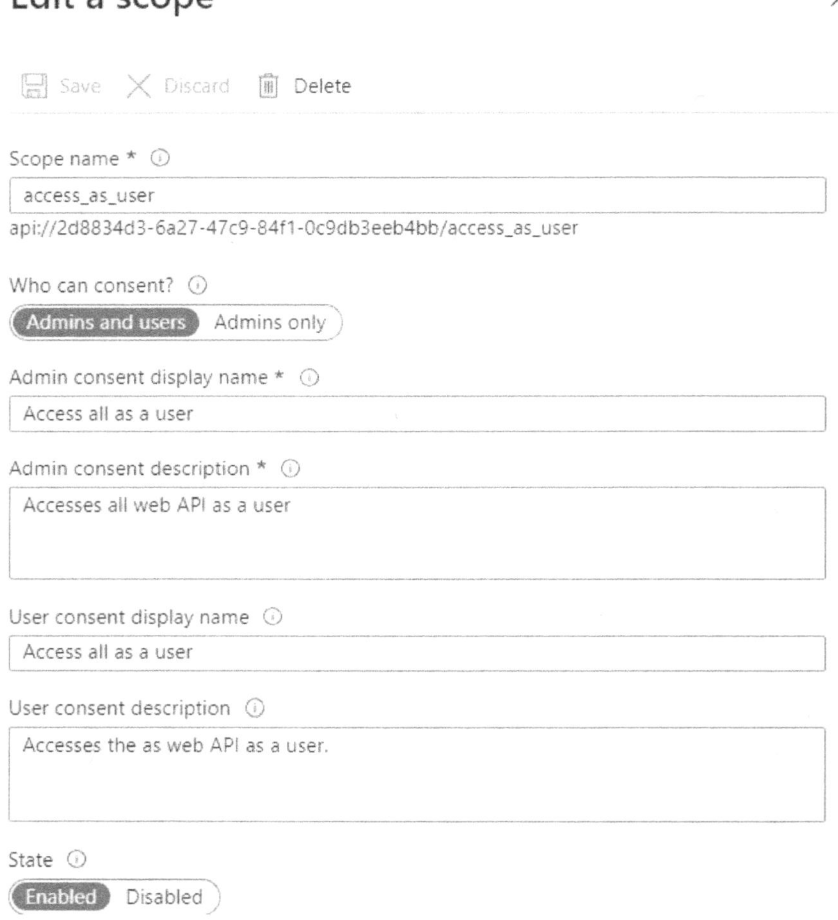

Figure 9.38: The access_as_user scope available for all users

The first step in securing a Web API using Azure was to create an application registration in AAD. The next topic will cover how you can implement security within a Web API in .NET.

IMPLEMENTING WEB API SECURITY

This section will focus on the details of how, programmatically, you can get the token and work with it. So, first, install NuGet, which does JWT validation using the Microsoft identity platform:

```
dotnet add package Microsoft.Identity.Web
```

In the Bootstrap folder, create the **SecuritySetup** class:

```
    public static class SecuritySetup
    {
        public static IServiceCollection AddSecurity(this
IServiceCollection services, IConfiguration configuration,
IWebHostEnvironment env)
        {
            services.
AddMicrosoftIdentityWebApiAuthentication(configuration);
            return services;
        }
    }
```

Then, in **Program.cs**, append this to **services**:

```
.AddSecurity()
```

The injected services are needed by the authorization middleware. So, add the following on an **app** to add authorization middleware:

```
    app.UseAuthentication();
    app.UseAuthorization();
```

This will be triggered on all endpoints decorated with the **[Authorize]** attribute. Make sure the preceding two lines are placed before **app.MapControllers();** or else the middleware will not be wired with your controllers.

Within **appsettings.json**, add the following configuration to link to your **AzureAd** security configuration:

```
  "AzureAd": {
    "Instance": "https://login.microsoftonline.com/",
    "ClientId": "2d8834d3-6a27-47c9-84f1-0c9db3eeb4ba",
    "TenantId": "ddd0fd18-f056-4b33-88cc-088c47b81f3e",
    "Audience": "api://2d8834d3-6a27-47c9-84f1-0c9db3eeb4bb"
  }
```

Lastly, add the **`Authorize`** attribute above each controller for any kind of security you choose:

```
[Authorize]
[ApiController]
[RequiredScope("access_as_user")]
[Route("[controller]")]
public class WeatherForecastController : ControllerBase
```

The **`Authorize`** attribute is essential for any type of security implementation. This attribute will perform the generic token validation, while **`[RequiredScope("access_as_user")]`** will check whether the **`access_as_user`** scope was included or not. What you now have is a secured API. If you try calling the **`WeatherForecast`** endpoints, you will get a **`401 - Unauthorised`** error.

> **NOTE**
>
> You can find the code used for this example at https://packt.link/ruj9o.

In the next section, you will learn how to generate a token through the token generator app and use it to securely access your API.

TOKEN GENERATOR APP

To call the API, you need to generate a token by creating a console application. Before you do that, however, you need to configure one more thing in your app registration. Your console application is considered a desktop app. So, when signing in, you need a redirect URI. This URI, returned with the code, is used to get the access token. To achieve this, perform the following steps:

1. From the left pane in AAD, select the **`Authentication`** option (**1**) to view all configurations with outside applications.

2. Next, click the **Add a platform** button (**2**) to configure a new application (token generator):

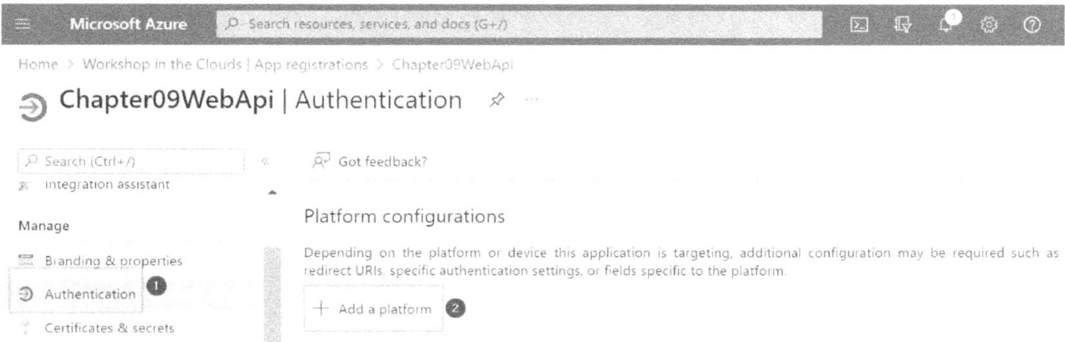

Figure 9.39: Authentication window with options to configure a new application

3. In the **Configure platforms** section, select the **Mobile and desktop applications** button (**3**) to register a console application token generator:

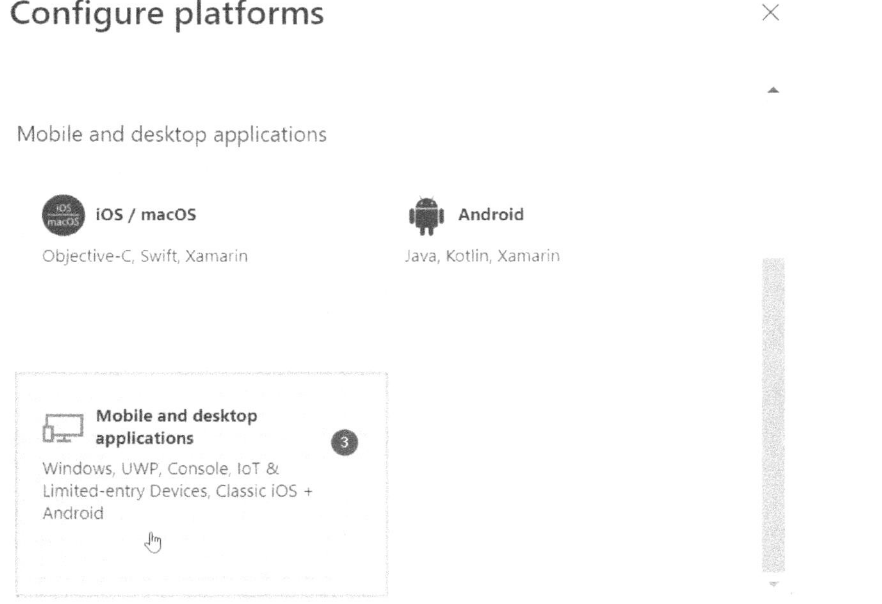

Figure 9.40: Selecting the Mobile and desktop applications platform for authentication

A new window will open on the screen.

4. Type your **Custom redirect URIs** that specify where you will return after the successful login to AAD when requesting the token. In this case, it doesn't matter so much. So, type any URL.

5. Then, click the **Configure** button (**4**):

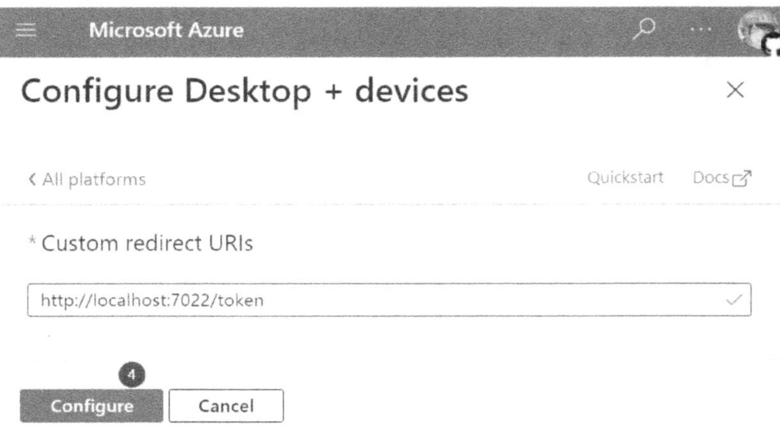

Figure 9.41: Configuring the redirect URI

That completes the configuration of AAD. Now that you have all the infrastructure for security, build a console application to generate an access token from AAD:

1. First, create a new project called **Chapter09.TokenGenerator**. It will allow you to generate authorization tokens needed to call your API.
2. Then, make it a console app on .NET Core to keep it simple and display a generated token.
3. Add **Microsoft.Identity.Client** by running the following command:

```
dotnet add package Microsoft.Identity.Client
```

   This will allow you to request a token later.

4. Next, in **Program.cs**, create a method to initialize an AAD application client. This will be used to prompt browser login, as if you were to log in to the Azure portal:

```
static IPublicClientApplication BuildAadClientApplication()
{
    const string clientId = "2d8834d3-6a27-47c9-84f1-0c9db3eeb4bb";
// Service
    const string tenantId = "ddd0fd18-f056-4b33-88cc-088c47b81f3e";
    const string redirectUri = "http://localhost:7022/token";
    string authority = string.Concat("https://login.microsoftonline.
com/", tenantId);

    var application = PublicClientApplicationBuilder.Create(clientId)
```

```
            .WithAuthority(authority)
            .WithRedirectUri(redirectUri)
            .Build();

        return application;
}
```

> **NOTE**
>
> The values used in the preceding code will differ, depending upon the AAD subscription.

As you can see, the application uses the **clientId** and **tenantId** configured in AAD.

5. Create another method to use the application that requires a user login on Azure to get an auth token:

```
static async Task<string>
GetTokenUsingAzurePortalAuth(IPublicClientApplication application)
{
```

6. Now, define the scopes you need:

```
            var scopes = new[] { $"api://{clientId}/{scope}" };
```

Replace **api://{clientId}/{scope}** with your own application ID URI if you are not using a default value.

7. Then, attempt to get a cached token:

```
            AuthenticationResult result;
            try
            {
                var accounts = await application.GetAccountsAsync();
                result = await application.AcquireTokenSilent(scopes,
accounts.FirstOrDefault()).ExecuteAsync();
            }
```

The cached token retrieval is required if the login was done earlier. If you haven't signed in before to get a token, you will need to log in to Azure AD:

```
            catch (MsalUiRequiredException ex)
            {
                result = await application.
```

```
AcquireTokenInteractive(scopes)
                    .WithClaims(ex.Claims)
                    .ExecuteAsync();
            }
```

8. Return the access token as the result of a logged-in user so that you can use it later to access your APIs:

```
            return result.AccessToken;
```

9. Now, call the two methods and print the result (using the minimal API):

```
var application = BuildAadClientApplication();
var token = await GetTokenUsingAzurePortalAuth(application);
Console.WriteLine($"Bearer {token}");
```

10. Finally, when you run the token app, it will ask you to sign in:

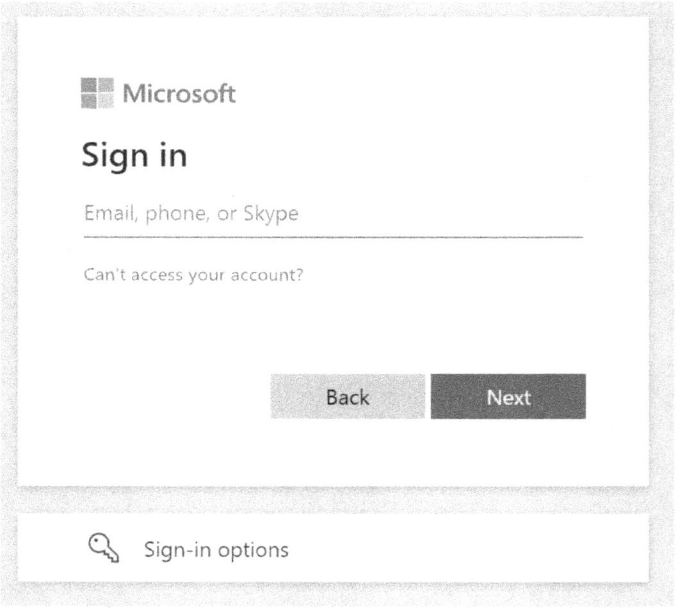

Figure 9.42: Sign-in request from Azure

A successful sign-in redirects you to a configured redirect URI with the following message:

```
Authentication complete. You can return to the application. Feel free
to close this browser tab.
```

You will see that the token will be returned in the console window:

Figure 9.43: Generated token from the app registration in the console app

Now, you can inspect the token using the https://jwt.io/ website. The following screen is displayed, showing two parts: **Encoded** and **Decoded**. The **Decoded** part is divided into the following sections:

- **HEADER**: This contains a type of token and the algorithm used to encrypt the token.
- **PAYLOAD**: The claims encoded within the token contain information, such as who requested the token and what access has been granted:

Figure 9.44: Encoded and decoded JWT version on the jwt.io website using your app registration

In this section, you learned how to secure an unsecured API. Security is not limited to just an authorization token. As a professional developer, you must be aware of the most common vulnerabilities in APIs. A list of the top 10 most common security issues is updated every four years based on the trends in the industry. This list is called the **Open Web Application Security Project** (**OWASP**) and can be reached at https://owasp.org/www-project-top-ten/.

In the next section, you will apply the changes needed for Swagger to work with the authorization token.

CONFIGURING SWAGGER AUTH

To pass an authorization header through Swagger, you will need to add some configuration. Follow these steps to do so:

1. In order to render an authorization button, add the following block of code inside the **SwaggerSetup** class, the **AddSwagger** method, and the **services.AddSwaggerGen(cfg =>** section:

```
                cfg.AddSecurityDefinition("Bearer", new
OpenApiSecurityScheme()
                {
                    Name = "Authorization",
                    Type = SecuritySchemeType.ApiKey,
                    Scheme = "Bearer",
                    BearerFormat = "JWT",
                    In = ParameterLocation.Header,
                    Description = $"Example: \"Bearer YOUR_TOKEN>\"",
                });
```

2. In order to forward the value of a bearer token with an authorization header, add the following code snippet:

```
                cfg.AddSecurityRequirement(new
OpenApiSecurityRequirement
                {
                    {
                        new OpenApiSecurityScheme
                        {
                            Reference = new OpenApiReference
                            {
                                Type = ReferenceType.SecurityScheme,
                                Id = "Bearer"
                            }
```

```
                                },
                                new string[] {}
                            }
                        });
```

3. When you navigate to https://localhost:7021/index.html, you will see that it now contains the **Authorize** button:

/swagger/v1/swagger.json

Figure 9.45: Swagger docs with Authorize button

4. Click the **Authorize** button to allow you to input the bearer token:

Available authorizations ×

Bearer (apiKey)

Example: "Bearer YOUR_TOKEN"

Name: Authorization

In: header

Value:

Bearer eyJ0eXAiOiJKV1QiLC

Authorize Close

Figure 9.46: Bearer token input after clicking the Authorize button

5. Now, send a request:

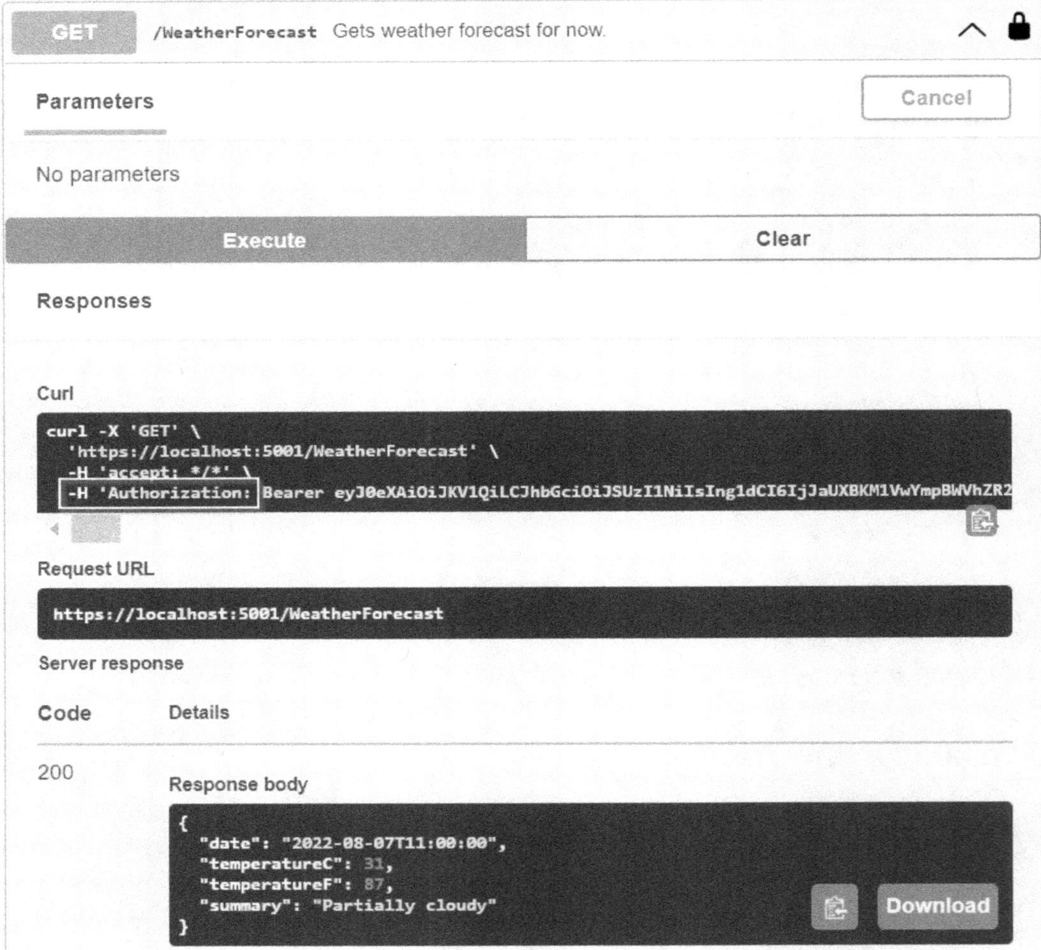

Figure 9.47: Swagger-generated request with a status of 200 generated in response

You will see that the authorization header is added, and the `ok` response (HTTP status code `200`) is returned.

In this section, you added some configuration to pass an authorization header through Swagger.

> **NOTE**
>
> You can find the code used for this example at https://packt.link/hMc2t.

If you make a mistake and your token validation fails, you will get either a **401 – unauthorized** or **403 – forbidden** status code returned (often without any details). Fixing this error might be a headache. However, it is not too difficult to get more information on what went wrong. The next section provides more details.

TROUBLESHOOTING TOKEN VALIDATION ERRORS

To simulate this scenario, try invalidating the client-id in **appsettings.json** by changing any single symbol (for example, the last letter to **b**). Run the request and see how the response is displayed as **401**, with nothing else appearing in the logs.

All the validations and incoming and outcoming requests can be tracked through a pipeline. All you must do is change the default minimum logged level from **info** to **Trace**. You can do this by replacing the **appsettings.development.json** file contents with the following:

```
{
  "Logging": {
    "LogLevel": {
      "Default": "Trace",
      "Microsoft": "Trace",
      "Microsoft.Hosting.Lifetime": "Trace"
    }
  }
}
```

Do not mix **appsettings.development.json** with **appsettings.json**. The former is used for configuration as a whole and the latter overrides the configuration but only in certain environments—development (local) in this case.

If you run the same request again, you will now see a verbose log in the console:

```
Audience validation failed. Audiences: 'api://2d8834d3-6a27-47c9-84f1-
0c9db3eeb4bb'. Did not match: validationParameters.ValidAudience:
'api://2d8834d3-6a27-47c9-84f1-0c9db3eeb4bc' or validationParameters.
ValidAudiences: 'null'.
```

Inspecting it deeper reveals the error as the following:

Audience validation failed; Audiences: 'api://2d8834d3-6a27-47c9-84f1-0c9db3eeb4bb'. Did not match validationParameters

This error indicates a mismatched audience configured in the JWT:

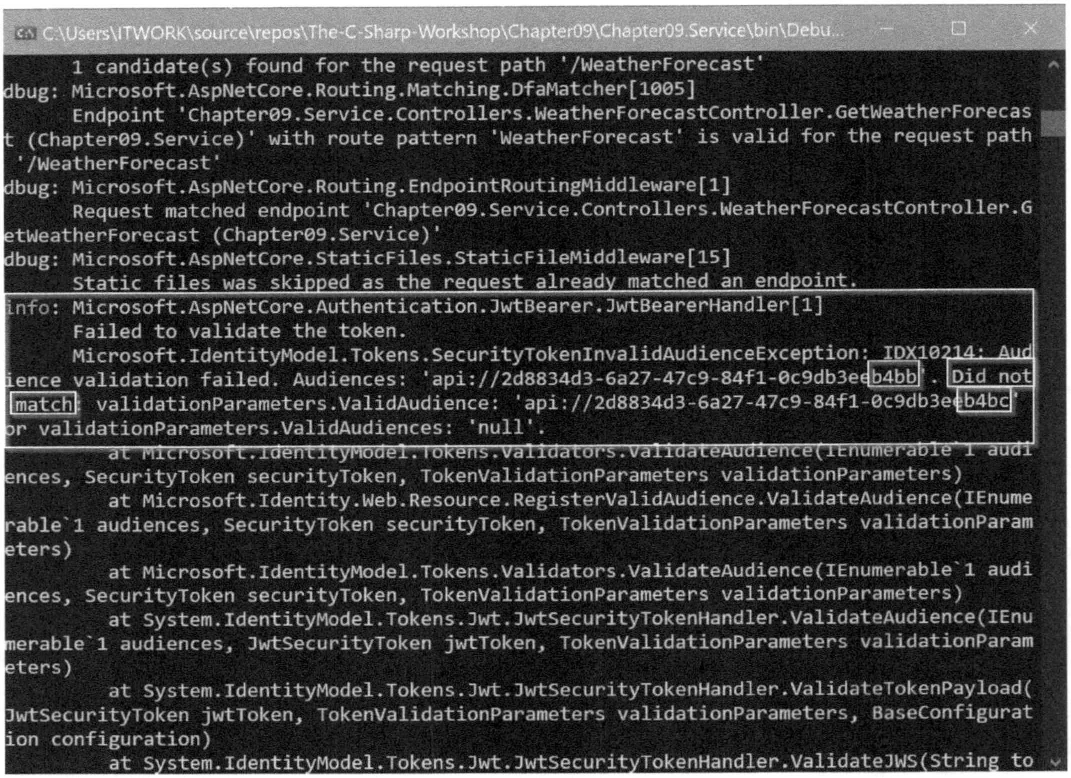

```
C:\Users\ITWORK\source\repos\The-C-Sharp-Workshop\Chapter09\Chapter09.Service\bin\Debu...
      1 candidate(s) found for the request path '/WeatherForecast'
dbug: Microsoft.AspNetCore.Routing.Matching.DfaMatcher[1005]
      Endpoint 'Chapter09.Service.Controllers.WeatherForecastController.GetWeatherForecas
t (Chapter09.Service)' with route pattern 'WeatherForecast' is valid for the request path
 '/WeatherForecast'
dbug: Microsoft.AspNetCore.Routing.EndpointRoutingMiddleware[1]
      Request matched endpoint 'Chapter09.Service.Controllers.WeatherForecastController.G
etWeatherForecast (Chapter09.Service)'
dbug: Microsoft.AspNetCore.StaticFiles.StaticFileMiddleware[15]
      Static files was skipped as the request already matched an endpoint.
info: Microsoft.AspNetCore.Authentication.JwtBearer.JwtBearerHandler[1]
      Failed to validate the token.
      Microsoft.IdentityModel.Tokens.SecurityTokenInvalidAudienceException: IDX10214: Aud
ience validation failed. Audiences: 'api://2d8834d3-6a27-47c9-84f1-0c9db3eeb4bb'. Did not
 match: validationParameters.ValidAudience: 'api://2d8834d3-6a27-47c9-84f1-0c9db3eeb4bc'
or validationParameters.ValidAudiences: 'null'.
         at Microsoft.IdentityModel.Tokens.Validators.ValidateAudience(IEnumerable`1 audi
ences, SecurityToken securityToken, TokenValidationParameters validationParameters)
         at Microsoft.Identity.Web.Resource.RegisterValidAudience.ValidateAudience(IEnume
rable`1 audiences, SecurityToken securityToken, TokenValidationParameters validationParam
eters)
         at Microsoft.IdentityModel.Tokens.Validators.ValidateAudience(IEnumerable`1 audi
ences, SecurityToken securityToken, TokenValidationParameters validationParameters)
         at System.IdentityModel.Tokens.Jwt.JwtSecurityTokenHandler.ValidateAudience(IEnu
merable`1 audiences, JwtSecurityToken jwtToken, TokenValidationParameters validationParam
eters)
         at System.IdentityModel.Tokens.Jwt.JwtSecurityTokenHandler.ValidateTokenPayload(
JwtSecurityToken jwtToken, TokenValidationParameters validationParameters, BaseConfigurat
ion configuration)
         at System.IdentityModel.Tokens.Jwt.JwtSecurityTokenHandler.ValidateJWS(String to
```

Figure 9.48: Token validation error with the error highlighted

Now it is time for you to learn about the SOA architecture where components of a system are hosted as separate services.

SERVICE-ORIENTED ARCHITECTURE

Software architecture has come a long way—evolving from monolithic to **Service-Oriented Architecture** (**SOA**). SOA is an architecture where major layers of applications are hosted as separate services. For example, there would be one or more Web APIs for data access, one or more Web APIs for business logic, and one or more client applications consuming it all. The flow would be like this: the client app calls the business Web API, which calls another business Web API or a data access Web API.

However, modern software architecture goes one step further to bring a more evolved architecture, called microservice architecture.

MICROSERVICE ARCHITECTURE

Microservice architecture is SOA with a single-responsibility principle applied. This means that, instead of service-as-a-layer, you now have hosted self-contained modules that have a single responsibility. A self-contained service has both data access and business logic layers. Instead of many services per layer, in this approach, you have many services per module.

The purpose of those self-contained modules is to allow multiple teams to work on different parts of the same system simultaneously without ever stepping on each other's toes. On top of that, parts in a system can be scaled and hosted independently and there is no single point of failure. Also, each team is free to use whatever technology stack they are most familiar with, as all the communication happens through HTTP calls.

This concludes the theoretical portion of this topic. In the following section, you will put all that you have learned into practice with an activity.

ACTIVITY 9.01: IMPLEMENTING THE FILE UPLOAD SERVICE USING MICROSERVICE ARCHITECTURE

A microservice should be self-contained and do just one thing. In this activity, you will sum up the steps needed for extracting a piece of code into a microservice that manages how you work with files through the web (delete, upload, and download). This should serve as an overall effective checklist of what needs to be done when creating a new microservice.

Perform the following steps to do this:

1. Create a new project. In this case, it will be a **`.NET Core Web API`** project on the .NET 6.0 framework.
2. Name it **`Chapter09.Activity.9.01`**.
3. Now, add the commonly used NuGet packages:
    - **`AutoMapper.Extensions.Microsoft.DependencyInjection`**
    - **`FluentValidation.AspNetCore`**
    - **`Hellang.Middleware.ProblemDetails`**
    - **`Microsoft.AspNetCore.Mvc.NewtonsoftJson`**

- **Microsoft.Identity.Web**
- **Swashbuckle.AspNetCore**

4. Next, include the Azure Blobs Client package as **Azure.Storage.Blobs**.
5. Create one or more controllers for communication with the Web API. In this case, you will move **FileController** to the **Controllers** folder.
6. In order to create one or more services for business logic, move **FilesService** to the **Services** folder and **FileServiceSetup** to the **Bootstrap** folder.
7. Then document API using XML docs and Swagger.
8. Update the **csproj** file to include XML docs.
9. Copy **SwaggerSetup** to the **Bootstrap** folder.
10. Configure **Controllers**. In this scenario, it will be a plain one-line **services.AddControllers()** under the **ControllersConfigurationSetup** class and the **AddControllersConfiguration** method.
11. Configure the problem details error mappings. In this case, there are no exceptions that you will explicitly handle. So, you will keep it as a one-liner within the **ExceptionMappingSetup** class and the **AddExceptionMappings** and **services.AddProblemDetails()** methods.
12. Secure the API.
13. Create AAD app registration for the new service. Refer to the *Application Registration* subsection in the *Securing the Web API* section.
14. Update the configuration of the new service based on the Azure AD app registration client, **tenant**, and **app** IDs.
15. Inject the needed services and configure the API pipeline.
16. Copy the **Program** class.
17. Since the **ConfigureServices** method contains extra services, you don't need to remove them. Leave the **Configure** method as is.
18. Run the service through Swagger and upload a test file. Don't forget to generate a bearer token first using the token generator app from the updated values learned earlier.

19. After that, try to get a test file that you just uploaded. You should see the status code **200**:
    - Get download link request:

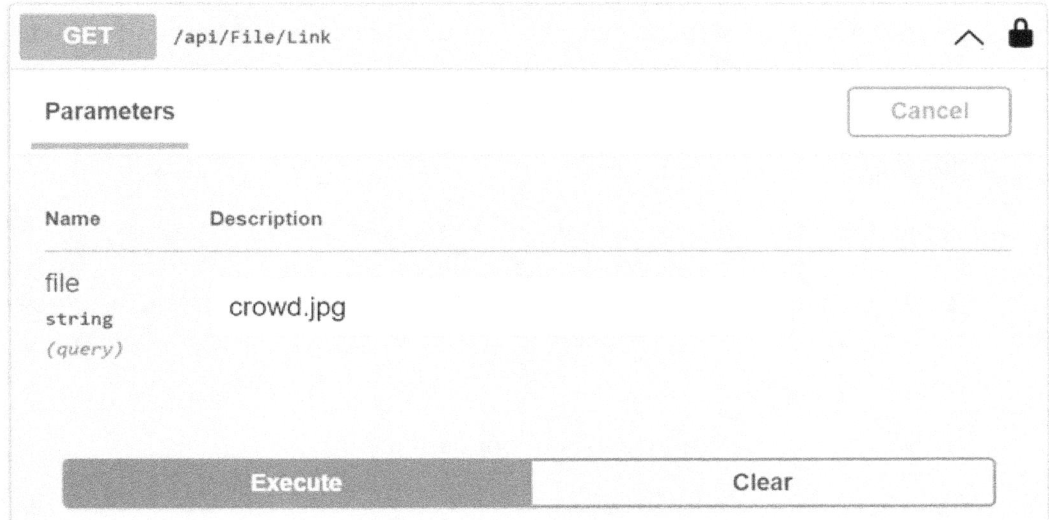

Figure 9.49: Get download link request in Swagger

- Get download link response:

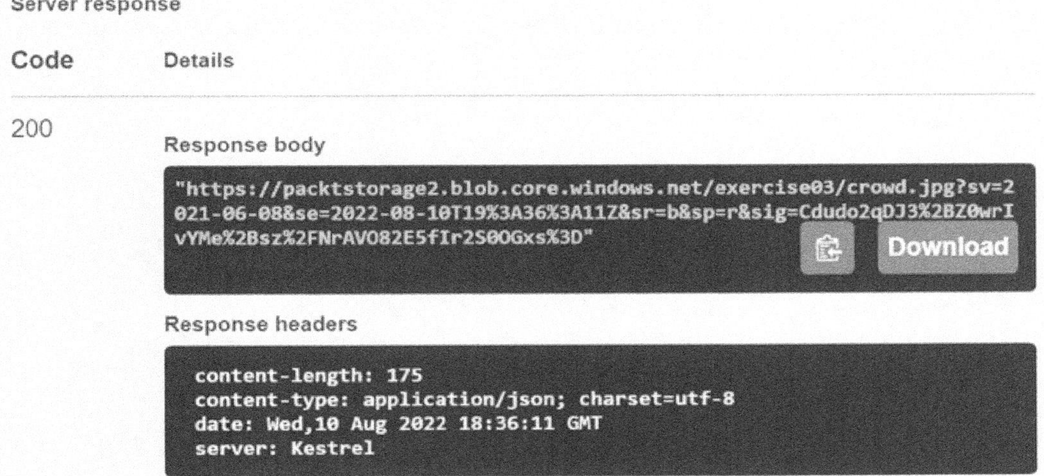

Figure 9.50: Get download link response in Swagger

> **NOTE**
>
> The solution to this activity can be found at https://packt.link/qclbF.

All the services that have been created so far require considerations such as hosting, scaling, and availability. In the following section, you will learn about serverless and Azure Functions.

AZURE FUNCTIONS

In the preceding section, you learned that microservice architecture is a self-contained service with both data access and business logic layers. With this approach, you have many services per module. However, working with microservices, especially at the start, might seem like a hassle. It might raise doubts such as the following:

- What does not big enough mean?
- Should you host on different servers or on the same machine?
- Is another cloud hosting model better?

These questions might be overwhelming. So, a simple way of calling your code through HTTP is by using Azure Functions. Azure Functions is a serverless solution that allows you to call your functions on the cloud. Serverless does not mean that there is no server; you just do not need to manage it by yourself. In this section, you will try to port **`CurrentTimeController`** from *Exercise 9.02* to an Azure Function.

> **NOTE**
>
> Before proceeding with the steps, install Azure Functions Core Tools first using the instructions here: https://docs.microsoft.com/en-us/azure/azure-functions/functions-run-local?tabs=v3%2Cwindows%2Ccsharp%2Cportal%2Cbash%2Ckeda#install-the-azure-functions-core-tools. Azure Functions Core Tools also requires the Azure CLI to be installed (if you want to publish an Azure Functions application and not on a server). Follow the instructions here: https://docs.microsoft.com/en-us/cli/azure/install-azure-cli-windows?tabs=azure-cli.

Perform the following steps to do so:

1. In VS Code, click the **Extenstions** icon (**1**).
2. Then search for **azure function** in the search text box (**2**).
3. Then, install the **Azure Functions** extension (**3**):

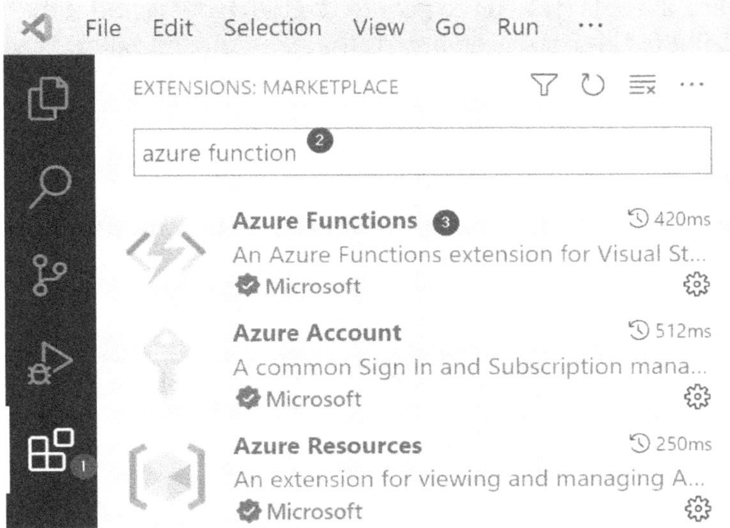

Figure 9.51: Searching for the Azure Functions extension in VS Code

A new Azure tab will appear on the left.

4. Click the new Azure tab.
5. On the new page, click the **Add** button (**1**).
6. Select the **Create Function...** option (**2**):

Figure 9.52: The new Azure Functions extension in VS Code with the Create Function... button

7. In the Create Function window, select **`HTTP trigger`**.
8. Enter the name **`GetCurrentTime.Get`**.
9. Name the project where it is held **`Pact.AzFunction`**.
10. On the last screen, select **`anonymous`**.

    At this point, there is no need to go into too much detail about this configuration. The key point to be considered here is that the function will be reachable publicly, through HTTP requests. A new project created through these steps will include the new Azure Function.

11. Now, navigate to the root of the new project folder to run the project.
12. Next, press **`F5`** or click the **`Start debugging to update this list`**... message:

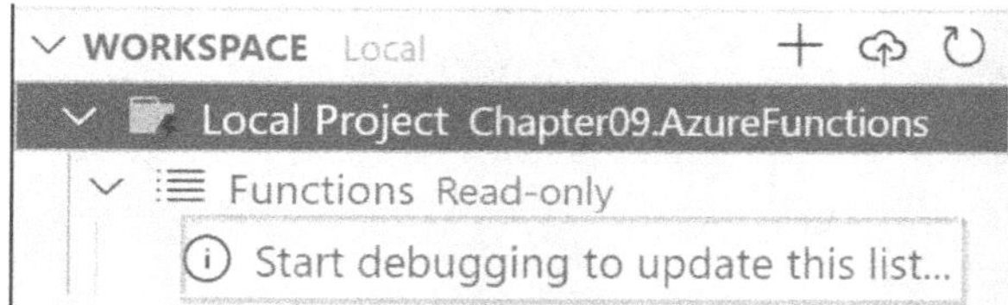

Figure 9.53: Azure Extension window with the to-be-built project

You will notice that upon a successful build, the message changes to the function name:

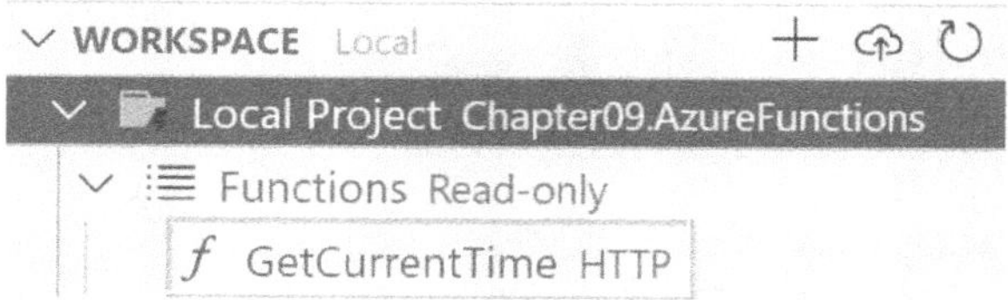

Figure 9.54: Azure Extension window with post-build project

The terminal output window, displayed at the bottom of VS Code, shows the following details:

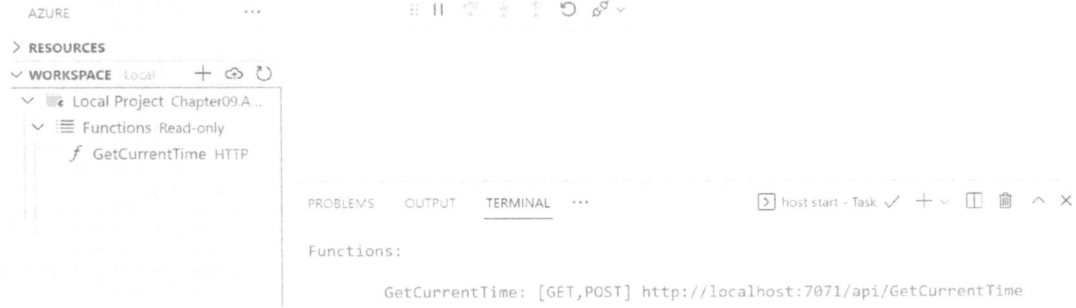

Figure 9.55: The terminal output after a successful build

13. Next, in VS Code Explorer, open **GetCurrentTime.cs**:
14. Note that in *Exercise 9.01*, you worked with the **GetCurrentTime** code. You will reuse the same code here:

```
namespace Pact.Function
{
    public static class GetCurrentTime
    {
        [Function("GetCurrentTime")]
        public static HttpResponseData
Run([HttpTrigger(AuthorizationLevel.Anonymous, "get", "post")]
HttpRequestData request,
            FunctionContext executionContext)
```

The template names are generated based on your configuration from before. An Azure Function is bound to an HTTP endpoint through the **[Function("GetCurrentTime")]** attribute.

Before you proceed, you might have noticed that, even though the function for getting the current time consumed a variable for **timezoneid**, there is no such variable here (yet). Unlike the previous REST APIs you created to pass parameters to an Azure Function, here you pass it through either a request body or query variables. The only problem here is that you will have to parse it yourself, as there are no bindings through attributes just like with the controller methods. The argument you need is just a simple string that can be passed as a query argument. This line parses the URI from the request and gets a **timezoneId** variable from the query string.

15. Use the **timezoneId** variable to get the current time in a specific zone:

```
        {
            var timezoneId = HttpUtility.ParseQueryString(request.
Url.Query).Get("timezoneId");
```

16. Next up is the business logic. So, use the **timezoneId** variable to get the current time in a specified time zone:

```
var timezoneInfo = TimeZoneInfo.FindSystemTimeZoneById(timezoneId);
            var time = TimeZoneInfo.ConvertTimeFromUtc(DateTime.
UtcNow, timezoneInfo);
```

17. Finally, serialize the results in **HTTP 200 Ok** as the **text/plain** content type:

```
var response = request.CreateResponse(HttpStatusCode.OK);
            response.Headers.Add("Content-Type", "text/plain;
charset=utf-8");
            response.WriteString(time.ToString());
            return response;
}
```

18. Run this code and navigate to **http://localhost:7071/api/GetCurrentTime?timezoneId=Central%20European%20Standard%20Time**.

    You will get the current time of that time zone, as follows:

```
2022-08-07 16:02:03
```

You have now grasped the workings of Azure Functions—a serverless solution to call your functions on the cloud.

It has been a long path through this book, but with the conclusion of this final activity, you have mastered all the concepts and skills required to create your own modern C# applications.

SUMMARY

In this chapter, you learned how to build your own REST Web API using the ASP.NET Core Web API template. You learned how to tackle the ever-growing complexity of configuration using bootstrap classes. You were introduced to the OpenAPI standard and Swagger, a tool used for calling an API to see whether it has successfully rendered the documentation. You also delved into mapping exceptions to specific HTTP status codes, along with how to map DTOs to domain objects and vice versa. In the second half of the chapter, you practiced securing the Web API using AAD, learned the concept of microservices, and created one yourself—both through a new dedicated Web API and through an Azure Function.

Knowing how to create and consume Web APIs is important because that's what most of the software development is all about. You either consume or create Web APIs at some point. Even if you don't have to create one yourself, grasping the ins and outs of it will help you as a professional developer.

This brings a close to *The C# Workshop*. Throughout this book, you have learned the basics of programming in C#, starting with simple programs that used arithmetic and logical operators, followed by the increasingly complex concepts of clean coding, delegates and lambdas, multithreading, client and server Web APIs, and Razor Pages applications.

This concludes the print copy of this book, but it is not the end of your journey. Visit the GitHub repository at https://packt.link/sezEm for bonus chapters—*Chapter 10, Automated Testing*, and *Chapter 11, Production-Ready C#: From Development to Deployment*—covering such topics as different forms of testing before you take an in-depth look at unit testing using Nunit (the most popular third-party testing library for C#), getting acquainted with Git and using GitHub to keep a remote backup of your code, enabling Continuous Deployment (CD) and deployment from your code to the cloud, studying the cloud using Microsoft Azure, in addition to learning how to use GitHub Actions to perform CI and CD to push application changes live in production.

HEY!

We are Jason Hales, Almantas Karpavicius, and Mateus Viegas the authors of this book. We really hope you enjoyed reading our book and found it useful for learning C#.

It would really help us (and other potential readers!) if you could leave a review on Amazon sharing your thoughts on *The C# Workshop*.

Go to the link https://packt.link/r/1800566492.

OR

Scan the QR code to leave your review.

Your review will help us to understand what's worked well in this book and what could be improved upon for future editions, so it really is appreciated.

Best wishes,

Jason Hales, Almantas Karpavicius, and Mateus Viegas

INDEX

A

B

C

D

E

F

G

H

I

K

L

M

N

O

P

R

S

T

V

W

www.ingramcontent.com/pod-product-compliance
Lightning Source LLC
LaVergne TN
LVHW081326050326
832903LV00024B/1047

* 9 7 8 1 8 0 0 5 6 6 4 9 1 *